**Annals of the International Society of Dynamic Games**
**Volume 5**

*Series Editor*
Tamer Başar

Annals of the International Society of Dynamic Games

# Advances in Dynamic Games and Applications

Jerzy A. Filar
Vladimir Gaitsgory
Koichi Mizukami

*Editors*

Birkhäuser
Boston · Basel · Berlin

Jerzy A. Filar
Centre for Industrial and
Applicable Mathematics
School of Mathematics
University of South Australia
Adelaide, SA 5000
Australia

Koichi Mizukami
Division of Mathematical
and Information Sciences
Faculty of Integrated Arts and Sciences
Hiroshima University
1-7-1, Kagamiyama
Higashi-Hiroshima City
739 Japan

Vladimir Gaitsgory
School of Mathematics
University of South Australia
Adelaide, SA 5000
Australia

**Library of Congress Cataloging-in-Publication Data**

Advances in dynamic games and applications / [edited by] Jerzy A. Filar, Vladimir Gaitsgory, Koichi Mizukami.
p. cm.—(Annals of the International Society of Dynamic Games ; v. 5)
Includes bibliographical references.
ISBN 0-8176-4002-9 (hardcover: alk. paper)
1. Game theory. I. Filar, Jerzy A., 1949– II. Gaĭtsgory, Vladimir Grigor'evich. III. Mizukami, Koichi, 1936– . IV. Series
QA269.A363 1999
519.3—dc21 99-30133

AMS Subject Classifications: 90C, 90D, 49M, 49N

Printed on acid-free paper.

ISBN 0-8176-4002-9
ISBN 3-7643-4002-9 SPIN 10636138

Typeset by The Bartlett Press, Inc., Marietta, GA.
Printed and bound by Sheridan Books, Inc., Ann Arbor, MI.
Printed in the United States of America.

9 8 7 6 5 4 3 2 1

# Contents

# Part V Applications

# Preface

Modern game theory has evolved enormously since its inception in the 1920s in the works of Borel and von Neumann and since publication in the 1940s of the seminal treatise "Theory of Games and Economic Behavior" by von Neumann and Morgenstern. The branch of game theory known as dynamic games is—to a significant extent—descended from the pioneering work on differential games done by Isaacs in the 1950s and 1960s. Since those early decades game theory has branched out in many directions, spanning such diverse disciplines as mathematics, economics, electrical and electronics engineering, operations research, computer science, theoretical ecology, environmental science, and even political science.

The papers in this volume reflect both the maturity and the vitality of modern day game theory in general, and of dynamic games, in particular. The maturity can be seen from the sophistication of the theorems, proofs, methods, and numerical algorithms contained in these articles. The vitality is manifested by the range of new ideas, new applications, the number of young researchers among the authors, and the expanding worldwide coverage of research centers and institutes where the contributions originated.

Most of the papers in the volume were presented at the 7th International Symposium on Dynamic Games and Applications that was held December 16–18, 1996, at Shonan Village Center in Kanagawa, Japan. The Symposium was attended by 93 participants from 21 countries spanning 4 continents. There was consensus among the participants that both the presentations and the papers included in the conference proceedings were of very high quality. All papers subsequently submitted for this volume were refereed at the highest professional level. The fact that there are only twenty-three papers at the end of this process bears testimony to the high standards set by the editorial board.

Since the contributing authors cover many diverse topics, it is hard to identify a few selected papers as representing the dominant research area in dynamic games today. However, it is possible to claim that the theme of the volume is the remarkable (and happy) blend between deep ideas and theory on the one hand, and the applications and numerical methods on the other hand. Whereas in some other subjects the division between the theory and applications causes a schism among the practitioners of these two aspects within a field, in dynamic games in late 1990s the theory and the applications reinforce each other in a most fruitful way. This synergy is strongly evident in the papers included in this volume.

The diversity of topics makes it difficult to classify the papers into neat, nonoverlapping cateories. However, for the convenience of readers the twenty-three papers have been organized into five cogent sections that are briefly described as follows.

- *Robust Control Design and $H^\infty$.* Papers included in this part deal with different issues related to design of robust controllers for systems with uncertain inputs. This topic represents a relatively recent and increasingly important branch of zero-sum game theory. Advances in this topic have provided both a unifying perspective and numerical methods for the design of robust controllers. In the first paper the authors (Altman, Başar, and Hovakimyan) consider a problem of design of information flow controllers in telecommunications networks. $H^\infty$ and LQG optimal controllers are analytically constructed, and their performances are illustrated using simulations. In the next paper Sasaki and Uchida construct sufficient optimality conditions for bilinear systems with $H^\infty$ type criterion. The applicability of these conditions is demonstrated on the example of an artificial rubber-muscle actuator controller design. Xiao and Başar use in their paper a differential game theoretic approach to obtain a set of necessary and sufficient conditions for the existence of a state-feedback controller under which a given $H^\infty$ bound (on disturbance attenuation) is achieved. These results are applied to a disturbance attenuation problem with control constraints. In the final paper of this section, Xu and Mizukami show that the results concerning the composite controller available for "standard" singularly perturbed (SP) systems ($A_{22}$ is invertible) also hold if a preliminary feedback can be designed first to yield an SP system with invertible $A_{22}$.
- *Pursuit-Evasion (P-E) Games.* This part includes papers devoted to both analytical and numerical aspects of pursuit-evasion games. This is a well-developed branch of game theory that dates back to the pioneering work of Steinhaus in the 1920s, and that has been sustained over the years by its obvious applicability both in the military context and in rescue and collision avoidance problems. The papers in this general area indicate both the maturity of the topic, and the fact that there are still many challenging problems that need to be solved. In particular, in the paper by Kovshov a geodesic parallel strategy (GPS) of pursuit is considered in a P-E game in which both pursuer and evader move on a surface of a unit sphere and can instantly change their directions. Main properties of GPS, including an estimate of the capture time, are described. On the other hand, the paper by Lachner, Breitner, and Pesch is devoted to a problem of how a "correct" driver (evader) can escape a collision with a "wrong" driver (pursuer), that is, a person driving on the wrong side of the road. Worst-case approaches based on P-E differential games are used, and an optimal collision avoidance strategy is derived. The paper by Lim examines a zero-sum game in which two agents of a team $R$ can optimally rendezvous while evading an enemy searcher $S$. The optimality of a special class

of strategies for $R$ is established, and the value of the game is found. The next paper in this section by Melikyan and Shinar discusses a method of singular characteristics that can be used for a construction of nonclassical solutions of Hamilton–Jacobi–Bellman–Isaacs equations arising in P-E Games. In the paper by Raivio and Ehtamo, a solution of a P-E game is decomposed into a sequence of iteratively solved optimal control problems. The convergence of the iterations is illustrated by a number of examples.

- *Coupled Dynamic and Stochastic Games.* Perhaps the most fundamental common feature of all game theoretic models is that of the players' fate being somehow coupled, so that a level of interdependence, which complicates the analysis of rational behavior, is created. While in the classic models this coupling occurred only directly in the payoff functions and/or dynamics of the model, nowadays researchers are exploring more sophisticated forms of coupling, or constraints, that are imposed on the previously solved models and that are intended to extend the applicability of game-theoretic methods to a wider and more realistic set of situations. Thus in a paper by Carlson and Haurie a situation is considered in which the state variables of individual players are coupled by convex constraints that are known a priori. These situations arise in environmental management. Among a number of interesting results, the authors establish the existence of an "averaging Nash equilibrium" in which the coupled state constraint is satisfied asymptotically. In a somewhat similar spirit, Altman and Shwartz consider noncooperative Markov games in which the players' costs must satisfy additional constraints that are specified a priori. This formulation is inspired by potential applications in telecommunications and flow control. Sufficient conditions for the existence of stationary equilibria are presented, and a link with coupled linear programs is established. In a different way, Altman, Başar, and Pan introduce games with coupled continuous and discrete dynamics. In particular, player 1 controls the continuous (deterministic) dynamics that, however, depend also on a state of a Markov process whose transition rates are controlled by player 2. Under the usual linear-quadratic and zero-sum assumptions, existence results are derived for a unique, game-theoretic saddle point. Numerical algorithms and variations on the assumptions about player 2 that are inspired by telecommunications applications are also included. In a paper by Nakagami, Kurano, and Yasuda the coupling of the players' fortunes is embedded in a given multivariate stochastic process (whose components are in one-to-one correspondence with the players), and in the associated stopping rules. In particular, each player can declare his desire to stop or continue the process at any time, and these individual declarations are summed up to arrive at the group decision to actually stop or continue the process. Player $p$'s individual payoff is then determined by the value of the $p$th component of the process. The authors derive conditions for the existence of Nash equilibria as well as their characterization via suitably constructed integral equations.

- *General Game Theoretic Developments*. This section contains five papers devoted primarily to the extensions and further analysis of well-established game theoretic concepts and models. In particular, refinement of Nash equilibria in games with perfect information is considered in Petrosjan's paper. It is proved that by introducing a "preference vector" for each player a unique (in the sense of payoffs) Nash equilibrium can be constructed. In the paper by Kimura, Sawasaki, and Tanaka it is shown that the existence of max-inf (or min-sup) implies the existence of the saddle-point value in a zero-sum game perturbed by constrained functions, as well as the existence of weak optimal solutions for the players. The paper by Murthy, Parthasarathy, and Sriparna demonstrates interconnections between the classical linear complimentary problem (LCP) and matrix games. The applicability of LCP to the solution of certain stochastic games is also demonstrated. The paper by Altman, Feinberg, and Shwartz examines stochastic games with the payoff being a linear combination of expected total rewards discounted on the infinite time horizon with different discount factors. Properties of optimal policies and, in particular, the existence (in the case of finite states-actions spaces) of an optimal pure Markov policy that becomes stationary from some epoch onward are established. Finally, in a paper by Küenle two-person nonzero-sum stochastic games with complete information and average cost criterion are treated. It is shown that there exists a quasi-stationary deterministic $2\epsilon$-equilibrium pair if in two related zero-sum games stationary, deterministic $\epsilon$-equilibrium pairs exist and a certain ergodicity property is satisfied.
- *Applications*. Last, but not least, the volume includes a section containing five papers that are devoted to game theoretic models arising from specific, and quite varied, applications. The importance of significant generic applications that stimulate the interest and imaginations of a new generation of game theorists cannot be overestimated. While the applications papers occasionally may lack powerful general results, that is compensated for by challenging the scholarly community with interesting and generic situations of conflict for which a general theory and successful analyses are yet to be developed. In particular, in the first paper of this section Dawid, Feichtinger, and Jørgensen model the conflict between a potential criminal offender and a law enforcement agency as a two-stage game with imperfect information. Relevant sequential equilibria are identified, and their dependence on parameters of the model is analysed. Bischi and Naimzada study stability properties of the Nash equilibrium of a discrete-time dynamical system on the plane representing a dynamic Cournot duopoly game. The loss of stability of the Nash equilibrium and the appearance of more complex attractors (as a certain parameter of the model changes its values) are described. Petrosjan and Zaccour construct the set of Nash equilibria for a multistage supergame model for the management of a downstream pollution dispute between two neighboring regions. A time-consistent procedure to share the cooperative outcome between the two players is devised. In the survey paper by de Palma transportation models are described. Stability of

the solutions using adjustment processes is also discussed. The final paper of this section by Sain, Spencer, Jr., Won, and Liberty deals with the problem of minimization of variance while meeting a certain mean cost level. Sufficient optimality conditions in the form of Hamilton–Jacobi–Bellman equations are derived. Application of results to the control of a representative civil structure under seismic disturbances is demonstrated.

We would like to conclude this preface by expressing our thanks and appreciation to many people who have worked so hard to make this volume of the Annals a success. In particular, we are indebted to all the members of the editorial board and the referees who gave generously of their time to ensure that the high quality of the Annals is maintained. We also acknowledge the help of the series editor, Professor Başar, for his helpful advice. Finally, we wish to acknowledge the essential contribution of Mrs. Angela McKay, the administrator of the Centre for Industrial and Applicable Mathematics at the University of South Australia. Angela maintained a manuscript database throughout the submission and review process and made sure that all submitted papers were processed as expeditiously as possible, given the constraints of rigorous refereeing by international experts. On behalf of the entire editorial board, we invite you to read and enjoy the many interesting contributions contained in this volume.

Adelaide, Australia

*Jerzy A. Filar*
*Vladimir Gaitsgory*

Hiroshima, Japan

*Koichi Mizukami*

# Contributors

*Eitan Altman*, INRIA BP 93, 2004 Route des Lucioles, 06902 Sophia Antipolis Cedex, France

*Tamar Başar*, Coordinated Science Laboratory, University of Illinois, 1308 West Main Street, Urbana, IL 61801, USA

*Gian Italo Bischi*, Instituto di Scienze Economiche, Università di Urbino, I-061029 Urbino, Italy

*Michael H. Breitner*, Technische Universität Clausthal, Institüt für Mathematik, D-38678, Clausthal-Zellerfeld, Germany

*Dean A. Carlson*, University of Toledo, Department of Mathematics, Toledo, OH 43606, USA

*Herbert Dawid*, University of Southern California, Department of Economics, Los Angeles, CA 90033, USA

*André de Palma*, Thema, University of Cergy-Pontoise, 33 Boulevard du Port, 95800 Cergy-Pontoise, France

*Harri Ehtamo*, Systems Analysis Laboratory, Helsinki University of Technology, PO Box 1100, 02015, Espoo, Finland

*Gustav Feichtinger*, Department of Operations Research and Systems Theory, Vienna University of Technology, A-1040 Vienna, Austria

*Eugene A. Feinberg*, Harriman School for Management and Policy, SUNY at Stony Brook, Stony Brook, NY 11794-3775, USA

*Alain B. Haurie*, Department of Management Studies, University of Geneva, 102 Carl-Vogt, CH-1211 Geneva, Switzerland; and GERAD-École des Hautes Etudes Cinnerciales, Montréal, Canada

*Naira Hovakimyan*, Institute of Mechanics, Armenian Academy of Sciences, Marshal Baghramian Street 24, Yerevan 375019, Armenia

*Steffen Jørgensen*, Department of Management, Odense University, DK-5230, Odense, Demnark

*Yutaka Kimura*, Department of Mathematics and Information Science, Graduate School of Science and Technology, Niigata University, 950-21, Niigata, Japan

*Masami Kurano*, Department of Mathematics, Faculty of Education, Chiba University, Chiba 263, Japan

*A.M. Kovshov*, Faculty of Applied Mathematics, St. Petersburg State University, 198904 Bibliotechnaja pl. 2., Petrodvorets, St. Petersburg, Russia

*Heinz-Uwe Küenle*, Institut für Mathematik, Brandenburgische Technische Universistät Cottbus, Postfach 10 13 44, 03013 Cottbus, Germany

*Rainer Lachner*, Technische Universität Clausthal, Institut für Mathematik, D-38678, Clausthal-Zellerfeld, Germany

*Stanley R. Liberty*, Department of Electrical Engineering, University of Nebraska-Lincoln, Lincoln, NE 68588, USA

*Wei Shi Lim*, Department of Decision Sciences, National University of Singapore, 10 Kent Ridge Crescent, Singapore 119260

*A.A. Melikyan*, Institute for Problems in Mechanics, Russian Academy of Sciences, Moscow, Russia

*Koichi Mizukami*, Faculty of Integrated Arts and Sciences, Hiroshima University, 1-7-1, Kagamiyama, Higashi-Hiroshima, Japan 739-8521

*G.S.R. Murthy*, Indian Statistical Institute, Street No. 8, Habsiguda, Hyderabad 500 007, India

*Jun-ichi Nakagami*, Department of Mathematics and Informatics, Faculty of Science, Chiba University, Chiba 263, Japan

*Ahmad Naimzada*, Unversità L. Bocconi, Via Sargattim, 1-20132 Milano, Italy

*Zigang Pan*, EE Department, Polytechnic University, Siz MetroTech Center, Brooklyn, NY 11201, USA

*T. Parthasarathy*, Indian Statistical Institute, 7, SJS Sansanwal Marg, New Delhi 110 016, India

*H. Josef Pesch*, Technische Universität Clausthal, Institut für Mathematik, D-38678, Clausthal-Zellerfeld, Germany

*Leon A. Petrosjan*, Faculty of Applied Mathematics, St. Petersburg State University 198904, Biblitechnaja pl. 2., Petrodvorets, St. Petersburg, Russia

*Tuomas Raivio*, Systems Analysis Laboratory, Helsinki University of Techology, PO Box 1100, 02015, Espoo, Finland

*Michael K. Sain*, Department of Electrical Engineering, University of Notre Dame, Notre Dame, IN 46556, USA

*B.F. Spencer, Jr.*, Department of Civil Engineering and Geological Sciences, University of Notre Dame, Notre Dame, IN 46556, USA

*Seigo Sasaki*, Department of Electrical Engineering, National Defense Academy, 1-10-20 Hashirimizu, Yokosuka, Kanagawa 239-8686, Japan

*Yoichi Sawasaki* Department of Mathematics and Information Science, Graduate School of Science and Technology, Niigata University, 95-021, Niigata, Japan

*Josef Shinar*, Faculty of Aerospace Engineering, Technion-Israel Institute of Technology, Technion City, Haifa 32000, Israel

*Adam Shwartz*, Electrical Engineering Department, Technion-Israel Institute of Technology, Technion City, Haifa 32000, Israel

*B. Sriparna*, Indian Statistical Institute, 7, SJS Sansanwal Marg, New Delhi 110 016, India

*Kensuke Tanaka*, Department of Mathematics, Faculty of Science, Niigata University, 950-21, Niigata, Japan

*Kenko Uchida*, Department of Electrical and Computer Engineering, Waseda University, 3-4-1 Okubo, Shinjuku, Tokyo 169, Japan

*Chang-Hee Won*, TT & C Section, ETRI, Taejon, Korea 305-600

*Mingqing Xiao*, Coordinated Science Laboratory, University of Illinois, 1308 West Main Street, Urbana, IL 61861, USA

*Hua Xu*, Graduate School of Systems Management, The University of Tsukuba, 3-29-1 Otsuka Bunkyoku, Tokyo, Japan 112-0012

*Masami Yasuda*, Department of Mathematics and Informatics, Faculty of Science, Chiba University, Chiba 263, Japan

*Georges Zaccour*, Gerad-École des Hautes Etudes Cinnerciales, 3000, Chemin de la Cote-Sainte-Catherine, Montreal, Quebec, Canada H3T 2A7

# PART I
# Robust Control Design and $H^\infty$

# Worst-Case Rate-Based Flow Control with an ARMA Model of the Available Bandwidth

Eitan Altman
INRIA B.P. 93
Sophia-Antipolis Cedex, France

Tamer Başar
University of Illinois
Coordinated Science Laboratory
Urbana, Illinois, USA

Naira Hovakimyan
Armenian Academy of Sciences
Institute of Mechanics
Yerevan, Armenia

**Abstract**

A central control problem that arises in high-speed networks is the control of the rate of flow of information into the network. A rate that is too high may result in congestion and hence in the degradation of performance measures: high delays within the network, high loss probabilities of information packets, and large delay variations; a rate that is too low may result in the under-utilization of the network and in low throughput. Most of the existing and future telecommunications networks therefore include dynamic flow control mechanisms, which have often been developed on growing available experience, using ad hoc techniques that did not come as a result of a control-theoretical study. This is due to the high complexity of the controlled systems, that are typically decentralized, have nonlinear dynamics, and may only use partial noisy delayed information. Some attempts have been made in recent years to use control theory to design flow controllers with, however, no explicit objective functions to be minimized; moreover, the class of control policies in existing theoretical work is quite restricted. In this paper we formulate explicitly some cost criteria to be minimized, related to performance measures mentioned above, including delays, throughputs, and loss probabilities. We present an approximating linearized model with quadratic cost, and follow two approaches to model interfering traffic and other unknown data: the $H^\infty$ approach, and LQG approach. We determine for both cases the optimal controllers, and illustrate the theoretical results using simulation.

## 1 Introduction

Congestion phenomena in high-speed networks are quite frequent, both on the Internet as well as in the ATM (Asynchronous Transfer Mode) both of which are widely used in today's high-speed telecommunications networks. Congestion results in the degradation of many important performance measures; it causes losses of packets, it increases delays, and decreases the transmission throughput. In order to protect the users from congestion, on one hand, and to allow for the efficient use of the network, on the other, flow control mechanisms have been introduced and adopted both in the Internet and in the ATM.

Several approaches to flow control in high-speed networks, based on some feedback on the state of the network, have been proposed: window flow control (e.g., [13]), credit-based control (see [15] and [16] and references therein), and rate-based flow control (see [9] and [16] and references therein). The last one has been selected by the ATM forum in 1994 as the main approach for flow control in the Available Bit Rate (ABR) type of service, see [1]. This class of service is designed to support applications that can adjust their input rate to the congestion state of the network (e.g., file transfers), as opposed to some real-time applications (such as high quality interactive video or audio applications) that typically use other higher priority services and to which a fixed amount of bandwidth is allocated. In networks that integrate several types of traffic, the flow control is used to adjust the rate of transmission of the first type of traffic to the time-varying bandwidth left over by the higher priority real-time applications. The ATM forum has not standardized the flow control mechanisms and does not wish to do so; its design and implementation are left to the switch designers.

We present in this paper two types of linear quadratic (LQ) models for designing optimal rate based control, so as to achieve optimal performance measures: LQG (Linear–Quadratic–Gaussian) and $H^\infty$. In both cases we allow for noisy delayed information. The objectives in the design of the controllers are to

i. minimize the variation of the bottleneck queue size around some desired level; this objective measure is directly related to the minimization of losses and maximization of throughput;
ii. obtain good tracking between input and output rates; this objective is related to the efficient use of the network; and
iii. minimize the jitter.

The relative importance (weight) of each one of these objectives is given as data in the design of the controller. The optimal controllers that are obtained are easy to implement: they turn out to be linear in the (estimated) state.

We consider a discrete time, linearized model for the dynamics, as introduced in [2] (which extends the models in [8] and [14]). A time unit corresponds to the time interval between two consecutive acquisitions of new information. According to [1], such information is available through RM (Resource Management) packets that travel from the source to the destination (forward packets) and then back from the destination to the source (backward packets). Information on the congestion (queue-length) and available bandwidth may be stamped onto these packets in the intermediate switches. We make the assumption that the RM packets are sent periodically. We model the interfering traffic, which may stem from higher priority real-time applications, by a general ARMA process. This is a generalization of the AR model for interfering traffic that was introduced in [3] and [4], which thus allows us to have a much larger class of models for the bandwidth available for the ABR traffic.

We describe briefly in the next section the linearized model for the state dynamics; detailed derivations of the nonlinear model as well as its linear approximation can be found in [2]. Simulations further justify this linearization; see [2].

We assume that the performance measures (such as throughputs, delays, and loss probabilities) are determined essentially by a bottleneck node. This assumption has both theoretical and experimental [7], [10] justification, and is often made in the literature in order to analyze or design controllers [2], [8], [12].

The control problem is solved in Sections 3 and 4 for the case when there is no delay in the measurements, and in Section 5 when the imperfect measurements are acquired with some delay, or alternatively when the control is updated more then once during a round trip. Simulation results are presented in Section 6, and some discussion on their interpretation is included in Section 7. The paper ends with the concluding remarks of Section 8, that summarize and highlight the advantages of the methodology and schemes introduced here.

## 2 The Model

We begin by providing the discrete-time model considered in this paper, where a time unit corresponds to a round trip delay. The more general model where a time unit corresponds to the period between consecutive instants of information acquisition, which may be smaller than a round trip delay, is formulated and studied later in Section 5. Let $q_n$ denote the queue length at a bottleneck link, and let $\mu_n$ denote the effective service rate available for traffic of the give source at that link at the beginning of the $n$th time slot. Let $x_n$ denote the source rate during the $n$th time slot. We thus consider a rate-based flow control where, based on the current and previous noisy information on $\mu_n$, $q_n$, and $x_n$ (to be made precise below), the controller updates the transmission rate. Hence, the queue length evolves according to

$$q_{n+1} = q_n + x_n - \mu_n \tag{1}$$

Since several sources with varying transmission rates may share the same bottleneck link, the service rate $\mu_n$ available to the controlled source may change over time in an unpredictable way. The other sources may be represented as some interference, which we model as a stable ARMA process:

$$\mu_n = \mu + \xi_n, \qquad \xi_n = \sum_{i=1}^{d} \ell_i \xi_{n-i} + \sum_{i=1}^{d} k_i \phi_{n-i}. \tag{2}$$

We assume that $\ell_i$'s, which will typically be obtained by some estimation/identification procedure (using, e.g., the results and framework of [11]), are such that (2) describes a stable system. The variable $\phi_n$ in (2) stands for disturbance, the nature of which will be described shortly.

At the end of step $n$, the source obtains a noisy estimate of the value of $\mu_n$, and a noisy estimate of the value of $q_n$. We use $\hat{\mu}_n$ to denote a measurement for the service rate at the bottleneck node, and $\hat{q}_n$ to denote a measurement for the bottleneck queue size. For our analysis, we assume that these measurements can be written as

$$\hat{\mu}_n = \mu_n + c v_n, \qquad \hat{q}_n = q_n + a w_n, \tag{3}$$

where $v_n$ and $w_n$ are disturbances whose nature will be described shortly, and $a$ and $c$ are some constants.

We introduce shifted versions of the variables $x_n$, $q_n$, $\hat{\mu}_n$, $\hat{q}_n$:

$$\tilde{x}_n := s_n - \mu, \qquad \tilde{q}_n := q_n - Q, \qquad \hat{\tilde{\mu}}_n := \hat{\mu}_n = \mu, \qquad \tilde{q}_n := \hat{q}_n - Q. \tag{4}$$

We further define the (scaled) variation in the input rate by $bu_n$, where $u_n$ will be called the control, and $b$ is some constant parameter. In terms of the new variables, we now have the following dynamics:

$$\tilde{q}_{n+1} = \tilde{q}_n + \tilde{x}_n - \xi_n, \tag{5}$$

$$\xi_n = \sum_{i=1}^{d} \ell_i \xi_{n-i} + \sum_{i=1}^{d} k_i \phi_{n-i}, \tag{6}$$

$$\tilde{x}_{n+1} = \tilde{x}_n + bu_n, \tag{7}$$

and the observations

$$\hat{\tilde{\mu}}_n = \xi_n + c v_n, \qquad \hat{\tilde{q}}_n = \tilde{q}_n + a w_n. \tag{8}$$

For each pair of integers $(k, K)$, with $K > k$, we introduce the following quadratic cost:

$$L_k^K = g\|\tilde{x} - \xi\|^2_{(k,K)} + \|\tilde{q}\|^2_{(k,K)} + \|u\|^2_{(k,K)}, \quad \text{where } g > 0,$$

$$\|u\|^2_{(k,K)} := \sum_{n=k}^{K} |u_n|^2, \quad \text{etc.} \tag{9}$$

This definition of the cost, over a period of length $K + 1 - k$ units, that is allowed to be infinite, allows us to quantify optimization criteria that have been used in previous control models [2]:

- The first term represents that quality with which the input rate tracks the available service rate, where $g$ is a positive weighting term. A high rate of tracking is known to be desirable; see, e.g., [8].
- The second term is a penalty for deviating from a desirable queue length. Setting some desired level of queue reflects the fact that we do not wish to have a large queue, so as to avoid losses, and on the other hand, we do not wish to have a low level of queue, since if the queue is empty and the input rate is lower than the service rate then there is a waste of potential throughput. We have not included any additional weighting on this term, as any such positive weight can be absorbed into the other variables.
- The last term stands for a penalty for high jitter, i.e., for high variability of (input) transmission rate, which is known to be undesirable; see [8].[1] In many proposed telecommunication networks, the source will pay more for higher burstiness (variability) of its input rate, since a highly variable input rate has typically a bad influence on other traffic. Again, as in the case of the second term, the weighting here has been normalized to 1, but any other weighting can be absorbed in the parameter $b$.

We initially allow the controller to be a function of all the past and present measurements and past actions:

$$u_n = f_n\left(\tilde{x}_m, \hat{\tilde{\mu}}_m, \hat{\tilde{q}}_m, u_{m-1}, m = k, \dots, n\right), \qquad n \geq k. \tag{10}$$

Later, in Section 5, we will consider the case where the dependence on $\hat{\tilde{\mu}}$ and $\hat{\tilde{q}}$ is with a delay of $\theta$ time units.

To obtain steady state solutions (controllers), we take the number of periods (time slots) in the problem to be sufficiently large; in particular, we let $k = -\infty$ and $K = \infty$. Over the "doubly-infinite" control horizon, we consider two types of models for the disturbances:

- **M1:** The Gaussian model. Here $v_n, w_n, \phi_n$ are taken as independent identically distributed Gaussian random variables with zero mean and unit variance. It then follows that the measurements $\tilde{\mu}_n$ and $\tilde{q}_n$ are unbiased. The cost to be minimized in this case is the "per time slot" expected value of the cost, written in precise mathematical terms as

$$J_G(f) = \limsup_{K\to\infty, k\to-\infty} \frac{1}{K+1-k} E\left[L_k^K(f;\phi, v, w)\right]. \tag{11}$$

- **M2:** The $H^\infty$ approach. In this case, $v_n$, $w_n$, and $\phi_n$ are taken as unknown (deterministic) disturbances. The cost to be minimized is the $\ell_2$ gain from the

[1] Note that if flows from several sources are multiplexed in a FIFO queue, and if the input rate of one source is highly variable, then also the output rate for that source tends to be highly variable, since packets from the other sources come in between the packets of that source.

disturbances to the cumulative cost $L_{-\infty}^{\infty}$, i.e.,

$$J_H(f) = \sup_{\{\phi_n, v_n, w_n\}_{n=-\infty}^{\infty}} \frac{L_{-\infty}^{\infty}(f; \phi, v, w)}{\|\phi, v, w\|^2}, \tag{12}$$

where

$$\|\phi, v, w\|^2 := \|\phi\|^2 + \|v\|^2 + \|w\|^2 = \sum_{n=-\infty}^{\infty} \left(|\phi_n|^2 + |v_n|^2 + |w_n|^2\right). \tag{13}$$

Here we also take the initial conditions for the state equations (5)–(7) in the infinite past to be *zero*, i.e.,

$$\lim_{n \to -\infty} \tilde{x}_n = 0, \qquad \lim_{n \to -\infty} \tilde{q}_n = 0, \qquad \lim_{n \to -\infty} \xi_n = 0. \tag{14}$$

Let us denote the square-root of the infimum of this cost function over all admissible controllers $\{f_n\}$ by $\gamma^*$:

$$\left(\gamma^*\right)^2 = \inf_{\{f_n\}_{n=-\infty}^{\infty}} J_H(f). \tag{15}$$

If there exists an actual minimizing control, say $f^*$, to the optimization problem above, then it can be shown [6] that, defining a soft-constrained cost function

$$L_\gamma(u; \phi, v, w) := L_{-\infty}^{\infty}(u; \phi, v, w) - \gamma^2 \|\phi, v, w\|^2,$$

to be viewed as the kernel of a two-player game, to be minimized by Player 1 (controlling $u$) and maximized by Player 2 (controlling $(\phi, v, w)$) $f^*$ has the property: $\sup_{\phi,v,w} L_{\gamma^*}(f^*; \phi, v, w) = \inf_f \sup_{\phi,v,w} L_{\gamma^*}(f; \phi, v, w)$. The quantity above is the upper value of the zero-sum dynamic game with kernel $L_{\gamma^*}$, which is in fact equal to zero. It can actually be shown that for any $\gamma \geq \gamma^*$, the upper value of the game with parameterized kernel $L_\gamma$ is zero, and for $\gamma < \gamma^*$, its upper value is infinite. Hence, $\gamma^*$ is the smallest positive scalar $\gamma$ for which the zero-sum game with kernel $L_\gamma$ has a finite upper value.

Instead of obtaining $f^*$ defined above, we will in fact solve a parameterized class of controllers, $\{f^\gamma, \gamma > \gamma^*\}$, where $f^\gamma$ is obtained from $\sup_{\phi,v,w} L_\gamma(f^\gamma; \phi, v, w) = \inf_f \sup_{\phi,v,w} L_\gamma(f; \phi, v, w)$. The controller $f^\gamma$ will clearly have the property that it ensures a performance level $\gamma^2$ for the index adopted for M2, i.e., the attenuation is bounded by

$$\mathcal{A}(f^\gamma; \phi, v, w) := \frac{\left\{L_{-\infty}^{\infty}(f^\gamma; \phi, v, w)\right\}^{1/2}}{\|\phi, v, w\|} \leq \gamma \qquad \text{for all } \phi, v, w. \tag{16}$$

It will turn out that the limit $\lim_{\gamma \to \infty} f^\gamma =: f^\infty$ is a well-defined controller, and solves uniquely the control problem with the Gaussian model, M1.

## 3 Complete Solution to the Problem with Model M2 and with an Additional Noise Perturbation in the Measurements

The optimal control problem formulated above can be solved by a suitable modification of the theory of discrete-time $H^\infty$-control developed in [6]. Toward this end, we first write (5)–(7) in standard state variable form. Introduce the $d$-dimensional vector $\eta_n := (\xi_{n-d+1}, \ldots, \xi_n)'$, that would be obtained from (6) by taking $k_1 = 1, k_i = 0, i \neq 1$, in terms of which (6) can be rewritten as a first order difference equation:

$$\eta_{n+1} = C\eta_n + \begin{pmatrix} 0 \\ \vdots \\ 1 \end{pmatrix} \phi_n, \quad \text{where} \quad C := \begin{pmatrix} 0 & 1 & 0 & 0 & \ldots & 0 \\ 0 & 0 & 1 & 0 & \ldots & 0 \\ . & . & . & . & \ldots & \\ \ell_d & \ell_{d-1} & \ell_{d-2} & & \ldots & \ell_1 \end{pmatrix}. \tag{17}$$

Then the actual $\xi_n$ generated by (6) is given by

$$\xi_n = (k_d, \ldots, k_1)\, \eta_n. \tag{18}$$

Further introducing the $(d+2)$-dimensional vector $z_n := \left(\tilde{x}_n, \tilde{q}_n, \eta_n'\right)'$, we can write (5)–(7) in the form:

$$z_{n+1} = Az_n + D\phi_{n+1} + Bu_n, \tag{19}$$

where

$$A = \left(\begin{array}{cc|ccc} 1 & 0 & 0 & \ldots\ 0 & 0 \\ 1 & 1 & -k_d & \ldots\ -k_2 & -k_1 \\ \hline 0 & & & C & \end{array}\right), \qquad D := \begin{pmatrix} 0 \\ \cdot \\ \cdot \\ \cdot \\ 0 \\ 1 \end{pmatrix}, \qquad B := \begin{pmatrix} b \\ 0 \\ \cdot \\ \cdot \\ \cdot \\ 0 \end{pmatrix}. \tag{20}$$

Now, the stability of the ARMA process $\xi_n$ generated by (2) is equivalent to the stability of the matrix $C$ in (17), which is captured in the following assumption required to hold throughout the remaining analysis:

- **A1.** All eigenvalues of $C$ are in the unit circle in the complex plane.

Under A1, we have the following fact,[2] which will be useful in the statement of our main result:

- **F1.** Under A1, the pair $(A, B)$ is stabilizable.

---

[2]This follows from the Hautus test by observing that the unstable eigenvalue 1 (of multiplicity 2) of the matrix $A$ is controllable, i.e., $\text{rank}(A - 1|B) = d + 2$.

As far as the measurements available to the controller go, we have two disturbance corrupted state measurements, given by (8), and one perfect measurement of the state, $\tilde{x}_n$. This is a "mixed" structure, that leads to a *singular* measurement equation, which is not allowed in standard theory. To circumvent this difficulty, we will first perturb $\tilde{x}_n$ with a small independent disturbance, to arrive at the measurement equation

$$\hat{\tilde{x}}_n = \tilde{x}_n + \epsilon \tilde{\tilde{v}}_n, \qquad 0 < \epsilon << 1, \tag{21}$$

and assume that now $\hat{\tilde{x}}_n$ is available for control purposes, instead of $\tilde{x}_n$. In the above, $\tilde{\tilde{v}}_n$ is the auxiliary disturbance introduced, and $\epsilon$ is a small positive quantity. This modification is only for mathematical convenience, which will in the end be made to vanish asymptotically by letting $\epsilon \to 0$, so as to recover the original information structure.

To introduce the modified measurement equation, let $y_n := \left(\hat{\tilde{x}}_n, \hat{\tilde{q}}_n, \hat{\tilde{\mu}}_n\right)'$, Then $y_n = Hz_n + E_\epsilon \tilde{v}_n$, where

$$H := \begin{pmatrix} 1\ 0\ 0 \dots 0\ 0 \\ 0\ 1\ 0 \dots 0\ 0 \\ 0\ 0\ 0 \dots 0\ 1 \end{pmatrix}, \quad E_\epsilon := \begin{pmatrix} \epsilon & 0 & 0 \\ 0 & a & 0 \\ 0 & 0 & c \end{pmatrix}, \quad \tilde{v} := \begin{pmatrix} \tilde{\tilde{v}}_n \\ w_n \\ v_n \end{pmatrix}. \tag{22}$$

Since $E_\epsilon E'_\epsilon > 0$ for all $\epsilon > 0$, this is a nonsingular measurement equation. Compatible with this modification we also modify the state equation (19), to write it as $z_{n+1} = Az_n + Bu_n + D_\epsilon \zeta_n$, where

$$D_\epsilon := \begin{pmatrix} 0 & 0 & \dots & 0 & 1 \\ \epsilon & 0 & \dots & 0 & 0 \end{pmatrix}, \qquad \zeta_n := \left(\phi_{n+1}, \tilde{\phi}_{n+1}\right)', \tag{23}$$

with $\tilde{\phi}_{n+1}$ being the new disturbance introduced.

To complete the description, we have to write the cost function (9) in terms of these new variables

$$L = \sum_{n=-\infty}^{\infty} \left\{ |z_n|_N^2 + |u_n|^2 \right\}, \tag{24}$$

where

$$N := \left(\begin{array}{cc|cccc} g & 0 & -gk_d & -gk_{d-1} & \dots & -gk_1 \\ 0 & 1 & 0 & 0 & \dots & 0 \\ \hline -gk_d & 0 & gk_d^2 & gk_dk_{d-1} & \dots & gk_dk_1 \\ -gk_{d-1} & 0 & gk_dk_{d-1} & gk_{d-1}^2 & \dots & gk_{d-1}k_1 \\ \cdot & \cdot & \cdot & \cdot & \dots & \cdot \\ \cdot & \cdot & \cdot & \cdot & \dots & \cdot \\ \cdot & \cdot & \cdot & \cdot & \dots & \cdot \\ -gk_1 & 0 & gk_dk_1 & gk_{d-1}k_1 & \dots & gk_1^2 \end{array}\right), \tag{25}$$

and the $H^\infty$ criterion (to be minimized) now becomes

$$\sup_{\tilde{v},\zeta} \frac{L}{\|\tilde{v},\zeta\|^2}, \tag{26}$$

the square-root of whose infimum over all admissible controllers we now denote by $\gamma^*(\epsilon)$. It is continuous at $\epsilon = 0$, and the quantity $\gamma^*(0)$ is equal to the $\gamma^*$ defined earlier (for the unmodified problem).

Three additional properties, related to the system matrices, are now worth noting, which we state below as facts:

- **F2** The pair $(A, H)$ is detectable.
- **F3** The pair $(A, N)$ is detectable.
- **F4** For each $\epsilon > 0$, the pair $(A, D_\epsilon)$ is controllable.

We are now in a position to state the solution to the modified $H^\infty$ control problem formulated above, directly from Başar and Bernhard [6, Chap. 6].

**Proposition 3.1.** *Let $\epsilon > 0$ be fixed, and* A1 *hold. Then, $\gamma^*(\epsilon)$ defined above is finite, and for all $\gamma > \gamma^*(\epsilon)$:*

(1) *There exists a minimal nonnegative-definite solution, $M_\epsilon$, to*

$$M = N + A'M\Lambda_\epsilon^{-1}A \tag{27}$$

*such that*

$$\gamma^2 I - D_\epsilon' M_\epsilon D_\epsilon > 0, \tag{28}$$

*where*

$$\Lambda_\epsilon := I + \left(BB' - \gamma^{-2}D_\epsilon D_\epsilon'\right)M. \tag{29}$$

(2) *There exists a minimal nonnegative-definite solution, $\Sigma_\epsilon$, to*

$$\Sigma = D_\epsilon D_\epsilon' + A\Sigma R_\epsilon^{-1}A' \tag{30}$$

*such that*

$$\gamma^2 I - N^{1/2}\Sigma_\epsilon N^{1/2} > 0, \tag{31}$$

*where*

$$R_\epsilon := I + \left(H'\left(E_\epsilon E_\epsilon'\right)^{-1}H - \gamma^{-2}N\right)\Sigma. \tag{32}$$

(3)

$$\Sigma_\epsilon^{1/2}M_\epsilon\Sigma_\epsilon^{1/2} < \gamma^2 I. \tag{33}$$

(4) *An $H^\infty$-controller that ensures the bound $\gamma^2$ in the index* (26) *is*

$$u_{n,\epsilon}^\gamma = f_\epsilon^\gamma\left(\hat{z}_{n|n}\right) = -B'M_\epsilon\Lambda_\epsilon^{-1}A\hat{z}_{n|n}, \tag{34}$$

*where $\hat{z}_{n|n}$ is generated by*

$$\hat{z}_{n|n} = \left(I + \Sigma_\epsilon H' \left(E_\epsilon E_\epsilon'\right)^{-1} H - \gamma^{-2}\Sigma_\epsilon M_\epsilon\right)^{-1} \left(\check{z}_n + \Sigma_\epsilon H' \left(E_\epsilon E_\epsilon'\right)^{-1} y_n\right), \tag{35}$$

$$\check{z}_{n+1} = A\check{z}_n + Bu_n + A\Sigma_\epsilon R_\epsilon^{-1}\left[\gamma^{-2} N \check{z}_n + H' \left(E_\epsilon E_\epsilon'\right)^{-1}(y_n - H\check{z}_n)\right]. \tag{36}$$

(5) *The controller* (34) *leads to a bounded input-bounded state stable system, i.e., for all bounded disturbances $\tilde{v}$, $\zeta$, the system state $z_n$ and the filter state $z_{n|n}$ remain bounded. Equivalently, the two matrices*

$$\left(I - BB'M_\epsilon\Lambda_\epsilon^{-1}\right) A \qquad \text{and} \qquad \left(I - H'\left(E_\epsilon E_\epsilon'\right)^{-1} H\Sigma_\epsilon R_\epsilon^{-1}\right) A'$$

*have all their eigenvalues in the unit circle.*

*If any one of the three properties* (1)–(3) *above does not hold, then $\gamma \le \gamma^*(\epsilon)$.*

**Remark 3.1.** The results of Proposition 3.1 are valid not only for all finite $\gamma$'s larger than $\gamma^*(\epsilon)$, but also as $\gamma \to \infty$. In this limiting case, which captures the solution to the $\epsilon$-perturbed problem under the Gaussian model (M1) of the disturbance, the constraints (28), (31), and (32) become irrevelant, and all other expressions admit "well-defined" forms, obtained by simply setting $\gamma^{-2} = 0$.

## 4 The Limiting Solution as $\epsilon \downarrow 0$

We next study the limit of the solution given in Proposition 3.1 as $\epsilon \downarrow 0$. For the limit to be well defined, it will be sufficient to show that equations and relationships (27)–(33) are well defined in the limit as $\epsilon \downarrow 0$. The analysis here is similar to that carried out in [4] for the case where interfering traffic was modeled by an AR process, with just $A$ and $N$ having different structures; it is reproduced here in full for the sake of completeness.

The first three of (27)–(33) depend continuously on $\epsilon \ge 0$, and hence in the limit (27) will be replaced by

$$M = N + A'M\Lambda_0^{-1}A, \tag{37}$$

where

$$\Lambda_0 := I + \text{diag}\left(b^2, 0, \ldots, 0, -\gamma^{-2}\right) M \tag{38}$$

and (37) will have to be solved under the scalar condition (as counterpart of (28)):

$$(0, \ldots, 0, 1)M(0, \ldots, 0, 1)' < \gamma^2, \tag{39}$$

which guarantees invertibility of (38). Again, for each $\gamma > \gamma^*(0)$, (37) admits a unique minimal nonnegative-definite solution, which also satisfies (39).

The limiting analysis of (30)–(32) is somewhat more complicated, since it involves the inverse of a matrix that becomes singular at $\epsilon = 0$ (which is $E_\epsilon E'_\epsilon$). However, this isolated signularity does not lead to any singularity for the solution of (30) as to be shown below. We will in fact show that in the limit as $\epsilon \downarrow 0$, the solution of (30) will be in the form

$$\Sigma_0 = \begin{pmatrix} 0 & \dots & 0 \\ \vdots & P & \\ 0 & & \end{pmatrix}, \tag{40}$$

where $P \geq 0$ is of dimensions $(d+1) \times (d+1)$. To see this, let us first rewrite $\Sigma R_\epsilon^{-1}$ as

$$\Sigma R_\epsilon^{-1} = \Sigma^{1/2} \left[ I + \Sigma^{1/2} H' \left( E_\epsilon E'_\epsilon \right)^{-1} H \Sigma^{1/2} - \gamma^{-2} \Sigma^{1/2} N \Sigma^{1/2} \right]^{-1} \Sigma^{1/2},$$

where $\Sigma^{1/2}$ denotes the unique nonnegative definite square root of $\Sigma$. Let the $[(d+1) \times (d+1)]$-dimensional lower block of $N$ be denoted by $\tilde{N}$, which is

$$\tilde{N} := \left( \begin{array}{c|cccc} 1 & 0 & 0 & \dots & 0 \\ \hline 0 & g k_d^2 & g k_d k_{d-1} & \dots & g k_d k_1 \\ 0 & g k_d k_{d-1} & g k_{d-1}^2 & \dots & g k_{d-1} k_1 \\ \cdot & \cdot & \cdot & \dots & \cdot \\ \cdot & \cdot & \cdot & \dots & \cdot \\ \cdot & \cdot & \cdot & \dots & \cdot \\ 0 & g k_d k_1 & g k_{d-1} k_1 & \dots & g k_1^2 \end{array} \right) \tag{41}$$

and suppose that $\Sigma$ is given by (40), where $P$ is arbitrary. Then, straightforward manipulations lead to

$$\Sigma^{1/2} N \Sigma^{1/2} = \text{blockdiag}\left(0,\, P^{1/2} \tilde{N} P^{1/2}\right),$$

$$\Sigma^{1/2} H' \left( E_\epsilon E'_\epsilon \right)^{-1} H \Sigma^{1/2} = \text{blockdiag}\left(0,\, P^{1/2} \tilde{H}' (\tilde{E}\tilde{E}')^{-1} \tilde{H} P^{1/2}\right),$$

where

$$\tilde{H} := \underbrace{\begin{pmatrix} 1 & 0 & \dots & 0 & 0 \\ 0 & 0 & \dots & 0 & 1 \end{pmatrix}}_{2\times(d+1)}, \qquad \tilde{E} := \underbrace{\begin{pmatrix} a & 0 \\ 0 & c \end{pmatrix}}_{2\times 2}. \tag{42}$$

Hence,

$$\begin{aligned} \Sigma R_\epsilon^{-1} &= \Sigma^{1/2} \left[ \text{blockdiag}\left(1,\, I + P^{1/2} \left( \tilde{H}' (\tilde{E}\tilde{E}')^{-1} \tilde{H} - \gamma^{-2} \tilde{N} \right) P^{1/2} \right) \right]^{-1} \Sigma^{1/2} \\ &= \text{blockdiag}\left(0,\, P^{1/2} \left[ I + P^{1/2} \left( \tilde{H}' (\tilde{E}\tilde{E}')^{-1} \tilde{H} - \gamma^{-2} \tilde{N} \right) P^{1/2} \right]^{-1} P^{1/2} \right) \\ &= \text{blockdiag}\,(0, \tilde{S}), \end{aligned} \tag{43}$$

which is independent of $\epsilon$. Pre- and post-multiplying this expression by $A$ and $A'$, we obtain

$$A\Sigma R_{\epsilon}^{-1}A' = \text{blockdiag}\left(0,\ \tilde{A}P^{1/2}\left[I + P^{1/2}\left(\tilde{H}'(\tilde{E}\tilde{E}')^{-1}\tilde{H} - \gamma^{-2}\tilde{N}\right)P^{1/2}\right]^{-1} \times P^{1/2}\tilde{A}'\right),$$

where $\tilde{A}$ is the $[(d+1)\times(d+1)]$-dimensional lower block of $A$, i.e.,

$$\tilde{A} := \left(\begin{array}{c|c} 1 & -k_d \ldots -k_2 - k_1 \\ \hline 0 & \\ \vdots & C \\ 0 & \end{array}\right). \tag{44}$$

Furthermore, since $\lim_{\epsilon\to 0} D_\epsilon D_\epsilon' = \text{diag}(0,\ldots,0,1)$, it readily follows that the structure (40) is consistent with (30). Introducing a $(d+1)$-dimensional vector $\tilde{D}$,

$$\tilde{D} := (0,\ldots,0,1)', \tag{45}$$

the equation to be satisfied by $P$ can now be written as

$$P = \tilde{D}\tilde{D}' + \tilde{A}P\left[I + \left(\tilde{H}'(\tilde{E}\tilde{E}')^{-1}\tilde{H} - \gamma^{-2}\tilde{N}\right)P\right]^{-1}\tilde{A}', \tag{46}$$

and (31) becomes equivalent to

$$\gamma^2 I - \tilde{N}^{1/2}P\tilde{N}^{1/2} > 0. \tag{47}$$

Finally, (33) simplifies to

$$P^{1/2}\tilde{M}P^{1/2} < \gamma^2 I, \tag{48}$$

where $\tilde{M}$ is the $[(d+1)\times(d+1)]$-dimensional lower block of $M$.

Hence, (27)–(29) are well-defined at $\epsilon = 0$, and (30)–(33) admit well-defined limits in the space of dimension $(d+1)$ instead of $(d+2)$.

To complete the limiting analysis, we still have to find the limiting form of the controller (34)–(36). Toward this end, write $z$, as $\epsilon \downarrow 0$, as

$$z = \left(\alpha, \beta'\right)', \tag{49}$$

where $\alpha$ is of dimension 1, and $\beta$ of dimension $d+1$. Compatibly, write (again as $\epsilon \downarrow 0$):

$$\hat{z}_{n|n} = \left(\hat{\alpha}_{n|n}, \hat{\beta}_{n|n}'\right)', \qquad \breve{z}_n = \left(\breve{\alpha}_n, \breve{\beta}_n'\right)'. \tag{50}$$

Then, using the earlier manipulations, and some straightforward extensions, it can be shown that $\hat{\alpha}_{n|n}, \hat{\beta}_{n|n}, \breve{\alpha}_n, \breve{\beta}_n$ are generated by (from (35)–(36), and using the notation introduced by (41)–(45)):

$$\hat{\alpha}_{n|n} = \breve{\alpha}_n, \tag{51}$$

$$\breve{\alpha}_{n+1} = \breve{\alpha}_n + bu_n, \tag{52}$$
$$\hat{\beta}_{n|n} = \left(I + P\tilde{H}'(\tilde{E}\tilde{E}')^{-1}\tilde{H} - \gamma^{-2}P\tilde{M}\right)^{-1}\left(\breve{\beta}_n + P\tilde{H}'(\tilde{E}\tilde{E}')^{-1}\tilde{y}_n\right), \tag{53}$$
$$\breve{\beta}_{n+1} = \tilde{A}\breve{\beta}_n + \tilde{A}\tilde{S}\left[\gamma^{-2}\tilde{N}\breve{\beta}_n + \tilde{H}'(\tilde{E}\tilde{E}')^{-1}\left(\tilde{y}_n - \tilde{H}\breve{\beta}_n\right)\right], \tag{54}$$
$$\tilde{y}_n := \left(\hat{\tilde{q}}_n, \hat{\tilde{\mu}}_n\right)'. \tag{55}$$

Clearly, (52) generates $\tilde{x}_n$ whose value is already available to the controller. Hence, the limiting controller is of dimension $d+1$, with the worst-case observer given by (53)–(54). The counterpart of (34) is then

$$u_n^\gamma = f^\gamma\left(x_n, \hat{\beta}_{n|n}\right) = -B'M_0\Lambda_0^{-1}A\begin{pmatrix}\tilde{x}_n \\ \hat{\beta}_{n|n}\end{pmatrix}, \tag{56}$$

and this achieves a level of disturbance attenuation no worse than $\gamma$, provided that $\gamma > \gamma^*$, where $\gamma^*$ is the optimum level of disturbance attenuation for the unperturbed problem, which can be computed as $\gamma^* = \lim_{\epsilon\downarrow 0}\gamma^*(\epsilon)$.

To see this, first note that for each fixed controller, the attenuation is monotonically nonincreasing with decreasing $\epsilon > 0$, and hence its inf sup value, $\gamma^*(\epsilon)$, also has the same property. Furthermore, since $\gamma^*(\epsilon) \geq 0$, the set $\{\gamma^*(\epsilon)\}_{\epsilon>0}$ has a limit as $\epsilon \downarrow 0$. If follows from the algebraic characterization of $\gamma^*(\epsilon)$, which was in terms of two Riccati equations and a spectral radius condition which are well defined at $\epsilon = 0$ (as discussed earlier), that its limit as $\epsilon \downarrow 0$ is indeed $\gamma^*(0)$. Furthermore, $\gamma^*(0) = \gamma^*$, because clearly $\gamma^* > \gamma^*(0)$ is not possible as this would imply $\gamma^* > \gamma^*(\epsilon)$ for some sufficiently small $\epsilon > 0$, contradicting the initial hypothesis that $\gamma^*$ corresponds to a more refined information for the controller than $\gamma^*(\epsilon)$; $\gamma^* < \gamma^*(0)$ is not possible either, because if $f^\gamma$ is a controller for the original problem, achieving a level of disturbance attenuation no worse than $\gamma$, with $\gamma^* < \gamma < \gamma^*(0)$, for the noise-perturbed problem with sufficiently small $\epsilon > 0$, still a performance level better than $\gamma^*(0)$ would be obtained using the same controller (not with noise perturbed measurement) by continuity of the controller with respect to the measurement, which would then violate the inequality $\gamma^*(0) \leq \gamma^*(\epsilon)$ for all $\epsilon > 0$. Now, finally, under the controller (56), a level of disturbance attenuation no worse than $\gamma > \gamma^*$ would be achieved because the controller $f_\epsilon^\gamma$, given by (34), is a continuous function of $\epsilon$, which is also well defined at $\epsilon = 0$, and given any $\hat{\gamma} > \gamma^*$, there exists $\epsilon_0 > 0$ such that for all $0 < \epsilon < \epsilon_0$, $\gamma^*(\epsilon) < \hat{\gamma}$, and

$$\sup_{\phi,v,w} L_{\hat{\gamma}}\left(f_\epsilon^{\hat{\gamma}}, \phi, v, w\right) = 0, \tag{57}$$

and by continuity at $\epsilon = 0$ (of also the concavity condition that asssures finiteness of the supremum in (57)),

$$\sup_{\phi,v,w} L_{\hat{\gamma}}\left(f^{\hat{\gamma}}, \phi, v, w\right) = 0,$$

where $f^{\hat{\gamma}}$ given by (56) is also $f_\epsilon^{\hat{\gamma}}\Big|_{\epsilon=0}$.

The result is now summarzied in the following theorem:

**Theorem 4.1.** *Consider the original control problem formulated in Section* 2, *under the $H^\infty$ model* (M2) *of the disturbance. Let Assumption* A1 *hold. Then $\gamma^*$ defined by* (15) *is finite, and for all $\gamma > \gamma^*$:*

(1) *There exists a minimal nonnegative-definite solution, $M_0$, to* (37) *such that* (39) *holds.*
(2) *There exists a minimal nonnegative-definite solution, $P$, to* (46) *such that* (47) *holds.*
(3) *Condition* (48) *holds.*
(4) *An $H^\infty$-controller that ensures the bound $\gamma^2$ in the index* (12) *is given by* (56).
(5) *The controller* (56) *leads to a bounded input-bounded state stable system* (*see Proposition* (3.1), *part* (5)).

*For a given $\gamma > 0$, if any one of the properties* (1)–(3) *above does not hold, then $\gamma \le \gamma^*$.*

We immediately have the following corollary to Theorem 4.1, obtained by letting $\gamma \to \infty$, which solves the original problem under the Gaussian interpretation (M1) of the disturbances (see Remark 3.1):

**Corollary 4.1.** *Under* A1, *the optimal control problem with the disturbance model* M1 *admits the solution:*

$$u_n^\infty = f^\infty\left(x_n, \hat{\beta}_{n|n}\right) = -B'\bar{M}(I + BB'\bar{M})^{-1}A\begin{pmatrix}\tilde{x}_n \\ \hat{\beta}_{n|n}\end{pmatrix} =: \mathcal{K}\begin{pmatrix}\tilde{x}_n \\ \hat{\beta}_{n|n}\end{pmatrix}, \tag{58}$$

*where $\bar{M}$ is the unique nonnegative-definite solution of*

$$\bar{M} = N + A'\bar{M}(I + BB'\bar{M})^{-1}A \tag{59}$$

*in the class of nonnegative-definite matrices, and*

$$\hat{\beta}_{n|n} = \left(I + \bar{P}\tilde{H}'(\tilde{E}\tilde{E}')^{-1}\tilde{H}\right)^{-1}\left(\breve{\beta}_n + \bar{P}\tilde{H}'(\tilde{E}\tilde{E}')^{-1}\tilde{y}_n\right), \tag{60}$$

$$\breve{\beta}_{n+1} = \tilde{A}\breve{\beta}_n + \tilde{A}\bar{S}\tilde{H}'(\tilde{E}\tilde{E}')^{-1}\left(\tilde{y}_n - \tilde{H}\breve{\beta}_n\right), \tag{61}$$

*where $\bar{S} := \bar{P}\left[I + \tilde{H}'(\tilde{E}\tilde{E}')^{-1}\tilde{H}\bar{P}\right]^{-1}$, and $\bar{P}$ is the unique nonnegative-definite solution of $\bar{P} = \tilde{D}\tilde{D}' + \tilde{A}\bar{S}\tilde{A}'$ in the class of nonnegative definite matrices.*

*Furthermore, the controller* (58) *leads to a stable system, and the optimal average (incremental) values of the three different additive terms in the cost* (11) *are given by (for $-\infty \le n \le \infty$)*

$$E\left[|\tilde{x}_n - \xi_n|^2\right] = \zeta_1'\Xi\zeta_1, \qquad E\left[|\tilde{q}_n|^2\right] = \zeta_2'\Xi\zeta_2,$$
$$E\left[|u_n|^2\right] = \mathcal{K}(\Xi - \text{blockdiag}\,(0, \bar{S}))\mathcal{K}', \tag{62}$$

*where $\Xi$ is the unique nonnegative definite solution of the linear matrix equation:*

$$\Xi = (A - B\mathcal{K})\Xi(A - B\mathcal{K})' + B\mathcal{K}\left[\text{blockdiag}\,(0, \bar{S})\right]\mathcal{K}'B' + DD', \tag{63}$$

*and $\zeta_1$ and $\zeta_2$ are $(d+2)$-dimensional vectors, defined by* $\zeta_1 := (1, 0, \ldots, 0, -1)'$ *and* $\zeta_2 := (0, 1, 0, \ldots, 0, )'$.

**Remark 4.1.** The expressions (62) can be related to other performance measures of interest, and in particular, to blocking probabilities. Let $\Phi$ denote the probability distribution of a standard Gaussian random variable. Then (62) enables us to use the following standard heuristic in order to approximate the blocking probabilities in steady state for a finite buffer of size $K >> Q$:

$$P(\text{blocking}) \sim P\left(q_n > K\right) = P\left(\tilde{q}_n > K - Q\right). \tag{64}$$

Since $\tilde{q}_n$ is Gausian with zero mean and second moment given by $\zeta_2'\Xi\zeta_2$ (see (62)), this leads to

$$P(\text{blocking}) \sim 1 - \Phi\left(\frac{K - Q}{\sqrt{\zeta_2'\Xi\zeta_2}}\right). \tag{65}$$

## 5 Delayed Measurements

In our basic model presented in Section 2, a time unit corresponded to a round trip delay. In applications where the end-to-end delays are quite large, however, it might be highly inefficient to wait for a whole round trip delay until the control is updated. In such cases, one may choose one of the following two options:

(i) Consider some intermediate nodes as virtual sources and destinations, and use a hop-by-hop flow control. In that case, one may use at each hop the optimal control schemes proposed in the previous sections.
(ii) Allow for updating of the control several times (say $\theta \geq 1$ times) during a round trip. In that case, a basic time unit corresponds to the round trip time divided by $(1 + \theta)$. In an ATM-type architecture where ABR (Available Bit Rate) traffic is controlled, this is realized by sending RM (Resource Management) packets that travel from the source to the destination (forward packets) and then back from the destination to the source (backward packets). Information on queue-length and available bandwidth are then stamped onto these packets at the intermediate switches.

One may also use a combination of the two possibilities above, and allow for a hop-by-hop control where, at each hop, control is updated at times, which are shorter than the round trip hop time.

We shall focus in this section on the second possibility above, and design the corresponding optimal controllers. The system can still be modeled by the dynamics described in Section 2. However, since the basic time unit is now shorter (by a factor of $1 + \theta$) than the time unit considered in the previous sections, the basic parameters of the model have to be rescaled accordingly. For example, $\mu$ appearing in (2) will now be one-($\theta + 1$)th of the earlier one, and so will $\mu_n$, $x_n$, and $u_n$. In terms of the new time unit, the problem fromulation will then be exactly as in Section 2—with one major difference. Now the measurements $\hat{\tilde{\mu}}$ and $\hat{\tilde{q}}$ are acquired (for control purposes) with a delay of $\theta \geq 1$ time units. Hence, the controller (10) will be of the form

$$u_n = f_n\left(\tilde{x}_m, \hat{\tilde{\mu}}_{m-\theta}, \hat{\tilde{q}}_{m-\theta}, u_{m-\theta}, m = k, \ldots, n\right), \tag{66}$$

where $k$ is actually taken to be $-\infty$. Here we could also have taken the dependence on $\tilde{x}_m$ to be with a delay of $\theta$ units, but since $\tilde{x}_m$ is generated by noise-free dynamics, this would not make any difference in the end result; that is, any performance that is achievable using $\tilde{x}_m$ could also be achieved using $\tilde{x}_{m-\theta}$.

We note that even though this structure was arrived at by increasing the frequency of control updates and by rescaling the parameters of the model (as discussed above), it can also be interpreted for a fixed time scale as the control having access to measurements with a delay of $\theta$ time units. In the framework of this interpretation, we denote the minimum $H^\infty$ performance level $\gamma^*(\epsilon)$ by $\gamma_\theta^*(\epsilon)$, to show explicitly its dependence on the delay factor $\theta$. When we vary $\theta$ (as we will do in the sequel) this will be done precisely in this framework, with the time-scale kept fixed. This now brings us to the point where we present the counterpart of Proposition 3.1 for the delayed measurements case. The small-noise perturbed problem is precisely the one formulated in Section 3, with the only difference being that now the control has access to $y_{m-\theta}$, $m \leq n$, at time $n$.

**Proposition 5.1.** *Consider the framework of Section* 3, *but with the controller having the structural form discussed above, with* $\theta \geq 1$. *Let* $\epsilon > 0$ *be fixed, and Assumption* A1 *hold. Then:*

(1) $\gamma_\theta^*(\epsilon)$ *is finite, and for each fixed* $\epsilon > 0$, *the ordering* $\theta_2 > \theta_1 \geq 1$ *implies that* $\gamma_{\theta_2}^*(\epsilon) \geq \gamma_{\theta_1}^*(\epsilon) \geq \gamma^*(\epsilon)$, *where* $\gamma^*(\epsilon)$ *is as defined in Proposition 3.1.*

(2) *An* $H^\infty$*-controller that ensures a bound* $\gamma^2 > \left(\gamma_\theta^*(\epsilon)\right)^2$ *in the index* (26) *is*

$$u_{n,\epsilon,\theta}^\gamma = f_{\epsilon,\theta}^\gamma\left(\hat{z}_{n|n-\theta}\right) = -B'M_\epsilon\Lambda_\epsilon^{-1}A\left(I - \gamma^{-2}\hat{\Sigma}_{\epsilon,\theta}M_\epsilon\right)^{-1}\hat{z}_{n|n-\theta}, \tag{67}$$

*where* $\hat{z}_{n|n-\theta} = \xi_n^{(n)}$, *with* $\xi_n^{(n)}$ *being the last step of the iteration* $\{\xi_k^{(n)}\}_{k=n-\theta+1}^n$ *generated by*

$$\xi_{k+1}^{(n)} = A\left(I - \gamma^{-2}\tilde{\Sigma}_{\epsilon,\theta}N\right)^{-1}\xi_k^{(n)} + Bu_k, \qquad \xi_{n-\theta+1}^{(n)} = \check{z}_{n-\theta+1}, \tag{68}$$

*and* $\{\check{z}_n\}_{n=0}^\infty$ *is as generated by* (36). *The matrix* $\tilde{\Sigma}_{\epsilon,\theta}$ *is the* $\theta$th *term generated by the recursion*

$$\tilde{\Sigma}_{\epsilon,,k+1} = A\tilde{\Sigma}_{\epsilon,k}\left(I - \gamma^{-2}N\tilde{\Sigma}_{\epsilon,k}\right)^{-1}A' + D_\epsilon D_\epsilon', \qquad \tilde{\Sigma}_{\epsilon,1} = \Sigma_\epsilon, \tag{69}$$

*satisfying the two spectral radius conditions*

$$N^{1/2}\tilde{\Sigma}_{\epsilon,k}N^{1/2} < \gamma^2 I, \qquad k = 1, \ldots, \theta, \tag{70}$$

$$M_\epsilon^{1/2}\tilde{\Sigma}_{\epsilon,k}M_\epsilon^{1/2} < \gamma^2 I, \qquad k = 1, \ldots, \theta + 1. \tag{71}$$

*The initializing matrix $\Sigma_\epsilon$ in* (69) *is a minimal nonnegative-definite solution of* (30), *satisfying* (31), *and all other terms, such as $M_\epsilon$, $\Lambda_\epsilon$, $D_\epsilon$, are defined in Proposition* 3.1.

*If any one of the conditions of Proposition* 3.1, *or* (70)–(71) *do not hold, then* $\gamma \leq \gamma_\theta^*(\epsilon)$.

**Proof.** The first statement above follows from the fact that in $H^\infty$-optimal control, less on-line information on the uncertainties cannot lead to improved optimum performance, which in this case is quantified by $\gamma_\theta^*(\epsilon)$. The other statements of the proposition follow directly from Theorem 6.6, p. 276, of the 2nd edition of [6]. Even though this theroem states the solution of only the finite-horizon case, its extension to the infinite-horizon case (which is dealt with here) follows by mimicking the arguments used in the nondelay case; see Theorem 5.5, p. 226, of the 2nd edition of [6]. □

**Remark 5.1.** The ordering of the $\gamma^*$'s in part (1) of Proposition 5.1 is valid, as discussed prior to the statement of the proposition, under the interpretation that the round-trip delay is fixed, and $\theta$ reflects the delay in the acquisition of the measurements. The problem of relevance, however, is the one where the number of control updates is increased (from 1 to $\theta + 1$) during each round trip. Under this interpretation, the time scaling will not be fixed, and we would then expect the $H^\infty$ performance level to be a nonincreasing function of $\theta$ (instead of being nondecreasing, as in the case of Proposition 5.1), because the control is applied more frequently on a given time interval.

To present the counterpart of Theorem 4.1 for the $\theta \geq 1$ case, we have to study the limiting process as $\epsilon \downarrow 0$. The analysis of Section 4 applies here intact, with the exception of the control law; we also have to determine the limiting expression for $\tilde{\Sigma}_{\epsilon,k}$.

Compatible with that of $\Sigma_0 = \lim_{\epsilon\downarrow 0}\Sigma_\epsilon$, it is not difficult to show that $\tilde{\Sigma}_{0,k} = \text{blockdiag}(0, \tilde{P}_k)$, where the $(d+1)$-dimensional square matrix $\tilde{P}_k$ is generated by the recursion

$$\tilde{P}_{k+1} = \tilde{A}\tilde{P}_k\left(I - \gamma^{-2}\tilde{N}\tilde{P}_k\right)^{-1}\tilde{A}' + \tilde{D}\tilde{D}', \qquad \tilde{P}_1 = P, \tag{72}$$

where $P$ is a minimal nonnegative-definite solution of (46), satisfying (47), and $\tilde{A}$, $\tilde{D}$ are as defined by (44) and (45), respectively. In view of this simplification, the two spectral radius conditions (70) and (71) become, respectively,

$$\tilde{N}^{1/2}\tilde{P}_k\tilde{N}^{1/2} < \gamma^2 I, \quad k = 1, \ldots, \theta, \qquad \text{and}$$

$$\tilde{M}^{1/2}\tilde{P}_k\tilde{M}^{1/2} < \gamma^2 I, \quad k = 1, \ldots, \theta + 1, \tag{73}$$

where $\tilde{N}$ was defined by (41) and $\tilde{M}$ in conjunction with (48).

To obtain the limiting form of the controller, we write (compatibly with (49) and (50)):

$$\hat{z}_{n|n-\theta} = \left(\hat{\alpha}_{n|n-\theta}, \hat{\beta}'_{n|n-\theta}\right)', \qquad \xi_k^{(n)} = \left(\tilde{\alpha}_k^{(n)}, \tilde{\beta}_k^{(n)'}\right)'. \tag{74}$$

Then, some manipulations yield

$$\begin{aligned} &\tilde{\alpha}_{k+1}^{(n)} = \tilde{\alpha}_k^{(n)} + bu_k, \qquad \tilde{\alpha}_{n-\theta+1}^{(n)} = \begin{cases} \tilde{x}_{n-\theta+1}, & n \geq \theta, \\ 0, & \text{else}, \end{cases} \\ \Rightarrow \quad &\tilde{\alpha}_k^{(n)} = \tilde{x}_k, \quad k = 0, 1, \ldots, \forall n \geq 0 \quad \Rightarrow \quad \tilde{\alpha}_{n|n-\theta} = \tilde{x}_n, \end{aligned} \tag{75}$$

and

$$\begin{aligned} \beta_{n|n-\theta} &= \tilde{\beta}_n^{(n)}, \\ \tilde{\beta}_{k+1}^{(n)} &= \underbrace{\begin{pmatrix} 0 & 1 & 0 & \ldots & 0 & 0 \\ 0 & 0 & 1 & \ldots & 0 & 0 \\ \cdot & \cdot & \cdot & \ldots & \cdot & \cdot \\ 0 & 0 & 0 & \ldots & 1 & 0 \\ 0 & 0 & 0 & \ldots & 0 & 1 \end{pmatrix}}_{(d+1)\times(d+2)} A\left(I - \gamma^{-2}\tilde{\Sigma}_{0,\theta}N\right)^{-1} \begin{pmatrix} \tilde{x}_k \\ \tilde{\beta}_k^{(n)} \end{pmatrix}, \\ \tilde{\beta}_{n-\theta+1}^{(n)} &= \begin{cases} \breve{\beta}_{n-\theta+1}, & n \geq \theta, \\ 0, & \text{else}, \end{cases} \end{aligned} \tag{76}$$

where $\breve{\beta}_n$, $n \geq 1$, is generated by (54). Note that in the delayed measurement case we do not have a separation between $\hat{\alpha}$ and $\hat{\beta}$ as clean as in the case with $\theta = 0$. The dynamics for $\tilde{\beta}^{(n)}$ above, and thereby those of $\hat{\beta}_{n|n-\theta}$ use also $\tilde{x}$ as an input, whereas (53) and (54) did not.

In view of these limiting expressions, the limiting expression for the controller (67), as $\epsilon \downarrow 0$, is

$$u_{n,\theta}^{\gamma} = f_{\theta}^{\gamma}\left(x_n, \hat{\beta}_{n|n-\theta}\right) = -B'M_0\Lambda_0^{-1}A\left(I - \gamma^{-2}\tilde{\Sigma}_{0,\theta}M_0\right)^{-1}\begin{pmatrix} \tilde{x}_n \\ \hat{\beta}_{n|n-\theta} \end{pmatrix}, \tag{77}$$

This now leads to the following theorem.

**Theorem 5.1.** *Consider the original control problem formulated in Section* 2, *under the $H^\infty$ model* (M2) *of the disturbance, and with delayed measurements, with the optimum $H^\infty$ performance level denoted by $\gamma_\theta^*$. Let Assumption* A1 *hold. Then $\gamma_\theta^*$ is finite, and for all $\gamma > \gamma_\theta^*$:*

(1) *Statements* (1)–(3) *of Theorem* 4.1 *hold.*
(2) *Conditions* (73) *hold, and the matrix sequence generated by* (72) *is well defined and nonnegative definite.*

(3) *An $H^\infty$-controller that ensures the bound $\gamma^2$ in* (12) *under $\theta$-delayed measurements is given by* (77). *For a given $\gamma > 0$, if any one of the conditions above does not hold, then $\gamma \leq \gamma_\theta^*$.*

By letting $\gamma \to \infty$ in Theorem 5.1, we arrive at the following counterpart of Corollary 4.1 under delayed measurements.

**Corollary 5.1.** *Under Assumption* A1, *the optimal control problem with the disturbance model* M1, *and $\theta$-delayed measurements, admits the unique solution*

$$u_{n,\theta}^{\infty} = f_\theta^\infty\left(x_n, \hat{\beta}_{n|n-\theta}\right) = -B'\bar{M}(I + BB"\bar{M})^{-1}A\begin{pmatrix} \tilde{x} \\ \hat{\beta}_{n|n-\theta} \end{pmatrix}, \tag{78}$$

*where $\bar{M}$ is the unique nonnegative-definite solution of* (59) *in the class of nonnegative-definite matrices, and*

$$\hat{\beta}_{n|n-\theta} = \tilde{\beta}_n^{(n)}, \qquad \tilde{\beta}_{k+1}^{(n)} = \tilde{A}\tilde{\beta}_k^{(n)}, \qquad \tilde{\beta}_{n-\theta+1}^{(n)} = \begin{cases} \breve{\beta}_{n-\theta+1}, & n \geq \theta, \\ 0, & else, \end{cases} \tag{79}$$

*where $\breve{\beta}_n$, $n \geq 1$, is generated by* (61), *and $\tilde{A}$ is as defined by* (44).

## 6 Simulation Results

First paralleling the simulations presented in [4], but for the genuine ARMA model of interfering traffic, we present in this section some simulation results using both the $H^\infty$ and Gaussian models, and with no delay in the measurements (or equivalently with control updated only once during each round trip); hence, this first set of simulations will constitute an illustration of the results of Section 4. Subsequently, we present simulations for the case of delayed information, thus illustrating the results of Section 5. For simulation purposes, the observation noises $v_n$ and $w_n$ are chosen as standard i.i.d. Gaussian sequences. In order to illustrate the effectiveness of the $H^\infty$ (worst-case) controller in the context of telecommunications, we present results obtained under different choices of perturbations (interfering traffic) $\phi_n$ (in (2)), such as:

(i) highly correlated perturbations: $\phi_n = \sin(0.2n)$;
(ii) i.i.d. Gaussian perturbations; and
(iii) i.i.d. uniformly distributed perturbations (over $[-1, 1]$).

We illustrate below how the design objectives influence the behavior of the system when driven by the optimal controller. More precisely, we shall focus on the influence of the parameters $g$ (see (9)) and $b$ (see (7)) on the system.

The parameter $g$ will influence the tracking between input and output rates, as can be seen from (5) and (9). The larger $g$ is, the more costly the deviations between the input and output rates will be. Indirectly, this may result in reducing the variations of the queue (see (5)).

The parameter $b$ is expected to influence the jitter, i.e., the variablility of the transmission rate $\tilde{x}$, which is given by (see (7)) $\hat{u}_n := bu_n = \tilde{x}_{n+1} - \tilde{x}_n$. It is easily seen from (7) and (9) that choosing a smaller value of $b$ results in a higher weight on deviations of $u$ from zero, and thus a higher weight on the variability of $\tilde{x}$. Thus, we would expect that smaller values for $b$ will be useful when designing controllers for video or voice traffic, where the transmission rate is typically required to be regular (i.e., low jitter). (Note that if one considers $\hat{u}_n$ as the control, instead of $u_n$, then the cost $L$ in (9) becomes $L = g\|\tilde{x} - \xi\|^2 + \|q\|^2 + b^{-2}\|u\|^2$. This shows that indeed lower values of $b$ result in higher weight on the jitter.)

Finally, if $g$ is small and $b$ is large, then the second term in (9) becomes important, that is, we pay more for deviations of the queue length from its target value $Q$. In practice, this part will be responsible for decreasing loss probabilities (that occur when queue sizes are large) and increasing throughputs (since we may lose potential throughput when the queue is empty, if, at that time, the input rate is lower than the output rate).

We shall examine three situations in the simulations below:
Case 1: The three different criteria are weighted equally in the cost $L$ to be minimized (see (7) and (9)): $g = b = 1$.
Case 2: $g = b = 0.1$. We thus place relatively less emphasis on the tracking between input and output rates and thus on the variability of the queue length; the objective that is most weighted is that of minimizing the jitter $\hat{u}$.
Case 3: $b = 10$, $g = 0.1$. The objective that is most weighted is that of minimizing the deviations of the queue size from $Q$.

To describe the interfering traffic, we have chosen an ARMA model of order 2 with $\ell_1 = 0.7$, $\ell_2 = -0.3$, $k_1 = 1$, $k_2 = 2$. The parameters $a$ and $c$ (appearing in (3)) were chosen to be $\sqrt{2}$. We chose $Q = 30$ and $\mu = 40$ units. (A unit corresponds to the number of packets transmitted during a round trip delay.) The duration of each simulation was 140 time units. The queues were initially empty. A steady state was reached in each case after 20 time units. The design of the controller is for a steady-state behavior. We present in Table 1, the performance (attenuation) for a transient period (first 100 units) as well as the steady-state period (last 100 units). In all cases, the performance is indeed considerably better at the steady state.

For each of the cases above, we considered both of the scenarios below:

- An almost optimal $H^\infty$ controller, with $\gamma$ very close to (slightly larger than) the optimal (smallest) value $\gamma^*$ (see (15)), so as to operate under stable conditions.
- Very high values of $\gamma$ ($\gamma = 100$), which resulted in a controller that is (almost) optimal for the Gaussian model.

In addition to the values obtained for the attenuation according to the simulations, which are presented in Table 1, we also present some figures (Figures 1–4) with each one depicting a set of four curves: the evolution of the queue size, the controlled input versus available (output) rate over the entire simulation interval as well as over a typical shorter period (a zoomed-in curve), and the evolution of

**Table 1**: Steady-state and transient attenuation levels in the absence of delay in measurements.

| | Parameters | | | | | Simulation results | |
|---|---|---|---|---|---|---|---|
| Sim. | $\gamma$ | $b$ | $g$ | pert. | obs. noise | ss. atten. | tr. atten. |
| 1 | 100 | 0.1 | 0.1 | sin | sin | 10.49 | 11.54 |
| 2 | $(\gamma^* = 12.03)$, 12.05 | 0.1 | 0.1 | sin | sin | 5.75 | 10.75 |
| 3 | 100 | 1 | 1 | sin | sin | 4.93 | 6.29 |
| 4 | $(\gamma^* = 5.17)$, 5.2 | 1 | 1 | sin | sin | 2.61 | 6.45 |
| 5 | 100 | 0.1 | 0.1 | sin | Gauss. | 7.90 | 8.87 |
| 6 | $(\gamma^* = 12.03)$, 12.05 | 0.1 | 0.1 | sin | Gauss. | 7.44 | 9.55 |
| 7 | 100 | 1 | 1 | sin | Gauss. | 3.33 | 4.67 |
| 8 | $(\gamma^* = 5.17)$, 5.2 | 1 | 1 | sin | Gauss. | 4.06 | 5.37 |
| 9 | 100 | 10 | 0.1 | sin | Gauss. | 2.27 | 3.57 |
| 10 | $(\gamma^* = 3.89)$, 4 | 10 | 0.1 | sin | Gauss. | 3.11 | 5.41 |
| 11 | 100 | 10 | 0.1 | sin | sin | 3.54 | 4.9 |
| 12 | $(\gamma^* = 3.89)$, 4 | 10 | 0.1 | sin | sin | 2.06 | 5.27 |
| 13 | 100 | 1 | 1 | Gauss. | Gauss. | 2.36 | 3.60 |
| 14 | $(\gamma^* = 5.17)$, 5.2 | 1 | 1 | Gauss. | Gauss. | 3.41 | 5.28 |
| 15 | 100 | 10 | 0.1 | Gauss. | Gauss. | 1.81 | 3.08 |
| 16 | $(\gamma^* = 3.89)$, 4 | 10 | 0.1 | Gauss. | Gauss. | 3.14 | 4.53 |
| 17 | 100 | 0.1 | 0.1 | Gauss. | Gauss. | 3.2 | 5.07 |
| 18 | $(\gamma^* = 12.03)$, 12.05 | 0.1 | 0.1 | Gauss. | Gauss. | 8.01 | 9.43 |
| 19 | 100 | 1 | 1 | Unif. | Unif. | 2.05 | 5.41 |
| 20 | $(\gamma^* = 5.17)$, 5.2 | 1 | 1 | Unif. | Unif.. | 3.77 | 7.66 |
| 21 | 100 | 0.1 | 0.1 | Unif. | Unif. | 3.19 | 7.34 |
| 22 | $(\gamma^* = 12.03)$, 12.05 | 0.1 | 0.1 | Unif. | Unif. | 6.56 | 13 |
| 23 | 100 | 10 | 0.1 | Unif. | Unif. | 1.49 | 4.51 |
| 24 | $(\gamma^* = 3.89)$, 4 | 10 | 0.1 | Unif. | Unif. | 3.17 | 6.18 |

the control. (Multiplying the control evolution by the parameter $b$ provides us with the jitter.)

In Table 2 we illustrate the effect of delay on steady-state attenuation. We chose for all simulations $b = 10$ and $g = 1$.

## 7 Conclusions Drawn from the Simulations

As expected, the curves below indicate that smaller values of $b$ and $g$ result in low jitter. Indeed, the jitter $(bu_n)$ is the smallest in Figure 1 (corresponding to simulation 2) for which $g = b = 0.1$; its values are between $-2$ and $+2$. In Figure 2 (corresponding to simulation 4), in which $g = b = 1$, the jitter has much higher oscillations, of amplitude 8. In Figure 3 (corresponding to simulation 6) in which $g = 0.1$, $b = 10$, it has oscillations of amplitude 10. (Note that for a fair comparison of all these figures, $u$ has to be multipied by $b$).

**Table 2**: Effect of the delay $\theta$ on attenuation.

| | | Parameters | | | Simulation results |
|---|---|---|---|---|---|
| Sim. | $\theta$ | $\gamma$ | pert. | obs. noise | ss. attenuation |
| a. | 2 | 100 | Gauss. | Gauss. | 2.45 |
| b. | 2 | 100 | Unif. | Unif. | 2.76 |
| c. | 2 | 100 | sin | Gauss. | 2.34 |
| d. | 2 | 100 | sin. | sin. | 3.09 |
| e. | 2 | $(\gamma^* = 9.5, 10.5)$ | Gauss. | Gauss. | 4.51 |
| f. | 2 | $(\gamma^* = 9.5, 10.5)$ | Unif. | Unif. | 7.20 |
| g. | 2 | $(\gamma^* = 9.5, 10.5)$ | sin. | Gauss. | 2.52 |
| h. | 2 | $(\gamma^* = 9.5, 10.5)$ | sin. | sin. | 2.91 |
| i. | 5 | 100 | Gauss. | Gauss. | 9.84 |
| j. | 5 | 100 | Unif. | Unif. | 9.77 |
| k. | 5 | 100 | sin. | Gauss. | 3.57 |
| l. | 8 | 100 | Gauss. | Gauss. | 16.60 |
| m. | 8 | 100 | Unif. | Unif. | 14.01 |
| n. | 8 | 100 | sin. | Gauss. | 7.62 |

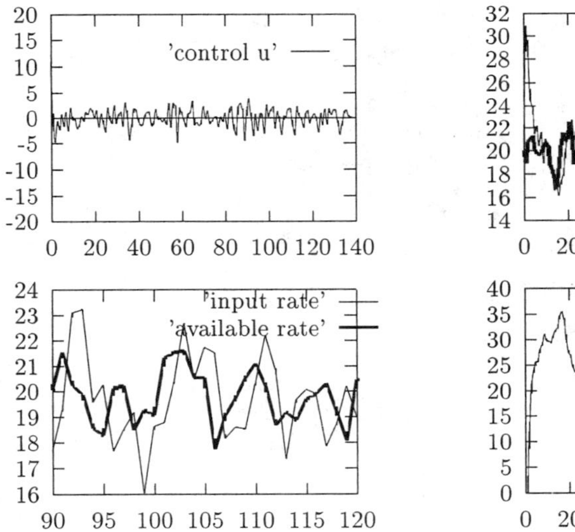


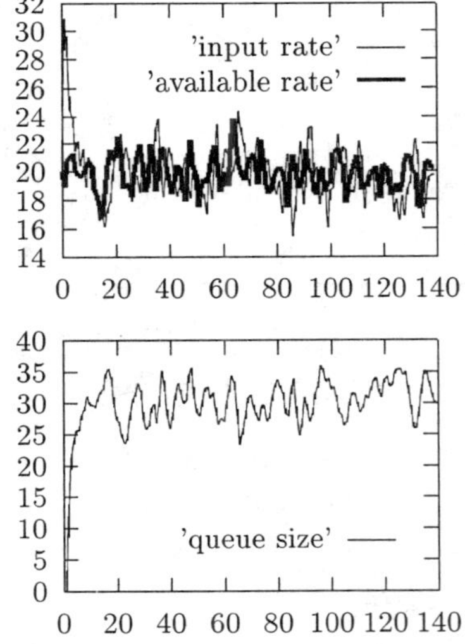


**Figure 1**: Simulation 6; $\gamma = 12.03$, unequally weighted costs $g = b = 0.1$, sinusoidal perturbations and Gaussian noise, $k_1 = k_2 = 1, l_1 = 0.7, l_2 = -0.3$

Large $b$ and relatively small $g$ result in the lowest queue lengths. Indeed, in Figure 3, for which $g = 0.1, b = 10$, the amplitude of the oscillations of the queue length is 5, whereas in Figure 1 ($g = b = 0.1$) the oscillations are of amplitude 9.

Finally, relatively large values of $g$ indeed lead to much better tracking of the input with respect to the output rate, as can be seen from Figures 2 and 4, where

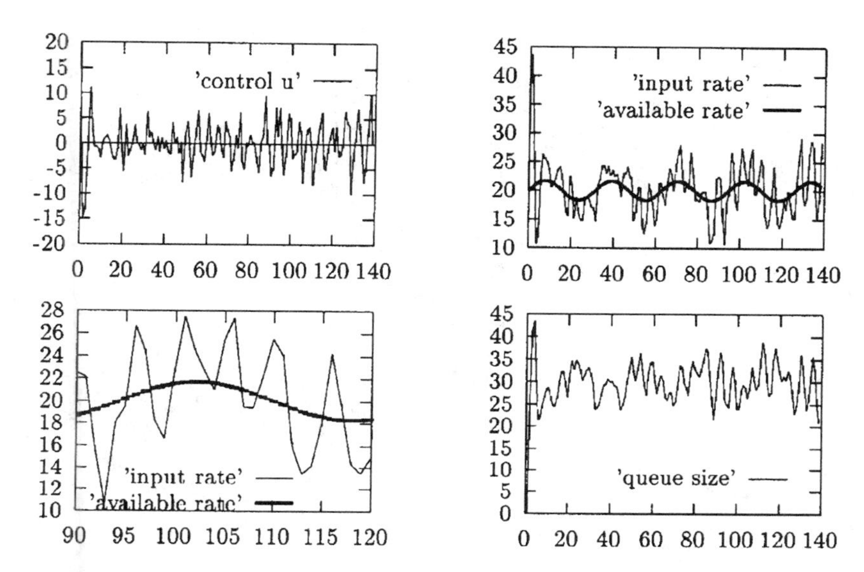


**Figure 2**: Simulation 8; $\gamma = 5.17$, equally weighted costs $g = b = 1$, sinusoidal perturbations and Gaussian noise, $k_1 = k_2 = 1$, $l_1 = 0.7$, $l_2 = -0.3$

$g = b = 1$. We observe from the figures that relatively large values of $g$ also lead to smaller variation in the queue length around its target level.

An interesting question that can be raised here is whether in specific applications one should choose an optimal $H^\infty$ controller (designed for the worst-case behavior of noise and perturbations) or a Gaussian one. When both the noise and the perturbations are Gaussian, we see from Table 1 (Simulations 13–18) that, as expected, the Gaussian controller performs better (leads to smaller attenuation). In the presence of non-Gaussian noise, however, the $H^\infty$ controller performs sometimes better and sometimes worse (note that we did not conduct any simulations under the worst-case noise, which is typically linear in the state). Indeed, in Simulations 1–4, we see that for the case of sinusoidal measurement perturbations and noise, the attenuation is better under an $H^\infty$ controller, whereas in the case of uniformly distributed perturbations and noise (Simulations 19–24) the situation is reversed.

When the measurements are delayed, we see in Table 2, as to be expected, that the attenuation increases with delay. As in the nondelayed case, we see that for the case of sinusoidal measurement perturbations and noise, the attenuation is better under an $H^\infty$ controller, whereas in the other cases the situation is reversed.

## 8 Concluding Remarks

We have presented in this paper a control-theoretic approach to designing optimal rate-based flow controller. Using the $H^\infty$-optimization based controller

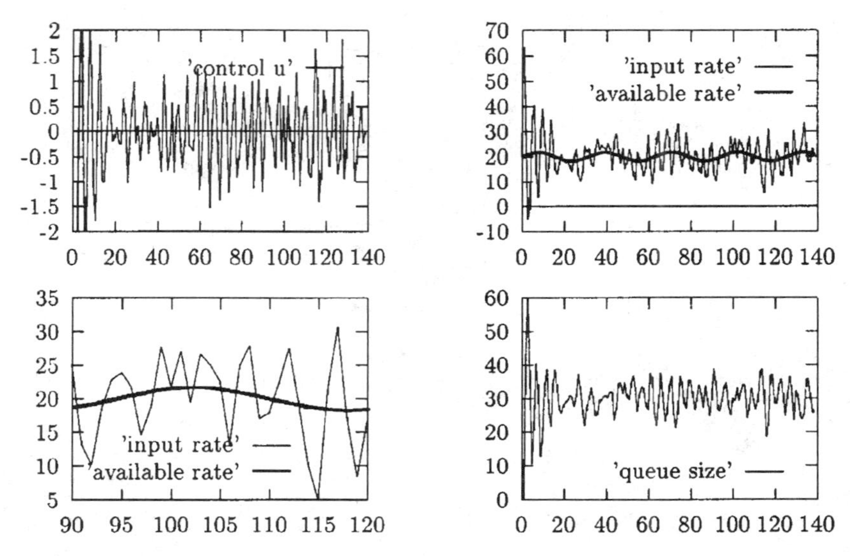


**Figure 3**: Simulation 10; $\gamma = 3.89$, unequally weighted costs $g = 0.1$, $b = 10$, sinusoidal perturbations and Gaussian noise, $k_1 = k_2 = 1$, $l_1 = 0.7$, $l_2 = -0.3$

design theory that was developed in the last decade, we were able to handle very general measurement noise and interfering traffic: these may be bursty, nonstationary, nonergodic, highly correlated, and even periodic. In spite of this generality in the description of the noise and interfering traffic, the resulting (optimal) controller is easy to implement, being linear in the (augmented) state of the system.

It turned out that the underlying control problem did not fall precisely into the standard $H^\infty$ setting with imperfect information, and to bring it to the standard setting, the perfect state measurements had to be perturbed by small noise. The limit of the solution thus obtained, as the perturbation noise vanishes, provided the optimal control for the original problem.

We further studied the problem of delayed information; due to the *certainty-equivalence* property of both $H^\infty$ and LQG optimal controllers (see [6]), it turned out that the problem with delayed information did not require an increase in the dimension of the state space (as is generally the case in optimal control problems with delayed information, see, e.g., [5]). The optimal controller thus remains simple, and easy to implement also in the presence of delay, and is thus appropriate for applications in high-speed networks.

The linear-quadratic control theoretic framework adopted in this paper offers an additional important advantage besides leading to a design that is *optimal* with respect to a performance index. It enables us to obtain *analytical expresssions for the performance measures*, as in (62), as well as expressions for expected blocking probabilities, as in Remark 4.1. This is especially important in Call Admission Control (CAC) of ABR-type traffic, since, when accepting a new call, the network has

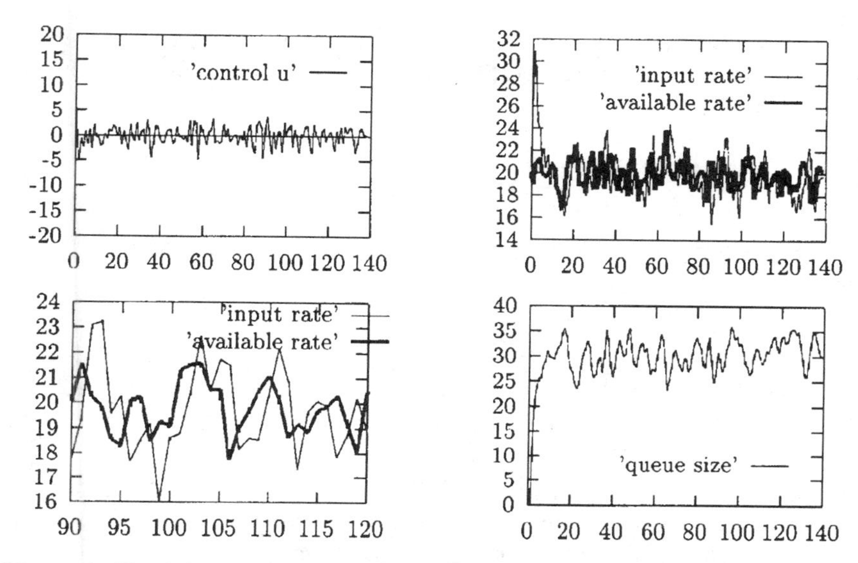


**Figure 4**: Simulation 13; $\gamma = 100$, equally weighted costs $g = b = 1$, Gaussian perturbations and noise, $k_1 = k_2 = 1$, $l_1 = 0.7$, $l_2 = -0.3$

to guarantee some Quality of Service which has been prenegotiated (and includes in particular an upper bound on the cell loss rate, a lower bound on the available transmission rate, etc.). To the best of our knowledge, there do not exist analytical expressions for performance measures for the control schemes proposed in [1], and this limits the possibility of the network to guarantee performance measures when using them.

The $H^\infty$ approach allows for the measurement noises to be state-dependent. This makes it possible, in particular, to consider quantized measurements. For instance, we may consider queue length information that is conveyed in a single congestion bit, that is turned on if and only if the queue length is beyond some threshold. We are currently investigating the computation of achievable performances for such scenarios.

Effectiveness and versatility of the control schemes developed here hinge on the modeling assumption that it is the source that computes the input rates, according to the information it acquires. A major limitation of this approach in the context of the control of ABR traffic in ATM is that, according to [1], the source receives through the RM packets only two bits of information on the status of the congestion (one is used to tell the source not to increase its rate, and the other one is used to tell the source to decrease its rate; if both bits are zero, then the source may increase its sending rate). An alternative possibility would be to let the switches compute the desired input rate from the source. The advantage is that the switch indeed possesses full (nondelayed) information about its own queue length and the available bandwidth. Derivation of the optimal control in this case will be similar to those carried out in this paper. A drawback of this approach is that, in the presence of many sources and destinations, it might require high computational effort in the switches.

## Acknowledgments

The research of the second author was supported by Grants ANI 98–13710, and NSF INT 98–04950.

## REFERENCES

[1] The ATM Forum Technical Committee. *Traffic Management Specification*, Version 4.0, 95-0013R8. October 1995.

[2] Altman, E., F. Baccelli, and J. C. Bolot. Discrete-Time Analysis of Adaptive Rate Control Mechanisms. In: *High Speed Networks and Their Performance* (H. G. Perros and Y. Viniotis, eds.). North-Holland, Amsterdam, pp. 121–140, 1994.

[3] Altman, E. and T. Başar. Optimal Rate Control for High Speed Telecommunication Networks. *Proceedings of the 34th IEEE Conference on Decision and Control*, pp. 1389–1394, New Orleans, Louisiana, December 13–15, 1995.

[4] Altman, E. and T. Başar. Optimal Rate Control for High Speed Telecommunication Networks: The Case of Delayed Information. *First Workshop on ATM Traffic Management, WATM'95*, IFIP, WG.6.2 Broadband Communication, pp. 115–122, Paris, France, December 6–8, 1995.

[5] Altman, E. and G. Koole. Control of a Random Walk with Noisy Delayed Information. *Systems and Control Letters*, **24**, 207–213, 1995.

[6] Başar, T. and P. Bernhard. *$H^\infty$-Optimal Control and Relaxed Minimax Design Problems: A Dynamic Game Approach*. Birkhäuser, Boston, MA, 1991 (2nd ed., 1995).

[7] Bolot, J.-C. End-to-End Delay and Loss Behavior in the Internet. *Proceedings of the ACM Sigcomm '93*, pp. 289–298, San Francisco, CA, Sept. 1993.

[8] Bolot, J.-C. and A. U. Shankar. Analysis of a Fluid Approximation to Flow Control Dynamics. *Proceedings of the IEEE INFOCOM '92*, Florence, Italy, pp. 2398–2407, May 1992.

[9] Bonomi, F. and K. W. Fendick. The Rate-Base Flow Control Framework for the Available Bit Rate ATM Service. *IEEE Network*, pp. 25–39, March/April 1995.

[10] Boxma, O. J. Sojourn Times in Cyclic Queues—The Influence of the Slowest Server. In: *Computer Performance and Reliability* (G. Iazeolla, P. J. Courtois, and O. J. Boxma (eds.). Elsevier Science, North-Holland, Amsterdam, pp. 13–24, 1988.

[11] Didinsky, G., Z. Pan, and T. Başar. Parameter Identification for Uncertain Plants Using $H^\infty$ Methods. *Automatica*, **31**, no. 9, 1227–1250, September 1995.

[12] Fendrick, K., M. A. Rodrigues, and A. Weiss. Analysis of a Rate-Based Control Strategy with Delayed Feedback. *Proceedings of the ACM SIGCOMM '92*, Baltimore, MD, pp. 136–148, August 1992.

[13] Jacobson, V. Congestion Avoidance and Control. *Proceedings of the ACM SIGCOMM '88*, Stanford, CA, pp. 314–329, August 1988.

[14] Keshav, S. A Control-Theoretic Approach to Flow Control. *Proceedings of the ACM SIGCOMM '91*, Zurich, Switzerland, September 1991.

[15] Kung, H. T. and R. Morris. Credit-Based Flow Control for ATM Networks. *IEEE Network*, pp. 40–48, March/April 1995.

[16] Ramakrishnan, K. K. and P. Newman. Integration of Rate and Credit Schemes for ATM Flow Control. *IEEE Network*, pp. 49–56, March/April 1995.

# $H_\infty$ Output Feedback Control Problems for Bilinear Systems

Seigo Sasaki
Department of Electrical Engineering
National Defense Academy
Yokosuka, Kanagawa 239-8686, Japan

Kenko Uchida
Department of Electrical and Computer Engineering
Waseda University
Shinjuku, Tokyo 169, Japan

## Abstract

In this paper, we consider $H_\infty$ output feedback control problems for bilinear systems, and present a design example of artifical rubber muscle actuator control system. First we derive two types of $H_\infty$ output feedback controller via the differential game approach. The controllers are characterized in terms of the solutions satisfying two Riccati inequalities depending on the state of generalized plant or controller. Second, we propose algorithms to solve the Riccati inequalities via specifying a domain of the state and solving constant coefficient Riccati inequalities. The proposed algorithms include an evaluation method for the domain of internal stability. Finally, we demonstrate efficiency of the proposed algorithm through a numerical example.

## 1 Introduction

Bilinear systems comprise perhaps the simplest class of nonlinear systems. However, the linearization of bilinear systems easily lose the essential nature of the problem for the systems. Moreover, bilinear systems are nonlinear systems that have a lot of practical applications in various fields. Many researchers have studied various aspects of bilinear systems for the past 30 years [11]. However, there is little research that considers exogenous inputs, e.g., disturbance, to bilinear systems. it is important for practical applications to consider the exogenous inputs for guaranteeing a good performance.

$H_\infty$ theory is a control theory that considers explicitly the exogenous inputs for guaranteeing a good performance [3]. Since the time-domain methodology has been developed for the linear $H_\infty$ control problem, $H_\infty$ theory has been generalized to nonlinear systems [7],[6]. In particular, the differential game theory [1] can generalize the linear $H_\infty$ theory in the time domain to the nonlinear $H_\infty$ theory.

For the linear $H_\infty$ control problems, we obtain the solution by solving the Riccati inequality. On the other hand, for the nonlinear $H_\infty$ control problems, we obtain the solution by solving the partial differential inequality called the "Hamilton–Jacobi–Isaacs (HJI) inequality." But until now, it seems to us that there does not exist a true effective method to solve the HJI inequality. It is just the same with the $H_\infty$ control problem for bilinear systems that are a special class of nonlinear systems. So far, there exists the formal tensor series solution of the HJI inequality for the bilinear $H_\infty$ state feedback control problem [2]. The work [13] solves finite-time horizon $H_\infty$ output feedback control problems for bilinear systems via the information state approach.

In this paper, we consider (infinite-time horizon) $H_\infty$ output feedback control problems for bilinear systems. Generalizing the approach of [10] that discusses the linear $H_\infty$ output feedback control problems, we derive two types of $H_\infty$ output feedback controller. To construct the controllers needs solutions satisfying two Riccati inequalities that depend on the state of a generalized plant or controller. But it is difficult to obtain the solutions satisfying the inequalities. We propose algorithms to obtain the solutions satisfying the Riccati inequalities via specifying a domain of the state and solving constant coefficient Riccati inequalities. The proposed algorithms include an evaluation method for the domain of internal stability. Finally, we demonstrate efficiency of the proposed algorithm through a design example of an artificial rubber muscle actuator control system.

The paper is organized as follows. Section 2 gives a statement of the bilinear $H_\infty$ output feedback control problem. Section 3 gives sufficient conditions for the existence of controllers solving the problem. Section 4 gives algorithms that obtain a solution satisfying the sufficient conditions. The final section demonstrates efficiency of the proposed algorithm through a numerical example.

**Notations.** For a vector $x$, $\|x\|$ denotes the Euclidean norm. For a matrix $X$, $\|X\|$ denotes the norm induced by the Euclidean norm (largest singular value). $\|\cdot\|_2$ denotes the norm of the space of square integrable signals, and is defined as

$$\|x\|_2 := \left(\int_0^\infty \|x(t)\|^2\, dt\right)^{\frac{1}{2}}, \qquad x \in L_2.$$

$L_{2\rho}$ denotes the set of bounded functions with $\|x(t)\| < \rho$ for all $t \in [0, \infty)$. For a real matrix $P$, $P > 0$ ($P \geq 0$) means $P$ is symmetric and positive (positive-semi)definite. $I$ denotes the identity matrix of appropriate dimensions.

## 2 Bilinear $H_\infty$ Control Problem

We consider a nonlinear system ($\Sigma$):

$$\dot{x} = Ax + B_1 w + B(x)u, \tag{1}$$

$$z = C_1 x \qquad\qquad + D_{12}u, \tag{2}$$

$$y = C_2 x + D_{21} w, \tag{3}$$

where

$$B(x) = B_2 + \{xN\}, \tag{4}$$

$$\{xN\} := \sum_{i=1}^{n} x_i N_i, \qquad N_i \in \mathbb{R}^{n\times r}, \quad i = 1, \ldots, n. \tag{5}$$

$x_i$ stands for the $i$th element of $x$.

$x(t) \in \mathbb{R}^n$ is the state, $u(t) \in \mathbb{R}^r$ is the control input, $y(t) \in \mathbb{R}^m$ is the measured output, $z(t) \in \mathbb{R}^q$ is the controlled output, and $w(t) \in \mathbb{R}^p$ is the exogenous input. $A$, $B_1$, $B_2$, $C_1$, $C_2$, $D_{12}$, $D_{21}$ are coefficient matrices of appropriate dimensions, and satisfy the following "Orthogonality Condition":

$$\text{(AO)} \qquad D_{12}^T [C_1 \quad D_{12}] = [0 \quad I], \tag{6}$$

$$D_{21} \left[B_1^T \quad D_{21}^T\right] = [0 \quad I]. \tag{7}$$

For given matrices $N_i$, $i$=1, ..., $n$, there exist matrices $M_j$, $j$=1, ..., $r$, such that

$$\{xN\}u = \{uM\}x, \tag{8}$$

for all $x$, $u$, where $\{uM\}$ is defined as

$$\{uM\} := \sum_{j=1}^{r} u_j M_j, \qquad M_j \in \mathbb{R}^{n\times n}, \quad j = 1, \ldots, r. \tag{9}$$

$u_j$ stands for the $j$th element of $u$.

For the system $(\Sigma)$, we consider an output feedback controller $(\Gamma)$ of the form

$$\dot{\xi} = \eta_1(\xi) + \eta_2(\xi) + \eta_3(\xi)y, \tag{10}$$

$$u = \theta_1(\xi), \tag{11}$$

where $\eta_1(\xi)$, $\eta_2(\xi)$, $\eta_3(\xi)$, $\theta_1(\xi)$ are sufficiently smooth functions with $\eta_1(0) = 0$, $\theta_1(0) = 0$.

For the closed-loop system $(\Sigma, \Gamma)$, an internal stability is defined as follows.

**Definition 2.1** (Internal Stability). Let $\Omega \subset \mathbb{R}^n \times \mathbb{R}^n$ be a domain that contains the equilibrum point $(0, 0)$. Consider the closed-loop system $(\Sigma, \Gamma)$ with $w = 0$. If the equilibrum point $(0, 0)$ is (locally) asymptotically stable and the solution $(x(t), \xi(t))$, starting in the domain $\Omega$, approaches the point $(0, 0)$ as $t \to \infty$, we say the closed-loop system $(\Sigma, \Gamma)$ is internally stable in the domain $\Omega$.

In the paper, we consider the following problem:

**Bilinear $H_\infty$ Control Problem [P].** Consider a system $(\Sigma)$. Given $\gamma(> 0)$, find an output feedback controller $(\Gamma)$ satisfying the following conditions $(P_1)$ and $(P_2)$, and characterize the domain $\Omega$ satisfying condition $(P_1)$:

$(P_1)$ The closed-loop system $(\Sigma, \Gamma)$ is internally stable in a domain $\Omega \subseteq \mathbb{R}^n \times \mathbb{R}^n$ that contains the equilibrium point $(0, 0)$.

($P_2$) Whenever $(x(0), \xi(0)) = (0, 0)$, there exists some $\rho(> 0)$ and $\|z\|_2 \leq \gamma \|w\|_2$ for all $w \in L_{2\rho}$.

## 3 Characterizations of Controllers

Now assume that the structures $\eta_1(\xi)$, $\eta_2(\xi)$, $\eta_3(\xi)$ of the controller $(\Gamma)$ are already designed. Then, the differentail game approach [7],[1] leads us to the fact that the problem [P] is solved by obtaining a solution $V(x_a)$ satisfying the HJI inequality

$$\min_{u=\theta_1(\xi)} \max_{w=w(x,\xi,u)} \left[ \frac{\partial V}{\partial x_a}(x_a)\{f_1(x_a)+f_2(x_a)w+f_3(x_a)u\}+\|z\|^2 - \gamma^2\|w\|^2 \right] \leq 0, \tag{12}$$

for the augmented system that consists of the system $(\Sigma)$ and the controller $(\Gamma)$, given as

$$\dot{x}_a = f_1(x_a) + f_2(x_a)w + f_3(x_a)u, \tag{13}$$

$$z = C_1 x + D_{12} u, \tag{14}$$

where

$$x_a = \begin{bmatrix} x \\ \xi \end{bmatrix}, \qquad f_1(x_a) = \begin{bmatrix} Ax \\ \eta_1(\xi) + \eta_3(\xi)C_2 x \end{bmatrix},$$

$$f_2(x_a) = \begin{bmatrix} B \\ \eta_3(\xi)D_{21} \end{bmatrix}, \qquad f_3(x_a) = \begin{bmatrix} B(x) \\ \eta_2(\xi) \end{bmatrix}.$$

It is not easy to obtain the solution $V(x_a)$ satisfying the HJI inequality (12). In this paper, by restricting a structure of the solution $V(x_a)$ to two particular types, we obtain two solutions presented in Theorems 3.1 and 3.2. The solution $V(x_a)$ for Theorems 3.1 has the structure

$$V(x_a) := \xi^T S \xi + \gamma^2 (x - \xi)^T Y^{-1} (x - \xi), \tag{15}$$

and the solution for Theorem 3.2 has the structure

$$V(x_a) := x^T X x + \gamma^2 (x - \xi)^T T^{-1} (x - \xi). \tag{16}$$

These structures of the solutions are the same as those of the linear case [10],[3]. We might think that the structures (15), (16) are introduced as the second-order approximation of $V(x_a)$ with respect to the linearization model in a neighborhood of the equilibrium point $(x, \xi) = (0, 0)$ of the bilinear system (1) and controller (10). This is not true, because the linearization model of the bilinear system (1) is generally useless because it lacks the control input. In this paper, we regard the state $x$ in the bilinear term as an unknown parameter, and the bilinear system (1) as a linear system with an unknown parameter. We can prove the following theorems with the same technique "completing the square" as in the linear case [10]:

**Theorem 3.1.** *Consider the system* $(\Sigma)$ *satisfying the condition* (AO). *If there exists a domain* $\Phi_1 \subset \mathbb{R}^n \times \mathbb{R}^n$ *that contains the origin, and for all* $(x, \xi) \in \Phi_1$, $(x, \xi) \neq (0, 0)$, *there exist* $S > 0$, $Y > 0$ *such that:*

(C11)

$$\begin{aligned} \mathrm{Eqn}_1(S, B(\xi)) := {} & S(A + \gamma^{-2} Y C_1^T C_1) + (A + \gamma^{-2} Y C_1^T C_1)^T S \\ & - S(B(\xi)B^T(\xi) - \gamma^{-2} Y C_2^T C_2 Y)S + C_1^T C_1 < 0, \end{aligned} \tag{17}$$

(C12)

$$\mathrm{Eqn}_2(Y) := Y A^T + AY - Y(C_2^T C_2 - \gamma^{-2} C_1^T C_1)Y + B_1 B_1^T < 0, \tag{18}$$

(C13)

$$\begin{aligned} & \xi^T \mathrm{Eqn}_1(S, B(\xi))\xi + \gamma^2 (x - \xi)^T Y^{-1} \\ & \quad \times [\mathrm{Eqn}_2(Y) + Y\{\underline{v}(\xi)M\}^T + \{\underline{v}(\xi)M\}Y]Y^{-1}(x - \xi) < 0, \end{aligned} \tag{19}$$

*where* $\underline{v}(\xi) := -B^T(\xi)S\xi$;

*then the controller* $(\Gamma)$ *solving the problem* [P] *is given as*

$$\dot{\underline{x}} = A\underline{x} - B(\underline{x})B^T(\underline{x})S\underline{x} + \gamma^{-2} Y C_1^T C_1 \underline{x} + Y C_2^T (y - C_2 \underline{x}), \tag{20}$$

$$u = -B^T(\underline{x})S\underline{x}. \tag{21}$$

*Moreover, the closed-loop system* $(\Sigma, \Gamma)$ *is internally stable in the maximum hyper-ellipsoid*

$$\Omega_1(\sigma_1) := \{(x, \xi) \mid \xi^T S \xi + \gamma^2 (x - \xi)^T Y^{-1} (x - \xi) \leq \sigma_1\}, \tag{22}$$

*that is contained in the domain* $\Phi_1$.

**Theorem 3.2.** *Consider the system* $(\Sigma)$ *satisfying the condition* (AO). *If there exists a domain* $\Phi_2 \subset \mathbb{R}^n \times \mathbb{R}^n$ *that contains the orgin, and for all* $(x, \xi) \in \Phi_2$, $(x, \xi) \neq (0, 0)$, *there exist* $X > 0$, $T > 0$ *such that:*

(C21)

$$\begin{aligned} \mathrm{Eqn}_3(X, B(x)) := {} & XA + A^T X - X(B(x)B^T(x)) \\ & - \gamma^{-2} B_1 B_1^T)X + C_1^T C_1 < 0, \end{aligned} \tag{23}$$

(C22)

$$\begin{aligned} \mathrm{Eqn}_4(T, B(x)) := {} & T(A + \gamma^{-2} B_1 B_1^T X)^T + (A + \gamma^{-2} B_1 B_1^T X)T \\ & - T(C_2^T C_2 - \gamma^{-2} X B(x) B^T(x) X)T + B_1 B_1^T < 0, \end{aligned} \tag{24}$$

(C23)

$$\begin{aligned} & x^T \mathrm{Eqn}_3(X, B(x))x + \gamma^2 (x - \xi)^T T^{-1} \\ & \quad \times [\mathrm{Eqn}_4(T, B(x)) + T\{\hat{v}(\xi)M\}^T + \{\hat{v}(\xi)M\}T]T^{-1}(x - \xi) \\ & \quad + \|B^T(x)Xx - B^T(\xi)X\xi\|^2 - \|B^T(x)X(x - \xi)\|^2 < 0, \end{aligned} \tag{25}$$

*where* $\hat{v}(\xi) := -B^T(\xi)X\xi$;

*then the controller* ($\Gamma$) *solving the problem* [P] *is given as*

$$\dot{\hat{x}} = A\hat{x} - B(\hat{x})B^T(\hat{x})X\hat{x} + \gamma^{-2}B_1B_1^TX\hat{x} + TC_2^T(y - C_2\hat{x}), \tag{26}$$

$$u = -B^T(\hat{x})X\hat{x}. \tag{27}$$

*Moreover, the closed-loop system* $(\Sigma, \Gamma)$ *is internally stable in the maximum hyper-ellipsoid*

$$\Omega_2(\sigma_2) := \{(x, \xi) \mid x^TXx + \gamma^2(x - \xi)^TT^{-1}(x - \xi) \leq \sigma_2\}, \tag{28}$$

*that is contained in the domain* $\Phi_2$.

**Remark 3.1.** Theorem 3.2 correspond to a case where we let $V(x) = x^TXx$ and $Q(x - \xi) = \gamma^2(x - \xi)^TT^{-1}(x - \xi)$ in Theorem 3.1 of [7]. In [7], however, there is not a result to which Theorem 3.1 corresponds. This paper gives another type of controller as shown for linear systems in [3] and [10].

**Remark 3.2.** We use condition (C23) (or (C13)) to obtain the largest domain possible. In an extreme neighborhood of the origin (e.g., in [7]), the higher-order terms of condition (C23) (or (C13)), with respect to the state, are contained in a "gap" of the inequality (24) (or (18)) and condition (C23) (or (C13)) becomes useless. The paper [7] does not discuss the evaluation of a domain in which a system in internally stable.

**Remark 3.3.** In Theorem 3.1, first we find a solution satisfying conditions (C11) and (C12), and construct the controller ($\Gamma$). Next, we use condition (C13) to evaluate the domain in which the closed-loop system $(\Sigma, \Gamma)$ is internally stable. In the same way, in Theorem 3.2, first we find a solution satisfying conditions (C21) and (C22), and construct the controller ($\Gamma$). Next, we use condition (C23) to evalutate the domain in which the closed-loop system $(\Sigma, \Gamma)$ is internally stable. Conditions (C11) (C12) and (C21) (C22) correspond to two Riccati inequalities in linear $H_\infty$ control problems [3], [10], repectively. We see these in the following algorthms. In Remark 4.3, we give more constructive characterization of domains $\Omega_1$ and $\Omega_2$.

## 4 Controller Synthesis Algorithms

In Section 3, we showed two theorems which present the sufficient conditions to obtain the output feedback controller ($\Gamma$) for the problem [P]. To construct the controller ($\Gamma$), from the sufficient conditions, we must solve the Riccate inequalities that depend on the state $x$ (or $\xi$) of the generalized plant (or the controller). In this section, we propose algorithms which consist of considering the admissible domain of the state $x$ (or $\xi$) and solving constant coefficient Riccati inequalities. The algorithms also give the domain $\Omega_1$ (or $\Omega_2$) in which the closed-loop system is internally stable. Algorithms 4.1 and 4.2 proposed here correspond to Theorems 3.1 and 3.2, respectively.

**Algorithm 4.1.**
**Step 1.** Set a domain $\Phi_\xi^{11} \subseteq \mathbb{R}^n$ that contains the origin. For all $\xi \in \Phi_\xi^{11}$, find a matrix $\underline{B}$ such that

$$\underline{BB}^T \leq B(\xi)B^T(\xi), \tag{29}$$

and $S > 0$, $Y > 0$, $\underline{\Delta}_1 > 0$, $\underline{\Delta}_2 > 0$ such that

$$\mathrm{Eqn}_1(S, \underline{B}) + \underline{\Delta}_1 = 0 \tag{30}$$

$$\mathrm{Eqn}_2(Y) + \underline{\Delta}_2 = 0. \tag{31}$$

At this time, the obtained $S$, $Y$ give a controller $(\Gamma)$ in the form of (20), (21).

**Step 2.** Find $\sigma_1$ such that

$$\Omega_1 = \{(x, \xi) \mid \xi^T S\xi + \gamma^2(x-\xi)^T Y^{-1}(x-\xi) \leq \sigma_1\} \subseteq \Phi_1. \tag{32}$$

$\Omega_1$ is a domain in which the closed-loop system $(\Sigma, \Gamma)$ is internally stable, where $\Phi_1$ is an admissible domain that contains $\Omega_1$, defined as

$$\Phi_1 := (\mathbb{R}^n \times \Phi_\xi^{11}) \cap \Phi^{12}. \tag{33}$$

Here $\Phi^{12}$ is defined as

$$\begin{aligned}\Phi^{12} := \{(x, \xi) \mid & -\xi^T[S(B(\xi)B^T(\xi) - \underline{BB}^T)S + \underline{\Delta}_1]\xi \\ & + \gamma^2(x-\xi)^T Y^{-1}[Y\{\underline{v}(\xi)M\}^T \\ & + \{(\underline{v}\xi)M\}Y - \underline{\Delta}_2]Y^{-1}(x-\xi) < 0\}.\end{aligned} \tag{34}$$

**Algorithm 4.2.**
**Step 1.** Set a domain $\Phi_x^{21} \subseteq \mathbb{R}^n$ that contains the origin. For all $x \in \Phi_x^{21}$, find matrices $\hat{B}_u$, $\hat{B}_l$ such that

$$\hat{B}_l\hat{B}_l^T \leq B(x)B^T(x) \leq \hat{B}_u\hat{B}_u^T, \tag{35}$$

and $X > 0$, $T > 0$, $\hat{\Delta}_1 > 0$, $\hat{\Delta}_2 > 0$ such that

$$\mathrm{Eqn}_3(X, \hat{B}_l) + \hat{\Delta}_1 = 0, \tag{36}$$

$$\mathrm{Eqn}_4(T, \hat{B}_u) + \hat{\Delta}_2 = 0. \tag{37}$$

At this time, the obtained $X$, $T$ give a controller $(\Gamma)$ in the form of (26), (27).

**Step 2.** Find $\sigma_2$ such that

$$\Omega_2 = \{(x, \xi) \mid x^T Xx + \gamma^2(x-\xi)^T T^{-1}(x-\xi) \leq \sigma_2\} \subseteq \Phi_2. \tag{38}$$

$\Omega_2$ is a domain in which the closed-loop system $(\Sigma, \Gamma)$ is internally stable, where $\Phi_2$ is an admissible domain that contains $\Omega_2$, defined as

$$\Phi_2 := (\Phi_x^{21} \times \mathbb{R}^n) \cap \Phi^{22}. \tag{39}$$

Here $\Phi^{22}$ is defined as

$$\Phi^{22} := \{(x, \xi) \mid -x^T[X(B(x)B^T(x) - \hat{B}_l\hat{B}_l^T)X + \hat{\Delta}_1]x$$

$$
\begin{aligned}
&- \gamma^2(x-\xi)^T X(\hat{B}_u \hat{B}_u^T - B(x)B^T(x))X(x-\xi) \\
&+ \gamma^2(x-\xi)^T T^{-1}[T\{\hat{v}(\xi)M\}^T + \{\hat{v}(\xi)M\}T - \hat{\Delta}_2]T^{-1}(x-\xi) \\
&+ \|B^T(x)Xx - B^T(\xi)X\xi\|^2 - \|B^T(x)X(x-\xi)\|^2 < 0\}.
\end{aligned} \tag{40}
$$

In Algorithms 4.1 and 4.2, (34) is given by substituting (17), (18), (30), and (31) into (19), and (40) is given by substituting (23), (24), (36), and (37) into (25).

**Remark 4.1.** A basic idea of the algorithms is to solve the constant coefficient Riccati inequalities instead of the Riccati inequalities that depend on the state $x$ (or $\xi$), by setting the admissible domain for $x$ (or $\xi$) and evaluting the nonlinear term with respect to $x$ (or $\xi$) in the admissible domain. It is a key point of the algorithms how large an admissible domain we set for $x$ (or $\xi$):

For example, in Algorithm 4.1, if there exists $\underline{B}$ such that $\underline{B}\,\underline{B}^T \leq B(\xi)B^T(\xi)$ for all $\xi \in \mathbb{R}^n$, then we get $\mathrm{Eqn}_1(S, B(\xi)) \leq \mathrm{Eqn}_1(S, \underline{B})$. Therefore, by solving the inequality $\mathrm{Eqn}_1(S, \underline{B}) < 0$, we can solve the inequality $\mathrm{Eqn}_1(S, B(\xi)) < 0$. Note that the smaller $\|\underline{B}\,\underline{B}^T\|$ is, then more difficult it is to obtain the solution; conversely, the larger $\|\underline{B}\,\underline{B}^T\|$ is, the smaller the admissible domain is (see Figure 1).

**Remark 4.2.** In Algorithm 4.1, we consider a domain of $\xi$ to solve the Riccati inequalities. In Algorithm 4.2, we consider a domain of $x$ to do so. At this time, we restrict a lower bound of the nonlinear term in Algorithm 4.1, and both upper and lower bounds in Algorithm 4.2. Therefore, in those algorithms, the structure (15) of solution $V(x_a)$ gives an easier task than another structure (16) does.

**Remark 4.3.** In Step 2, to obtain a domain in which the system is internally stable (i.e., to get $\sigma_1$ in Algorithm 4.1, or $\sigma_2$ in Algorithm 4.2), we solve a nonlinear optimization problem given as

$$
\sigma_1 = \min_{(x,\xi)\in\partial\Phi_1} \left[\xi^T S\xi + \gamma^2(x-\xi)^T Y^{-1}(x-\xi)\right], \tag{41}
$$

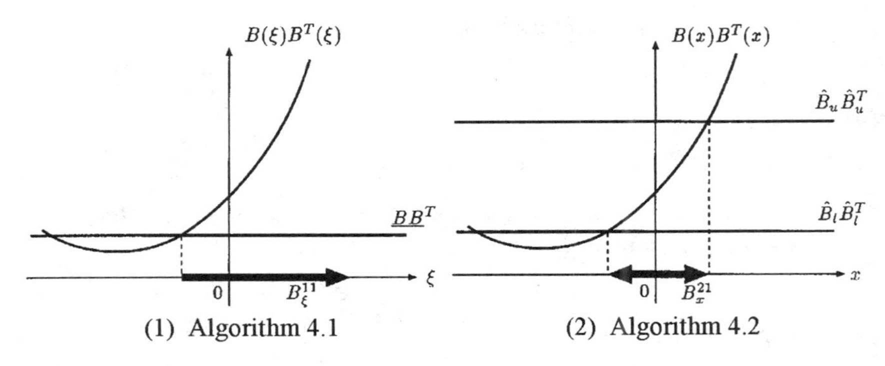


**Figure 1**: Bilinear term versus admissible domain.

or

$$\sigma_2 = \min_{(x,\xi)\in\partial\Phi_2} \left[x^T X x + \gamma^2 (x-\xi)^T T^{-1}(x-\xi)\right]. \tag{42}$$

Section 5.2.2 will discuss the problem in detail.

# 5 Numerical Example

In this section, we present a design example of the artificial rubber muscle actuator control system. We demonstrate the efficiency of the proposed algorithm through a numerical example. We model a single-link manipulator with paralleled artificial rubber muscle actuators as a bilinear system, and construct a nonlinear $H_\infty$ output feedback controller which controls the joint angle.

## 5.1 Bilinear Model of Rubber Muscle Actuators

We consider a single-link manipulator with paralleled artificial rubber muscle actuators as shown in Figure 2. By transforming the difference in shrinking forces of the actuators into a rotation force, we use a pair of actuators like a human muscle. On the basis of [8] and [9], we model it as a bilinear system of the form

$$\dot{x}_n = A_n x_n + (B_n + x_{n1}N_{n1} + x_{n2}N_{n2})u, \tag{43}$$

$$y_n = C_n x_n, \tag{44}$$

where $x_n$, $u$, $A_n$, $B_n$, $C_n$, $N_{n1}$, and $N_{n2}$ are given as

$$x_n = \begin{bmatrix} \theta \\ \dot{\theta} \end{bmatrix}, \qquad u = \begin{bmatrix} u_f \\ u_e \end{bmatrix},$$

$$A_n = \begin{bmatrix} 0 & 1 \\ 0 & 0 \end{bmatrix}, \qquad B_n = \frac{r}{I}\begin{bmatrix} 0 & 0 \\ 1 & -1 \end{bmatrix}, \qquad C_n = \begin{bmatrix} 1 & 0 \end{bmatrix},$$

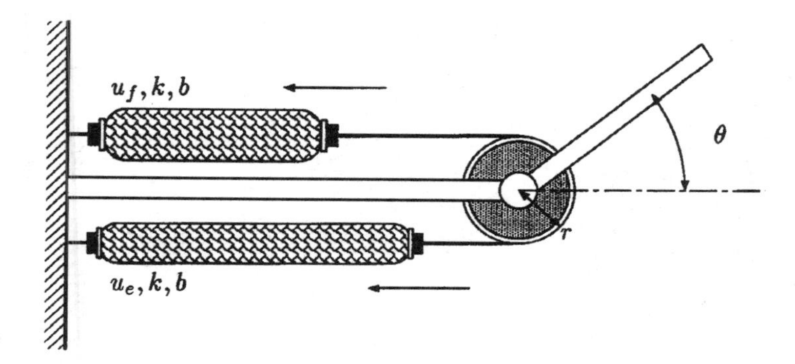


**Figure 2**: Single-link manipulator with paralleled artificial rubber muscle actuators.

$$N_{n1} = -\frac{r}{I}\begin{bmatrix} 0 & 0 \\ k & k \end{bmatrix}, \qquad N_{n2} = -\frac{r}{I}\begin{bmatrix} 0 & 0 \\ b & b \end{bmatrix}.$$

Here $\theta$ denotes a joint angle, $u_f$, $u_e$ shrinking-forces, $I$ an inertial moment, $r$ a radius of joint part, and $k$, $b$ constant coefficients. The following discussion uses $I/r = 0.03, k = 0.2, b = 0.05$ [8].

### 5.2 Control System Design

We design a control system for the bilinear model (43) and (44). First, we construct a generalized plant meeting our specifications. Then, for the generalized plant, we construct an $H_\infty$ output feedback controller, and at the same time, we evaluate a domain of stability.

#### 5.2.1 Generalized Plant Design

We shall design a control system meeting the following specifications:

(S1) A joint angle $\theta$ should track output signals of a reference model ($w_1 \to z_1$).
(S2) Sensitivity from process noises to a control input should be reduced ($w_2 \to z_2$).

Thus, we obtain a generalized plant (Figure 3) of the form

$$\frac{d}{dt}\begin{bmatrix} x_n \\ x_p \end{bmatrix} = \begin{bmatrix} A_n & 0 \\ 0 & -a_p \end{bmatrix}\begin{bmatrix} x_n \\ x_p \end{bmatrix} + \begin{bmatrix} 0 & 1 & 0 \\ b_p & 0 & 0 \end{bmatrix} w$$
$$+ \left( \begin{bmatrix} B_n \\ 0 \end{bmatrix} + x_n \begin{bmatrix} N_n \\ 0 \end{bmatrix} + x_p \begin{bmatrix} 0 \\ 0 \end{bmatrix} \right) u, \qquad (45)$$

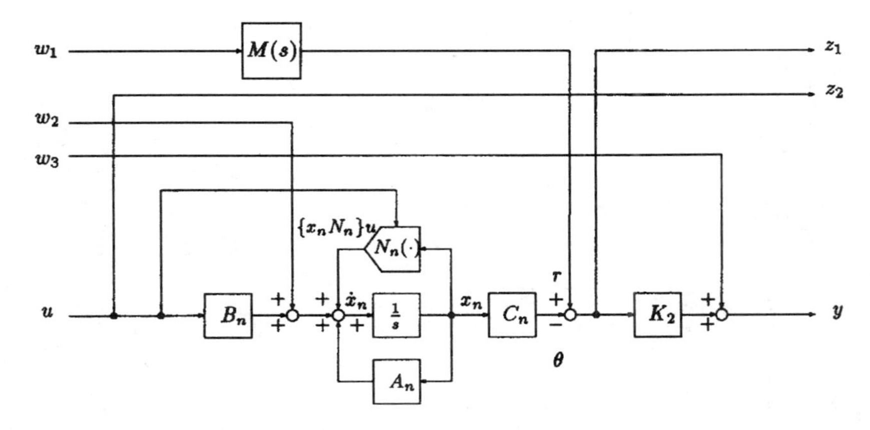


**Figure 3**: Generalized plant.

$$\begin{bmatrix} z_1 \\ z_2 \end{bmatrix} = \begin{bmatrix} -C_n & 1 \\ 0 & 0 \end{bmatrix} \begin{bmatrix} x_n \\ x_p \end{bmatrix} + \begin{bmatrix} 0 \\ 1 \end{bmatrix} u, \tag{46}$$

$$y = \begin{bmatrix} -K_2 C_n & K_2 \end{bmatrix} \begin{bmatrix} x_n \\ x_p \end{bmatrix} + \begin{bmatrix} +0 & 0 & 1 \end{bmatrix} w, \tag{47}$$

where $x_p$ denotes a state of the reference model $M(s)$ of the form

$$M(s) = \frac{b_p}{s + a_p} = \frac{1}{s+1}.$$

The model $M(s)$ does not reflect on any particular actual problem. We consider an external input $w_3$ only for the orthongonality condition to hold. Correspondingly, we add a gain block $K_2 = 10$ to the nominal plant in order to attenuate an influence of $w_3$ on constructing a controller.

### 5.2.2 Controller Design and Evaluation of Domain of Stability

We construct an output feedback controller by using Algorithm 4.1 given in Section 4. The following numerical calculations use MATLAB/LMITOOL [4]/Optimization Toolbox [5].

**Algorithm 5.1.**
**Step 1.** We shall construct a controller. The bilinear term of (45) gives

$$B(\xi) = \begin{bmatrix} B_n \\ 0 \end{bmatrix} + \xi_1 \begin{bmatrix} N_{n1} \\ 0 \end{bmatrix} + \xi_2 \begin{bmatrix} N_{n2} \\ 0 \end{bmatrix} + \xi_3 \begin{bmatrix} 0 \\ 0 \end{bmatrix},$$

where $\xi$ denotes a state of the controller. Figures 4 and 5 illustrate maximum and minimum eigenvalues of $B(\xi)B^T(\xi)$ as functions of states of the controller $(\xi_1, \xi_2) \in ([-100, 100], [-100, 100])$, respectively. From these figures, we easily

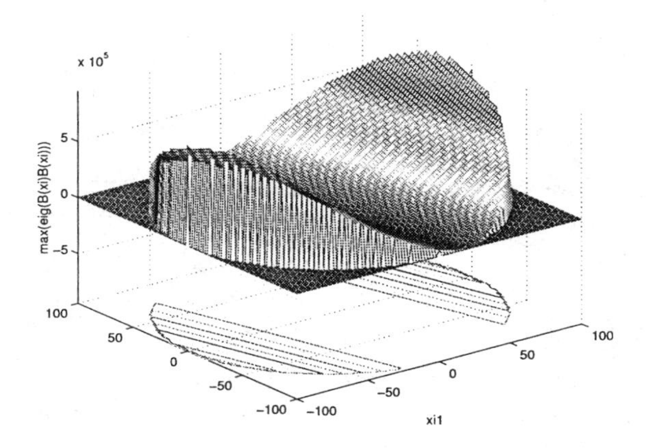


**Figure 4**: Max-eigenvalues of $B(\xi)B^T(\xi)$.

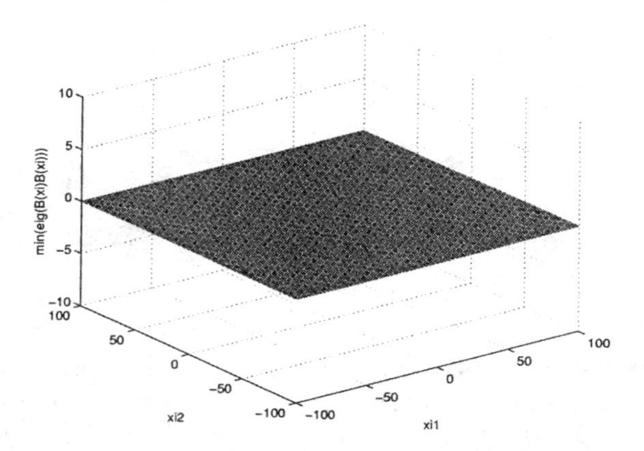


**Figure 5**: Min-eigenvalues of $B(\xi)B^T(\xi)$.

obtain an inequality $B(0)B^T(0) \leq B(\xi)B^T(\xi)$ for all $\xi \in \mathbb{R}^3$. Thus, the matrix $\underline{BB^T}$, which satisfies

$$\begin{bmatrix} 0 & 0 & 0 \\ 0 & 2222.2 & 0 \\ 0 & 0 & 0 \end{bmatrix} = \underline{BB^T} < B(0)B^T(0),$$

assures us that the inequality $\underline{BB^T} < B(\xi)B^T(\xi)$ holds globally, i.e., $\Phi_\xi^{11} = \mathbb{R}^3$. By using this matrix $\underline{BB^T}$, we obtain solutions

$$S = \begin{bmatrix} 1.0828 & 0.11208 & -0.97144 \\ 0.11208 & 0.015301 & -0.097308 \\ -0.97144 & -0.097308 & 0.87464 \end{bmatrix}, \tag{48}$$

$$Y = \begin{bmatrix} 1.6254 & 0.46810 & 1.3757 \\ 0.46810 & 0.61939 & 0.29732 \\ 1.3757 & 0.29732 & 1.3460 \end{bmatrix}, \tag{49}$$

that satisfy Riccati equations (30) and (31) with $\gamma = 0.8$, and a controller.

**Step 2.** We shall evaluate a domain of stability $\Omega_1(\sigma_1)$, which is done by finding $\sigma_1$. To obtain $\sigma_1$ (or a lower bound of $\sigma_1$), we solve two contrained nonlinear optimization problems

$$\min_{(x,\xi)\in\partial B_r} V(x,\xi) \leq \sigma_1 = \min_{(x,\xi)\in\partial\Phi_1} V(x,\xi),$$

where $V(x,\xi) = \xi^T S\xi + \gamma^2(x-\xi)^T Y^{-1}(x-\xi)$, and $B_r$ is a maximum hyperball contained in $\Phi_1 = (\mathbb{R}^3 \times \mathbb{R}^3) \cap \Phi^{12}$. Then we obtain an inequality (see Figure 6)

$$4.0057 \times 10^{-5} < \sigma_1 = 1.1177.$$

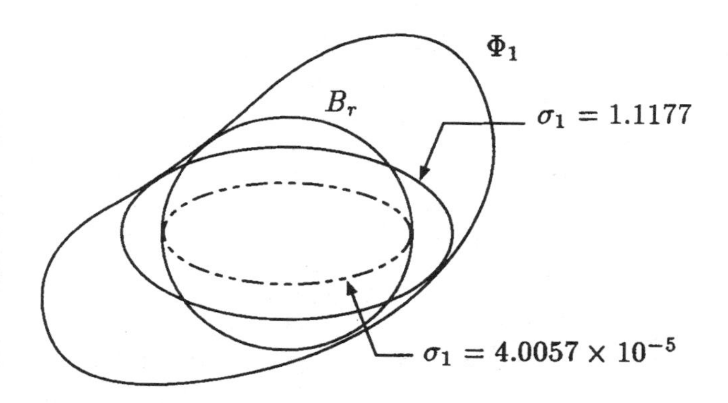


**Figure 6**: Domain of stability.

Here, if the value $\sigma_1 = 1.1177$ is a local optimal solution, then the value is only an upper bound of the true solution. In this problem, we obtain a fairly small lower bound because there is a difference of $1 \sim 10^5$ between magnitudes of the eigenvalues of $S$ and $Y$.

### 5.3 Computer Simulations

We show the efficiency of the obtained controller through computer simulations. Consider four cases: Case 1 (R1 + D1), Case 2 (R1 + D2), Case 3 (R2 + D1), and Case 4 (R2 + D2), where reference input $w_1$ and process noise $w_2$ are given as follows:

$$\begin{array}{llll} \text{R1}: & w_1 = 1(0 \le t \le 8), & \text{D1}: & w_2 = 0, \\ \text{R2}: & w_1 = \sin(0.5\pi t)\ (0 \le t \le 8), & \text{D2}: & w_2 = -0.3(1.5 \le t \le 8). \end{array}$$

Results for the four cases are shown in Figures 2.7–2.10. Figures 2.7 and 2.9 show that a joint angle $\theta$ tracks reference output signals $r$, and so means that the obtained controller satisfies the specification (S1). Figures 2.8 and 2.10 show that the obtaining controller attenuates influence of process noises $w_2$ and satisfies the specification (S2).

## 6 Conclusion

We considered $H_\infty$ output feedback control problems for bilinear systems, and presented a design example of an artificial rubber muscle actuator control system. We derived two types of $H_\infty$ output feedback controller via the differential game approach, and proposed algorithms to construct the controllers based on Riccati inequalities. We also demonstrated the efficiency of the proposed algorithm. The algorithm has also been examined through other numerical examples. Details can be found in [12].

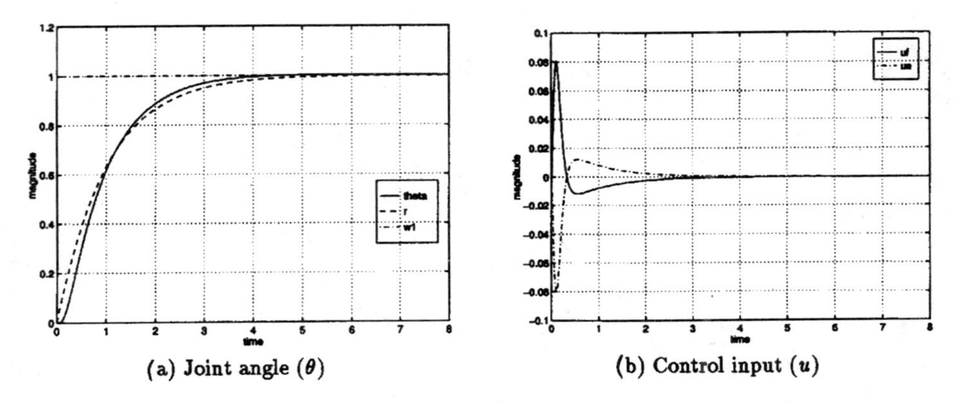


(a) Joint angle ($\theta$) (b) Control input ($u$)

**Figure 7**: Case 1 (R1 + D1).

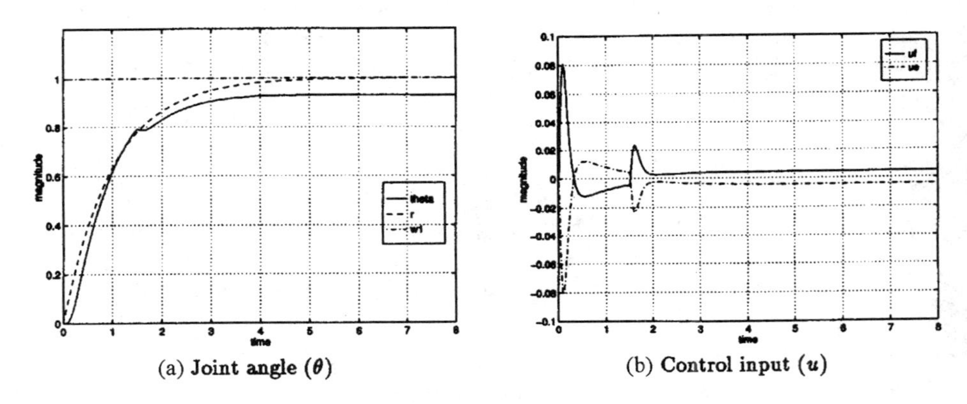


(a) Joint angle ($\theta$) (b) Control input ($u$)

**Figure 8**: Case 2 (R1 + D2).

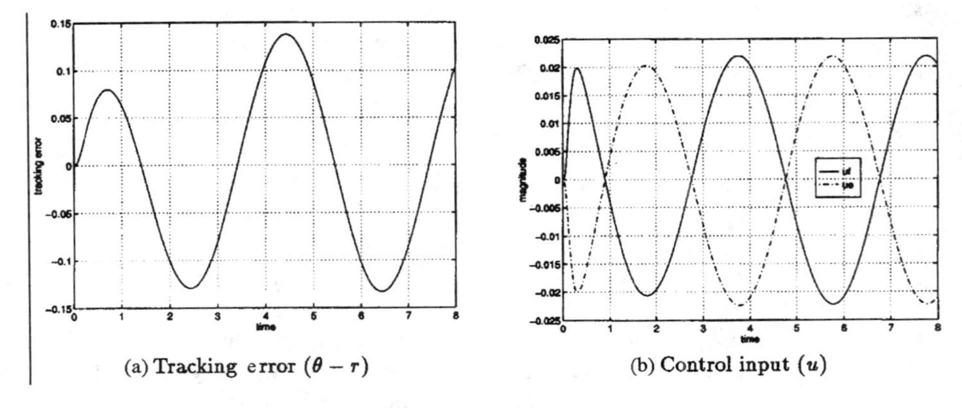


(a) Tracking error ($\theta - r$) (b) Control input ($u$)

**Figure 9**: Case 3 (R2 + D1).

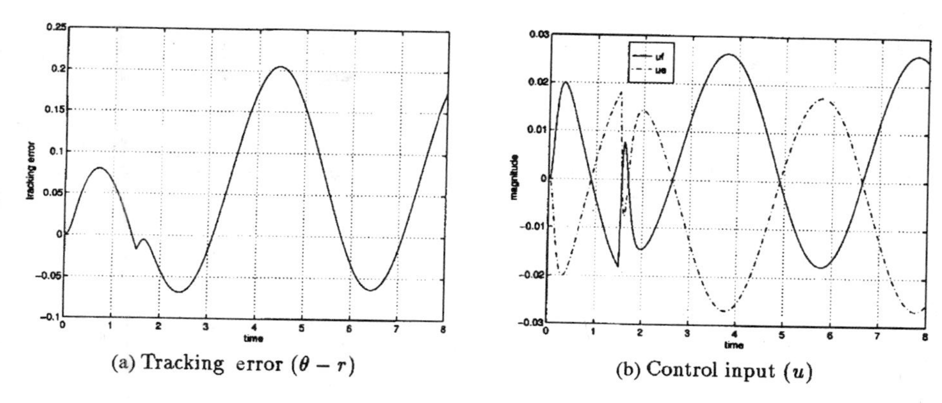


(a) Tracking error $(\theta - r)$

(b) Control input $(u)$

**Figure 10**: Case 4 (R2 + D2).

## REFERENCES


[1] Başar, T. and P. Bernhard. *$H_\infty$-Optimal Control and Related Minimax Design Problems*. Birkhäuser, Boston, 2nd ed., 1995.

[2] Chen, B. S. and S. Tenqchen. $H_\infty$ Control Design in Bilinear Systems: A Tensor Formal Series Approach. *Systems & Control Letters*, **23**, 15–16, 1994.

[3] Doyle, J. C., K. Glover, P. P. Khargonekar, and B. A. Francis. State-space Solutions to Standard $H_2$ and $H_\infty$ Control Problems. *IEEE Transactions on Automatic Control*, **34**, No. 8, 831–847, 1989.

[4] Ghaoui, L. E., F. Delebecque, and R. Nikoukhah. LMITOOL: *A User-Friendly Interface for LMI Optimatization, User's Guide*, 1995.

[5] Grace, A. *Optimization Toolbox User's Guide*. The MathWorks, 1994.

[6] Imura, J., T. Sugie, and T. Yoshikawa. Internal Stability and $L_2$ gain of Nonlinear Systems—Derivation of Bounded Real Condition. *Transactions of SICE*, **29**, No. 6 659–667, 1993. In Japanese.

[7] Isidori, A. $H_\infty$ Control via Measurement Feedback for Affine Nonlinear Systems. *International Journal of Robust and Nonlinear Control*, **4**, 553–574, 1994.

[8] Ito, K. and T. Tsuji. The Bilinear Characteristics of Muscle-Skeletomotor System and the Application to Prosthesis Control. *Transactions of IEE Japan*, **105-C**, No. 10, 201–209, 1985. In Japanese.

[9] Kato, A. et al. The Bilinear Model of Muscle-Skeletomotor System and the Impedance Characteristics of Rubber Artifical Muscle. *Proceedings of the 5th Annual of Conference of the Robotics Society of Japan*, pp. 657–658, 1987. In Japanese.

[10] Kojima, A., M. Fujita, K. Uchida, and E. Shimemura. Linear Quadratic Differential Game and $H_\infty$ Control—A Direct Approach Based on Completing the Square. *Transactions of SICE*, **28**, No. 5, 570–577, 1992. In Japanese.

[11] Mohler, R. R. *Application to Bilinear Control*, Vol. 2 of *Nonlinear Systems*. Prentice Hall, Englewood Cliffs, NJ, 1991.

[12] Sasaki, S. and K. Uchida. Synthesis of $H_\infty$ Output Feedback Control for Bilinear Systems. *Proceedings of the 35th IEEE Conference on Decision and Control*, pp. 3282–3287, Kobe, Japan, 1996.

[13] Teolis, C. A., S. Yuliar, M. R. James, and J. S. Baras. Robust $H_\infty$ Output Feedback Control for Bilinear Systems. *Proceedings of the 33rd Conference on Decision and Control*, pp. 1421–1426, Lake Buena, 1994.

# $H^\infty$ Control of a Class of Infinite-Dimensional Linear Systems with Nonlinear Outputs

Mingqing Xiao and Tamer Başar
University of Illinois at Urbana-Champaign
Coordinated Science Laboratory
Urbana, Illinois, USA

**Abstract**

For a class of infinite-dimensional linear systems with nonlinear outputs, and using a differential game-theoretic approach, we obtain a set of necessary and sufficient conditions for the existence of a state-feedback controller under which a given $H^\infty$ bound (on disturbance attenuation) is achieved. Characterization of such a controller is given, and the result is applied to a disturbance attenuation problem with control constraints.

## 1 Introduction

In a standard infinite-dimensional linear $H^\infty$-control problem, we have the input–output relationship

$$\begin{aligned} \dot{x}(t) &= Ax(t) + Bu(t) + Dw(t), \qquad t \in (t_0, t_f], \\ x(t_0) &= x_0, \\ z(t) &= C_1 x(t) + D_{12} u(t), \end{aligned} \tag{1}$$

where $A$ is the infinitesimal generator of a $C_0$-semigroup $e^{At}$ on a real separable Hilbert space $X$ (state space), and $B \in \mathcal{L}(U, X)$, $D \in \mathcal{L}(W, X)$ are linear operators defined on real separable Hilbert spaces $U$ (control space), and $W$ (disturbance space), respectively. Let $Z$ be the controlled output space (another real Hilbert space) and $C_1 \in \mathcal{L}(X, Z)$, $D_{12} \in \mathcal{L}(U, Z)$. Let us use the notation $(\cdot, \cdot)_X$ to denote the inner product on $X$, and $\|\cdot\|_X$ to denote the norm of an element out of $X$, and adopt the natural counterparts of this convention for the other Hilbert spaces $U$, $W$, and $Z$. Under the standard hypothesis, $D_{12}^*[C_1, D_{12}] = [0, I]$, where $^*$ denotes the adjoint, we have

$$(z, z)_Z = (C_1 x, C_1 x)_Z + (u, u)_U. \tag{2}$$

Introduce the quadratic cost

$$L_{x_0}(u_{[t_0,t_f]}, w_{[t_0,t_f]}) = \int_{t_0}^{t_f} (z(\tau), z(\tau))_Z \, d\tau + (Q_f x(t_f), x(t_f))_X$$

$$= \int_{t_0}^{t_f} ((C_1x(\tau), C_1x(\tau))_Z + (u(\tau), u(\tau))_U)\, d\tau$$
$$+ (Q_f x(t_f), x(t_f))_X, \tag{3}$$

where $Q_f \in \mathcal{L}(X, X)$ with $Q_f = Q_f^* \geq 0$. For every fixed $u \in L^2([t_0, t_f]; U)$, $L_{x_0}$ induces a natural mapping of $L^2([t_0, t_f]; W)$ into nonnegative reals which we denote by $\mathcal{F}_\mu$. The minimax ($H^\infty$-optimal) controller design problem with state feedback is one of finding a state-feedback controller $\hat{\mu}$ that minimizes the index

$$\sup_{w \in L^2([t_0, t_f]; W)} \rho(\mu, w), \tag{4}$$

where

$$\rho(\mu, w) := \left\{ \frac{\mathcal{F}_\mu{}^{1/2}}{\|w\|_2} \right\} \quad \text{and} \quad \|w\|_2 := \left\{ \int_{t_0}^{t_f} \|w(t)\|_W^2 dt \right\}^{1/2}. \tag{5}$$

If this minimum value (whenever it exists) is denoted by $\gamma^*$, then a related problem is the following: Given a $\gamma > \gamma^*$, find a state-feedback controller, $\hat{\mu}_\gamma$, such that

$$\inf_{\mu \in \mathcal{M}_{CL}} \sup_w \rho(\mu, w) < \gamma. \tag{6}$$

Implicit in these definitions is the assumption that the initial state of the system is zero (i.e., $x_0 = 0$), so that $\mathcal{F}_\mu$ would be associated with a linear gain from $w$ to $z$, if $\mu$ is chosen to be linear. Another point to mention is that, as shown in [8][1] the study of such $H^\infty$-optimal control problems is equivalent to the study of a differential game with dynamics (1) and kernel function

$$J_\gamma(t_0, x_0; u, w) = \int_{t_0}^{t_f} (\|C_1x(t)\|_Z^2 + \|u(t)\|_U^2 - \gamma^2 \|w(t)\|_W^2)\, dt + (Q_f x(t_f), x(t_f))_X, \tag{7}$$

which is to be minimized with respect to the controller $u(\cdot) = \mu(\cdot, x(\cdot))$, and maximized with respect to $w$. This equivalence is as far as the upper value of this differential game goes, though one can show that under some appropriate conditions, for all $\gamma > \gamma^*$, this differential game has a saddle point, while for $\gamma < \gamma^*$, its upper value is infinite. Such a result is true regardless of whether $x_0 = 0$ or not; see [23] for the infinite-dimensional case, and [8] for the finite-dimensional case.

The fundamental $H^\infty$-control problem for finite-dimensional linear systems was originally formulated by Zames [33], and first studied in the frequency domain [17]. Later, it was discovered that time-domain methods based on state-space

[1] Even though this reference covers only the finite-dimensional case, the same arguments used there apply to the infinite-dimensional case as well, to prove this equivalence under some appropriate assumptions. For more details, see [34].

representations lead to simpler and more complete derivations for a large class of systems [8], [14], [20], [27], [29], [31], [32].

The fact that the $H^\infty$-optimal control problem is also a differential game was expounded by Başar and Bernhard [8], the techniques of which are now widely used in dealing with $H^\infty$-optimal control problems in finite- or infinite-dimensional spaces. The differential game whose performance index is given by (7) (or more generally by $J_\gamma(t_0, 0; \mu, w) = \mathcal{F}_\mu - \gamma^2\|w\|_2^2$) is called the soft-constrained game associated with the disturbance attenuation problem, because in this game there is no hard bound on the disturbance $w$. It turns out that for a large class of systems, the zero-sum dynamic game has equal upper and lower values (whenever they are both finite), which makes the theory on saddle-point solutions of linear-quadratic dynamic games directly applicable to such classes of worst-case design problems.

In the infinite-dimensional case, the state-space approach was first used to treat the singular $H^\infty$-control problem for the Pritchard–Salamon class, in connection with stability radius optimization [28]. The results obtained in this framework were then generalized to a larger class of linear systems [18], [19]. For the boundary control problems we refer the reader to [5], [23], [24], [25]. As far as we know, in the infinite-dimensional case, almost all systems considered heretofore were linear and the cost functions (or kernel functions in terms of the terminology of game theory) were *quadratic*. The main advantage of dealing with quadratic cost functions is that they lead to linear optimal controllers: $\mu(x) = Fx$, under which it is relatively easy to prove the stability of the feedback system. In spite of such nice features, a quadratic cost function also has its limitations. It cannot handle, for example, constraints on control, or the state, which can, however, be incorporated into an unconstrained optimization framework by choosing discontinuous weighting functions.

Motivated by these considerations, we consider in this paper, a class of (infinite-dimensional) systems with *nonlinear* regulated outputs, defined as follows:

$$z(t) = f(x(t), u(t)), \qquad (x(\cdot), u(\cdot)) \in L^2([t_0, t_f]; X) \times L^2([t_0, t_f]; U), \tag{8}$$

where $f : X \times U \to Z$ is nonlinear with respect to $x$ or $u$. This nonlinearity in the output translates into a cost function of the form:

$$L_{x_0}(u, w) = \int_{t_0}^{t_f} \{g(x(t)) + h(u(t))\}\, dt + \phi_0(x(t_f)), \tag{9}$$

where $g$, $h$, and $\phi_0$ satisfy certain regularity conditions listed in the next section. The cost function (9) covers a large class of practical problems, such as problems with control constraints, state constraints, as well as problems where the state is weighted differently in different regions. Of course, the quadratic cost function (7) is a special case of (9). The basic approach to the problem here is game-theoretic, which associates a particular differential game with the disturbance attenuation problem. Here we only consider the finite horizon case, with the extension to the infinite-horizon problem to be dealt with in another forthcoming paper. Under a rather weak assumption on the infinitesimal generator $A$ (which is that $A$ generates a $C_0$-semigroup on $X$), we develop a general theory for existence and characterization of disturbance attenuating time-varying state-feedback controllers for $H^\infty$ problems in infinite dimensions. The study leads to original results not only in the infinite-dimensional case, but also for finite-dimensional systems with nonlinear outputs. Two examples included in this paper serve to illustrate the theory. The paper ends with two Appendices, containing some auxiliary results used in the main body of the paper.

## 2 General Formulation of the Problem and the Main Result

### 2.1 Problem Formulation

Consider a class of uncertain dynamical systems governed by the set of evolution equations

$$\begin{aligned} \dot{x}(t) &= Ax(t) + Bu(t) + Dw(t), \qquad t_0 < t \leq t_f, \\ x(t_0) &= x_0, \\ z(t) &= f(x(t), u(t)), \end{aligned} \tag{10}$$

where $x(t) \in X$ is the state, $u(t) \in U$ the control, $w(t) \in W$ the disturbance, and $z(t) \in Z$ the controlled output, with $X, U, W, Z$ being real separable Hilbert spaces. $A$ is the infinitesimal generator of a $C_0$-semigroup $e^{At}$ on $X$; $B \in \mathcal{L}(U, X)$, $D \in \mathcal{L}(W, X)$ are linear operators defined on $U$ and $W$, respectively; and $f : X \times U \to Z$ is such that

$$\|f(x, u)\|_Z^2 = g(x) + h(u), \qquad (x, u) \in X \times U, \tag{11}$$

where $g$ and $h$ satisfy certain regularity conditions, which will be made precise shortly. Here and throughout in the sequel we shall use the asterisk symbol to denote dual operators and $(\cdot, \cdot)$ to denote inner products on appropriate Hilbert spaces. Denote the cost function as

$$L_{x_0}(u, w) = \int_{t_0}^{t_f} \{g(x(t)) + h(u(t))\}\, dt + \phi_0(x(t_f)), \tag{12}$$

which we wish to minimize under worst choices for the unknown disturbance. For every $\gamma \in \mathbb{R}^+$, let $J_\gamma : [t_0, t_f] \times X \times L^2([t_0, t_f]; U) \times L^2([t_0, t_f], W) \to \mathbb{R}$ be given by

$$J_\gamma(t, x, u, w) = \int_t^{t_f} \{g(x(s)) + h(u(s)) - \gamma^2 \|w(s)\|^2\}\, ds + \phi_0(x(t_f)), \tag{13}$$

where $t \to x(t)$ is the solution of (18) (in the "mild" sense).

We now make the notion of an admissible control policy (or law) precise.

**Definition 2.1.** An admissible feedback control policy for system (10) is a (nonlinear, multivalued) mapping $F : [t_0, t_f] \times X \to U$ such that for every $x_0 \in X$ and $w \in L^2([t_0, t_f]; W)$ the Cauchy problem

$$\dot{x}(t) \in Ax + BFx + Dw, \qquad t \in (t_0, t_f], \quad x(t_0) = x_0,$$

has at least one mild solution.

**Remark 2.1.** Note that in accordance with the definition above, an admissible controller has access to instantaneous perfect measurements of the state of the system (10).

**Remark 2.2.** The reason for allowing $F$ to be multivalued is because various terms in the cost function $L$, such as $g$, $h$, $\phi_0$, are not necessarily differentiable, thus making the value function also nondifferentiable. Because of this, the optimal control is generally expressed in terms of subdifferentials, which generally contain more than one element.

We shall denote by $\mathcal{M}_{CL}$ the class of admissible feedback control policies for system (18), as per Definition 2.1. Also, by a possible abuse of notation, we will write $L_{x_0}(\mu, w)$ for (20) when $u$ is generated by the control policy $\mu$.

**Definition 2.2.** System (10) is $\gamma$-robust if there exists a $\mu \in \mathcal{M}_{CL}$ and a nonnegative function $U : X \to \mathbb{R}$ such that for some $\epsilon \in (0, \gamma)$

$$L_{x_0}(\mu, w) \le \epsilon^2 \int_{t_0}^{t_f} \|w(t)\|_W^2 \, dt + U(x_0), \qquad \forall (x_0, w) \in X \times L^2([t_0, t_f]; W). \tag{14}$$

**Remark 2.3.** In the above definition, the initial condition $x_0$ plays a role since the system (18) becomes nonlinear when a nonlinear feedback control is applied. In the quadratic cost function case, a possible choice for $U$ is $U(x_0) = \alpha \|x_0\|_X^2$, where $\alpha > 0$ is a constant positive parameter. Note that in this case $U(0) = 0$, and hence the above definition subsumes the case of the standard linear $H^\infty$-optimal control problem (see, e.g., [8], [9], [18], and [19]). If there is some a priori information on $x_0$, such as it having a nominal value $\bar{x}_0$, then $U(x_0)$ could be picked as $U(x_0) = \alpha \|x_0 - \bar{x}_0\|_X^2$.

The differential game associated with this problem is defined by kernel (30), whose upper value is the quantity of interest:

$$\inf_{\mu \in \mathcal{M}_{CL}} \sup_{w \in L^2([t_0, t_f], W)} J_\gamma(t_0, x_0; \mu, w). \tag{15}$$

It will in fact be shown that if this upper value is finite for some $\gamma = \gamma^0$, then the game has a value for all $\gamma > \gamma^0$, which is attained by some state-feedback (saddle-point) policies ($u(t) = \hat{\mu}(t, x(t))$, $w(t) = \hat{\nu}(t, x(t))$), where the feedback policy $\nu(\cdot, \cdot)$ for the disturbance is defined in a way analogous to that of $\mu$. This result will be established under the following assumptions on $g$,$h$, and $\phi_0$:

**Assumption 2.1.**

(A1) $g : X \to \mathbb{R}$ and $\phi_0 : X \to \mathbb{R}$ are proper nonnegative lower semicontinuous convex functions.

(A2) $h : U \to \overline{\mathbb{R}}$ is a lower semicontinuous nonnegative convex function, nonidentically $+\infty$.

(A3)

$$\int_{t_0}^{t_f} h(u_1(t))\,dt = \int_{t_0}^{t_f} h(u_2(t))\,dt \text{ implies } \int_{t_0}^{t_f} \|u_1(t)\|_U^2\,dt = \int_{t_0}^{t_f} \|u_2(t)\|_U^2\,dt$$

and there exists a proper convex, nondecreasing function $\omega : \mathbb{R}^+ \to \mathbb{R}^+$ with the properties $\omega(0) = 0, \lim_{r\to+\infty} \omega(r^2)/r = +\infty$ such that $h(u) \geq \omega(\|u\|_U^2)$.

(A4) $g$ is locally Lipschitz, and there exist constants $C_1$ and $C_2$ such that

$$\sup\{\|\xi\| : \xi \in \partial g(x)\} \leq C_1\|x\| + C_2.$$

(A5) The function $u \to h(u)$ is continuous and $u \to \partial h(u)$ is continuous and bounded on every bounded subset of $U$ and $u \to h^*(u)$ is Fréchet differentiable, where $h^*$ is the conjugate function of $h$, i.e., $h^*(u) = \sup_{v\in U}\{(u, v) - h(v)\}$.

(A6) $\phi_0 : X \to \mathbb{R}$ is convex, locally Lipschitz.

**Remark 2.4.** Assumptions (A1)–(A6) are automatically satisfied when the output is the standard output defined in (18), (19).

Let us denote by $\Phi$ a class of functions $\{\varphi\}$ satisfying the following properties:

(1) $\varphi : [t_0, t_f] \times X \to \mathbb{R}$;
(2) for every $s \in [t_0, t_f]$, $\varphi(s, \cdot)$ is convex and continuous on $X$;
(3) for every $x \in W^{1,2}([t_0, t_f]; X)$, $t \to \varphi(t, x(t))$ is absolutely continuous and the following rule of differentiation holds:

$$\frac{d}{dt}\varphi(t, x(t)) = \varphi_t(t, x(t)) + (\eta(t, x(t)), \dot{x}(t)) \qquad \text{a.e.} \quad t \in [t_0, t_f],$$

where $\eta \in \partial\varphi$, where $\partial\varphi$ denotes the subdifferential of $\varphi$; i.e.,

$$\partial\varphi(x_0) = \{y \in X : \varphi(x_0) \leq \varphi(h) + (y, x_0 - h), \ \forall h \in X\}$$

and $W^{1,2}([t_0, t_f]; X)$ is the typical Sobolev space (see, e.g., [1]).

## 2.2 Statement of the Main Result

Under conditions (A1)–(A3) and (A5), let the following Hamilton–Jacobi–Isaacs equation admit a solution $\varphi_\gamma \in \Phi$ for $\eta \in \partial\varphi_\gamma$ for a given $\gamma > 0$:

$$\begin{aligned} &\frac{\partial\varphi_\gamma}{\partial s}(s, x) + (Ax, \eta(x)) - h^*(-B^*\eta(x)) + g(x) + \frac{\gamma^{-2}}{4}\|D^*\eta(x)\|_W^2 = 0, \\ &\quad \text{a.e.} \quad s \in (t_0, t_f), \\ &\varphi_\gamma(t_f, h) = \phi_0(h), \qquad \forall h \in X. \end{aligned} \tag{16}$$

Furthermore, let the following properties hold:

(i) $x \to \eta(x)$ is continuous;
(ii) $D(\eta) \supseteq D(A)$ and $\|\eta(x)\|_X \le c_1\|x\|_X + c_2$ for all $x \in D(A)$, for some constants $c_1 > 0$, $c_2 > 0$.

Then, the zero-sum differential game (2.6) admits a feedback saddle-point solution, which is given by

$$\hat{\mu}(t;x(t)) = \partial h^*(-B^*\eta(x(t))), \qquad \hat{\mu}(\cdot;x(\cdot)) \in L^2([t_0,t_f];U) \cap C([t_0,t_f];U),$$
$$\hat{\nu}(t;x(t)), = \frac{\gamma^{-2}}{2} D^*\eta(x(t)), \qquad \hat{\nu}(\cdot;x(\cdot)) \in L^2([t_0,t_f];W) \cap C([t_0,t_f];W), \tag{17}$$

and $\varphi_\gamma(t,x)$ is the value of the game. Moreover, (14) holds with $\epsilon = \gamma$.

Conversely, under conditions (A1)–(A6), let the $\gamma$-robustness property hold and let

$$\varphi_\gamma(t,x) := \inf_{\mu \in \mathcal{M}_{CL}} \sup_{w \in L^2([t_0,t_f];W)} J_\gamma(t,x,\mu,w).$$

Then, $\varphi_\gamma(t,x)$ is finite for any $x_0 \in X$, and $\varphi_\gamma \in \Phi$ and satisfies (16) almost everywhere with respect to $s \in [t_0,t_f]$. Moreover, if (i) and (ii) above are satisfied, then (17) is an admissible optimal feedback control policy.

In the next two sections, we will present several theorems, which will capture separately the necessity and sufficiency parts of the main result above.

## 3 Proof of the Main Result: Necessity

We first consider the following sup inf problem, which constitutes the lower value of the game:

$$\varphi_\gamma(t_0,x_0) := \sup_{w \in L^2([t_0,t_f];W)} \inf_{u \in L^2([t_0,t_f];U)} J_\gamma(t_0,x_0;u,w). \tag{18}$$

**Theorem 3.1.** *Let the assumptions* (A1)–(A3) *hold. Suppose that for a given* $\gamma > 0$*, the* $\gamma$*-robustness property holds for system (18). Then:*

(i) *(18) has at least one solution* $(\hat{u},\hat{w})$*;*
(ii) *There exists a function* $p \in C([t_0,t_f];X)$ *such that*

$$\dot{\hat{x}}(t) = A\hat{x}(t) + B\hat{u}(t) + D\hat{w}(t), \tag{19}$$
$$\dot{p}(t) \in -A^*p(t) + \partial g(\hat{x}(t)), \tag{20}$$
$$B^*p(t) \in \partial h(\hat{u}(t)), \tag{21}$$
$$p(t_f) + \partial\phi_0(\hat{x}(t_f)) \ni 0, \tag{22}$$
$$\hat{x}(t_0) = x_0, \tag{23}$$
$$D^*p(t) = -2\gamma^2\hat{w}(t), \tag{24}$$

*where solutions in* (19), (20) *are both defined in the "mild" sense.*

(iii) $(\hat{u}, \hat{w})$ *is an open-loop saddle-point solution of the differential game* $J_\gamma(t_0, x_0; u, w)$ *on the space* $L^2([t_0, t_f]; U) \times L^2([t_0, t_f]; W)$.

**Proof.** (i) Under the hypotheses of the theorem, there exist a $\mu \in \mathcal{M}_{CL}$ and $\epsilon \in (0, \gamma)$ such that

$$\int_{t_0}^{t_f} (g(x(t)) + h(\mu(t, x(t))))\, dt + \phi_0(x(t_f))\} \leq \epsilon^2 \int_{t_0}^{t_f} \|w(t)\|_W^2\, dt + U(x_0)$$

for all $(x_0, w) \in X \times L^2([t_0, t_f]; W)$ and for some nonnegative function $U : X \to \mathbb{R}$. This yields that $\sup_w J_\gamma(t_0, x_0; \mu, w)$ is finite for a given $x_0 \in X$. Since $J_\gamma(t_0, x_0; \cdot, w)$ is convex, lower semicontinuous, coercive, and $\not\equiv \infty$, it has a unique minimum, say $u^w(t)$. Let $\Psi(w) = -\inf_{L^2([t_0,t_f];U)} J_\gamma(t_0, x_0; u, w)$. Then there exists $\Gamma : L^2([t_0, t_f]; W) \to L^2([t_0, t_f]; U)$ such that

$$\begin{aligned} \Psi(w) = -J_\gamma(t_0, x_0; u^w, w) &= \gamma^2 \|w\|_2^2 - L_{x_0}(\Gamma w, w) \\ &\geq (\gamma^2 - \epsilon^2)\|w\|_2^2 - U(x_0). \end{aligned}$$

Let $\{w_k\}_{k=1}^\infty \subset L^2([t_0, t_f]; W)$ be a minimizing sequence,

$$\Psi(w_k) \to \inf_{w \in L^2([t_0,t_f];W)} \Psi(w).$$

Note that $\{w_k\}_{k=1}^\infty$ is bounded in $L^2([t_0, t_f]; W)$. Hence $\exists \hat{w} \in L^2([t_0, t_f]; W)$ such that

$$w_k \to \hat{w} \quad \text{weakly in } L^2([t_0, t_f]; W).$$

Clearly, we have

$$L_{x_0}(\Gamma w_k, w_k) \leq \int_{t_0}^{t_f} (g(x_k(t)) + h(\Gamma \hat{w}(t)))\, dt + \phi_0(x_k(t_f)), \tag{25}$$

where $x_k(t)$ satisfies

$$\begin{aligned} \dot{x}_k(t) &= Ax_k(t) + B\Gamma \hat{w}(t) + Dw_k(t), \\ x_k(t_0) &= x_0. \end{aligned}$$

It is not difficult to see that

$$x_k \to x_{\hat{w}} \quad \text{in } C([t_0, t_f]; X)$$

by the Arzela–Ascoli theorem (see Appendix II), and that $(x_{\hat{w}}, \Gamma\hat{w}, \hat{w})$ satisfies (19). Assumption (A3) and (25) further imply that $\Gamma w_k$ is bounded, and thus $\exists \overline{u} \in L^2([t_0, t_f]; U)$ such that

$$\Gamma w_k \to \overline{u} \quad \text{weakly on } L^2([t_0, t_f]; U).$$

Let us introduce

$$\dot{x}^{w_k}(t) = Ax_k^{w_k}(t) + B\Gamma w_k(t) + Dw_k(t),$$
$$x_{w_k}(t_0) = x_0.$$

Again by the Arzela–Ascoli theorem, there exists a subsequence of $x^{w_k}$ (to be denoted by $x^{w_k}$ again) that converges to $x \in C([t_0, t_f]; X)$. Now taking the limit on both sides of (25), we have

$$L_{x_0}(\overline{u}, \hat{w}) \leq \int_{t_0}^{t_f} (g(x_{\hat{w}}(t)) + h(\Gamma\hat{w}(t)))\, dt + \phi_0(x_{\hat{w}}) = \Psi(\hat{w}).$$

Therefore

$$\Gamma\hat{w} = \overline{u}, \qquad x = x_{\hat{w}}.$$

By assumption (A3), we have

$$\int_{t_0}^{t_f} \|\Gamma\hat{w}\|_U^2\, dt = \int_{t_0}^{t_f} \|\overline{u}\|_U^2\, dt$$

and hence

$$\Gamma w_k \to \Gamma w \quad \text{strongly in } L^2([t_0, t_f]; U).$$

Thus $\Gamma$ is weakly lower semicontinuous on $L^2([t_0, t_f]; w)$; therefore, $\sup_w\{\Psi(w), w \in L^2([t_0, t_f]; W)\}$ has at least one solution. Clearly $(\Gamma\hat{w}, \hat{w})$ is a solution of (18).

(ii) Let $(\hat{u}, \hat{w}) = (\Gamma\hat{w}, \hat{w})$ be a solution of (18). Note that

$$\inf_u J_\gamma(t_0, x_0; u, \hat{w}) = J_\gamma(t_0, x_0; \hat{u}, \hat{w}).$$

By Theorem 1.1 of Chapter 4 of [6], there exists a function $p \in C([t_0, t_f]; X)$ such that (19)–(23) are true. Now we claim (24). Let $w \in L^2([t_0, t_f], W)$ be arbitrary but fixed, and let $x \in C([t_0, t_f]; X)$ be such that

$$\dot{x}(t) = Ax(t) + B\hat{u}(t) + Dw(t), \qquad x(t_0) = x_0. \tag{26}$$

Let $q(t) \in \partial g(\hat{x}(t))$. From the definitions of $\partial g$, $\partial h$, and $\partial\phi_0$, and (22), we have

$$g(\hat{x}(t)) \leq g(x(t)) + (\hat{x}(t) - x(t), q(t)), \tag{27}$$
$$\phi_f(\hat{x}(t_f)) \leq \phi_f(x(t_f)) + (\hat{x}(t_f) - x(t_f), -p(t_f)). \tag{28}$$

Hence

$$\begin{aligned} J_\gamma(t_0, x_0; \hat{u}, \hat{w}) \leq J_\gamma(t_0, x_0; \hat{u}, w) &+ \int_{t_0}^{t_f} (\hat{x}(t) - x(t), q(t))\, dt \\ &+ (\hat{x}(t_f) - x(t_f), -p(t_f)) \\ &+ \gamma^2(\|w\|^2_{L^2([t_0,t_f],W)} - \|\hat{w}\|^2_{L^2([t_0,t_f],W)}). \end{aligned}$$

By using (19) and (26), we have

$$\int_{t_0}^{t_f} (\hat{x}(t) - x(t), q(t))\, dt = \int_{t_0}^{t_f} \left( \int_{t_0}^{t} e^{A(t-s)} D(\hat{w}(s) - w(s))\, ds, q(t) \right) dt.$$

Applying Fubini's theorem to interchange the order of integration and using (20) yields

$$\begin{aligned}\int_{t_0}^{t_f}(\hat{x}(t)-x(t),q(t))\,dt &= \int_{t_0}^{t_f}\left(D(\hat{w}(s)-w(s)),\int_s^{t_f}e^{A^*(t-s)}q(t)\,dt\right)ds\\ &= \int_{t_0}^{t_f}(D(\hat{w}(s)-w(s)),e^{A^*(t_f-s)}p(t_f)-p(s))\,ds\\ &= \int_{t_0}^{t_f}(\hat{w}(s)-w(s),D^*e^{A^*(t_f-s)}p(t_f)-D^*p(s))_W\,ds.\end{aligned}$$

Using (19) and (26) again, we arrive at

$$(\hat{x}(t_f)-x(t_f),-p(t_f))=\int_{t_0}^{t_f}(e^{A(t_f-s)}D(\hat{w}(s)-w(s)),-p(t_f))\,ds,$$

and hence

$$\begin{aligned}&\int_{t_0}^{t_f}(\hat{x}(t)-x(t),q(t))\,dt+(\hat{x}(t_f)-x(t_f),-p(t_f))\\ &\qquad+\gamma^2(\|w\|^2_{L^2([t_0,t_f],W)}-\|\hat{w}\|^2_{L^2([t_0,t_f],W)})\\ &=\int_{t_0}^{t_f}(\hat{w}(s)-w(s),-D^*p(s))_W+\gamma^2(\|w\|^2_{L^2([t_0,t_f],W)}-\|\hat{w}\|^2_{L^2([t_0,t_f],W)})\\ &=\gamma^2\left\{\int_{t_0}^{t_f}\|w(s)+\frac{\gamma^{-2}}{2}D^*p(s)\|^2_W ds-\int_{t_0}^{t_f}\|\hat{w}(s)+\frac{\gamma^{-2}}{2}D^*p(s)\|^2_W\,ds\right\}.\end{aligned}$$

Therefore

$$\begin{aligned}J_\gamma(t_0,x_0;\hat{u},\hat{w})\le\ & J_\gamma(t_0,x_0;\hat{u},w)\\ &+\gamma^2\left\{\int_{t_0}^{t_f}\|w(s)+\frac{\gamma^{-2}}{2}D^*p(s)\|^2_W\,ds\right.\\ &\left.-\int_{t_0}^{t_f}\|\hat{w}(s)+\frac{\gamma^{-2}}{2}D^*p(s)\|^2_W\,ds\right\}\end{aligned}$$

Suppose $\hat{w}(s)\neq-(\gamma^{-2}/2)D^*p(s)$, and choose $w(s)=-(\gamma^{-2}/2)D^*p(s)$. Then we have

$$J_\gamma(t_0,x_0;\hat{u},\hat{w})<J_\gamma(t_0,x_0;\hat{u},w)$$

which contradicts our initial hypothesis that $(\hat{u},\hat{w})$ is a solution of (18).

(iii) To prove that $(\hat{u},\hat{w})$ is also a saddle point of $J_\gamma(t_0,x_0;u,w)$, we need to show that $\forall(u,w)\in L^2([t_0,t_f],U)\times L^2([t_0,t_f],W)$ the following is true:

$$J_\gamma(t_0,x_0;\hat{u},w)\le J_\gamma(t_0,x_0;\hat{u},\hat{w})\le J_\gamma(t_0,x_0;u,\hat{w}).$$

In (i) we have already proven that the left-hand side inequality is true. Now we show that the above right-hand side inequality also holds. Fix $(y,v)\in C([t_0,t_f];X)\times$

$L^2([t_0, t_f], U)$ such that

$$\dot{y}(t) = Ay(t) + Bv(t) + Dw(t), \qquad y(t_0) = \tilde{x},$$

where $\tilde{x} \in X$. Let $q(t) \in \partial g(\hat{x}(t))$; then (20) and (21) yield

$$\int_{t_0}^{t_f} (h(\hat{u}(t)) + g(\hat{x}(t)))\, dt \leq \int_{t_0}^{t_f} (\hat{u}(t) - v(t), B^* p(t))_U\, dt + \int_{t_0}^{t_f} (\hat{x}(t) - y(t), q(t))_X\, dt. \tag{29}$$

We also note

$$\int_{t_0}^{t_f} (\hat{x}(t) - y(t), q(t))_X\, dt = \int_{t_0}^{t_f} (\hat{w}(t) - w(t), D^* e^{A^*(t_f - t)} p(t_f) - D^* p(t))_W\, dt$$
$$+ \int_{t_0}^{t_f} (\hat{u}(t) - u(t), B^* e^{A^*(t_f - t)} p(t_f) - B^* p(t))_U\, ds$$
$$+ \int_{t_0}^{t_f} (e^{A(t - t_0)}(x_0 - \tilde{x}), q(t))_X\, dt.$$

Since

$$\int_{t_0}^{t_f} (e^{A(t - t_0)}(x_0 - \tilde{x}), q(t))_X\, dt = \left( x_0 - \tilde{x}, \int_{t_0}^{t_f} p(t_f) - p(t_0) \right)_X,$$

and (22) provides

$$(\hat{x}(t_f) - y(t_f), -p(t_f))_X = (x_0 - \tilde{x}, e^{A^*(t_f - t_0)} p(t_f))_X$$
$$+ \int_{t_0}^{t_f} (e^{A(t_f - t)} D(\hat{w}(t) - w(t), -p(t_f))_X\, ds$$
$$+ \int_{t_0}^{t_f} (e^{A(t_f - t)} B(\hat{u}(t) - v(t)), -p(t_f))_X\, dt,$$

(29) becomes

$$J_\gamma(t_0, x_0, \hat{u}, \hat{w}) - J_\gamma(t_0, \tilde{x}, v, w) \leq (x_0 - \tilde{x}, -p(t_0))_X + \gamma^2 \int_{t_0}^{t_f} \|w(t) + \frac{\gamma^{-2}}{2} D^* p(t)\|_W^2\, dt. \tag{30}$$

Letting $\tilde{x} = x_0$, $w = -(\gamma^{-2}/2) D^* p(t)$, the conclusion follows. □

**Corollary 3.1.** *Under the hypotheses of Theorem* 3.1, *we have*

$$\{-p(t_0)\} = \partial \varphi_\gamma(t_0, x_0).$$

**Proof.** From (80) of Theorem 3.1, we have

$$\varphi_\gamma(t_0, x_0) - \inf_v J_\gamma(t_0, x_0, v, w) \leq (x_0 - \tilde{x}, -p(t_0)) + \gamma^2 \int_{t_0}^{t_f} \left\| w(t) + \frac{\gamma^{-2}}{2} D^* p(t) \right\|_W^2 dt. \tag{31}$$

Let $w(t) = \hat{w}(t)$; then we get

$$\varphi_\gamma(t_0, x_0) - \inf_v J_\gamma(t_0, x_0, v, \hat{w}) \leq (x_0 - \tilde{x}, -p(t_0)).$$

Since

$$\inf_v J_\gamma(t_0, x_0, v, \hat{w}) \leq \sup_w \inf_v J_\gamma(t_0, x_0, v, w),$$

we have

$$\varphi_\gamma(t_0, x_0) - \varphi_\gamma(t_0, \tilde{x}) \leq (x_0 - \tilde{x}, -p(t_0)), \tag{32}$$

i.e., $-\frac{1}{2}p(t_0) \in \partial\varphi(t_0, x_0)$. To complete the proof, let us define $Px_0 = \{-p(t_0)\}$ and next show that $P$ is maximal monotone. Let $y_0 \in X$ be arbitrary but fixed. We claim that the following equation

$$\lambda x_1 + Px_1 = y_0$$

has at least one solution for some $\lambda$. Let us introduce the sup inf problem

$$\sup_{w \in L^2([t_0,t_f];W)} \inf_{(u,x_0)\in L^2([t_0,t_f];U)\times X} \left\{ \int_{t_0}^{t_f} (g(x(t)) + h(u(t)) - \gamma^2 \|w(t)\|_W^2)\, dt \right.$$
$$\left. + \phi_0(x(t_f)) - \frac{\lambda^2}{2} \left\| x_0 - \frac{1}{\lambda^2} y_0 \right\|^2 \right\},$$

where $x(t)$, $x_0$ satisfy (10) and $\lambda \in \mathbb{R}^+$. We claim that when $\lambda$ is sufficiently large, the above sup inf admits a finite solution.

By Theorem 1.1 of Chapter 4 of [6], for every $w \in L^2([t_0, t_f]; W)$, a unique solution for the minimization problem is given by

$$B^* p^w(t) \in \partial h(u^w(t)), \tag{33}$$
$$\dot{p}(t) \in -A^* p(t) + \partial g(x^w(t)), \tag{34}$$
$$p^w(t_f) + \partial\phi_0(x^w(t_f)) i 0, \tag{35}$$
$$p^w(t_0) = \lambda^2 x_0 - y_0. \tag{36}$$

Because of the $\gamma$-robustness assumption, when $\lambda$ is sufficiently large, we have

$$\int_{t_0}^{t_f} (g(x^w(t)) + h(u^w(t)) - \gamma^2 \|w(t)\|_W^2)\, dt + \phi_0(x^w(t_f)) - \frac{\lambda^2}{2} \left\| x_0 - \frac{1}{\lambda^2} y_0 \right\|_X^2$$
$$\leq -(\gamma^2 - \epsilon^2) \int_{t_0}^{t_f} \|w(t)\|_W^2\, dt - \frac{\lambda^2}{2} \left\| x_0 - \frac{1}{\lambda^2} y_0 \right\|_X^2 + U(x_0)$$
$$\leq -(\gamma^2 - \epsilon^2) \int_{t_0}^{t_f} \|w(t)\|_W^2\, dt.$$

Thus, the above sup inf problem has a solution $(x_0, \overline{u}, \overline{w})$. For such $x_0$, let $(\hat{u}, \hat{w})$ be a solution of the lower value game (18). By letting $w = \hat{w}$ in (33)–(35), we have $\hat{w}(t) = -(\gamma^{-2}/2)D^* p^{\hat{w}}(t)$, and hence

$$p^{\hat{w}}(t_0) \in Px_0.$$

Note that (32) yields $Px_0 = \lambda^2 x_0 - y_0$, and therefore completes the proof. □

**Remark 3.1.** More generally, it is easy to see that $\{-p(t)\} = \partial\varphi_\gamma(t, x(t))$ is true for $t_0 \leq t \leq t_f$.

**Remark 3.2.** Since $B^* p(t) \in \partial h(\hat{u}(t))$ implies $\hat{u}(t) \in \partial h^*(B^* p(t))$, using Remark 3.1, we have the following relationship between the output and the optimal controller:

$$\hat{u}(t) \in \partial h^*(B^*(-\partial\varphi_\gamma(t, \hat{x}(t)))).$$

As mentioned before, the control we are mainly interested in is the state-feedback control. Later we shall see that the $H^\infty$-optimal feedback control for system (10) with cost function (12) is in fact

$$\hat{\mu}(t, x(t)) \in \partial h^*(B^*(-\partial\varphi_\gamma(t, x(t)))).$$

The following lemma establishes some important properties of the lower value function $\varphi_\gamma(t, x)$, defined by (18) with $(t_0, x_0)$ replaced by an arbitrary $(t, x)$:

**Lemma 3.1.** *Suppose that for a given $\gamma > 0$, the $\gamma$-robustness property holds. Then:*

(i) *for every $x \in D(A)$, $s \to \varphi_\gamma(s, x)$ is Lipschitz on $[t_0, t_f]$;*
(ii) *for every $s \in [t_0, t_f]$, $x \to \varphi_\gamma(s, x)$ is Lipschitz on every bounded subset of $X$; and*
(iii) *$t \to \varphi_\gamma(t, x(t))$ is absolutely continuous and for $x \in W^{1,2}([t_0, t_f]; X)$ we have*

$$\frac{d}{dt}\varphi_\gamma(t, x(t)) = \varphi_{\gamma t}(t, x(t)) + (\eta(t, x(t)), \dot{x}(t)), \qquad a.e. \quad t \in (t_0, t_f), \tag{37}$$

*where $\eta(t, x(t)) \in \partial\varphi_\gamma(t, x(t))$.*

**Proof.** (i) Recall that

$$\varphi_\gamma(t_0, x_0) = \sup_w \inf_u J_\gamma(t_0, x_0; u, w),$$

and note that the $\gamma$-robustness property guarantees $\varphi_\gamma(t_0, x_0)$ to be finite for any $x_0 \in X$. As in the proof of Theorem 2.1, for each $t \geq 0$, and fixed $x \in X$, we introduce the mapping $\Gamma_x^t : W \to U$ defined by

$$\varphi_\gamma(t, x) = \sup_w J_\gamma(t, x; \Gamma_x^t w, w) = J_\gamma(t, x; \Gamma_x^t \hat{w}_{[t,t_f]}, \hat{w}_{[t,t_f]}),$$

assuming at the last step the existence of a maximizing $w$, without any loss of generality. Let $u_{[t_1,t_f]} \in L^2([t_0, t_f]; U)$ be the open-loop control defined by $u_{[t_1,t_f]} = \Gamma_{x_0}^{t_1} \hat{w}_{[t_1,t_f]}$. Note that by our assumption on $h$, $\exists u_0 \in U$ such that $h(u_0) < +\infty$.

Define

$$\overline{u} = \begin{cases} u_0, & t_0 \le t < t_1, \\ \Gamma_{x_0}^{t_1} \hat{w}_{[t_1,t_f]}, & t_1 \le t \le t_f. \end{cases}$$

Then we have

$$\begin{aligned}\varphi_\gamma(t_0, x_0) - \varphi_\gamma(t_1, x_0) &\le J_\gamma(t_0, x_0; \overline{u}, \hat{w}_{[t_0,t_f]}) - J_\gamma(t_1, x_0; \Gamma_{x_0}^{t_1} w_{[t_1,t_f]}, \hat{w}_{[t_0,t_f]}) \\ &= \int_{t_0}^{t_1} (g(x^{\overline{u}}(s)) + h(\overline{u}) - \gamma^2 \|\hat{w}_{[t_0,t_f]}\|(s))\, ds \\ &+ \int_{t_1}^{t_f} (g(x^{\overline{u}}(s)) - g(x^{\Gamma_{x_0}^{t_1} w_{[t_1,t_f]}}(s)))\, ds \\ &+ \phi_0(x^{\overline{u}}(t_f) - \phi_0(x^{\Gamma_{x_0}^{t_1} w_{[t_1,t_f]}}(t_f)).\end{aligned}$$

Note the following inequalities:

$$\begin{aligned}\|x^{\overline{u}}(t) - x^{u_{[t_0,t_f]}}(t)\| &\le \|e^{A(t-t_0)}x_0 - e^{A(t-t_1)}x_0\| \\ &+ \int_{t_0}^{t_1} \|e^{A(t-s)}(Bu_0 + D\hat{w}_{[t_0,t_1]})\|\, ds\end{aligned}$$

and

$$\|e^{A(t-t_0)}x_0 - e^{A(t-t_1)}x_0\| \le C\|Ax_0\||t_1 - t_0|.$$

Since $g$, $\phi_0$ are locally Lipschitz, the result follows.

(ii) Let $x_0, x_1 \in X$. Without any loss of generality, assume that $\varphi_\gamma(t_0, x_0) \ge \varphi_\gamma(t_0, x_1)$. Introduce again the mapping $\Gamma_{x_1}^{t_0} : W \to U$ such that

$$\varphi_\gamma(t_0, x_1) = \sup_w J_\gamma(t_0, x_1; \Gamma_{x_1}^{t_0} w, w) = J_\gamma(t_0, x_1; \Gamma_{x_1}^{t_0} \hat{w}_0, \hat{w}_0).$$

Now note the inequality

$$\begin{aligned}\varphi_\gamma(t_0, x_0) - \varphi_\gamma(t_0, x_1) &\le J_\gamma(t_0, x_0; \Gamma_{x_1}^{t_0} \hat{w}_0, \hat{w}_0) - J_\gamma(t_0, x_1; \Gamma_{x_1}^{t_0} \hat{w}_0, \hat{w}_0) \\ &= \int_{t_0}^{t_1} (g(x^{x_0}(s)) - g(x^{x_1}(s))\, ds \\ &+ \phi_0(x^{x_1}(t_f)) - \phi_0(x^{x_0}(t_f)).\end{aligned}$$

Clearly

$$\|x^{x_1}(t) - x^{x_0}(t)\| \le \|e^{A(t-t_0)}(x_0 - x_1)\| \le C\|x_0 - x_1\|.$$

Again since $g$, $\phi_0$ are locally Lipschitz, the desired result readily follows.

(iii) Let $x_s$ be the mild solution of

$$\begin{aligned}\dot{x}_s(t) &= Ax_s(t) + Bu_s(t) + Dw_s(t), \qquad s \le t \le t_f, \\ x_s(s) &= x,\end{aligned}$$

where $(u_s, w_s) \in L^2(s, t_f; U) \times L^2(s, t; W)$ and $x \in X$. Introduce

$$\varphi_\gamma(s, x) := \sup_{w \in L^2([s,t_f];W)} \inf_{u \in L^2([s,t_f];U)} J_\gamma(s, x; u_s, w_s).$$

Let $(\hat{u}_s, \hat{w}_s)$ be a solution for $\varphi_\gamma(s, x)$ such that

$$\varphi_\gamma(s, x) = \int_s^{t_f} (g(\hat{x}_s) + h(\hat{u}_s) - \gamma^2 \|\hat{w}_s\|_W)\, ds + \phi_0(\hat{x}_s(t_f)).$$

Note that $t \to g(\hat{x}_s(t)) + h(\hat{u}_s(t)) - \gamma^2 \|\hat{w}_s(t)\|$ is continuous by assumption (A5). Let $s \in [t_0, t_f]$ be such that the function $t \to \varphi_\gamma(t, x_s(t))$ is differentiable at $t = s$. If $x \in W^{1,2}([t_0, t_f]; X)$, then

$$\begin{aligned} \frac{d}{dt}\varphi_\gamma(t, x_s(t))|_{t=s} &= \lim_{t\to s}(\varphi_\gamma(t, x_s(s)) - \varphi_\gamma(s, x_s(s)))(t-s)^{-1} \\ &\quad + \lim_{t\to s}(\varphi_\gamma(t, x_s(t)) - \varphi_\gamma(t, x_s(s))(t-s)^{-1} \end{aligned}$$

and there exists $\eta \in \varphi_\gamma(x)$ such that

$$\lim_{t\to s}(\varphi_\gamma(t, x_s(t)) - \varphi_\gamma(t, x_s(s))(t-s)^{-1} = (\eta, \dot{x}_s(t))$$

by virtue of (ii) and the mean value property (cf. [2] or [6]). □

**Remark 3.3.** The preceding lemma infers that system (10) with cost function (12) being $\gamma$-robust implies $\varphi_\gamma \in \Phi$ (see definition of $\Phi$ in Section 2).

**Remark 3.4.** Lemma 3.2(i) implies that for $x \in D(A)$, $\varphi_\gamma(\cdot, x)$ is absolutely continuous on $[t_0, t_f]$, and thus is almost everywhere differentiable on $[t_0, t_f]$.

**Theorem 3.2.** *$\gamma$-robustness property holds for system* (10) *with cost* (12). *Then, there exists $\eta \in \partial\varphi_\gamma$ such that $\forall x \in D(A)$ the following equation holds:*

$$\begin{aligned} &\frac{\partial \varphi_\gamma}{\partial s}(s, x) + (Ax, \eta(s, x)) - h^*(-B^*\eta(s, x)) \\ &\qquad + g(x) + \frac{\gamma^{-2}}{4}\|D^*\eta(s, x)\|_W^2 = 0, \\ &\qquad\qquad\qquad\qquad\qquad\qquad a.e.\ s \in (t_0, t_f), \qquad (38) \\ &\varphi_\gamma(t_f, h) = \phi_0(h), \qquad \forall h \in X. \end{aligned}$$

**Proof.** For each fixed $x \in D(A)$ we may assume the existence of a mapping $\Gamma_x : W \to U$ such that

$$\begin{aligned} \varphi_\gamma(s, x) &= \sup_{L^2([s,t_f];W)} \inf_{L^2([s,t_f];U)} J_\gamma(s, x; u_s, w_s) = \sup_{L^2([s,t_f];W)} J_\gamma(s, x; \Gamma_x w_s, w_s) \\ &= J_\gamma(s, x; \Gamma_x \hat{w}_s, \hat{w}_s). \end{aligned}$$

Let us consider the following system, where we henceforth drop the subscript on $\Gamma$:

$$\dot{x}_s(t) = Ax_s(t) + B\Gamma\hat{w}_s(t) + D\hat{w}_s(t), \qquad s \le t \le t_f, \tag{39}$$
$$x_s(s) = x.$$

Since the pair $(\Gamma\hat{w}_s(t), \hat{w}_s(t))$ is continuous on $[t_0, t_f]$ by Theorem 3.1, and since (39) implies

$$x_s(t) = e^{A(t-s)}x + \int_s^t e^{A(t-\tau)}(B\Gamma\hat{w}_s(\tau) + D\hat{w}_s(\tau))\,d\tau,$$

and $x \in D(A)$, we have

$$\lim_{t\to s} \frac{x_s(t) - x}{t - s} = Ax_s(s) + B\Gamma\hat{w}(s) + D\hat{w}_s(s). \tag{40}$$

Note that

$$\frac{d\varphi_\gamma}{dt}(t, x_s(t)) + g(x_s(t) + h(\Gamma\hat{w}_s(t)) - \gamma^2\|\hat{w}_s(t)\|^2) = 0, \qquad \text{a.e.} \quad t \in [t_0, t_f], \tag{41}$$

and

$$\frac{d\varphi_\gamma}{dt}\,|_{t=s} = \varphi_{\gamma t}(s, x) + (\eta(s, x), \dot{x}(s)), \qquad \text{a.e.} \quad t \in [t_0, t_f], \tag{42}$$

where $\eta(s, x) \in \partial\varphi_\gamma(s, x)$. Combining (40), (41), and (42) yields

$$\begin{aligned}\varphi_{\gamma t}(s, x) &+ (\eta(s, x), Ax + B\Gamma\hat{w}_s(s) + D\hat{w}_s(s)) \\ &+ g(x) + h(\Gamma\hat{w}_s(s)) - \gamma^2\|\hat{w}_s(s)\|_W^2 = 0.\end{aligned} \tag{43}$$

Since

$$h(\Gamma\hat{w}_s) + h^*(-B^*\eta) = (-B\Gamma\hat{w}_s, \eta),$$

by virtue of Corollary 3.2 we may take $p(s) = -\eta(s, x) \in \varphi_\gamma(s, x)$, and under this choice (43) becomes

$$\varphi_{\gamma s}(s, x) + (\eta(s, x), Ax) - h^*(-B^*\eta(s, x)) + \frac{\gamma^{-2}}{4}\|D^*\eta(s, x)\|_W^2 = 0.$$

The proof of the theorem is thus complete. □

## 4 Proof of the Main Result: Sufficiency

The following theorem provides a sufficient condition for existence of a solution of (10) when a nonlinear feedback control policy is used:

**Theorem 4.1.** *Let $\varphi$ be a proper lower semicontinuous convex function: $X \to \mathbb{R}$, $D(A) \cap D(\partial\varphi) \neq \phi$, and $\partial\varphi : X \to X$ satisfy*

$$\sup\{\|\eta\|,\ \eta \in \partial\varphi(x)\} \le C_1\|x\|_X + C_2, \tag{44}$$

*where $C_1$, $C_2$ are constants. Then the Cauchy problem*

$$\dot{x}(t) \in Ax(t) + B\partial\varphi(x(t)) + Dw(t), \qquad x(t_0) = x_0, \tag{45}$$

*has at least one mild solution over $[t_0, t_f]$ for every $w \in L^2([t_0, t_f]; W)$.*

**Proof.** By the assumptions on $\varphi$, $\partial\varphi$ is a maximal monotone operator. Define $\varphi_\lambda : X \to \overline{\mathbb{R}}$ as follows:

$$\varphi_\lambda(y) = \|y - J_\lambda y\|^2/2\lambda + \varphi(J_\lambda y), \qquad \lambda > 0, \tag{46}$$

where $J_\lambda = (I + \lambda\partial\varphi)^{-1}$ is a nonexpansive single-valued operator from $X$ to $D(\partial\varphi)$ where $D(\partial\varphi)$ denotes the valid domain for $\partial\varphi$. $\varphi_\lambda$ is Fréchet differentiable and $\nabla\varphi_\lambda = (\partial\varphi)_\lambda = \lambda^{-1}(I - J_\lambda)$ is Lipschitzian. Consider the approximation equation

$$\dot{x}_\lambda(t) = Ax_\lambda(t) + B\nabla\varphi_\lambda(x_\lambda(t)) + Dw(t), \qquad x_\lambda(t_0) = x_0, \tag{47}$$

which is equivalent to the following integral equation:

$$x_\lambda(t) = e^{A(t-t_0)}x_0 + \int_{t_0}^{t} e^{A(t-s)}(B\nabla\varphi_\lambda(x_\lambda(s)) + Dw(s))\, ds =: (Qx_\lambda)(t).$$

Set $X_1 = C([t_0, t_f]; X)$. It follows from the nonexpansive property of $J_\lambda$ by induction that

$$\|Q^n(x_\lambda^1) - Q^n(x_\lambda^2)\|_{X_1} \leq \frac{C\|B\|^n(t_f - t_0)^n}{\lambda^n n!}\|x_\lambda^1 - x_\lambda^2\|_{X_1}, \qquad n = 1, 2, \ldots,$$

and thus $Q$ has a unique fixed point $x_\lambda$ in $X_1$, which is the mild solution of (47) for each $\lambda > 0$. By the assumption made on $\partial\varphi$, we have

$$\|\nabla\varphi_\lambda(x_\lambda)\| \leq C_3\|x_\lambda\| + C_4\,,$$

and in view of Gronwall's lemma (see Appendix II) it is readily seen that $\{x_\lambda\}$ is uniformly bounded on $[t_0, t_f]$. The Arzela–Ascoli theorem infers that there is a subsequence $\{x_{\lambda_n}\}$ of $\{x_\lambda\}$ that is uniformly convergent to a continuous function $x = x(t)$ on $[t_0, t_f]$ as $\lambda_n \to 0$. Since $\nabla\varphi_\lambda(x_\lambda) \in \partial\varphi((I + \lambda\partial\varphi)^{-1}x_\lambda)$, $\nabla\varphi_\lambda(x_\lambda)$ is also uniformly bounded, and furthermore there exists a subsequence $\nabla\varphi_{\lambda_n}(x_{\lambda_n})$ and $y \in X$ such that $\nabla\varphi_{\lambda_n}(x_{\lambda_n}) \to y$ weakly on $X$ and $y \in \partial\varphi(x)$. Note that $x_\lambda$ is equicontinuous on $[t_0, t_f]$, which implies that there exists a subsequence of $\nabla\varphi_{\lambda_n}(x_{\lambda_n})$ which strongly converges to $y \in \partial\varphi(x)$. Thus $x = x(t)$ is a mild solution of (45), which completes the proof. □

We now return to the cost function $J_\gamma$, and consider the following problem: Given a fixed but arbitrary $w \in L^2([t_0, t_f]; W)$, minimize

$$J_\gamma(t_0, x_0; u, w) = \int_{t_0}^{t_f} \{g(x) + h(u) - \gamma^2\|w\|_W^2\}\, ds + \phi_0(x(t_f)) \tag{48}$$

over all $u \in L^2([t_0, t_f]; U)$. Define

$$\psi^w(x_0) := \inf_u J_\gamma(t_0, x_0; u, w). \tag{49}$$

**Corollary 4.1.** *Suppose that:*

(i) $\sup\{\|\xi\| : \xi \in \partial g(x)\} \le C_1\|x\| + C_2$; *and*
(ii) $u \to h^*(u)$ *is Fréchet differentiable on* $U$.

*Then there exists an optimal feedback control* $\mu^w(t, x(t))$ *for* (49), *given by*

$$\mu^w(t, x(t)) \in \partial h^*(B^*(-\partial\psi^w(x(t)))).$$

**Proof.** Since $J_\gamma(t_0, x_0; \cdot, w)$ is convex, lower semicontinuous, coercive, and $\not\equiv \infty$, it has at least one minimum, say $\mu^w(t)$. In this case we know that there exists a function $p^w(t) \in C([t_0, t_f]; X)$ such that

$$\dot{p}^w(t) \in -A^* p^w(t) + \partial g(x(t)), \tag{50}$$
$$Bp^w(t) \in \partial h(u^w(t)), \tag{51}$$
$$p^w(t_f) + \phi_0(x(t_f)) \ni 0, \tag{52}$$

(50) implies that

$$p^w(t) = e^{-A^*(t-t_f)}p(t_f) - \int_t^{t_f} e^{-A^*(t-t_f)}q(x(t))\,dt,$$

where $q(x(t)) \in \partial g(x(t))$. By the assumption on $\partial g$, we have

$$\|p^w(t)\| \le \overline{C}_1\|x\| + \overline{C}_2.$$

Since $\partial\psi^w(x(t)) = \{-p^w(t)\}$ (cf. [6]),

$$\sup\{\|\zeta\| : \zeta \in \partial\psi^w(x)\} \le C_3\|x\| + C_4. \tag{53}$$

From (51) we have

$$u^w(t) = \partial h^*(Bp^w(t))$$

since $h$ is lower semicontinuous. Hence $\exists\eta \in \partial\psi^w(x(t))$ such that

$$\mu^w(t) = \partial h^*(-B\eta(x(t))).$$

From (53) and Theorem 4.1, we know that the feedback control

$$\mu^w(t, x(t)) = \partial h^*(-B\eta(x(t)))$$

is admissible in the sense of Definition 2.1. Note that when this feedback control is used, $p^w$ still satisfies (50)–(52), and hence, it is also an optimal feedback control. This completes the proof of the corollary. □

We are now in a position to study the inf sup problem, i.e., the upper value of the game. Recall that $\mathcal{M}_{CL}$ is defined as the class of (nonlinear) feedback controllers $\mu : X \to U$, and in terms of this notation, and for $s \in (t_0, t_f)$ and $x \in X$, let

$$\varphi_\gamma(s, x) = \inf_{\mu\in\mathcal{M}_{CL}} \sup_{w\in L^2([t_0,t_f];W)} J_\gamma(s, x; \mu, w). \tag{54}$$

**Theorem 4.2.** *Suppose that $u \to h^*(u)$ is Fréchet differentiable and for $x \in D(A)$ the following equation admits a solution $\varphi_\gamma \in \Phi$ for $\eta \in \partial\varphi_\gamma$:*

$$\frac{\partial \varphi_\gamma}{\partial s}(s,x) + (Ax, \eta(x)) - h^*(-B^*\eta(x)) + g(x) + \frac{\gamma^{-2}}{4}\|D^*\eta(x)\|_W^2 = 0, \qquad \text{a.e.} \quad s \in (t_0, t_f), \tag{55}$$

$$\varphi_\gamma(t_f, h) = \phi_0(h), \qquad \forall h \in X.$$

*Moreover, assume that:*

(i) *$x \to \eta(x)$ is continuous; and*
(ii) *$D(\eta) \supseteq D(A)$ and $\|\eta(x)\|_X \le c_1\|x\|_X + c_2$ for all $x \in D(A)$.*

*Then the differential game (54) admits a feedback saddle-point solution, which is given by*

$$\hat{\mu}(t; x(t)) = \partial h^*(-B\eta(x(t))),$$
$$\hat{\nu}(t; x(t)) = \frac{\gamma^{-2}}{2} D^*\eta(x(t)),$$

*and $\varphi_\gamma(s,x)$ is the optimal value of (54). In particular*

$$\varphi_\gamma(t_0, x_0) = \inf_{\mu \in \mathcal{M}_{CL}} \sup_{w \in L^2([t_0,t_f];W)} J_\gamma(t_0, x_0; \mu, w).$$

**Proof.** Since system (10) (when the initial condition is $x(s) = x$) has just a mild solution, $x(t)$ is not necessarily differentiable; thus we consider its Yosida approximation (cf. [26]):

$$\dot{x}_n(t) = A_n x_n(t) + Bu_n(t) + Dw_n(t), \qquad x_n(s) = x, \tag{56}$$

where $A_n = n^2 R(n;A) - nI$ is the Yosida approximation of $A$ and $R(n;A) = (nI - A)^{-1}$, $w_n \in C^1([t_0,t_f];W) : w_n(\cdot) \to w(\cdot)$ and $u_n \in C^1([t_0,t_f];U) : u_n(\cdot) \to u(\cdot)$. Since $A_n \in L(X)$, (56) has a unique solution $x_n \in W^{1,2}([t_0,t_f];X) \cap L^2([t_0,t_f];D(A))$ for each sufficiently large positive integer $n$ and $x_n(t) \to x(t)$ uniformly on $X$ and $x(t)$ is the mild solution of (10). In (55), let $x = x_n(t)$, $s = t$; then we have

$$\varphi_{\gamma t}(t, x_n(t)) + (A_n x_n(t), \eta(x_n(t))) - h^*(-B^*\eta(x_n(t))) + g(x_n(t)) + \frac{\gamma^{-2}}{4}\|D^*\eta(x_n(t))\|_W^2 = 0.$$

In view of $\varphi_\gamma \in \Phi$ and (55),

$$\frac{d\varphi_\gamma}{dt}(t, x_n(t)) - (Bu_n + Dw_n, \eta(x_n(t))) - h^*(-B^*\eta(x_n(t))) + g(x_n(t)) + \frac{\gamma^{-2}}{4}\|D^*\eta(x_n(t))\|_W^2 = 0. \tag{57}$$

Let $\overline{u}_n = \overline{u}(t; x_n(t)) = \partial h^*(-B^*\eta(x_n(t)))$, and according to Theorem 4.1, $\overline{u}(t; x_n(t))$ is an admissible feedback control policy for system (56). Thus we have

$$h(\overline{u}_n) + h^*(-B^*\eta(x_n(t))) = (-B\overline{u}_n, \eta(x_n(t))),$$

and hence (57) becomes

$$\frac{d\varphi_\gamma}{dt}(t; x_n(t)) + (B\overline{u}_n - Bu_n, \eta(x_n(t))) + h(\overline{u}) - h(u)$$
$$(g(x_n(t)) + h(u(t)) - \gamma^2\|w(t)\|_W^2) + \gamma^2\|w(t) - \frac{\gamma^{-2}}{2}D^*\eta(x(t))\|_W^2 = 0.$$

Integrating the above equation from $t = s$ to $t = t_f$, and letting $n \to \infty$, yields

$$\int_s^{t_f} \{(B\overline{u}(t; x(t)) - Bu(t), \eta(x(t))) + h(\overline{u}(t; x(t))) - h(u(t))\}\, dt$$
$$+\gamma^2 \int_s^{t_f} \left\| w(t) - \frac{\gamma^{-2}}{2}D^*\eta(x(t))\right\|_W^2 dt + J_\gamma(s, x; u, w) = \varphi_\gamma(t_0, x_0). \quad (58)$$

Since $\overline{u} \in \partial h^*(-B^*\eta(x))$, it follows that

$$(B\overline{u}(t; x(t)) - Bu(t), \eta(x(t))) + h(\overline{u}(t; x(t))) - h(u(t)) \leq 0.$$

Minimizing $J_\gamma(s, x; \cdot, w)$ over $u$, we have

$$\int_s^{t_f} \{(B\overline{u}(t; x(t)) - Bu(t), \eta(x(t))) + h(\overline{u}(t; x(t))) - h(u(t))\}\, dt = 0.$$

Maximizing $J_\gamma(s, x; u, \cdot)$ over $w$, we have

$$w(t) = \frac{\gamma^{-2}}{2}D^*\eta(x(t)).$$

Let

$$\mu(t; x(t)) = \partial h^*(-B^*\eta(x(t))),$$
$$\nu(t; x(t)) = \frac{\gamma^{-2}}{2}D^*\eta(x(t)).$$

Then, for any $(u, w) \in L^2([s, t_f]; U) \times L^2([s, t_f]; W)$, we have

$$J_\gamma(s, x; \mu, w) \leq J_\gamma(s, x; \mu, \nu) \leq J_\gamma(s, x; u, \nu)$$

and therefore it is readily seen that $(\mu, \nu)$ is a saddle-point solution and

$$\varphi_\gamma(s, x) = \inf_{\mu\in\mathcal{M}_{CL}} \sup_w J_\gamma(s, x; \mu, w) = \sup_w \inf_{\mu\in\mathcal{M}_{CL}} J_\gamma(s, x; \mu, w).$$

In particular, letting $s = t_0$ and $x = x_0$, we have

$$\varphi_\gamma(t_0, x_0) = \inf_{\mu\in\mathcal{M}_{CL}} \sup_w J_\gamma(t_0, x_0; \mu, w) = \sup_w \inf_{\mu\in\mathcal{M}_{CL}} J_\gamma(t_0, x_0; \mu, w). \qquad \square$$

**Corollary 4.2.** *Under the hypotheses of Theorem 4.3, for any $(x_0, w) \in X \times L^2([t_0, t_f]; W)$ the following holds:*

$$\int_{t_0}^{t_f} \{g(x(t)) + h(\mu(x(t)))\} \, dt + \phi_0(x(t_f)) \leq \gamma^2 \int_{t_0}^{t_f} \|w(t)\|_W^2 \, dt + \varphi_\gamma(t_0, x_0),$$

*where $\mu(x(t)) = \partial h^*(-B\eta(x(t)))$.*

**Proof.** This is a direct consequence of (57). □

**Corollary 4.3.** *Under the hypotheses of Theorem 4.3, with $\varphi_\gamma(t_0, x_0) = 0$, we have*

$$\inf_{\mu \in \mathcal{M}_{CL}} \sup_w \rho(\mu, w) \leq \gamma^2,$$

*where $\rho$ was defined by (5).*

**Proof.** Let

$$\omega(u(t)) := \int_{t_0}^{t_f} \{(B\overline{u}(t; x(t)) - Bu(t), \eta(x(t))) + h(\overline{u}(t; x(t))) - h(u(t))\} \, dt,$$

where

$$\overline{u}(t; x(t)) = \partial h^*(-B^*\eta(x(t))).$$

According to (57), we have

$$\sup_w \rho(u, w) = \rho(u, \overline{w}) = \gamma^2 - \gamma^2 \frac{\omega(u(t))}{\int_{t_0}^{t_f} \|w(t)\|_W^2 \, dt}.$$

Since $\omega(u) \geq 0$ for any $u \in L^2([t_0, t_f]; U)$, we have

$$\inf_{\mu \in \mathcal{M}_{CL}} \sup_w \rho(\mu, w) \leq \gamma^2.$$ □

## 5 Application to a Class of Problems with Control Constraints

In this section, we consider the $H^\infty$-optimal control for system (1) under control constraints. Let $U_0$ be a closed convex subset of $U$, and let the system be described by

$$\begin{aligned} \dot{x}(t) &= Ax(t) + Bu(t) + Dw(t), \qquad x(t_0) = x_0, \\ z(t) &= Cx(t) + R^{1/2}u(t), \end{aligned} \tag{59}$$

where $C \in L(X, X)$, $R \in L(U, U)$ with $R^*[C, R] = [0, I]$.

The $\gamma$-robustness property with the control constraint $U_0$ amounts to the existence of $\mu^c \in \mathcal{M}_{CL}^c \subset \mathcal{M}_{CL}$ and a nonnegative function $V : X \to \mathbb{R}$ such that

$$\int_{t_0}^{t_f} \{\|C(x(t))\|_Z^2 + \|\mu^c(t, x(t))\|_U^2\} \, dt + \phi_0(x(t_f)) \leq \epsilon^2 \int_{t_0}^{t_f} \|w(t)\|_W^2 \, dt + V(x_0), \tag{60}$$

where $0 < \epsilon < \gamma$, and $\mathcal{M}_{CL}^{c}$ represents a subclass of admissible feedback controls in $\mathcal{M}_{CL}$ with $\mu^{c}(t; x(t)) = F^{c}x(t) \in U_0$ for all $w \in L^2([t_0, t_f], W)$ and $x_0 \in X$, $F^c : [t_0, t_f] \times X \to U_0$.

Let $I_{U_0}$ be the indicator function of $U_0$, i.e.,

$$I_{U_0}(u) = \begin{cases} 0 & \text{if } u \in U_0, \\ +\infty & \text{if } u \notin U_0. \end{cases}$$

Define $P_{U_0} : U \to U_0$ by $P_{U_0}(u) = u_0$ where $u_0 \in U_0$ is such that

$$\inf_{v \in U_0} \|v - u\|_U = \|u_0 - u\|_U.$$

From the Best Approximation Theorem (see, e.g., Theorem 2.3 of [2]), we know that $P_{U_0}$ is well defined. If $U_0$ is a subspace of $U$, $P_{U_0}$ is just the projection operator of $U$ on $U_0$. Let $h(u) = \|u\|_U^2 + I_{U_0}(u)$. Now we can restate the $\gamma$-robustness property with control constraints as follows: There exists a control $\mu \in \mathcal{M}_{CL}$ (not necessarily assuming that $\mu$ takes values in $U_0$) and a nonnegative function $V : X \to \mathbb{R}$ such that

$$\int_{t_0}^{t_f} \{\|C(x(t))\|_Z^2 + h(u(t))\}\, dt + \phi_0(x(t_f)) \le \epsilon^2 \int_{t_0}^{t_f} \|w(t)\|_W^2\, dt + V(x_0) \quad (61)$$

holds for any $(x_0, w) \in X \times L^2([t_0, t_f]; W)$ where $0 < \epsilon < \gamma$.

If (61) is satisfied we know that in fact $u \in U_0$. Note that

$$h^*(u) = \tfrac{1}{4}(\|u\|_U^2 - \|u - P_{U_0}(u)\|_U^2) \quad (62)$$

and hence that $\partial h^*(u)$ consists of a single element $P_{U_0}$ (hence it is Gâteaux differentiable); in fact, $\partial h^*$ is Fréchet differentiable (see Appendix I). This now brings us to the following theorem:

**Theorem 5.1.** *Suppose that the following equation admits a solution $\varphi_\gamma \in \Phi$ for $\eta(s, x) \in \partial\varphi_\gamma(s, x)$:*

$$\begin{aligned} &\frac{\partial \varphi_\gamma}{\partial s}(s, x) + (Ax, \eta(s, x)) - h^*(-B^*\eta(s, x)) \\ &\qquad + (Cx, Cx) + \frac{\gamma^{-2}}{4}\|D^*\eta(s, x)\|_W^2 = 0, \\ &\qquad\qquad a.e. \quad s \in (t_0, t_f), \\ &\varphi_\gamma(t_f, h) = \phi_0(h), \qquad \forall h \in X, \end{aligned} \quad (63)$$

*and also that:*

(i) *$x \to \eta(\cdot, x)$ is continuous; and*
(ii) *$D(\eta) \supseteq D(A)$ and $\|\eta(t, x)\|_X \le c_1\|x\|_X + c_2$ for all $x \in D(A)$.*

*Then we have*

$$\int_{t_0}^{t_f} \{\|C(x(t))\|_Z^2 + h(\mu(t))\}\, dt + \phi_0(x(t_f)) \le \gamma^2 \int_{t_0}^{t_f} \|w(t)\|_W^2\, dt + V(x_0), \quad (64)$$

*where* $\mu(t, x(t)) = P_{U_0}(-B^*\eta(t, x(t)))$*, which is an admissible feedback control for system* (59).

*Conversely, if the* $\gamma$*-robustness property with control constraints (i.e.,* (61)*) holds, then* $\varphi_\gamma(s, x)$ *defined as below satisfies* (63)*:*

$$\varphi_\gamma(t_0, x_0) = \inf_{\mu \in \mathcal{M}_{CL}} \sup_{w}, J_\gamma(t_0, x_0; \mu, w), \tag{65}$$

*where*

$$J_\gamma(t_0, x_0; \mu, w) = \int_{t_0}^{t_f} \{\|C(x(t))\|_Z^2 + h(\mu(t; x(t))) - \gamma^2 \|w(t)\|_W^2\}\, dt + \phi_0(x(t_f)).$$

When the control constraint set $U_0$ is a subspace of the control space $U$, we can obtain a more explicit expression. Note that in this case, $P_{U_0}$ is a linear operator from $U$ to $U_0$. Thus we can easily verify:

(i) $D(\varphi_\gamma)(t, \cdot) = X$; and
(ii) $x \to \partial\varphi_\gamma(t, x)$ is linear; provided that the $\gamma$-robustness property with control constraints holds and $\varphi_\gamma$ is given by (65).

For convenience, let $\phi_f \equiv 0$. Since $\varphi_\gamma$ is maximal monotone on $X$, and $\partial\varphi_\gamma$ is self-adjoint on $X$ for every $t \in [t_0, t_f]$, we have

$$\varphi_\gamma(t, x) = (Z_\gamma(t)x, x), \qquad \forall x \in X, \qquad \partial\varphi_\gamma(t) = 2Z_\gamma(t),$$

where $P_{U_0}$ is the projection operator of $U$ on $U_0$. In view of this, (63) now leads to

$$\begin{aligned} \dot{Z}_\gamma(t) + A^* Z_\gamma(t) + Z_\gamma(t)A - Z_\gamma(t)\left(2BP_{U_0}B^* - BP_{U_0}^* P_{U_0}B^* - \frac{1}{\gamma^2}DD^*\right)Z_\gamma(t) \\ +C^*C = 0, \\ Z_\gamma(t_f) = 0. \end{aligned} \tag{66}$$

Let

$$K_{x_0}(u, w) := \int_{t_0}^{t_f} \{\|C(x(t))\|_Z^2 + h(u(t))\}\, dt$$

and

$$\rho(u, w) := K_0(u, w) \Big/ \int_{t_0}^{t_f} \|w(t)\|_W^2\, dt.$$

**Corollary 5.1.** *Given* $\gamma > 0$,

$$\sup_w \inf_u \rho(u, w) < \gamma^2$$

*if and only if* (66) *admits a unique mild solution. In this case, the optimal feedback control is given by*

$$\underline{\mu}^*(t, x(t)) = -P_{U_0}B^* Z_\gamma(t)x(t), \qquad 0 \le t \le t_f,$$

*and the worst disturbance is (given in state-feedback form)*

$$\underline{\nu}^*(t, x(t)) = \gamma^{-2} D^* Z_\gamma(t) x(t), \qquad 0 \le t \le t_f.$$

*Moreover,*

$$\sup_w \inf_u \rho(u, w) < \gamma^2 \quad \textit{implies} \quad \inf_{\mathcal{M}_{CL}} \sup_w \rho(\mu, w) = \sup_w \inf_u \rho(u, w) < \gamma^2.$$

**Proof.** If (66) admits a mild solution, and for $x \in X$ we let

$$\varphi_\gamma(t, x) = (Z_\gamma(t)x, x),$$

then $\varphi_\gamma$ satisfies (63) and verifies conditions (i) and (ii) of Theorem (5.1); hence we have

$$\int_{t_0}^{t_f} \{\|C(x(t))\|_Z^2 + h(\mu(x(t)))\}\, dt \le \gamma^2 \int_{t_0}^{t_f} \|w(t)\|_W^2\, dt, \quad \forall w \in L^2([t_0, t_f]; W), \tag{67}$$

where $\mu(t, x(t)) = P_{U_0}(-B^* Z_\gamma(t) x(t))$ and $x$ can be viewed as the solution of

$$\dot{x} = \left(A - \left(B P_{U_0} B^* Z_\gamma - \frac{1}{\gamma^2} D D^* Z_\gamma\right)\right)x + D\left(w - \frac{1}{\gamma^2} D^* Z_\gamma x\right), x(t_0) = 0.$$

Hence, immediately

$$\int_{t_0}^{t_f} \|x(t)\|_X^2 \le c_0 \int_{t_0}^{t_f} \left\| w - \frac{1}{\gamma^2} D^* Z_\gamma x \right\|_W^2 dt. \tag{68}$$

Combining (67) and (68) we have

$$\sup_w \inf_u \rho(u, w) < \gamma^2. \tag{69}$$

Conversely, if (69) is true for a given $\gamma$, according to definition of $\rho$ we know that there exists a $\mu \in \mathcal{M}_{CL}$ such that

$$\int_{t_0}^{t_f} (\|Cx(t)\|_W^2 + h(\mu(x(t))))\, dt \le (\gamma^2 - \delta^2) \int_{t_0}^{t_f} \|w(t)\|_W^2\, dt,$$

where $\delta : 0 < \delta < \gamma$. Thus we can define

$$\varphi_\gamma(t, x) = \sup_w \inf_u \{K_0(u, w) - \gamma^2 \|w\|_2^2\}.$$

Following the discussion preceding Corollary 5.2, we know that $\partial\varphi_\gamma$ is the solution of (66). Next we claim

$$\sup_w \inf_u \rho(u, w) = \inf_{\mathcal{M}_{CL}} \sup_w \rho(\mu, w).$$

Since $\inf_{\mathcal{M}_{CL}} \sup_w \rho(\mu, w) < \gamma^2$, (66) admits a unique mild solution. Again let $\varphi_\gamma(t, x) = (Z_\gamma(t)x, x)$. By (57), for $(u, w) \in L^2([t_0, t_f]; U_0) \times L^2([t_0, t_f]; W)$, we have

$$\int_{t_0}^{t_f} \{(-B P_{U_0} B^* Z_\gamma(t) x(t) - Bu(t), Z_\gamma(t) x(t))$$

$$+ h(-P_{U_0}B^*Z_\gamma(t)x(t)) - h(u(t))\} \, dt$$
$$+ \gamma^2 \int_{t_0}^{t_f} \|w(t) - \gamma^{-2} D^* Z_\gamma(t)x(t)\|_W^2 \, dt + J_\gamma(t_0, 0; u, w) = 0, \quad (70)$$

and thus

$$\sup_w \rho(u, w) = \gamma^2 - \frac{\omega(u(t))}{\int_{t_0}^{t_f} \|w(t)\|_W^2 \, dt},$$

where

$$\omega(u(t)) = \int_{t_0}^{t_f} \{(BP_{U_0}B^*Z_\gamma(t)x(t) + Bu(t), Z_\gamma(t)x(t))$$
$$- h(-P_{U_0}B^*Z_\gamma(t)x(t)) + h(u(t))\} \, dt \ge 0 \quad (71)$$

and $x$ satisfies

$$\dot{x} = \left(A - \left(BP_{U_0}B^*Z_\gamma - \frac{1}{\gamma^2}DD^*Z_\gamma\right)\right)x + B(u - P_{U_0}B^*Z_\gamma x), \qquad x(t_0) = 0. \quad (72)$$

Similar to (68) we have

$$\int_{t_0}^{t_f} \|x(t)\|_X^2 \, dt \le c_1 \int_{t_0}^{t_f} \|u - P_{U_0}B^*Z_\gamma x\|_U^2 \, dt. \quad (73)$$

and (71) and (73) imply that

$$\inf_{\mathcal{M}_{CL}} \sup_w \rho(u, w) < \gamma^2.$$

Note that

$$\inf_{\mathcal{M}_{CL}} \sup_w \rho(u, w) \ge \sup_w \inf_u \rho(u, w).$$

Therefore we can conclude that

$$\inf_{\mathcal{M}_{CL}} \sup_w \rho(u, w) = \sup_w \inf_u \rho(u, w).$$

□

**Remark 5.1.** Two extreme cases are:

(i) $U_0 = \{0\}$ in which case $P_{U_0} = 0$; and
(ii) $U_0 = U$ in which case $P_{U_0} = I$.

In both cases (66) is consistent with the known results in the finite-dimensional case (see [8]). Another important application arises when $U_0$ is a finite-dimensional space, which is a realistic situation, arising due to implementation constraints.

## 6 Examples

**Example 6.1.** Let $\Omega$ be an open bounded subset of $\mathbb{R}^n$ with regular boundary $\partial\Omega$. Consider the state equation:

$$\begin{aligned}
\frac{\partial^2 x(t,\xi)}{\partial t^2}(t,\xi) &= \Delta_\xi x(x,\xi) + (Bu(t,\cdot))(\xi) + (Dw(t,\cdot))(\xi) \quad \text{in } (0,T)\times\Omega,\\
x(t,\xi) &= 0 \quad \text{on } (0,T)\times\partial\Omega, \qquad (74)\\
x(0,\xi) &= x_0(\xi) \quad \text{in } \Omega,\\
\frac{\partial x}{\partial t}(0,\xi) &= x_1(\xi) \quad \text{in } \Omega,
\end{aligned}$$

where $\Delta_\xi$ denotes the standard Laplacian operator. Set $X = H_0^1(\Omega) \oplus L^2(\Omega)$ and $U = W = L^2(\Omega)$ and denote by $Y = \begin{bmatrix} x^0 \\ x^1 \end{bmatrix}$ a generic element of $X$. Define the inner product on $X$ by

$$\left(\begin{bmatrix} x^0 \\ x^1 \end{bmatrix}, \begin{bmatrix} z^0 \\ z^1 \end{bmatrix}\right) = \int_\Omega (\nabla_\xi x^0 \cdot \nabla_\xi z^0 + x^1 z^1)\, d\xi.$$

Let $\Lambda$ be the self-adjoint positive operator on $L^2(\Omega)$, defined by

$$D(\Lambda) = H^2(\Omega) \cap H_0^1(\Omega), \qquad \Lambda x = -\Delta_\xi x.$$

Then we have

$$(X, Z) = (\sqrt{\Lambda} x^0, \sqrt{\Lambda} z^0) + (x^1, z^1).$$

Define the linear operator $A$ on $H$:

$$\begin{aligned}
AY &= \begin{bmatrix} 0 & 1 \\ -\Lambda & 0 \end{bmatrix} \begin{bmatrix} x^0 \\ x^1 \end{bmatrix}, \qquad \forall X \in D(A),\\
D(A) &= H^2(\Omega) \cap H_0^1(\Omega) \oplus H_0^1(\Omega).
\end{aligned}$$

We know that $A$ is the infinitesimal generator of a contraction group in $X$ because $A^* = -A$ (cf. the Stone theorem, see, e.g., [26]). For simplicity, we may define $B \in L(U, X)$ and $D \in L(W, X)$ as

$$Bu = \begin{bmatrix} 0 \\ u \end{bmatrix}, \quad u \in U, \qquad Dw = \begin{bmatrix} 0 \\ w \end{bmatrix}, \quad w \in W.$$

Setting

$$x^0(t) = x(t,\cdot), \qquad x^1(t) = \frac{\partial x}{\partial t}(t,\cdot), \qquad u(t) = u(t,\cdot), \qquad w(t) = w(t,\cdot),$$

the state equation (74) can be written as

$$\dot{Y} = AY + Bu + Dw, \qquad Y(0) = Y_0,$$

where

$$Y(t) = \begin{bmatrix} x^0(t) \\ x^1(t) \end{bmatrix}, \qquad Y_0 = \begin{bmatrix} x^0 \\ x^1 \end{bmatrix}.$$

A natural choice for the function $g$ is

$$g(x(t,\cdot)) = \begin{cases} |\nabla_\xi x(t,\cdot)|^2 + \left|\dfrac{\partial x}{\partial t}(t,\cdot)\right|^2, & x(t,\cdot)\dfrac{\partial x}{\partial t}(t,\cdot) \geq 0, \\ |\nabla_\xi x(t,\cdot)|^2, & x(t,\cdot)\dfrac{\partial x}{\partial t}(t,\cdot) \leq 0. \end{cases}$$

Let the $\gamma$-parametrized cost function be

$$\begin{aligned} J_\gamma(t, Y_0; u, w) = & \int_t^T \int_\Omega (g(x(t,\xi)) + |u(t,\xi)|^2 - \gamma^2 |w(t,\xi))\, dt\, d\xi \\ & + \int_\Omega \left( |\nabla_\xi x(T,\xi)|^2 + \left|\frac{\partial x}{\partial t}(T,\xi)\right|^2 \right) d\xi. \end{aligned}$$

Introducing the value function

$$\begin{aligned} \varphi_\gamma(t, Y_0) &= \sup_w \inf_u J_\gamma(t, Y_0; u, w) \\ &= \inf_{\mu \in M_{CL}} \sup_w J_\gamma(t, Y_0; u, w), \end{aligned}$$

it is not difficult to see that $D(\varphi_\gamma) = X$ and $Y_0 \to \partial\varphi_\gamma(t, Y_0)$ is linear. Therefore we may have

$$\varphi_\gamma(t, Y_0) = (P_\gamma(t)Y_0, Y_0)_H, \qquad P_\gamma(t) = \partial\varphi_\gamma(t, Y_0),$$

where $P_\gamma$ satisfies

$$\begin{aligned} & \dot{P}_\gamma(t) + \begin{bmatrix} 0 & -1 \\ \Lambda & 0 \end{bmatrix} P_\gamma + P_\gamma(t) \begin{bmatrix} 0 & 1 \\ -\Lambda & 0 \end{bmatrix} \\ & \quad - P_\gamma(t) \begin{bmatrix} 0 & 0 \\ 0 & 1-\gamma^{-2} \end{bmatrix} P_\gamma(t) \in \left\{ \begin{bmatrix} 1 & 0 \\ 0 & 1 \end{bmatrix}, \begin{bmatrix} 1 & 0 \\ 0 & 0 \end{bmatrix} \right\}, \\ & P_\gamma(T) = \begin{bmatrix} 1 & 0 \\ 0 & 1 \end{bmatrix}. \end{aligned}$$

Let $P_\gamma^1(t)$ and $P_\gamma^2(t)$ be the solutions of the above, when, respectively, the first matrix and the second matrix on the right-hand side is taken. Such solutions exist when $\gamma > 1$ because the equation in these cases is the standard Riccati equation corresponding to the wave state equation [10, p. 145, Theorem 2.1]. We can represent $P_\gamma^i$ as

$$P_\gamma^i = \begin{bmatrix} p_{11}^i & p_{12}^i \\ p_{21}^i & p_{22}^i \end{bmatrix}, \qquad i = 1, 2,$$

where

$$p_{11}^i \in \mathcal{L}(H_0^1(\Omega)), \qquad p_{12}^i \in \mathcal{L}(L^2(\Omega); H_0^1(\Omega)),$$
$$p_{21}^i \in \mathcal{L}(H_0^1(\Omega); L^2(\Omega)), \quad p_{22}^i \in \mathcal{L}(L^2(\Omega)),$$

where in fact the following identities hold:

$$(p_{11}^i)^* = \Lambda p_{11}^i \Lambda^{-1}, \qquad (p_{12}^i)^* = p_{21}^i \Lambda^{-1}, \qquad (p_{21}^i)^* = \Lambda p_{12}^i, \qquad (p_{22}^i)^* = p_{22}.$$

For $\gamma > 1$, the optimal state-feedback control policy is

$$\underline{\mu}(t, x(t,\xi)) = \begin{cases} -(p_{21}^1(t)x(t,\cdot))(\xi) + \left(p_{22}^1(t)\dfrac{\partial x}{\partial t}(t,\cdot)\right)(\xi), & x(t,\xi)\dfrac{\partial x}{\partial t}(t,\xi) \ge 0, \\ -(p_{21}^2(t)x(t,\cdot))(\xi) + \left(p_{22}^2(t)\dfrac{\partial x}{\partial t}(t,\cdot)\right)(\xi), & x(t,\xi)\dfrac{\partial x}{\partial t}(t,\xi) \le 0, \end{cases}$$

and the worst disturbance (in state-feedback form) is

$$\underline{\nu}(t, x(t,\xi)) = \begin{cases} \gamma^{-2}\left\{(p_{21}^1(t)x(t,\cdot))(\xi) + \left(p_{22}^1(t)\dfrac{\partial x}{\partial t}(t,\cdot)\right)(\xi)\right\} & x(t,\xi)\dfrac{\partial x}{\partial t}(t,\xi) \ge 0, \\ \gamma^{-2}\left\{(p_{21}^2(t)x(t,\cdot))(\xi) + \left(p_{22}^2(t)\dfrac{\partial x}{\partial t}(t,\cdot)\right)(\xi)\right\} & x(t,\xi)\dfrac{\partial x}{\partial t}(t,\xi) \le 0. \end{cases}$$

**Example 6.2.** Let $X = Z = U = W = L^2(\Omega)$ and $Q = \Omega \times (0, t_f)$, and consider the following system :

$$\begin{aligned} \frac{\partial y}{\partial t} &= (\Delta + c)y + u + w \quad \text{in } (x,t) \in Q, \\ y &= 0 \quad \text{in } \partial\Omega \times (0, t_f), \\ y(x,0) &= y_0(x) \quad \text{in}\Omega, \end{aligned} \tag{75}$$

where $\Delta$ is the Laplacian operator with respect to $x \in \mathbb{R}^n$. We denote by $A$ the linear self-adjoint operator in $X$, defined by:

$$\begin{cases} Ay = \Delta y + cy, & \forall y \in D(A), \\ D(A) = H^2(\Omega) \cap H_0^1(\Omega). \end{cases}$$

Hence (75) can be written in the abstract form (10). Let $K$ be a nontrivial closed convex cone in $X$ with vertex at the origin and $y_1 \in X$. The state constraint is $X_0 = y_1 + K$. Let $U_0 := \{v \in U | y(t_f; v) \in X_0\}$, and

$$\phi_0(y) = \int_\Omega |y(x) - y_1(x)|^2\, dx + I_{X_0}(y),$$

where

$$I_{X_0}(y) = \begin{cases} 0 & \text{if } y \in X_0, \\ +\infty & \text{if } y \notin X_0. \end{cases}$$

Let the cost function be

$$L_{y_0}(u, w) = \int_0^{t_f} \int_\Omega \{|y(t,x) - z_d(x,t)|^2 + |u(t,x)|^2\}\, dx\, dt + \phi_0(y(t_f)),$$

where $z_d \in L^2([0, t_f]; X)$. Note that the subdifferential of $\phi_0$ is

$$\partial\phi_0(y) = y - y_1 + N_{X_0}(y),$$

where $N_{X_0}(y)$ is the normal cone of $X_0$ at $y$. For a given $\gamma$, if there exists a controller $\mu$ and a nonnegative function $U(\cdot)$ such that

$$L_{y_0}(\mu, w)! \leq! \epsilon^2 \int_0^{t_f} \int_\Omega |w(t,x)|^2\, dx\, dt + U(y_0), \quad \forall (y_0, w) \in X \times L^2([0, t_f]; W),$$

holds for $0 < \epsilon < \gamma$, then according to Theorem 3.1, the open-loop representation of the optimal controller is

$$u(t) = -B^* p(t) = -p(t),$$

where $p \in C([0, t_f]; X)$ satisfies

$$\begin{aligned} &\dot{p} = -Ap + y^* - z_d, \\ &(p(t_f), \overline{y} - y^*(t_f)) \geq (y_1 - y^*(t_f), \overline{y} - y^*(t_f)), \qquad \forall \overline{y} \in K, \\ &\dot{y}^* = Ay^* + p - \frac{1}{\gamma^2} p, \\ &y^*(0) = y_0. \end{aligned} \tag{76}$$

In fact (76) is equivalent to saying that $y^*(t_f) = P_{X_0}(y_1 - p(t_f))$, where $P_{X_0}$ is defined by

$$P_{X_0} : X \to X_0, \qquad \inf_{z \in X_0} \|z - y\|_X = \|P_{X_0}(y) - y\|_X.$$

If we define

$$\begin{aligned} \varphi_\gamma(s, y) = &\sup_{L^2([0,t_f];W)} \inf_{L^2([0,t_f];U)} \int_s^{t_f} \int_\Omega \{|y(t,x) \\ &- z_d(t,x)|^2 + |u(t,x)|^2\}\, dx\, dt + \phi_0(y(t_f)) \\ &- \gamma^2 \int_s^{t_f} \int_\Omega |w(t,x)|^2\, dx\, dt \end{aligned}$$

we know from Theorems 3.1 and 3.3 that $-\frac{1}{2}p(t) \in \partial\varphi_\gamma(t, y(t))$, and $\varphi_\gamma$ satisfies the equation

$$\begin{aligned} &\frac{\partial \varphi_\gamma}{\partial s}(s, y(s,x)) + \int_\Omega \nabla_x y(s,x) \cdot \nabla_x p(s,x)\, dx \\ &\quad - \int_\Omega |(-p(s,x))|^2\, dx + \int_\Omega |y(s,x) - z_d(s,x)|^2\, dx \\ &\quad + \frac{\gamma^{-2}}{4} \int_\Omega |p(s,x)|^2\, dx = 0, \qquad \forall y(s,\cdot) \in H_0^1(\Omega), \\ &\qquad\qquad\qquad\qquad\qquad\qquad a.e. \quad s \in (t_0, t_f), \\ &\varphi_\gamma(t_f, h) = \phi_0(h), \qquad \forall h \in X. \end{aligned} \tag{77}$$

If (77) admits a solution $\varphi_\gamma \in \Phi$ for a given $\gamma > 0$, and (i)–(ii) in Theorem 4.3 are also satisfied for some $\eta \in \partial\varphi_\gamma$, then the state-feedback $H^\infty$-optimal control is given by

$$\mu(t, y(t,x)) = -\eta(t; y(t,x)), \qquad \text{a.e.} \quad x \in \Omega, \quad t \in [0, t_f],$$

and the worst disturbance is (given in state-feedback form)

$$\nu(t; y(t,x)) = \frac{\gamma^{-2}}{2}\eta(t; y(t,x)), \qquad \text{a.e.} \quad x \in \Omega, \quad t \in [0, t_f].$$

Moreover, we have

$$\int_0^{t_f}\int_\Omega \{|y(t,x) - z_d(x,t)|^2 + |\mu(t; y(t,x))|^2\}\, dx\, dt + \int_\Omega |y(x) - y_1(x)|^2\, dx$$
$$\leq \gamma^2 \int_0^{t_f}\int_\Omega |w(t,x)|^2\, dx\, dt + \varphi_\gamma(0, y_0)$$

for any $(y_0, w) \in X \times L^2([0, t_f]; W)$.

## 7 Conclusions

In this paper, we have studied the $H^\infty$-optimal control problem for infinite-dimensional systems where the cost functions are not necessarily quadratic. By making use of the property that studying the $H^\infty$-optimal control problem is equivalent to studying the upper value of a differential game with kernel function $J_\gamma(x_0; u, w) = L_{x_0}(u, w) - \gamma^2\|w\|_2^2$, for $\gamma > \gamma^0$ for some $\gamma^0 > 0$, we have obtained a characterization of disturbance-attenuating controllers, parametrized by $\gamma$.

We have studied here only the finite-horizon case, where the Arzela–Ascoli theorem was used. In the infinite-horizon case, since $[t_0, \infty)$ is not a compact set in $\mathbb{R}$, further assumptions on the system are needed, such as $A$ generating a compact $C_0$-semigroup. This extension will be considered in a future paper. Also under investigation are the cases where the state information is sampled, and where (finite) dimensionality constraints are imposed on the controller. Another line of research would be to consider extensions to the nonlinear dynamics case, in which case the upper or lower value functions will not be smooth (almost everywhere differentiable) as in the present problem. The Hamilton–Jacobi–Isaacs (HJI) partial differential equations associated with such systems generally do not admit classical solutions, and hence one way to approach these problems is via the concept of viscosity solutions. This would provide an appropriate framework in which to study solutions to HJI equations for nonlinear infinite-dimensional systems, and thus the associated disturbance attenuating controllers. For some representative prior work on this general topic, we refer the reader to [3], [8], [11], [12], [15], [16], [22], and [30].

## Appendix I

*Let $X$, $Y$ be real normed linear spaces, and $f : X \to Y$. $f$ is Fréchet differentiable at $x \in X$ if and only if it is Gâteaux differentiable at $x$ and the following limit*

$$\lim_{t\to 0} \frac{1}{t}[f(x+tu) - f(x)] = [Df(x)]u$$

*is uniform on the unit sphere of $X$. If we denote the Fréchet derivative of $f$ by $f'$, then*

$$Df[x] = f'[x].$$

**Proof.** Suppose that for any $\varepsilon > 0$ there exists $\delta > 0$ such that if $|t| < \delta$,

$$\left\| \frac{1}{t}[f(x+tu) - f(x)] - [Df(x)]u \right\| < \varepsilon$$

holds for all $u$ on the unit sphere. Let $v = tu$; when $\|v\| < \delta$ we have

$$\|f(x+v) - f(x) - [Df(x)]v\| < \varepsilon\|v\|.$$

Since $Df(x) \in \mathcal{L}(X, Y)$, $f'(x)$ exists and $Df[x] = f'[x]$.

Conversely, assume that $f$ is Fréchet differentiable at $x$. Then for any $\varepsilon > 0$, there is a $\delta > 0$ such that if $\|v\| < \delta$, we have

$$\|f(x+v) - f(x) - f'(x)v\| < \varepsilon\|v\|.$$

For $\|u\| = 1$, and with $\|t\| < \delta$, the above inequality yields

$$\left\| \frac{1}{t}[f(x+tu) - f(x)] - f'(x)u \right\| < \varepsilon\|u\| = \varepsilon.$$

Therefore $(1/t)[f(x+tu) - f(x)]$ is uniformly convergent to $f'[x]u$. Clearly, it is also the Gâteaux derivative of $f$ at $x$. □

## Appendix II

We provide in this appendix, for the sake of completeness, precise statements for two results used in the main body of the paper: the Arzela–Ascoli theorem and Gronwall's lemma (inequality):

**Arzela–Ascoli Theorem.** *Let $M$ be a compact metric space. If $\mathcal{F}$ is a uniformly bounded equicontinuous family in $C(M)$, then every sequence of functions in $\mathcal{F}$ contains a uniformly convergent subsequence.*

**Gronwall's Lemma.** *Let*

$$u : [a, b] \to [0, \infty),$$
$$v : [a, b] \to \mathbb{R}^n,$$

be continuous functions and let $C$ be a constant. Then if

$$v(t) \leq C + \int_a^t v(s)u(s)\,ds$$

for $t \in [a, b]$, it follows that

$$v(t) \leq C \exp\Big( \int_a^t u(s)\,ds \Big)$$

*for* $t \in [a, b]$.

## Acknowledgments

This research was supported in part by the U.S. Department of Energy under Grant DOE-DEFG-02-94ER13939, and in part by the National Science Foundation under Grant ECS-93-12807.

## REFERENCES

[1] Adams, R. A. *Sobolev Space*. Academic Press, New York 1975.

[2] Aubin, J. P. *An Introduction to Nonlinear Analysis*. Springer-Verlag, New York, 1993.

[3] Ball, J. A. and J. W. Helton. Viscosity Solutions of Hamilton–Jacobi Equations Arising in Nonlinear $H^\infty$ Control. Preprint, 1994.

[4] Barbu, V. The $H^\infty$-Problem with Control Constraints. *SIAM J.Control and Optimization*, **32**, 952–964, 1994.

[5] Barbu, V. $H^\infty$ Boundary Control with State-Feedback: The Hyperbolic Case. *SIAM J.Control and Optimization*, **33**, 684–701, 1995.

[6] Barbu, V. and Th. Precupanu. *Convexity and Optimization in Banach Space*. Reidel, Dordrecht, 1986.

[7] Barbu, V. and G. Da Prato. *Hamilton–Jacobi Equations in Hilbert Spaces*. Reidel, Dordrecht, 1983.

[8] Başar, T. and P. Bernhard. *$H^\infty$-Optimal Control and Related Minimax Design Problems*. Birkhäuser, Boston, 2nd ed, 1995.

[9] Bensoussan, A. and P. Bernhard. On the Standard Problem of $H^\infty$-Optimal Control for Infinite-Dimensional Systems. *Proceedings of the Conference on Control and Identification of Partial Differential Equation*, South Hadlley, MA, 117–140, 1992.

[10] Bensoussan, A., G. D. Prato, M. C. Delfour, and S. K. Mitter. *Representation and Control of Infinite-Dimensional Systems*, Volume II. Birkhäuser, Boston, 1992.

[11] Crandall, M. C., H. Ishii, and P. L. Lions. User's Guide to Viscosity Solutions of Second Order Partial Differential Equations. *Bulletin American Mathematical Society*, **27**, 1–67, 1992.

[12] Crandall, M. C. and P. L. Lions. Viscosity Solutions of Hamilton–Jacobi Equations. *Transactions of the American Mathematical Society*, **277**, 1–42, 1983.

[13] Curtain, R. F. and H. J. Zwart. *An Introduction to Infinite-Dimensional Linear Systems Theory*. Springer-Verlag, New York, 1995.

[14] Doyle, J., K. Glover, P. Khargonekar, and B. Francis. State-Space Solutions to Standard $H_2$ and $H^\infty$-Control Problem. *IEEE Transactions on Automatic Control*, **34**, 831–847, 1989.

[15] Evans, L. C. and P. E. Souganidis. Differential Games and Representation Formulas for Solutions of Hamilton–Jacobi Equations. *Indiana University Mathematics Journal*, **33**, 773–797, 1984.

[16] Fleming, W. H. and H. M. Soner. *Controlled Markov Processes and Viscosity Solutions*. Springer-Verlag, New York, 1993.

[17] Francis, B. A. *A Course in $H^\infty$-Control Theory*. Springer-Verlag, New York, 1987.

[18] Ichikawa, A. *Differential games and $H^\infty$-problems*. Presented at the MTNS, Kobe, Japan, 1991.

[19] Keulen, B., M. Peters, and R. Curtain. $H_\infty$-Control with State-Feedback: The Infinite-Dimensional Case. *Journal of Mathematical Systems, Estimation, and Control*, **3**, 1–39, 1993

[20] Khargonekar, P. P., I. R. Petersen, and M. A. Rotea. $H^\infty$-Optimal Control with State Feedback. *IEEE Transactions on Automatic Control*, **33**, 786–788, 1988.

[21] Lions, J. L. *Optimal Control of Systems Governed by Partial Differential Equations*. Springer-Verlag, Berlin, 1971.

[22] Lions, J. L. *Generalized Solutions of Hamilton–Jacobi Equations*. Pitman, Boston, 1982.

[23] McMillan, C. and R. Triggiani. Min-Max Game Theory and Algebraic Riccati Equations for Boundary Control Problems with Continuous Input-Solution Map. Part II: the General Case. *Applied Mathematics Optimization*, **29**, 1–65, 1994.

[24] McMillan, C. and R. Triggiani. Min-Max Game Theory and Algebraic Riccati Equations for Boundary Control Problems with Analytic Semigroups: The Stable Case. Preprint, 1995.

[25] McMillan, C. and R. Triggiani. Algebratic Riccati Equations Arising in Game Theory and in $H_\infty$-Control Problems for a Class of Abstract Systems. *Differential Equations with Applications to Mathematical Physics*, pp. 239–247, 1993.

[26] Pazy, A. *Semigroup of Linear Operators and Applications to Partial Differential Equations*. Springer-Verlag, New York, 1983.

[27] Petersen, I. R. Disturbance attenuation and $H^\infty$ -optimization: A Design Method Based on the Algebraic Riccati Equation. *IEEE Transactions on Automatic Control*, **32**, 427–429, 1987.

[28] Pritchard, A. J. and S. Townley. Robustness Optimization for Abstract, Uncertain Control Systems: Unbounded Inputs and Perturbations. *Proceedings of IFAC Symposium on Distributed Parameter Systems* (El Jai, Amouroux, eds.), pp. 117–121, 1990.

[29] Scherer, G. $H^\infty$-Control by State-Feedback for Plants with Zeros on the Imaginary Axis. Preprint, 1992.

[30] Soravia, P. $H^\infty$ Control of Nonlinear Systems: Differential Games and Viscosity Solutions. *SIAM Journal on Control and Optimization*, **34**, 1071–1097, 1996.

[31] Stoorvogel, A. A. *The $H^\infty$-Control Problem: A State-Space Approach.* Prentice Hall, New York, 1992.

[32] Tadmor, G. Worst-Case Design in the Time Domain. The Maximum Principle and the Standard $H^\infty$-Problem. *MCSS*, **3**, 301–324, 1990.

[33] Zames, G. Feedback and Optimal Sensitivity: Model Reference Transformation, Multiplicative Seminorms and Approximate Inverses. *IEEE Transactions on Automatic Control* **AC-26**, 301–320, 1981.

[34] Xiao, M. and T. Başar. Solutions to Generalized Riccati Evolution Equations and $H^\infty$-Optimal Control Problems on Hilbert Spaces. Preprint, 1997.

[35] Zhao, Y. The Global Attractor of Infinite-dimensional Dynamical Systems Governed by a Class of Nonlinear Parabolic Variational Inequalities and Associated Control Problems. *Applicable Analysis*, **54**, 163–180, 1994.

# Nonstandard Extension of $H^{\infty}$-Optimal Control for Singularly Perturbed Systems

Hua Xu
The University of Tsukuba
Graduate School of Systems Management
Bunkyo-ku, Tokyo, Japan

Koichi Mizukami
Hiroshima University
Faculty of Integrated Arts and Sciences
Higashi-Hiroshima, Japan

**Abstract**

In this paper, we consider a nonstandard extension of the $H^{\infty}$-optimal control problem for singularly perturbed systems and relax the nonsingularity conditions made in the paper of Pan and Başar [11]. We prove that all the results concerning the composite controller in the paper above still hold true even if the system is a nonstandard singularly perturbed system. We show that the composite optimal controller, which guarantees a disturbance attenuation level larger than the upper bound of the full-order system when $\varepsilon$ is sufficiently small, can be constructed very simply by only revising a slow controller. This paper also provides a concise procedure to construct the composite controller. Some numerical examples, which can not be solved using the method of the above paper, are presented to illustrate the theoretical results.

## 1 Introduction

It is well known that some large-scale systems possess two-time-scale property. In general, such large-scale systems can be described by singularly perturbed models (Kokotović et al. [10]). Using the singular perturbation methodology, one can partition a singularly perturbed system into two reduced-order subsystems which are independent of the small singular perturbation parameter $\varepsilon > 0$. Various control schemes based on the subsystems, such as composite optimal control, stochastic control, observer-based control, and so on, have been developed in the literature. In a recent paper, Pan and Başar [11] have studied the $H^{\infty}$-optimal control problem of singularly perturbed systems with perfect state measurements by making use of a differential game-theoretic approach. It is shown that as the singular perturbation parameter $\varepsilon$ approaches zero, the optimal disturbance attenuation level for the full-order system converges to a value that is bounded above by the maximum of the optimal disturbance attenuation levels for the slow and fast subsystems. Furthermore, for a disturbance attenuation level $\gamma$ which is larger than the upper bound, one can always construct the composite controllers such that the disturbance attenuation level $\gamma$ is attained for the full-order systems when $\varepsilon$ is sufficiently small.

In this paper, we will consider a nonstandard extension of the $H^\infty$-optimal control problem for singularly perturbed systems. Using an alternative approach, we will relax the nonsingularity conditions existing in the paper of Pan and Başar [11]. As we have known, the basic assumption A3 in the paper of Pan and Başar [11], that is, the matrices $A_{22}(t)$ and $Q_{22}(t)$ are invertible for all $t \in [0, t_f]$, plays an important role in the study of the problem. Both the concavity analysis of the related slow cost function and the derivations of various complicated formulas depend very much on the inverses of $A_{22}(t)$ and $Q_{22}(t)$. In the literature (Khalil [7]), a singularly perturbed system with an invertible matrix $A_{22}(t)$ is usually referred as a standard singularly perturbed system. Otherwise, it is called a nonstandard singularly perturbed system. The purpose of the present paper is to prove that all the results concerning the composite controller in the paper of Pan and Başar [11] still hold true even if the system is a nonstandard singularly perturbed system. A useful concavity analysis (sufficient conditions) of the related slow cost function, especially, the relationship between the concavity of the related slow cost function and the disturbance attenuation level $\gamma$ in the cost function, is obtained from the analysis of the soft-constrained differential game of the fast subsystem. It is worth noting that the corresponding analysis in the standard case depends on the inverses of $A_{22}(t)$ and $Q_{22}(t)$. We also show that the composite optimal controller, which guarantees a disturbance attenuation level larger than the upper bound of the full-order system when $\varepsilon$ is sufficiently small, can be constructed simply by only revising a slow controller which is obtained from the associated soft-constrained differential game of the slow subsystem. Our basic thought is to view the slow subsystem as a special kind of descriptor system. Although the basic idea of Pan and Başar [11] is borrowed in the reasoning of this paper, a lot of technical details are different. Moreover, since the paper is valid for both standard and nonstandard singularly perturbed systems, it will extend considerably the domain of application of the $H^\infty$-optimal control for singularly perturbed systems.

Besides Pan and Başar [11],[12], there are also several papers dealing with the $H^\infty$-optimal control problem for singularly perturbed systems (see, for instance, Fridman [5],[6], Drăgan [3],[4] and Khalil and Chen [8]). Among them, Fridman [5],[6] proposed a method on the basis of the exact decomposition of the full-order Riccati equations to the reduced-order slow and fast equations, which is also valid for nonstandard singularly perturbed systems.

## 2 Problem Formulation

Consider a linear time-varying singularly perturbed system

$$\dot{x}_1 = A_{11}x_1 + A_{12}x_2 + B_1u + D_1w, \qquad x_1(0) = 0, \tag{1a}$$

$$\varepsilon \dot{x}_2 = A_{21}x_1 + A_{22}x_2 + B_2u + D_2w, \qquad x_2(0) = 0, \tag{1b}$$

and the quadratic cost function

$$L(u, w) = x^T(t_f)Q_f x(t_f) + \int_0^{t_f} [x^T(t)Q(t)x(t) + u^T(t)u(t)]\, dt, \tag{2}$$

where $\varepsilon$ is a small positive parameter, $x^T := (x_1^T, x_2^T)$ is the $n$-dimensional state vector, with $x_1$ of dimension $n_1$ and $x_2$ of dimension $n_2 := n - n_1$, $u$ is the $m$-dimensional control, and $w$ is the $p$-dimensional disturbance. All matrices above are of appropriate dimensions with $A_{ij}(t)$, $Q_{ij}(t)$, $B_i(t)$, $D_i(t)$ $(i, j = 1, 2)$ being continuously differentiable in $t \geq 0$, and

$$Q(t) = Q^T(t) = \begin{bmatrix} Q_{11}(t) & Q_{12}(t) \\ Q_{21}(t) & Q_{22}(t) \end{bmatrix} \geq 0,$$

$$Q_f = Q_f^T = \begin{bmatrix} Q_{f11} & \varepsilon Q_{f12} \\ \varepsilon Q_{f21} & \varepsilon Q_{f22} \end{bmatrix} \geq 0. \tag{3}$$

The system (1) is said to be in the standard form if the matrix $A_{22}(t)$ is nonsingular, $\forall t \in [0, t_f]$. Otherwise, it is called the nonstandard singularly perturbed system.

The $H^\infty$ optimal control problem for the full-order system (1) is well known (Başar and Bernhard, [1]). For each $\varepsilon > 0$, the $H^\infty$-optimal performance $\gamma^*(\varepsilon)$ is determined as the infimum of those $\gamma$'s under which the generalized Riccati differential equation

$$\dot{\tilde{Z}} + A_\varepsilon^T(t)\tilde{Z} + \tilde{Z}A_\varepsilon(t) - \tilde{Z}S_{\varepsilon\gamma}(t)\tilde{Z} + Q(t) = 0, \qquad \tilde{Z}(t_f) = Q_f, \tag{4}$$

has a bounded solution on $[0, t_f]$, where

$$A_\varepsilon(t) = \begin{bmatrix} A_{11}(t) & A_{12}(t) \\ A_{21}(t)/\varepsilon & A_{22}(t)/\varepsilon \end{bmatrix}, \qquad A(t) = \begin{bmatrix} A_{11}(t) & A_{12}(t) \\ A_{21}(t) & A_{22}(t) \end{bmatrix}, \tag{5a}$$

$$B_\varepsilon(t) = \begin{bmatrix} B_1(t) \\ B_2(t)/\varepsilon \end{bmatrix}, \qquad B(t) = \begin{bmatrix} B_1(t) \\ B_2(t) \end{bmatrix}, \tag{5b}$$

$$D_\varepsilon(t) = \begin{bmatrix} D_1(t) \\ D_2(t)/\varepsilon \end{bmatrix}, \qquad D(t) = \begin{bmatrix} D_1(t) \\ D_2(t) \end{bmatrix}, \tag{5c}$$

and

$$S_{\varepsilon\gamma}(t) = B_\varepsilon(t)B_\varepsilon^T(t) - \frac{1}{\gamma^2}D_\varepsilon(t)D_\varepsilon^T(t),$$

$$S_\gamma(t) = B(t)B^T(t) - \frac{1}{\gamma^2}D(t)D^T(t), \tag{6}$$

with the $ij$th block of $S_\gamma$ denoted henceforth by $S_{ij\gamma}$, $i, j = 1, 2$, and $S_{ij\gamma}^T = S_{ji\gamma}$, $i \neq j$.

In order to overcome the computation difficulties caused by high dimensionality and numerical stiffness, a composite design method has been provided by Pan and Başar [11], from which an upper bound $\bar{\gamma}$ for $\gamma^*(\varepsilon)$ has been found on the basis of the slow subsystem and the fast subsystem decomposition. Moreover, when $\varepsilon$ is sufficiently small, a composite controller can always be constructed which guarantees a disturbance attenuation level that is larger than the upper bound of the full-order system. All the results above are obtained on the basic assumption that the matrices $A_{22}(t)$ and $Q_{22}(t)$ are invertible for all $t \in [0, t_f]$. In this paper, we shall derive the same results as the above without using the inverses of $A_{22}(t)$ and $Q_{22}(t)$.

## 3 Decomposition and Solutions of Slow and Fast Subproblems

In this section, we first decompose the full-order system into a slow subsystem and a fast subsystem, and establish the associated soft-constrained slow and fast games. Then, we solve the resulted slow and fast $H^\infty$ optimal control problems, respectively.

### 3.1 The Slow and Fast Subsystems and the Associated Soft-Constrained Games

The slow subsystem is formed by neglecting the fast modes, which is equivalent to letting $\varepsilon = 0$ in (1),

$$E\dot{x}_s = Ax_s + Bu_s + Dw_s, \qquad Ex_s(0) = 0, \tag{7}$$

where $x_s(t) = [x_{1s}^T(t)\ \ x_{2s}^T(t)]^T$, $E = \operatorname{diag}[I_{n_1}, 0]$ and $A$, $B$, $D$ are defined in (5). The corresponding reduced (slow) cost function is

$$L_s = x_{1s}^T(t_f)Q_{f11}x_{1s}(t_f) + \int_0^{t_f} [x_s^T(t)Q(t)x_s(t) + u_s^T(t)u_s(t)]\,dt, \tag{8}$$

**Remark 3.1.** We do not use the inverse of $A_{22}(t)$, which does not exist in a nonstandard case, to eliminate the variable $x_{2s}$ in (7)–(8). The slow subsystem (7) is formulated as a descriptor system (Dai [2]) which may display an impulse phenomenon in the solution if $A_{22}(t)$ is singular.

The associated soft-constrained slow game is defined by the descriptor equation (7) and the cost function

$$L_{\gamma s} = L_s - \gamma^2 \int_0^{t_f} w_s^T(t)w_s(t)\,dt. \tag{9}$$

On the other hand, the fast subsystem is defined as in Pan and Başar [11].

$$\frac{d}{d\tau}x_{2f}^t = A_{22}(t)x_{2f}^t + B_2(t)u_f^t + D_2(t)w_f^t, \qquad x_{2f}^t(0) = x_{2f}(t), \quad t \in [0, t_f], \tag{10}$$

where $\tau = (t' - t)/\varepsilon$, $t$ is taken to be frozen and $t'$ to vary on the same scale as $t$, $x_{2f} = x_2 - x_{2s}$, $u_f = u - u_s$, and $w_f = w - w_s$. The associated soft-constrained fast game is defined by the system equation (10) and the cost function

$$L^t_{\gamma f} = \int_0^\infty [x^{tT}_{2f} Q_{22}(t) x^t_{2f} + u^{tT}_f u^t_f - \gamma^2 w^{tT}_f w^t_f] d\tau. \tag{11}$$

## 3.2 The Solution of the Associated Soft-Constrained Fast Game

The associated soft-constrained fast game is a standard infinite horizon differential game of state space systems for each fixed $t \in [0, t_f]$.

**Assumption.** The triplet $(A_{22}, B_2, C_2)$ is stabilizable and observable for every $t \in [0, t_f]$, where $Q_{22} = C_2^T C_2$.

As is well known, given the generalized algebraic Riccati equation

$$A^T_{22}(t) Z_f + Z_f A_{22}(t) - Z_f S_{22\gamma}(t) Z_f + Q_{22}(t) = 0, \tag{12}$$

the upper value of the associated soft-constrained fast game is finite if, and only if, the GARE (12) admits a positive definite solution (Başar and Bernhard [1]).

Let

$$\gamma^t_f = \inf\{\gamma > 0 | \text{the GARE (12) has a minimal positive definite solution}\}$$

and

$$\gamma_f = \sup\{\gamma^t_f | t \in [0, t_f]\}.$$

Then, any disturbance attenuation level $\gamma > \gamma_f$ can be guaranteed by a feedback controller

$$u^*_{f\gamma}(t) = -B^T_2(t) Z_{f\gamma}(t) x_{2f}(t), \tag{13}$$

where $Z_{f\gamma}(t)$ is the minimal positive definite solution of (12).

## 3.3 The Solution of the Associated Soft-Constrained Slow Game

Associated with the GARE (12) is the $2n_2 \times 2n_2$ Hamiltonian matrix

$$H_\gamma(t) = \begin{bmatrix} A_{22}(t) & -S_{22\gamma}(t) \\ -Q_{22}(t) & -A^T_{22}(t) \end{bmatrix}. \tag{14}$$

Let

$$\gamma_0 := \max\{\gamma > 0 | \text{the Hamiltonian matrix } H_\gamma(t) \text{ is singular at a time instant } t' \in [0, t_f]\}.$$

**Remark 3.2.** When $\gamma = \infty$, the matrix $H_\infty(t)$ of (14) is nonsingular if and only if the matrices $[A_{22}(t)\ B_2(t)]$ and $[A^T_{22}(t)\ C^T_2(t)]$ are of full row rank for all $t \in [0, t_f]$(Wang et al., [13]). Therefore, it is also necessary and sufficient for $\gamma_0 < \infty$.

Let us further introduce a reduced-order generalized Riccati differential equation

$$\dot{Z}_s + A_0^T Z_s + Z_s A_0 - Z_s S_{0\gamma} Z_s + Q_0 = 0, \qquad Z_s(t_f) = Q_{f11}. \tag{15}$$

The coefficient matrices of the GRDE (15) are obtained from the formula

$$\begin{bmatrix} A_0(t) & -S_{0\gamma}(t) \\ -Q_0(t) & -A_0^T(t) \end{bmatrix} = T_1(t) - T_2(t) T_4^{-1}(t) T_3(t), \tag{16}$$

where

$$T_1(t) = \begin{bmatrix} A_{11}(t) & -S_{11\gamma}(t) \\ -Q_{11}(t) & -A_{11}^T(t) \end{bmatrix}, \qquad T_2(t) = \begin{bmatrix} A_{12}(t) & -S_{12\gamma}(t) \\ -Q_{12}(t) & -A_{21}^T(t) \end{bmatrix}, \tag{17a}$$

$$T_3(t) = \begin{bmatrix} A_{21}(t) & -S_{12\gamma}^T(t) \\ -Q_{12}^T(t) & -A_{12}^T(t) \end{bmatrix}, \qquad T_4(t) = \begin{bmatrix} A_{22}(t) & -S_{22\gamma}(t) \\ -Q_{22}(t) & -A_{22}^T(t) \end{bmatrix}. \tag{17b}$$

Note that $T_4(t) = H_\gamma(t)$ is nonsingular for every $\gamma > \gamma_0$ and $t \in [0, t_f]$.

**Remark 3.3.** The advantage of the formula (16) lies in that the GRDE (15) is obtained without using the inverses of $A_{22}(t)$ and $Q_{22}(t)$. Therefore, the formula (16) is valid for a nonstandard singularly perturbed system while the corresponding formulas of Pan and Başar [11] (see (3.15)–(3.21) and (3.34)–(3.39) for comparison) are only useful in the standard case. It can be further shown that, details of which are omitted, the GRDE (15) is identical with the GRDE (3.19) or (3.38) of Pan and Başar [11] in the standard case. This fact implies that the formula (16) includes the corresponding formulas of Pan and Başar [11] as a special case.

The explicit expressions for $A_0(t)$, $S_{0\gamma}(t)$, and $Q_0(t)$ are also available which will be used in the later analyses

$$A_0 = A_{11} + N_{10} A_{21} + S_{12\gamma} N_{20}^T + N_{10} S_{22\gamma} N_{20}^T, \tag{18a}$$

$$S_{0\gamma} = S_{11\gamma} + N_{10} S_{12\gamma}^T + S_{12\gamma} N_{10}^T + N_{10} S_{22\gamma} N_{10}^T, \tag{18b}$$

$$Q_0 = Q_{11} - N_{20} A_{21} - A_{21}^T N_{20}^T - N_{20} S_{22\gamma} N_{20}^T, \tag{18c}$$

$$N_{10} = -\hat{A}_{120} \hat{A}_{220}^{-1}, \qquad N_{20} = \hat{Q}_{120} \hat{A}_{220}^{-1}, \tag{18d}$$

$$\hat{A}_{120} = A_{12} - S_{12\gamma} Z_{22\gamma}, \quad \hat{A}_{220} = A_{22} - S_{22\gamma} Z_{22\gamma}, \quad \hat{Q}_{120} = Q_{12} + A_{21}^T Z_{22\gamma}, \tag{18e}$$

where $Z_{22\gamma}$ is any real solution of a generalized algebraic Riccati-like equation

$$A_{22}^T(t) Z_{22\gamma} + Z_{22\gamma}^T A_{22}(t) - Z_{22\gamma}^T S_{22\gamma}(t) Z_{22\gamma} + Q_{22}(t) = 0. \tag{19}$$

Note that the GARLE (19) is different from the GARE (12). They are equivalent only for those symmetric solutions.

Let

$$\gamma_s := \inf\{\gamma' > \gamma_0 | \forall \gamma \geq \gamma',$$

the GRDE (15) has a bounded nonnegative definite solution over $[0, t_f]\}$.

**Proposition 3.1.** *Consider the associated soft-constrained slow game* (7)–(8). *Any disturbance attenuation level $\gamma > \bar{\gamma} := \max\{\gamma_s, \gamma_f\}$ can be guaranteed by the linear feedback controller*

$$u_{s\gamma}^*(t) = -[B_1^T(t)\ B_2^T(t)] \begin{bmatrix} Z_{s\gamma}(t) & 0 \\ Z_{m\gamma}(t) & Z_{22\gamma}(t) \end{bmatrix} \begin{bmatrix} x_{1s}(t) \\ x_{2s}(t) \end{bmatrix}, \tag{20}$$

*where $Z_{s\gamma}(t)$ is the bounded nonnegative definite solution of the GRDE* (15), *$Z_{22\gamma}(t)$ is the real solution of the GARLE* (19) *and $Z_{m\gamma}(t)$ is given by*

$$Z_{m\gamma}(t) = -N_{20}^T(t) + N_{10}^T(t) Z_{s\gamma}(t). \tag{21}$$

**Proof.** Since $\bar{\gamma} := \max\{\gamma_s, \gamma_f\}$, the GRDE (15) has the bounded nonnegative definite solution $Z_{s\gamma}(t)$ and the GARE (12) has the minimal positive definite solution $Z_{f\gamma}(t)$ for every $\gamma > \bar{\gamma}$ and $t \in [0, t_f]$. However, comparing the GARLE (19) with the GARE (12), we know that the GARLE (19) also has a real solution (e.g., $Z_{22\gamma}(t) = Z_{f\gamma}(t)$) for every $\gamma > \bar{\gamma}$ and $t \in [0, t_f]$. After extending the results of Xu and Mizukami [14], we can prove that the existences of the solution $Z_{s\gamma}(t)$ on $[0, t_f]$ and the solution $Z_{22\gamma}(t)$ for every $t \in [0, t_f]$ are necessary and sufficient for the existence of a real bounded solution $Z(t)$ to the following generalized Riccati differential equation:

$$\text{(i) } E\dot{Z} + A^T(t)Z + Z^T A(t) - Z^T S_\gamma(t) Z + Q(t) = 0, \quad Z(t_f) = E Q_f E, \tag{22a}$$

$$\text{(ii) } EZ = ZE, \tag{22b}$$

where

$$Z(t) = \begin{bmatrix} Z_{s\gamma}(t) & 0 \\ Z_{m\gamma}(t) & Z_{22\gamma}(t) \end{bmatrix}$$

is nonnegative definite in the sense that $x_s^T E Z(t) x_s \geq 0$ for any $Ex_s \neq 0$. Using the GRDE (22), a standard completion of squares shows that

$$u_{s\gamma}^*(t) = -B^T(t) Z(t) x_s(t), \tag{23a}$$

$$w_{s\gamma}^*(t) = \frac{1}{\gamma^2} D^T(t) Z(t) x_s(t), \tag{23b}$$

constitutes a linear feedback saddle-point solution for the associated soft-constrained slow game (7)–(8). Therefore, the linear feedback controller (20) guarantees a disturbance attenuation level $\gamma > \bar{\gamma}$. □

**Remark 3.4.** $\gamma > \bar{\gamma}$ is only sufficient, not necessary, for the existence of a linear feedback saddle-point solution to the associated soft-constrained slow game because the quantity $\gamma_f$ is determined by the GARE (12) rather than the GARLE (19).

However, taking into account the associated soft-constrained fast game, we arrive at the following results. For every $\gamma > \bar{\gamma}$, both the slow game and the fast game have linear feedback saddle-point solutions. For every $\gamma < \bar{\gamma}$, either the slow game or the fast game has an unbounded upper value.

**Remark 3.5.** The corresponding quantity $\bar{\gamma}$ in Pan and Başar [11] cannot be obtained without using the inverses of $A_{22}(t)$ and $Q_{22}(t)$. Instead of using the inverses of $A_{22}(t)$ and $Q_{22}(t)$, we arrive at $\bar{\gamma}$ through the simultaneous analyses of the slow and fast games.

## 4 The Linear Feedback Composite Controller

In this section, we will construct a composite controller for the problem. As stated in the proof of Proposition 1, the linear feedback controller (20) which guarantees a disturbance attenuation level $\gamma > \bar{\gamma}$ for the slow subsystem is not unique. We now select a special linear feedback controller for the slow subsystem, that is,

$$u_{s\gamma}^{*}(t) = -[B_1^T(t)\ \ B_2^T(t)] \begin{bmatrix} Z_{s\gamma}(t) & 0 \\ Z_{m\gamma}(t) & Z_{f\gamma}(t) \end{bmatrix} \begin{bmatrix} x_{1s}(t) \\ x_{2s}(t) \end{bmatrix} = -B^T(t)K(t)x_s(t), \tag{24}$$

where $Z_{f\gamma}(t)$ is the minimal positive definite and strictly feedback stabilizing solution of (12) for every $t \in [0, t_f]$. $Z_{m\gamma}(t)$ is the corresponding one obtained when $Z_{22\gamma}(t) = Z_{f\gamma}(t)$ in (21). Then, the linear feedback composite controller is constructed as follows:

$$u_{c\gamma}^{*}(t) = u_{s\gamma}^{*}(t) + u_{f\gamma}^{*}(t) = -[B_1^T(t)\ \ B_2^T(t)] \begin{bmatrix} Z_{s\gamma}(t) & 0 \\ Z_{m\gamma}(t) & Z_{f\gamma}(t) \end{bmatrix} \begin{bmatrix} x_1(t) \\ x_2(t) \end{bmatrix}, \tag{25}$$

where $x_1(t) \approx x_{1s}(t)$ and $x_2(t) \approx x_{2s}(t) + x_{2f}(t)$.

It is worth noting that $u_{c\gamma}^{*}$ is obtained very simply by only revising the slow controller (20). More precisely, we achieve $u_{c\gamma}^{*}(t)$ by selecting $Z_{22\gamma} = Z_{f\gamma}$ and changing $x_{1s}(t)$ to $x_1(t)$, and $x_{2s}(t)$ to $x_2(t)$.

We now provide the same results as Pan and Başar [11] on the composite controller, but under rather weak assumptions, for the finite horizon case.

In order to study the finite horizon case, we adopt the boundary layer method (see Yackel and Kokotović, [16]) for the generalized Riccati differential equation (4). Let $\tilde{Z}$ be partitioned as follows:

$$\tilde{Z}(t,\varepsilon) = \begin{bmatrix} \tilde{Z}_{11}(t,\varepsilon) & \varepsilon\tilde{Z}_{12}(t,\varepsilon) \\ \varepsilon\tilde{Z}_{21}(t,\varepsilon) & \varepsilon\tilde{Z}_{22}(t,\varepsilon) \end{bmatrix}, \qquad \tilde{Z}_{12}(t,\varepsilon) = \tilde{Z}_{21}^T(t,\varepsilon). \tag{26}$$

Substituting it into (4) yields

$$\dot{\tilde{Z}}_{11} + A_{11}^T\tilde{Z}_{11} + \tilde{Z}_{11}A_{11} + A_{21}^T\tilde{Z}_{21} + \tilde{Z}_{21}^T A_{21} - \tilde{Z}_{11}S_{11\gamma}\tilde{Z}_{11} - \tilde{Z}_{21}S_{22\gamma}\tilde{Z}_{21} - \tilde{Z}_{11}S_{12\gamma}\tilde{Z}_{21} - \tilde{Z}_{21}^T S_{12\gamma}^T\tilde{Z}_{11} + Q_{11} = 0, \quad \tilde{Z}_{11}(t_f) = Q_{f11}, \tag{27a}$$

$$\varepsilon\dot{\tilde{Z}}_{21} + \varepsilon\tilde{Z}_{21}A_{11} + \tilde{Z}_{22}A_{21} + A_{12}^T\tilde{Z}_{11} + A_{22}^T\tilde{Z}_{21} - \varepsilon\tilde{Z}_{21}S_{11\gamma}\tilde{Z}_{11} - \varepsilon\tilde{Z}_{21}S_{12\gamma}^T\tilde{Z}_{21} - \tilde{Z}_{22}S_{12\gamma}^T\tilde{Z}_{11} - \tilde{Z}_{22}S_{22\gamma}\tilde{Z}_{21} + Q_{12}^T = 0, \quad \tilde{Z}_{21}(t_f) = Q_{f12}^T, \tag{27b}$$

$$\varepsilon\dot{\tilde{Z}}_{22} + A_{22}^T\tilde{Z}_{22} + \tilde{Z}_{22}A_{22} + \varepsilon A_{12}^T\tilde{Z}_{21}^T + \varepsilon\tilde{Z}_{21}A_{12} - \tilde{Z}_{22}S_{22\gamma}\tilde{Z}_{22} - \varepsilon\tilde{Z}_{22}S_{12\gamma}^T\tilde{Z}_{21}^T - \varepsilon\tilde{Z}_{21}S_{12\gamma}\tilde{Z}_{22} - \varepsilon^2\tilde{Z}_{21}S_{11\gamma}\tilde{Z}_{21}^T + Q_{22} = 0, \quad \tilde{Z}_{22}(t_f) = Q_{f22}. \tag{27c}$$

Furthermore, take

$$\tilde{Z}_{11}(t,\varepsilon) = Z_{11}(t,\varepsilon) + P_{11}(\tau,\varepsilon), \tag{28a}$$
$$\tilde{Z}_{21}(t,\varepsilon) = Z_{21}(t,\varepsilon) + P_{21}(\tau,\varepsilon), \tag{28b}$$
$$\tilde{Z}_{22}(t,\varepsilon) = Z_{22}(t,\varepsilon) + P_{22}(\tau,\varepsilon), \tag{28c}$$

where $\tau = (t_f - t)/\varepsilon$. $(Z_{11}, Z_{21}, Z_{22})$ are the outer solutions of (27), and $(P_{11}, P_{21}, P_{22})$ are the boundary layer correction solutions of (27), which are significant only near $t = t_f$, that is, near $\tau = 0$, and are required to have the property

$$\lim_{\tau\to\infty} P_{ij}(\tau,\varepsilon) = 0, \qquad i, j = 1, 2. \tag{29}$$

Now, substituting $(Z_{11}, Z_{21}, Z_{22})$ into (27) and setting $\varepsilon = 0$ give the following zero-order differential and algebraic equations:

$$\dot{Z}_{11}^{(0)} + Z_{11}^{(0)}A_{11} + Z_{21}^{(0)T}A_{21} + A_{11}^T Z_{11}^{(0)} + A_{21}^T Z_{21}^{(0)} - Z_{11}^{(0)}S_{11\gamma}Z_{11}^{(0)} - Z_{21}^{(0)T}S_{12\gamma}^T Z_{11}^{(0)} - Z_{11}^{(0)}S_{12\gamma}Z_{21}^{(0)} - Z_{21}^{(0)T}S_{22\gamma}Z_{21}^{(0)} + Q_{11} = 0, \quad Z_{11}^{(0)}(t_f) = Q_{f11}, \tag{30a}$$

$$Z_{22}^{(0)}A_{21} + A_{12}^T Z_{11}^{(0)} + A_{22}^T Z_{21}^{(0)} - Z_{22}^{(0)}S_{12\gamma}^T Z_{11}^{(0)} - Z_{22}^{(0)}S_{22\gamma}Z_{21}^{(0)} + Q_{12}^T = 0, \tag{30b}$$

$$Z_{22}^{(0)}A_{22} + A_{22}^T Z_{22}^{(0)} - Z_{22}^{(0)}S_{22\gamma}Z_{22}^{(0)} + Q_{22} = 0. \tag{30c}$$

Moreover, the boundary layer correction solutions $(P_{11}, P_{21}, P_{22})$ satisfy the following differntial equations:

$$\frac{dP_{11}(\tau,\varepsilon)}{d\tau} = \varepsilon\dot{Z}_{11} - \varepsilon\dot{\tilde{Z}}_{11}, \tag{31a}$$
$$\frac{dP_{21}(\tau,\varepsilon)}{d\tau} = \varepsilon\dot{Z}_{21} - \varepsilon\dot{\tilde{Z}}_{21}, \tag{31b}$$
$$\frac{dP_{11}(\tau,\varepsilon)}{d\tau} = \varepsilon\dot{Z}_{22} - \varepsilon\dot{\tilde{Z}}_{22}. \tag{31c}$$

Note that $(Z_{11}, Z_{21}, Z_{22})$ and $(\tilde{Z}_{11}, \tilde{Z}_{21}, \tilde{Z}_{22})$ satisfy the same differential equations (27), but with different end conditions (see Yackel and Kokotović, [16]). Therefore, substituting $(Z_{11}, Z_{21}, Z_{22})$ and $(\tilde{Z}_{11}, \tilde{Z}_{21}, \tilde{Z}_{22})$ into (31) and setting $\varepsilon = 0$ give the following zero-order differential equations:

$$\frac{dP_{11}^{(0)}}{d\tau} = 0, \qquad P_{11}^{(0)}(0) = 0, \tag{32a}$$

$$\begin{aligned}\frac{dP_{21}^{(0)}}{d\tau} &= A_{22}^T(t_f)P_{21}^{(0)} + P_{22}^{(0)}A_{21}(t_f) - P_{22}^{(0)}S_{12\gamma}^T(t_f)Z_{11}^{(0)}(t_f)\\ &\quad - P_{22}^{(0)}S_{22\gamma}(t_f)Z_{21}^{(0)}(t_f) - Z_{22}^{(0)}(t_f)S_{22\gamma}(t_f)P_{21}^{(0)} - P_{22}^{(0)}S_{22\gamma}(t_f)P_{21}^{(0)},\\ P_{21}^{(0)}(0) &= Q_{f21} - Z_{21}^{(0)}(t_f),\end{aligned} \tag{32b}$$

$$\begin{aligned}\frac{dP_{22}^{(0)}}{d\tau} &= P_{22}^{(0)}A_{22}(t_f) + A_{22}^T(t_f)P_{22}^{(0)} - Z_{22}^{(0)}(t_f)S_{22\gamma}(t_f)P_{22}^{(0)}\\ &\quad - P_{22}^{(0)}S_{22\gamma}(t_f)Z_{22}^{(0)}(t_f) - P_{22}^{(0)}S_{22\gamma}(t_f)P_{22}^{(0)},\\ P_{22}^{(0)}(0) &= Q_{f22} - Z_{22}^{(0)}(t_f).\end{aligned} \tag{32c}$$

**Theorem 4.1.** *Consider the singularly perturbed system* (1)–(2). *Let the matrix triplet* $(A_{22}(t), B_2(t), C_2(t))$ *be controllable and observable for each* $t \in [0, t_f]$, *and the following condition hold:* $Q_{f22} \le Z_{f\gamma}(t_f)$, *where* $Z_{f\gamma}(t_f)$ *is the solution to* (12) *at* $t = t_f$ *with* $\gamma$ *fixed. Then:*

(1) $\gamma^*(\varepsilon) \le \bar{\gamma}$, *asympotically as* $\varepsilon \to 0$, *where* $\bar{\gamma}$, *as defined in Proposition* 1, *is finite.*

(2) $\forall\gamma > \bar{\gamma}$, $\exists\varepsilon_\gamma > 0$ *such that* $\forall\varepsilon \in [0, \varepsilon_\gamma)$, *the GRDE* (4) *admits a bounded nonnegative definite solution, and consequently, the game has a finite upper value. Furthermore, the solution to the GRDE* (4) *can be approximated by*

$$\tilde{Z}(t) = \begin{bmatrix} Z_{s\gamma}(t) + O(\varepsilon) & \varepsilon(Z_{m\gamma}^T(t) + P_{21}^{(0)T}(\tau)) + O(\varepsilon^2) \\ \varepsilon(Z_{m\gamma}(t) + P_{21}^{(0)}(\tau)) + O(\varepsilon^2) & \varepsilon(Z_{f\gamma}(t) + P_{22}^{(0)}(\tau)) + O(\varepsilon^2) \end{bmatrix}, \tag{33}$$

*for all* $t \in [0, t_f]$, *where* $P_{21}^{(0)}(\tau)$, $P_{22}^{(0)}(\tau)$ *are the solutions to* (32b)–(32c), *and has the property* $P_{21}^{(0)}(\tau) \to 0$, $P_{22}^{(0)}(\tau) \to 0$ *as* $\tau \to \infty$.

(3) $\forall\gamma > \bar{\gamma}$, *if we apply the composite controller* $u_{c\gamma}^*(t)$ *to the system, then* $\exists\varepsilon_\gamma' > 0$ *such that* $\forall\varepsilon \in [0, \varepsilon_\gamma')$, *the disturbance attenuation level* $\gamma$ *is attained for the full-order system.*

**Proof.** After decomposing the solution of (27) into the outer solution and the boundary layer correction solution, we prove the theorem by following Yackel and Kokotović [16] and Pan and Başar [11]. For more details, see Xu and Mizukami [15]. □

## 5 Design Procedure of Composite Controller and Examples

For a disturbance attenuation level $\gamma$ larger than the upper bound, the composite controller can be constructed by following a very concise procedure.

**Step 1.** Calculate $A_0(t)$, $S_{0\gamma}(t)$, $Q_0(t)$ by using the formula (16).

**Step 2.** Find the corresponding solutions $Z_{f\gamma}(t)$, $Z_{s\gamma}(t)$ of the GARE (12) and the GRDE (15).

**Step 3.** Calculate $Z_{m\gamma}(t)$ by using the formula (21), or, more conveniently, using the formula

$$Z_{m\gamma}(t) = L_2(t) - Z_{f\gamma}(t)L_1(t), \tag{34}$$

where

$$\begin{bmatrix} L_1(t) \\ L_2(t) \end{bmatrix} = -T_4^{-1}(t)T_3(t)\begin{bmatrix} I_{n_1} \\ Z_{s\gamma}(t) \end{bmatrix}, \tag{35}$$

and $T_4(t)$, $T_3(t)$ are defined in (17b). Note that the formula (21) is equivalent to the formula (34). The detail is omitted here.

**Step 4.** Construct the composite controller (25).

The main part (Step 2) of the above design procedure involves solving two reduced-order differential or algebraic Riccati equations. Since these two Riccati equations are independent with each other, parallel computations to find the solutions are possible. That is an advantage of this paper.

In this paper, only the finite horizon problem is treated. The same result is also available for the infinite horizon problem (Xu and Mizukami [15]). To the end of this section, we only present two numerical examples for the infinite horizon case. These two examples do not satisfy the basic assumption of Pan and Başar [11]. Hence, they cannot be solved directly using their method.

**Example 5.1.** Consider the first nonstandard singularly perturbed system

$$\begin{bmatrix} \dot{x}_1 \\ \varepsilon\dot{x}_2 \end{bmatrix} = \begin{bmatrix} 1 & 2 \\ 2 & 0 \end{bmatrix}\begin{bmatrix} x_1 \\ x_2 \end{bmatrix} + \begin{bmatrix} 2 \\ 1 \end{bmatrix} u + \begin{bmatrix} 1 \\ 3 \end{bmatrix} w, \tag{36}$$

with the performance index

$$L_\gamma = \int_0^\infty (2x_1^2 + 2x_1x_2 + x_2^2 + u^2 - \gamma^2 w^2)\,dt. \tag{37}$$

For this example, we obtain two quantities, $\gamma_{s\infty} = \gamma_{f\infty} = 3$, hence, $\bar{\gamma}_\infty = 3$. We can also compute the minimax disturbance attenuation level $\gamma_\infty^*(\varepsilon)$ of the full-order system (36)–(37) for different values of $\varepsilon$ as shown in Table 1. Now, we choose $\gamma = 3.5 > \max\{\gamma_{s\infty}, \gamma_{f\infty}\}$ to design the composite controller. Then, we obtain

$$Z_{s\gamma} = 0.61373, \qquad Z_{f\gamma} = 1.94145,$$
$$u_{c\gamma}^*(t) = -9.03037x_1 - 1.94145x_2.$$

Table 1.

| $\varepsilon$ | 1 | 0.5 | 0.1 | 0.01 | 0.001 | 0.0001 |
|---|---|---|---|---|---|---|
| $\gamma^*$ | 1.3918 | 1.5591 | 2.1357 | 2.7657 | 2.9671 | 2.9965 |
| $\gamma_c^*$ | 3.3277 | 3.3277 | 3.3277 | 3.3277 | 3.3277 | 3.3277 |

After applying the composite controller $u_{c\gamma}^*(t)$ to the full-order system (36)–(37), we arrive at the corresponding disturbance attenuation bound $\gamma_c^*$, which is also tabulated in Table 1.

**Example 5.2.** Consider the second nonstandard singularly perturbed system

$$\begin{bmatrix} \dot{x}_1 \\ \varepsilon\dot{x}_2 \end{bmatrix} = \begin{bmatrix} 2 & 1 \\ -1 & 0 \end{bmatrix} \begin{bmatrix} x_1 \\ x_2 \end{bmatrix} + \begin{bmatrix} 2 \\ 2 \end{bmatrix} u + \begin{bmatrix} 1 \\ 3 \end{bmatrix} w, \tag{38}$$

with the performance index,

$$L_\gamma = \int_0^\infty (2x_1^2 + 2x_1x_2 + 3x_2^2 + u^2 - \gamma^2 w^2)\, dt. \tag{39}$$

For this example, we obtain two quantities, $\gamma_{s\infty} = 3.7749$, $\gamma_{f\infty} = 1.5$, hence, $\bar{\gamma}_\infty = 3.7749$. We can also compute the minimax disturbance attenuation level $\gamma_\infty^*(\varepsilon)$ of the full-order system (38)–(39) for different values of $\varepsilon$ as shown in Table 2. In order to design the composite controller, we choose $\gamma = 4 > \max\{\gamma_{s\infty}, \gamma_{f\infty}\}$. Then, we obtain

$$Z_{s\gamma} = 131.20882, \qquad Z_{f\gamma} = 0.934199,$$
$$u_{c\gamma}^*(t) = -53.13037x_1 - 1.868398x_2.$$

After applying the composite controller $u_{c\gamma}^*(t)$ to the full-order system (38)–(39), we arrive at the corresponding disturbance attenuation bound $\gamma_c^*$, which is also tabulated in Table 2.

Besides the Examples 1 and 2 given above, we also apply the present method to the examples in Pan and Başar [11], which are standard singularly perturbed systems. The same results about $\gamma_c^*$, $u_{c\gamma}^*$ are obtained. This fact further verifies that the method of this paper is a unified one and includes the method of Pan and Başar [11] as a special case.

Table 2.

| $\varepsilon$ | 1 | 0.5 | 0.1 | 0.01 | 0.001 | 0.0001 |
|---|---|---|---|---|---|---|
| $\gamma^*$ | 1.9583 | 2.3170 | 3.2183 | 3.7050 | 3.7677 | 3.7742 |
| $\gamma_c^*$ | 3.9822 | 3.9822 | 3.9822 | 3.9822 | 3.9822 | 3.9822 |

## 6 Conclusions

In this paper, we have considered the extension of the $H^\infty$-optimal control problem for singularly perturbed systems to include the nonstandard case. We show that the same results of Pan and Başar [11] concerning the composite controller still hold true even if the system is a nonstandard singularly perturbed system. The upper bound of the optimal disturbance attenuation level is obtained from the simultaneous analyses of the slow and the fast games. It is worth noting that the corresponding analysis in the standard case depends on the inverses of $A_{22}(t)$ and $Q_{22}(t)$. We also show that the composite optimal controllers, which guarantee a disturbance attenuation level larger than the upper bound of the full-order system when $\varepsilon$ is sufficiently small, can be constructed simply by only revising the slow controller. A concise procedure to construct the composite controller for an achievable performance level of the full-order system is given which permits parallel computations to the related Riccati equations. Since the paper is valid for both standard and nonstandard singularly perturbed systems, it has extended considerably the domain of application of the $H^\infty$-optimal control for singularly perturbed systems.

## Acknowledgments

This work is supported by the Ministry of ESSC of Japan under Grant-in-Aid for EYS 08750538 and was presented at the 7th International Symposium on Dynamic Games and Applications, Shonan Village Center, Kanagawa, Japan, December 16–18, 1996.

## REFERENCES

[1] Başar, T. and P. Bernhard. *$H^\infty$-Optimal Control and Related Minimax Design Problems: A Dynamic Game Approach*. Birkhäuser, Boston, 1995.

[2] Dai, L. *Singular Control Systems*. Lecture Notes in Control and Information Sciences (M. Thoma and A. Wyner, eds.). Springer-Verlag, Berlin, 1998.

[3] Drăgan, V. Asymptotic Expansions for Game-Theoretic Riccati Equations and Stabilization with Disturbance Attenuation for Singularly Perturbed Systems. *Systems & Control Letters*, **20**, 455–463, 1993.

[4] Drăgan, V. $H_\infty$-Norms and Disturbance Attenuation for Systems with Fast Transients. *IEEE Transactions on Automatic Control*, **AC-41**, 747–750, 1996.

[5] Fridman, E. Exact Decomposition of Linear Singularly Perturbed $H^\infty$-Optimal Control Problem. *Kybernetika*, **31**, 589–597, 1995.

[6] Fridman, E. Near-Optimal $H^\infty$ Control of Linear Singularly Perturbed Systems. *IEEE Transactions on Automatic Control*, **AC-41**, 236–240, 1996.

[7] Khalil, H. K. Feedback Control of Nonstandard Singularly Perturbed Systems. *IEEE Transactions on Automatic Control*, **AC-34**, 1052–1060, 1989.

[8] Khalil, H. K. and F. Chen. $H^\infty$-Control of Two-Time-Scale Systems. *Systems & Control Letters*, **19**, 35–42, 1992.

[9] Kokotović, P. V. and R. A. Yackel. Singular Perturbation of Linear Regulators: Basic Theorems. *IEEE Transactions on Automatic Control*, **AC-17**, 29–37, 1972.

[10] Kokotović, P. V., H. K. Khalil, and J. O'Reilly. *Singular Perturbation Methods in Control: Analysis and Design*. Academic Press, New York, 1986.

[11] Pan, Z. and T. Başar. $H^\infty$-Optimal Control for Singularly Perturbed Systems. Part I: Perfect State Measurements. *Automatica*, **29**, 401–423, 1993.

[12] Pan, Z. and T. Başar. Time-Scale Separation and Robust Controller Design for Uncertain Nonlinear Singularly Perturbed Systems under Perfect State Measurements. *International Journal of Robust and Nonlinear Control*, **6**, 585–608, 1996.

[13] Wang, Y. Y., S. J. Shi and Z. J. Zhang. A Descriptor-System Approach to Singular Perturbation of Linear Regulators. *IEEE Transactions on Automatic Control*, **AC-33**, 370–373, 1988.

[14] Xu, H. and K. Mizukami. On the Isaacs Equation of Differential Games for Descriptor Systems. *Journal of Optimization Theory and Applications*, **83**, 405–419, 1994.

[15] Xu, H. and K. Mizukami. Nonstandard Extension of $H^\infty$-Optimal Control for Singularly Perturbed Systems, *Proceedings of the 7th International Symposium on Dynamic Games and Applications*, **2**, 931–948, 1996.

[16] Yackel, R. A. and P. V. Kokotović. A Boundary Layer Method for the Matrix Riccati Equation. *IEEE Transactions on Automatic Control*, **AC-18**, 17–24, 1973.

# PART II
# Pursuit-Evasion (P-E) Games

# Geodesic Parallel Pursuit Strategy in a Simple Motion Pursuit Game on the Sphere

A. M. Kovshov
Saint Petersburg State University
Faculty of Applied Mathematics
St. Petersburg, Russia

**Abstract**

The two-person differential simple pursuit game on the sphere is considered and the following problems are discussed. What is the fastest strategy for the pursuer? What does the Appolonius circle on the sphere look like? Is there universal pursuit strategy on the sphere? Can the evader avoid the collision with the pursuer? The well-known pursuit Π-strategy for the same game on the plane is extended to the sphere by two different ways. A variant of the extension is discussed, the differences between games on the plane and on the sphere are discovered, and the proofs of the properties of the strategy on the sphere are obtained.

## 1 Introduction

This paper is about one of the *parallel pursuit strategies* on the sphere. The parallel pursuit strategies on the sphere are the variants of the well-studied strategy of parallel pursuit (Π-strategy) for simple motion pursuit games on the plane, that has been investigated and described by L. A. Petrosjan in [1].

There are at least two ways to transfer the Π-strategy to the sphere. Each way produces a new strategy on the sphere and these strategies are not the same. Both strategies are named *parallel* because they are derived from the parallel pursuit strategy on the plane. One of them was denoted by the $\Pi_1$-strategy and the other one by the $\Pi_2$-strategy.

The full name of $\Pi_2$-strategy is the *relative parallel strategy* because this strategy conserves the bearing of the pursuer relative to the evader. For example, if the evader moves along the equator, and at some instant the direction from the evader to the pursuer is a direction to the north, then the pursuer will be north of the evader until the end of the game. The $\Pi_1$-strategy was named the *geodesic parallel strategy*. This strategy prescribes the pursuer to move along geodesic lines. For example, if the evader moves along a geodesic line all the time, then the pursuer moves along another geodesic line toward the intersection of these geodesic lines, such that both players come to the intersection point simultaneously.

The strict definition of the $\Pi_2$-strategy and the description of some its properties have been given in [2]. This paper contains similar information about the $\Pi_1$-strategy.

## 2 Definition of the Geodesic Parallel Strategy

Let two mobile points $P$ and $E$ lie on the unit radius sphere. Point $E$ is an evader, and point $P$ is a pursuer. The evader and the pursuer are the players. Both players can move on the sphere and can instantly change the direction of their motion. The absolute value of the evader's velocity is $\sigma$, the absolute value of the pursuer's velocity is $\rho$, and $\sigma < \rho$. Let us consider the game in which the goal of the pursuer is to capture the evader, and the goal of the evader is to avoid collision with the pursuer. Assume that the evader is captured by the pursuer at instant $t$ if the corresponding coordinates of both players are equal at this instant. During the game the players know their own coordinates and the coordinates of the other player at any instant. Moreover, the pursuer knows the evader's velocity vector all the time. Further, we will not say that the game takes place on the sphere exactly with unit radius, but it will be implied.

Denote by $\vec{v}_E$ the evader's velocity vector and by $\vec{v}_P$ the pursuer's velocity vector. Obviously these vectors are the tangent vectors to the sphere at the points $P$ and $E$, respectively. (See Figure 1.)

Associate these velocity vectors with the large circles $C_E$ and $C_P$, in such a way that the large circle $C_E$ contains point $E$ and the vector $\vec{v}_E$ is a tangent vector to $C_E$, respectively, the large circle $C_P$ contains point $P$ and the vector $\vec{v}_P$ is a tangent vector to $C_P$. (The radius of the large circle is equal to the sphere's radius.)

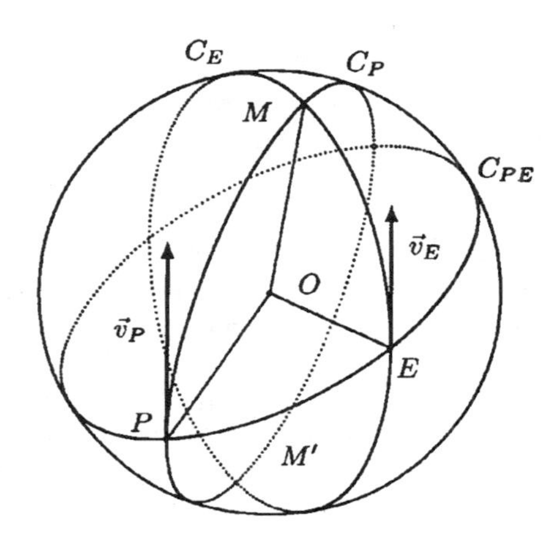


**Figure 1**: Players' position.

Construct the third large circle $C_{PE}$ through the points $P$ and $E$. This circle is the intersection of the sphere and a plane that contains three points $P$, $E$, and $O$, where $O$ is the center of the sphere. This plane can be constructed by only one way when the points $P$, $E$, and $O$ don't lie on the same line, i.e., points $P$ and $E$ are different and not diametrically opposite. If at least one of the large circles $C_E$ and $C_P$ does not coincide with the large circle $C_{PE}$, then $C_E$ and $C_P$ have exactly two common points.

As is known, large circles on the sphere are the geodesic lines. If an arc of some large circle has length not greater than $\pi$, then we will say that this arc is a *geodesic arc*. It is clear that such an arc is the shortest curve on the sphere that joins their boundary points and its length is equal to the *geodesic distance* between these points.

Let the evader move along $C_E$ and the pursuer move along $C_P$, where $C_P$ and $C_E$ do not coincide. Let, at some instant $t$, both players find themselves at one of two intersection points of $C_P$ and $C_E$. This meeting point was named *the center of pursuit*. We will denote it by $M$.

**Proposition 2.1.**

*For any players' position, when points $P$ and $E$ are not the same and not diametrically opposite, and for any direction of the evader's motion along the arbitrary large circle $C_E$ which does not coincide with the large circle $C_{PE}$, there exists a large circle $C_P$ such that the pursuer moving along $C_P$ will come to the point $M$ simultaneously with the evader, where point $M$ is one of the intersection points of $C_E$ and $C_P$, moreover the arcs $PE$, $PM$, and $EM$ constitute a spherical triangle i.e., the length of each arc is not greater than $\pi$ (See Figure 2.).*

**Proof.** Introduce designations:

$\psi$—the angle between arcs $PE$ and $EM$;
$\phi$—the angle between arcs $PE$ and $PM$;

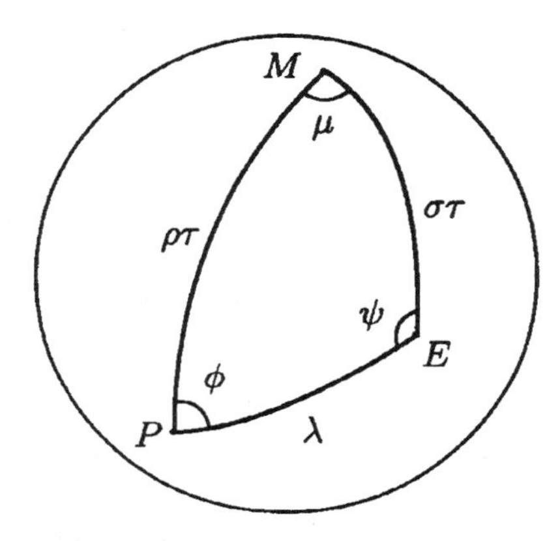


**Figure 2**: Spherical triangle.

$\mu$—the angle between arcs $EM$ and $PM$;
$\lambda$—the length of the arc $PE$; and
$\tau$—the time needed for players to go from the start points $P$ and $E$ to the common finishing point $M$. Consequently, the length of arc $EM$ equals $\sigma\tau$ and the length of arc $PM$ equals $\rho\tau$.

We suppose that all angles are between zero and $\pi$. Evidently, $\psi \in [0, \pi]$ corresponds to only half of the possible directions of the evader's motion, but if the direction of the evader's motion is such that angle $\psi < 0$, then the other angles $\phi$ and $\mu$ are less than zero too. Consequently, we can consider only absolute values of the angles.

It is easy to show that for any angle $\psi$ there exist angle $\phi$ and a segment of time $[0, \tau]$ such that Proposition 2.1 is true. The value of time $\tau$ is a root of the well-known equation for spherical triangles

$$\cos\rho\tau = \cos\lambda\cos\sigma\tau + \sin\lambda\sin\sigma\tau\cos\psi. \tag{1}$$

Transform (1) into the form

$$\cos\psi = \frac{\cos\rho\tau - \cos\lambda\cos\sigma\tau}{\sin\lambda\sin\sigma\tau}. \tag{2}$$

We can find $\tau$ from (2) for extreme values of $\psi$:

$$1 = \frac{\cos\rho\tau - \cos\lambda\cos\sigma\tau}{\sin\lambda\sin\sigma\tau}, \qquad -1 = \frac{\cos\rho\tau - \cos\lambda\cos\sigma\tau}{\sin\lambda\sin\sigma\tau}. \tag{3}$$

From the first equation of (3) we have $\cos\rho\tau = \cos(\lambda - \sigma\tau)$. Then, as far as we know, $\rho\tau < \pi$ and $\lambda - \sigma\tau < \pi$ and we have

$$\tau = \frac{\lambda}{\rho + \sigma}. \tag{4}$$

From the second equation of (3) we have $\cos\rho\tau = \cos(\lambda + \sigma\tau)$. This equation has two solutions. The first takes place when $\lambda + \sigma\tau \leq \pi$ and the second takes place when $\lambda + \sigma\tau > \pi$.

In the first case, when $\tau \leq (\pi - \lambda)/\sigma$, we have $\rho\tau = \lambda + \sigma\tau$, hence

$$\tau = \frac{\lambda}{\rho - \sigma}. \tag{5}$$

Obviously this solution needs a $\lambda$ that satisfies the inequality:
$\lambda/(\rho - \sigma) \leq (\pi - \lambda)/\sigma$, that is, $\lambda \leq \pi(\rho - \sigma)/\rho$.

In the second case, when $\tau > (\pi - \lambda)/\sigma$, we have $2\pi - \rho\tau = \lambda + \sigma\tau$, hence

$$\tau = \frac{2\pi - \lambda}{\rho + \sigma}. \tag{6}$$

Evidently for this solution it is necessary that $\lambda$ satisfies the inequality:
$(2\pi - \lambda)/(\rho + \sigma) > (\pi - \lambda)/\sigma$, that is, $\lambda > \pi(\rho - \sigma)/\rho$.

Thus we have found $\tau$ for extreme values of $\psi$ for all $\lambda$: $\lambda \in ]0, \pi[$. Obviously the right side of (2) is a continuous function of $\tau$. Consequently the right side of (2) increases from $-1$ to 1 while $\tau$ increases from the value given by (4) to the value given by (5) or (6). Thus for all admissible $\lambda$ and $\psi$ there exists at least one root $\tau$ of (1).

The proposition is proved. □

The other angles $\phi$ and $\mu$ can be obtained by the formulas

$$\cos\phi = \frac{\cos\sigma\tau - \cos\lambda\cos\rho\tau}{\sin\lambda\sin\rho\tau}, \tag{7}$$

$$\cos\mu = \frac{\cos\lambda - \cos\rho\tau\cos\sigma\tau}{\sin\rho\tau\sin\sigma\tau}. \tag{8}$$

The denominators of the right sides of (7) and (8) are not equal to zero because $0 < \sigma\tau < \rho\tau < \pi$ and $0 < \lambda < \pi$. Consequently, (7) and (8) are correct for all $\tau$.

Producing more detailed calculations, and after consideration of the singular cases when $\psi = 0$ and $\psi = \pi$, we find the correspondence between $\lambda$ and the range of variation of $\tau$:

$$\frac{\lambda}{\rho+\sigma} \le \tau \le \frac{\lambda}{\rho-\sigma}, \qquad \text{when} \quad 0 < \lambda \le \frac{\rho-\sigma}{\rho}\pi, \tag{9}$$

$$\frac{\lambda}{\rho+\sigma} \le \tau \le \frac{2\pi-\lambda}{\rho+\sigma}, \qquad \text{when} \quad \frac{\rho-\sigma}{\rho}\pi \le \lambda < \pi. \tag{10}$$

**Proposition 2.2.** *Equation* (1) *gives one-to-one correspondence between* $\psi$ *from* $[0, \pi]$ *and* $\tau$ *from* (9) *and* (10).

**Proof.** Since $\cos\psi$ is a continuous monotone function on $[0, \pi]$ and the right side of (2) is a continuous bounded function of $\tau$ on intervals given by (9)–(10), then it is sufficient to prove that the right side of (2) is a monotone function.

Find the derivative function from the right side of (2) with respect to $\tau$ and assume that the derivative is equal to zero. After calculations we obtain

$$\frac{\sigma}{\rho}\left(\frac{\cos\lambda - \cos\rho\tau\cos\sigma\tau}{\sin\rho\tau\sin\sigma\tau}\right) = 1.$$

From (8) it follows that this equation is equivalent to the following:

$$\cos\mu = \frac{\rho}{\sigma}.$$

The last expression has a contradiction because $\rho > \sigma > 0$ and $\cos\mu \le 1$. Consequently, the derivative function has no roots.

That proves the proposition. □

Since the relation between $\tau$ and $\psi$ is one-to-one, then we can consider the relation between $\phi$ and $\psi$ as a function $\phi(\psi)$. This is the function that defines the $\Pi_1$-strategy.

**Proposition 2.3.** (Fastest Property of the $\Pi_1$-Strategy). *If the evader moves along a fixed large circle with constant speed, then the pursuer can't capture him before the instant of time* $t = \tau$ *given by the* $\Pi_1$*-strategy.*

**Proof.** Assume that the opposite is true, and that the pursuer can capture the evader at some point $M_1$, that lies on the geodesic arc $EM$, where $E$ is a start position of the evader and $M$ is a center of pursuit given by the $\Pi_1$-strategy.

From this it follows that the pursuer can reach point $M$ from point $M_1$ faster than the evader because $\rho > \sigma$. Consequently, there exists some trajectory $PM_1M$, such that moving along this trajectory the pursuer can reach point $M$ in less time than if he moves along geodesic arc $PM$, where $P$ is a start point of the pursuer. But it is impossible because the geodesic arc is the shortest curve joining points $P$ and $M$.

The proposition is proved. □

**Corollary.** *Let the pursuer move along geodesic arc $PM_1$ and let the evader move along another geodesic arc $EM_1$. If the pursuer comes to point $M_1$ earlier than the evader, then the center of pursuit $M_*$ lies on geodesic arc $EM_1$ between points $E$ and $M_1$. If the pursuer comes to point $M_1$ later than the evader, then the center of pursuit $M_*$ lies on the extension of geodesic arc $EM_1$ so that point $M_1$ finds itself on geodesic arc $EM_*$ between points $E$ and $M_*$.*

## 3 Two-Jointed Outrunning

Let us consider the players in some initial position $P_0$, $E_0$ at the start time $t_0 = 0$. For any direction of the evader's motion given by angle $\psi \in [0, \pi]$ there exists a center of pursuit $M(\psi; P_0, E_0)$. For some initial position $P_0$, $E_0$, and some angle $\psi$, let us consider the center of pursuit $M$.

Let the evader move from $E_0$ to $M$, not along the geodesic arc $E_0M$ but along some two-jointed polygonal line consisting of two geodesic arcs $E_0E_\theta$ and $E_\theta M$. Assume that the geodesic distance between $E_0$ and $E_\theta$ is equal to $\sigma\theta$ and denote by $a$ the geodesic distance between $E_\theta$ and $M$.

Let the pursuer use the $\Pi_1$-strategy on the time interval $[0, \theta]$. Evidently, the pursuer's trajectory on this time interval is a geodesic arc $P_0P_\theta$. Let the pursuer move from point $P_\theta$ to point $M$ by the shortest way along geodesic arc $P_\theta M$. The geodesic distance between $P_0$ and $P_\theta$ is equal to $\rho\theta$. Denote by $b$ the geodesic distance between $P_\theta$ and $M$. We will find the conditions that are necessary for the evader to come to point $M$ earlier than the pursuer. For this we need

$$\frac{a}{\sigma} < \frac{b}{\rho}. \tag{11}$$

**Proposition 3.1.** *At point $M$ the evader can outrun the pursuer who uses the $\Pi_1$-strategy if the evader's deviation $\alpha$ from the geodesic arc $E_0M$ causes the pursuer's deviation $\beta$: $\beta > \alpha$ from the geodesic arc $P_0M$.*

**Proof.** Let $E(t)$ and $P(t)$ be the location points of the players at instant $t$. It is clear that

$$a = \sigma\tau - \sigma \int_0^\theta \cos\alpha(t)\,dt, \qquad b = \rho t - \rho \int_0^\theta \cos\beta(t)\,dt, \tag{12}$$

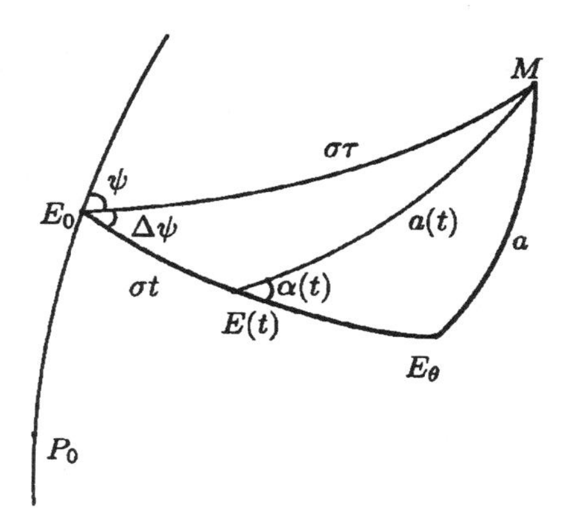


**Figure 3**: Two-jointed trajectory.

where $\alpha(t)$ is an angle between the geodesic arcs $E(t)E_\theta$ and $E(t)M$ (see Figure 3.); and $\beta(t)$ is an angle between the geodesic arcs $P(t)P_\theta$ and $P(t)M$. Substituting (12) into (11) we obtain

$$\int_0^\theta \big(\cos\beta(t) - \cos\alpha(t)\big)\, dt < 0. \tag{13}$$

Evidently, functions $\alpha(t)$ and $\beta(t)$ are continuous on $[0, \theta]$. From this it follows that if $\alpha(0) > \beta(0)$, then there exists such a $\delta > 0$ that $\alpha(t) < \beta(t)$ for any $t \in [0, \delta]$. Consequently, for all $\theta$: $\theta < \delta$ the inequality (13) is true.

We have thus shown that if the conditions of the proposition are satisfied, then there exists such a trajectory $E_0 E_\theta M$ that the evader moving along this trajectory can outrun the pursuer at point $M$ (See Figure 4.).

The proposition is proved. □

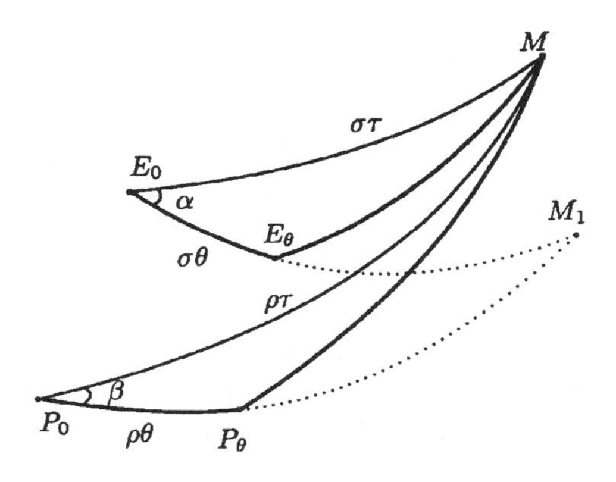


**Figure 4**: Two-jointed outrunning.

Let us find how $\beta$ depends on $\alpha$. As far as $\alpha = |\Delta\psi|, \beta = |\Delta\phi|$ we will obtain the relation between $\psi$ and $\phi$. Unfortunately, it is impossible to find function $\phi(\psi)$ in explicit form, but we can obtain the derivative $d\phi/d\psi$.

Indeed

$$\frac{d\cos\phi}{d\tau} = -\sin\phi \cdot \frac{d\phi}{d\tau}, \qquad \frac{d\cos\psi}{d\tau} = -\sin\psi \cdot \frac{d\psi}{d\tau},$$

hence

$$\frac{d\phi}{d\psi} = \frac{d\cos\phi}{d\tau}\left(\frac{d\cos\psi}{d\tau}\right)^{-1}\frac{\sin\psi}{\sin\phi}. \tag{14}$$

Differentiating (2) we obtain

$$\begin{aligned}\frac{d\cos\psi}{d\tau} &= \frac{\sin\lambda\sin\sigma\tau(-\rho\sin\rho\tau+\sigma\cos\lambda\sin\sigma\tau)-\sigma\sin\lambda\cos\sigma\tau(\cos\rho\tau-\cos\lambda\cos\sigma\tau)}{\sin^2\lambda\sin^2\sigma\tau} \\ &= \frac{-\rho\sin\lambda\sin\sigma\tau\sin\rho\tau+\sigma\sin\lambda(\cos\lambda-\cos\sigma\tau\cos\rho\tau)}{\sin^2\lambda\sin^2\sigma\tau}.\end{aligned}$$

Substituting (8) into the last formula

$$\frac{d\cos\psi}{d\tau} = \sin\sigma\tau\sin\rho\tau\sin\lambda \cdot \frac{-\rho+\sigma\cos\mu(\tau)}{\sin^2\lambda\sin^2\sigma\tau} = \frac{\sin\rho\tau}{\sin\sigma\tau}\cdot\frac{\sigma\cos\mu(\tau)-\rho}{\sin\lambda}. \tag{15}$$

In such a way we can find a formula for $d\cos\phi/d\tau$ from (7):

$$\frac{d\cos\phi}{d\tau} = \frac{\sin\sigma\tau}{\sin\rho\tau}\cdot\frac{\rho\cos\mu(\tau)-\sigma}{\sin\lambda}. \tag{16}$$

After substitution of (15) and (16) into (14) obtain

$$\frac{d\phi}{d\psi} = \frac{\rho\cos\mu(\tau)-\sigma}{\sigma\cos\mu(\tau)-\rho}\cdot\frac{\sin^2\sigma\tau\cdot\sin\psi}{\sin^2\rho\tau\cdot\sin\phi}.$$

By the well-known formula for sines of spherical triangles, and applying it to the spherical triangle $P_0E_0M$, we have

$$\frac{\sin\psi}{\sin\rho\tau} = \frac{\sin\phi}{\sin\sigma\tau}, \tag{17}$$

hence

$$\frac{d\phi}{d\psi} = \frac{\rho\cos\mu(\tau)-\sigma}{\sigma\cos\mu(\tau)-\rho}\cdot\frac{\sin\sigma\tau}{\sin\rho\tau}. \tag{18}$$

Let us investigate $d\phi/d\psi$. If, for a given start distance $\lambda_0$, the inequality $|d\phi/d\psi| > 1$ is true for some $\psi$, then there exists such a $|\Delta\phi|$ that the conditions of Proposition 3.1 are satisfied and the evader can outrun the pursuer at the center of pursuit $M(\psi)$.

We will look for such a $\lambda$, for which the opposite inequality is true:

$$\left|\frac{d\phi}{d\psi}\right| \le 1. \tag{19}$$

From (18) inequality (19) is equivalent to the following:

$$\left|\frac{\sigma - \rho\cos\mu}{\rho - \sigma\cos\mu}\right| \le \frac{\sin\rho\tau}{\sin\sigma\tau}. \tag{20}$$

The right-hand side of (20) is greater than 0 as far as $0 < \sigma\tau < \rho\tau \le \pi$. Let us investigate the right-hand side of (20) as a function with respect to $\tau$. Denote it by $S(\tau)$. When $\tau = 0$ we can assume that

$$S(\tau) = \lim_{\tau\to 0} S(\tau) = \lim_{\tau\to 0} \frac{\sin\rho\tau}{\sin\sigma\tau} = \frac{\rho}{\sigma}.$$

Let us find such intervals on which $S(\tau)$ is a monotone function. Solve the equation

$$\frac{d}{d\tau}S(\tau) = 0. \tag{21}$$

Find the derivative

$$\frac{d}{d\tau}S(\tau) = \frac{d}{d\tau}\left(\frac{\sin\rho\tau}{\sin\sigma\tau}\right) = \frac{-(\sigma+\rho)\sin(\rho-\sigma)\tau - (\sigma-\rho)\sin(\rho+\sigma)\tau}{2\sin^2\sigma\tau}. \tag{22}$$

From (21) and (22)

$$\frac{\sin(\rho-\sigma)\tau}{(\rho-\sigma)\tau} = \frac{\sin(\rho+\sigma)\tau}{(\rho+\sigma)\tau}. \tag{23}$$

Thus it is necessary to solve the equation $f(x_1) = f(x_2)$, where $f(x) = (\sin x)/x$, and $x_1 = (\rho-\sigma)\tau$, $x_2 = (\rho+\sigma)\tau$. Note that since $0 < \tau \le \pi/\rho$ and $0 < \sigma < \rho$, then $0 < x_1 < \pi$ and $0 < x_2 < 2\pi$.

Evidently, we can't obtain the roots of (23) in explicit form, but we can find the intervals such that each of them contains only a root. It is clear that if on some interval $I$ function $f(x)$ is monotonous, then there are no roots of (23) on $I$. Investigate the monotonicity of $f(x)$

$$\frac{d\,f(x)}{dx} = \frac{d}{dx}\left(\frac{\sin x}{x}\right) = \frac{x\cos x - \sin x}{x^2}.$$

This derivative function is equal to zero when $x$ satisfies the equation $\operatorname{tg} x - x = 0$, where $x \ne 0$, $x \ne \pi/2 + \pi k$, $k = 0, 1, 2, 3\ldots$.

Find the intervals that contain positive roots of $\operatorname{tg} x - x = 0$. The derivative of $\operatorname{tg} x - x$ is nonnegative for all $x$:

$$\frac{d}{d\tau}(\operatorname{tg} x - x) = \frac{1}{\cos^2 x} - 1 = \operatorname{tg}^2 x \ge 0,$$

and equal to zero only in points $\pi k$, where $k = 0, 1, 2, \ldots$, therefore we can say that function $\operatorname{tg} x - x$ monotonically increases on the domain of definition.

Evidently, function $\operatorname{tg} x - x$ monotonically increases from 0 to $+\infty$ on $]0, \pi/2[$. As far as this function is equal to zero when $x = 0$, then there are no roots when $x$: $0 < x < \pi/2$. On $]\pi/2, \pi[$ the function $\operatorname{tg} x - x$ also monotonically increases from $-\infty$ to $-\pi$, consequently, this function is negative for all $x \in ]\pi/2, \pi[$ and there are no roots on this interval. When $x \in ]\pi, 3\pi/2[$ the function $\operatorname{tg} x - x$ monotonically increases from $-\pi$ to $+\infty$, consequently, our function has exactly one root on this interval. There are no roots on $]3\pi/2, 2\pi[$ for the same reason as on $]\pi/2, \pi[$.

Since function $\operatorname{tg} x - x$ has only one root on the interval $[0, 2\pi]$ then its generating function $f(x) = \sin x/x$ changes their monotonicity on $[0, 2\pi]$ only once, and it happens between $\pi$ and $3\pi/2$. Consequently, function $f(x)$ is monotonous on $[0, \pi]$.

It is easy to show that $f(x)$ decreases on $[0, \pi]$. (For example, $f'(\pi/2) = -4/\pi^2 < 0$.) Hence for all $x_1, x_2 \in [0, \pi]$: $x_1 < x_2$ the inequality $f(x_1) > f(x_2)$ is fulfilled. Consequently, there are no such $x_1, x_2 \in [0, \pi]$ that the equation $f(x_1) = f(x_2)$ is true.

But this equation has no roots when $x_1 \in [0, \pi]$, and $x_2 \in [\pi, 2\pi]$ because $f(x_1) > 0$ and $f(x_2) < 0$. The single root exists only when $x_1$ and $x_2$ are in $[\pi, 2\pi]$, but as we showed before, $x_1$ is always less than $\pi$, therefore this root is not interesting.

Thus there are no such $\tau$ that (23) is true, consequently, $dS/d\tau$ is not equal to zero everywhere in the domain of the definition. From this it follows that the function $S(\tau)$ is monotonous. It is easy to show that $S(\tau)$ is monotonically decreasing. Calculate the sign of $S'(\tau)$ when $\tau = \pi/(2 \cdot (\rho - \sigma))$. In formula (22) the denominator of the fraction is always negative. After substitution of $\tau = \pi/(2 \cdot (\rho - \sigma))$ into the numerator we obtain

$$-(\rho + \sigma) - (\rho - \sigma) \sin\left(\frac{\rho + \sigma}{\rho - \sigma} \cdot \frac{\pi}{2}\right).$$

The value of the last expression is negative because $\rho + \sigma > \rho - \sigma > 0$ and the absolute value of sine is no greater than one.

Thus function $S(\tau)$ is monotonous on $[0, \pi/\rho]$ and decreases from $\rho/\sigma$ to 0.

Now let us consider the left side of inequality (20). The denominator is always positive because $\rho > \sigma$ and $|\cos \mu| \leq 1$. The nominator is positive when $\cos \mu < \sigma/\rho$ and negative when $\cos \mu > \sigma/\rho$. Therefore when $\mu$: $\arccos(\sigma/\rho) \leq \mu \leq \pi$, inequality (20) takes the form

$$\frac{\sigma - \rho \cos \mu}{\rho - \sigma \cos \mu} \leq \frac{\sin \rho\tau}{\sin \sigma\tau}.$$

Let us obtain the range of values of the left side. Denote the left side by $C^+(\mu)$. Investigate the monotonicity of $C^+(\mu)$. Take the derivative

$$\frac{dC^+(\mu)}{d\mu} = \frac{d}{d\mu}\left(\frac{\sigma - \rho \cos \mu}{\rho - \sigma \cos \mu}\right) = \frac{\rho^2 - \sigma^2}{(\rho - \sigma \cos \mu)^2} \sin \mu.$$

The last expression is equal to zero when $\mu = \pi$ and positive for all other $\mu$ from the considered interval $[\arccos(\sigma/\rho), \pi]$, because for such $\mu$ the sine is positive and the multiplier of the sine is positive too from $\rho > \sigma$. Thus function $C^+(\mu)$ increases from 0 to 1 while $\mu$ increases from $\arccos(\sigma/\rho)$ to $\pi$.

Consider the inequality (20) when $\sigma/\rho < \cos\mu \le 1$:

$$\frac{\rho\cos\mu - \sigma}{\rho - \sigma\cos\mu} \le \frac{\sin\rho\tau}{\sin\sigma\tau}.$$

Denote the left side by $C^-(\mu)$ and take the derivative

$$\frac{dC^-(\mu)}{d\mu} = \frac{d}{d\mu}\left(\frac{\rho\cos\mu - \sigma}{\rho - \sigma\cos\mu}\right) = -\frac{\rho^2 - \sigma^2}{(\rho - \sigma\cos\mu)^2} \cdot \sin\mu \le 0.$$

The last expression is equal to zero only when $\mu = 0$. Consequently, function $dC^-(\mu)$ decreases from 1 to 0 while $\mu$ increases from 0 to $\arccos(\sigma/\rho)$.

We have thus shown that

$$\left|\frac{\sigma - \rho\cos\mu(\tau)}{\rho - \sigma\cos\mu(\tau)}\right| \le 1 \qquad \text{for all} \quad \mu \in [0, \pi],$$

and the left side is equal to one only when $\mu = 0$ or $\mu = \pi$.

Now we will find such a $\tau$ that the right side of (20) is not less than one:

$$1 \le \frac{\sin\rho\tau}{\sin\sigma\tau}. \tag{24}$$

In that case, the inequality (20) is true.

Since the numerator and denominator in (24) are greater than zero, then (24) is equivalent to the following inequality: $\sin\rho\tau - \sin\sigma\tau \ge 0$ hence

$$\sin\frac{\rho - \sigma}{2}\tau \cdot \cos\frac{\rho + \sigma}{2}\tau \ge 0.$$

As far as $0 < \tau \le \pi/\rho$ then $\sin(\tau(\rho - \sigma)/2) > 0$. It remains to solve

$$\cos\frac{\rho + \sigma}{2}\tau \ge 0, \qquad \text{from this we obtain} \qquad \frac{\rho + \sigma}{2}\tau \le \frac{\pi}{2},$$

hence

$$\tau \le \frac{\pi}{\rho + \sigma}. \tag{25}$$

There are no other solutions because $0 < \tau \le \pi/\rho$, hence $0 < \tau(\rho + \sigma)/2 < \pi$. Evidently, on $[0, \pi]$ it is only solution.

We have thus shown that when $\tau$ is such that

$$\tau \in \left[0, \frac{\pi}{\rho + \sigma}\right], \tag{26}$$

then inequality (20) is true.

**Proposition 3.2.** *If at the start instant the geodesic distance $\lambda$ between players is less than or equal to $[(\rho - \sigma)/(\rho + \sigma)]\pi$, then for any $\psi \in [0, \pi]$ the evader* *moving along any two-jointed polygonal line cannot outrun the pursuer at the center of pursuit $M(\psi; \lambda_0)$.*

**Proof.** It is sufficient to prove that if $\lambda$ satisfies the conditions of the proposition then $\tau$ satisfies the inequality (26).

After substitution $[(\rho - \sigma)/(\rho + \sigma)]\pi$ into (9) we obtain

$$\frac{\rho - \sigma}{(\rho + \sigma)^2}\pi \leq \tau \leq \frac{\pi}{\rho + \sigma}.$$

That proves the proposition. □

**Notation.** It is easy to show that if $\lambda$: $[(\rho - \sigma)/(\rho + \sigma)]\pi < \lambda < \pi$, then there exists such a $\psi$ that $\tau(\psi)$ does not satisfy inequality (26).

## 4 Condition of the $\Pi_1$-Loops Inclusion

As was shown before in Proposition 2.1, for each players' position $P_0$, $E_0$ and for any angle $\psi$ of the evader's motion direction there exists a center of pursuit $M(\psi; P_0, E_0)$. Let us prove the following proposition:

**Proposition 4.1.** *Relation $M(\psi; P_0, E_0)$ gives a continuous closed curve on the sphere.*

**Proof.** Evidently, the dependency $\tau$ on $\psi$ given by equality (1) is a continuous function. Also we know that $\psi \in [-\pi, \pi]$ and $\tau(\psi) = \tau(-\psi)$, consequently, $\tau(\pi) = \tau(-\pi)$. As $\sigma\tau < \pi$ for all $\tau$ then each pair $\tau$, $\psi$ produces a center of pursuit $M(\psi)$ and each $M(\psi)$ corresponds to one and only one pair $\tau$, $\psi$. Thus the curve of all possible centers of pursuit is continuous, closed, and without self-intersections.

The proposition is proved. □

**Definition.** Let the continuous closed curve, consisting of possible centers of pursuit for some initial players' position $P_0$, $E_0$, be named by the *strategy loop* or $\Pi_1$-*loop*. Denote it by $L_{\Pi_1}(P_0, E_0)$.

The $\Pi_1$-loop separates the sphere on two sides. By *scope of the* $\Pi_1$-*loop* we will mean that part of the sphere that contains the evader's location point $E_0$. Denote the scope of $L_{\Pi_1}(P_0, E_0)$ by $D_{\Pi_1}$. When we say about the scope corresponding to the players position $E(t)$, $P(t)$ at the instant of time $t$ we will write $D_{\Pi_1}(t)$.

Assume that the evader moves along the geodesic arc $E_\theta M$ to the center of pursuit $M$ from the time instant $\theta$ . We will say that $L_{\Pi_1}(\theta)$ is a *seizing loop* if for all possible $M_\theta$ and for all $t$: $\theta < t$ the following expression $D_{\Pi_1}(\theta) \supset D_{\Pi_1}(t)$ is true.

**Proposition 4.2.** *If the players' initial position $E_0 P_0$ is such that for all $M$: $M \in L_{\Pi_1}(0)$ there does not exist a two-jointed polygonal line $E_0 E_\theta M$, along which the evader can outrun the pursuer at point $M$, then the $\Pi_1$-loop $L_{\Pi_1}(0)$ is a seizing loop.*

**Proof.** Let the evader move along some geodesic arc $E_0M$, where $M$ is the center of pursuit. Suppose that the proposition is not true. In that case, there exists such an instant $t$: $t = \theta > 0$, that $D_{\Pi_1}(\theta) \not\subset D_{\Pi_1}(0)$. Consequently, there exists such a point $M_1 \in L_{\Pi_1}(\theta)$, that $M_1 \notin D_{\Pi_1}(0)$. From this it follows that the geodesic arc $E(\theta)M_1$ certainly intersects the $\Pi_1$-loop $L_{\Pi_1}(0)$ at some point $M_0$. As $M_0$ lies on the geodesic arc $E(\theta)M_1$, and $M_1$ is a center of pursuit when $t = \theta$, then from the fastest property of the $\Pi_1$-strategy it follows that the evader, moving from $E(\theta)$ to $M_1$ along the geodesic arc will came to a point $M_0$ before his capture by the pursuer even if the last one moves from $P(\theta)$ to $M_0$ the shortest way.

We have thus shown that if the evader moves from $E_0$ to $E(\theta)$, and after that to $M_0$, then he achieves $M_0$ in less time than the pursuer, who uses the $\Pi_1$-strategy, does. But $M_0 \in L_{\Pi_1}(0)$, consequently, the evader moving along the geodesic arcs $E_0E(\theta)$ and $E(\theta)M_0$, can reach $M_0$ before the pursuer will find himself there. Thus we have a contradiction with the conditions of the proposition, consequently, our assumption is not true.

The proposition is proved. □

**Proposition 4.3.** *The geodesic arc joining points $E_0$ and $M$: $M \in L_{\Pi_1}(0)$, does not contain other points from $L_{\Pi_1}(0)$ except for $M$.*

**Proof.** The proof directly follows from the fastest property. □

**Corollary.** *If some point $A$ lies on the geodesic arc that joins $E(t)$ and point $M_t$: $M_t \in L_{\Pi_1}(t)$, then $A \in D_{\Pi_1}(t)$. Moreover, $A \in L_{\Pi_1}(t)$ if and only if $A = M_t$.*

**Proposition 4.4.** *If for the initial position $P_0$, $E_0$ there exists at least one point $M$ on the $\Pi_1$-loop $L_{\Pi_1}(0)$, such that the evader moving along some two-jointed polygonal line $E_0E_\theta M$ can outrun the pursuer at point $M$, then $L_{\Pi_1}(0)$ is not a seizing loop.*

**Proof.** When the evader moves along the geodesic arc $E(\theta)M$, the pursuer using the $\Pi_1$-strategy moves along the geodesic arc $P(\theta)M_\theta$, where $M_\theta$ is a center of pursuit corresponding to the players' position $P(\theta)$, $E(\theta)$ and the direction of the evader's motion along the geodesic arc $E(\theta)M$. From the corollary after Proposition 2.3 point $M$ lies on the geodesic arc $E(\theta)M_\theta$. Since $M_\theta \in L_{\Pi_1}(\theta)$, then from the corollary after Proposition 4.3 we have $M \in D_{\Pi_1}(\theta)$ and $M \notin L_{\Pi_1}(\theta)$. Consequently, there exists such an $\varepsilon$-neighborhood with center $M$, that all points of the $\varepsilon$-neighborhood belong to $D_{\Pi_1}(\theta)$.

From the other side, $M \in L_{\Pi_1}(0)$ and $L_{\Pi_1}(0)$ is a boundary of $D_{\Pi_1}(0)$, consequently, any $\varepsilon$-neighborhood with the center at point $M$ contains such a point that do not belong to $D_{\Pi_1}(0)$. Thus, there exist such points from an $\varepsilon$-neighborhood with center $M$, that belong to $D_{\Pi_1}(\theta)$ and don't belong to $D_{\Pi_1}(0)$.

From this it follows: $D_{\Pi_1}(\theta) \not\subset D_{\Pi_1}(0)$, consequently, $L_{\Pi_1}(0)$ is not a seizing loop.

The proposition is proved. □

**Definition.** We will say that *the condition of the* $\Pi_1$*-loops inclusion is completed* for some players' initial position, if for any evaders' motion the following expression $D_{\Pi_1}(t_1) \supset D_{\Pi_1}(t_2)$ is true for all pairs $t_1, t_2$: $t_1 < t_2$.

**Notation.** If the players' initial position is such that for any evaders' trajectory and for any time instant $t$ the $\Pi_1$-loop $L_{\Pi_1}(t)$ is a seizing loop, then the condition of the $\Pi_1$-loops inclusion is completed. It directly follows from the definitions of the $\Pi_1$-loops inclusion and the seizing loop.

**Proposition 4.5.** *If from the start instant* $t = 0$ *the evader moves from the initial location point* $E_0$ *along the geodesic arc* $E_0M$ *to the center of pursuit* $M$ *and reaches* $M$ *at instant* $t = \tau$ *simultaneously with the pursuer that moves from the initial location point* $P_0$ *and uses the* $\Pi_1$*-strategy, then if* $\tau$ *satisfies inequality* (25)*, that is,* $\tau \leq \pi/(\rho + \sigma)$*, then the geodesic distance* $\lambda(t)$ *between the players monotonically decreases during their motion from initial position to meeting point* $M$.

**Proof.** Let us consider two large circles on the sphere. Denote the angle between the planes of these circles by $\mu$ and let $\mu \in [0, \pi]$. Denote the intersection points of circles by $M$ and $M'$. We will consider only two halves of our circles. These half-circles construct two angles in points $M$ and $M'$ and the values of these angles are equal to $\mu$. Locate on the first half-circle the movable point $P(s)$ and locate on the other half-circle a movable point $E(s)$, where $s$: $s \in [0, \pi/\rho]$. Let $P$ and $E$ depend on $s$ in such a way that the geodesic distance between $P$ and $M$ is equal to $\rho s$, and the geodesic distance between $E$ and $M$ is equal to $\sigma s$.

Since the geodesic arcs $E(s)M$ and $P(s)M$ are no longer than $\pi$, then these arcs compose with the geodesic arc $E(s)P(s)$ the spherical triangle. Denote by $\lambda(s)$ the geodesic distance between $P(s)$ and $E(s)$.

Examine the dependence of $\lambda$ on $s$. By the cosine formula of spherical triangles, applying it to the spherical triangle $P(s)E(s)M$, we have

$$\cos\lambda = \cos\rho s \cos\sigma s + \sin\rho s \sin\sigma s \cos\mu. \tag{27}$$

This expression can be transformed to the following:

$$\cos\lambda = \frac{1}{2}(1 - \cos\mu)\cos(\rho+\sigma)s + \frac{1}{2}(1+\cos\mu)\cos(\rho-\sigma)s.$$

Find the intervals of monotonicity for function $\lambda(s)$. Take the derivative of $\cos\lambda$ with respect to $s$

$$\frac{d\cos\lambda}{ds} = -\frac{\rho+\sigma}{2}(1-\cos\mu)\sin(\rho+\sigma)s - \frac{\rho-\sigma}{2}(1+\cos\mu)\sin(\rho-\sigma)s.$$

Equate the derivative to 0. We obtain

$$\frac{\rho+\sigma}{2}(\cos\mu - 1)\sin(\rho+\sigma)s = \frac{\rho-\sigma}{2}(1+\cos\mu)\sin(\rho-\sigma)s.$$

The trivial root $s = 0$ is not interesting because it corresponds to the extreme value of $s$.

When $\mu = 0$ there is a root $s = \pi/(\rho - \sigma)$, but this root is not from the admissible interval $[0, \pi/\rho]$.

When $\mu = \pi$, there is a root $s = \pi/(\rho + \sigma)$.

Look for the other roots when $0 < \mu < \pi$. Transform the last formula into the form

$$\frac{\sin(\rho + \sigma)s}{\sin(\rho - \sigma)s} = \frac{\rho - \sigma}{\rho + \sigma} \cdot \frac{\cos \mu + 1}{\cos \mu - 1}. \tag{28}$$

The first fraction on the right side of (28) is positive as $0 < \sigma < \rho$, the second one is negative because the nominator is positive and the denominator is negative, consequently, the right side of (28) is negative.

In the left side of (28) the denominator is always positive when $0 < s \leq \pi/\rho$, as the argument of sine in the denominator for such $s$ is less than $\pi$ and greater than 0. Consequently, to satisfy (28) the nominator on the left side should be negative. Hence the argument of sine in the nominator should be greater than $\pi$ and less than $2\pi$. Thus the derivative $d \cos \lambda/ds$ has roots on the interval $]0, \pi/\rho[$ only when $(\rho + \sigma)s > \pi$. Consequently, when

$$0 < s \leq \frac{\pi}{\rho + \sigma}, \tag{29}$$

function $\cos \lambda(s)$ is monotonous. Show that this function is monotonically decreasing on interval (29). Substitute $s = \pi/(\rho + \sigma)$ into $d \cos \lambda/ds$:

$$\left.\frac{d \cos \lambda}{ds}\right|_{s=\pi/(\rho+\sigma)} = -\frac{\rho - \sigma}{2}(1 + \cos \mu) \sin \frac{\rho - \sigma}{\rho + \sigma}\pi.$$

The last expression is negative when $0 \leq \mu < \pi$.

When $\mu = \pi$ the derivative

$$\left.\frac{d \cos \lambda}{ds}\right|_{\mu=\pi} = -(\rho + \sigma) \sin(\rho + \sigma)s$$

takes negative values when $s \in ]0, \pi/(\rho + \sigma)[$. Since $d \cos \lambda(s)/ds = -\sin \lambda \cdot d\lambda/ds$ and $\sin \lambda > 0$ when $0 < \lambda < \pi$, then the derivatives of $\lambda$ and $\cos \lambda$ with respect to $s$ have opposite signs, hence $\lambda(s)$ increases when $s$ increases from 0 to $\pi/(\rho+\sigma)$. Consequently, for all $\mu$ the inequality $\lambda(s_1) \geq \lambda(s_2)$ is true, where $s_1$ and $s_2$ are such that $s_1, s_2 \in [0, \pi/(\rho + \sigma)]$ and $s_1 > s_2$.

Consider the players at the initial position $E_0$, $P_0$. If the evader chooses such a geodesic line to move along that time interval $\tau$, in which he will be captured by the evader, is not greater than $\pi/(\rho + \sigma)$, then assuming $s = \tau - t$ we obtain $\pi/(\rho + \sigma) \geq s \geq 0$ and $s$ monotonically decreases when $t$ increases.

Since for $s \in [0, \pi/(\rho + \sigma)]$ the geodesic distance $\lambda(s)$ between the players monotonically increases, then when $t$ increases from 0 to $\tau$, where $\tau \leq \pi/(\rho+\sigma)$, value of $\lambda\big(s(t)\big)$ monotonically decreases.

The proposition is proved. □

**Corollary.** *If the start players' position is such that the geodesic distance $\lambda$ between players satisfies the inequality $\lambda < \pi(\rho - \sigma)/(\rho + \sigma)$, then $\lambda$ decreases while the evader moves along any geodesic arc.*

**Proposition 4.6.** *If at the start instant of time the geodesic distance $\lambda$ between players is such that*

$$\lambda \leq \frac{\rho - \sigma}{\rho + \sigma} \pi, \tag{30}$$

*and if the evader moves along any polygonal line consisting of geodesic arcs when the pursuer uses the $\Pi$-strategy, then for any time instant $t$ the corresponding $\Pi$-loop $L_{\Pi_1}(t)$ is a seizing loop for any possible center of pursuit.*

**Proof.** From the corollary after Proposition 4.5 it follows that the geodesic distance $\lambda(t)$ between players is less than $\pi(\rho - \sigma)/(\rho + \sigma)$ all the time during the game. From the Proposition 3.2 it follows that for any $t$ there are no such $M \in L_{\Pi_1}(t)$ that the evader, moving along any two-jointed polygonal line, can outrun the pursuer at $M$. From the Proposition 4.2 it follows that for all $t$ the $\Pi_1$-loop $L_{\Pi_1}(t)$ is a seizing loop for all $M \in L_{\Pi_1}(t)$.

The proposition is proved. □

**Proposition 4.7.** (Inclusion Theorem). *The inequality* (30) *is a condition of the* $\Pi_1$-*loop inclusion.*

**Proof.** In Proposition 4.6 we have proved the $\Pi_1$-loop inclusion for any polygonal evader's trajectory. Evidently, all continuous trajectories may be approximated by polygonal trajectories with any precision. This fact gives hope that the inequality (30) is a condition of the $\Pi_1$-loop inclusion not only for polygonal trajectories but for continuous ones as well.

Unfortunately, a full strict proof of this theorem obtained by the author is too long to be included in this paper, it needs about 25 pages. You can find an exact proof in the second paper of [3]. □

**Corollary.** *If $\lambda$ satisfies the inequality* (30) *then the pursuer using the* $\Pi_1$-*strategy can capture the evader in time* $t_*$:

$$t_* \leq \frac{\lambda}{(\rho - \sigma)}.$$

## 5 Summary

We discussed only a small part of the parallel strategy properties. But we have discussed the main properties of the $\Pi_1$-strategy. They are the fastest property and the condition of the $\Pi_1$-loops inclusion. On the plane, the condition of inclusion is always complete. On the sphere, this condition is complete only if the distance between players is not large. Indeed the influence of the sphere's curvature to the game is small when $\lambda$ is very small relative to the radius of the sphere. When

$\lambda > \pi(\rho - \sigma)/(\rho + \sigma)$ the curvature influence has the other quality and the game receives a few new features.

## REFERENCES

[1] Petrosjan, L. A. *Differential Games of Pursuit.* World Scientific, Singapore, 1993.

[2] Kovshov, A. M. *The Simple Pursuit by a Few Objects on the Multidimensional Sphere. International Year-Book of Game Theory and Applications.* Nova Science, New-York, vol. **2**, pp. 27–36, 1996.

[3] Kovshov, A. M. Parallel Strategies in Pursuit Game on the Sphere—Candidate Dissertation (Ph.D.) on Physics and Mathematics. St. Petersburg State University, Registered by Russian Center of Scientific and Technical Information (VNTC) 960827 Entrance number k/8047. St. Petersburg, 122 pp., 1996. (Russian)

[4] Kovshov A. M. Two-Person Pursuit Game on the Half-Sphere—International Conference on Interval and Computer-Algebraic Methods in Science and Engineering INTERVAL'94, Abstracts. St. Petersburg, pp. 148–149, 1994.

# Real-Time Collision Avoidance: Differential Game, Numerical Solution, and Synthesis of Strategies

Rainer Lachner, Michael H. Breitner, and H. Josef Pesch
Technische Universität Clausthal
Institut für Mathematik
Clausthal-Zellerfeld, Germany

**Abstract**

Contemporary developments of on-board systems for automatic or semi-automatic driving include car collision avoidance systems. For this purpose two approaches based on pursuit-evasion differential games are compared. On a freeway a correct driver (evader) is faced with a wrong-way driver (pursuer), i.e., a person driving on the wrong side of the road. The correct driver tries to avoid collision against all possible maneuvers of the wrong-way driver and additionally tries to stay on the freeway. The representation of the optimal collision avoidance behavior along many optimal paths is used to synthesize an optimal collision avoidance strategy by means of neural networks. Examples of simulations that prove a satisfactory performance of the real-time collision avoidance scheme are presented.

## 1 Introduction

More than 3,000 cases where people drive on the wrong side of the road are officially registered on German freeways every year, but the actual number is estimated to be three to five times higher. On average, 75 people are injured and 16 are killed per 100 cases, see [12], [13], and [19]. Present on-board systems control, among others, the engine (automotive speed control), braking system (automotive anti-lock braking), and steering system (power steering). Current position and velocity vectors of the car and all neighboring cars on the freeway can be measured by on-board sensors using radar or ultrasonic waves. Alternatively, freeway mounted systems that transmit their measurements to on-board systems of the cars around will be used in the near future; see [5]. Because of drunkenness, fatigue, or panic, future maneuvers of wrong-way drivers are generally unpredictable. Because of this uncertainty on-board collision avoidance systems must:

[1]This paper is dedicated to Professor Dr. Dr. h. c. Roland Bulirsch on the occasion of his 65th birthday.

(1) detect reliably and as early as possible other cars on a collision course;
(2) warn drivers immediately;
(3) compute optimal steering and velocity control strategies in real-time against all possible maneuvers of all neighboring cars;
(4) advise drivers on monitors or wind-screen head-up displays or adopt steering and velocity control of the cars.

## 2 Hierarchical Formulation Based on the Game of Two Cars

We restrict ourselves to a collision avoidance problem for two cars: one is driven by the correct driver C, the other one by the wrong-way driver W. For this collision avoidance problem the kinematic equations can be sufficiently realistically modeled by

$$\dot{x}_W = v_W \sin\phi_W, \tag{1}$$

$$\dot{x}_C = v_C \sin\phi_C, \tag{2}$$

$$\dot{y} = v_W \cos\phi_W - v_C \cos\phi_C, \tag{3}$$

$$\dot{v}_C = b_C \eta_C, \tag{4}$$

$$\dot{\phi}_W = w_W(v_W) u_W, \tag{5}$$

$$\dot{\phi}_C = w_C(v_C) u_C, \tag{6}$$

see Figure 1. The subscripts W and C of the notations refer to the wrong-way driver W and the correct driver C, respectively. The independent variable $t$ denotes time, the state variables $x_W, x_C$ denote the distances of W and C from the left-hand side of the freeway, and $y$ denotes the distance between W and C orthogonal to the $x$-direction. The state variables $\phi_W$, $\phi_C$ denote the driving directions of W and C and $v_W$ and $v_C$ are the velocities of W and C. The control variables $u_W$, $u_C$ denote the turn rates of W and C, and $\eta_C$, the velocity change rate. Without oversimplification $v_W$ is taken as constant and the maximum angular velocities $w_W(v_W)$ and $w_C(v_C)$ are prescribed depending on the type of car. The kinematic constraints of bounded radii of curvature are taken into account by the control variable inequality constraints

$$-1 \le u_W \le +1, \tag{7}$$

$$-1 \le u_C \le +1. \tag{8}$$

Steering $u_W = \pm 1$ and $u_C = \pm 1$ means executing an extreme right/left turn for W and C, respectively. The kinematic constraint of acceleration and deceleration for C is taken into account by the control variable inequality constraint

$$-0.1 \le \eta_C \le +1, \tag{9}$$

where $\eta_C = 1$ means maximum braking, and $\eta_C = -0.1$ means maximum acceleration. In order to stay on the freeway, both W and C must obey the state

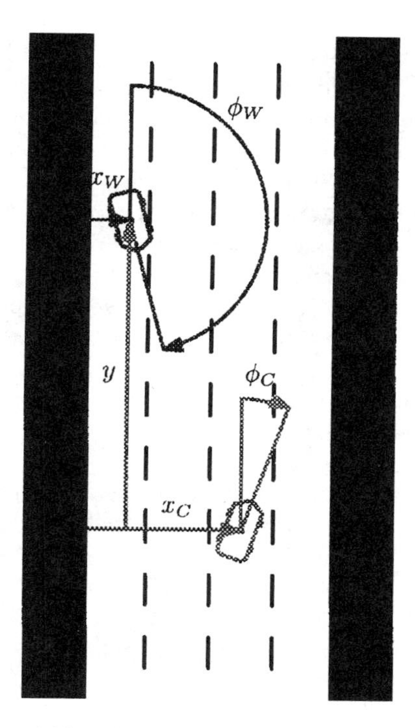


**Figure 1**: State variables of the collision avoidance problem for two cars.

constraints

$$0\text{m} \le x_W \le 15\text{m}, \tag{10}$$

$$0\text{m} \le x_C \le 15\text{m}. \tag{11}$$

The goal of the correct driver C is to avoid collision against all possible maneuvers of W. It can be formulated as

$$\min_{\gamma_W \in \Gamma_W} \max_{\gamma_C \in \Gamma_C} \min_{t_0 \le t < \infty} d(z(t)) \quad \text{w.r.t.} \quad z(t_0) = z_0 \tag{12}$$

with the vector $z := (x_W, x_C, y, v_C, \phi_W, \phi_C)^\top$ of state variables, with (feedback) strategies $\gamma_W(z)$ and $\gamma_C(z)$, with sets $\Gamma_W$ and $\Gamma_C$ of all admissible strategies, and with Euclidean distance $d(z)$ between W and C. The control variables are determined by $u_W(z) := \gamma_W(z)$ and $(u_C(z), \eta_C(z))^\top := \gamma_C(z)$. The strategy $\gamma_W$ is admissible, if the constraints (7) and (10) are fulfilled for all $t \in [t_0, t_f]$, and the strategy $\gamma_C$ is admissible, if the constraints (8), (9), and (11) are fulfilled for all $t \in [t_0, t_f]$. The collision avoidance maneuver starts at the time $t_0$ at $z(t_0) = z_0$, when the correct driver C notices W. The collision avoidance maneuver ends at the terminal time $t_f < \infty$ when the minimum distance between W and C is attained.

The idea of analyzing collision avoidance problems in a differential game framework dates back to Isaacs' work [16] and [17] in the early 1950s, unclassified and published in the mid 1960s [18]. The theory of collision avoidance, e.g., between cars, aircraft, or ships, can be embedded in the theory of pursuit and evasion. Omitting state constraints (10) and (11) and assuming constant velocity $v_C$, the collision avoidance problem under consideration is the famous *game of two cars* introduced in [17] and [18]. If additionally the correct driver C needs not obey the constraint of bounded curvature, i.e., $u_C \in ]-\infty, \infty[$, the famous *homicidal chauffeur game* arises. Although the essential number of state variables is only three for the game of two cars and only two for the homicidal chauffeur game, their full solution is extraordinarily baffling. The state-space is cut by diverse singular manifolds, e.g., barrier, universal, and dispersal manifolds, which subdivide the state-space into several subregions; see Figure 2. Neither the game of two cars nor the homicidal chauffeur game can be solved fully, i.e., optimal strategies $\gamma_W^*(z)$ and $\gamma_C^*(z)$ cannot be calculated explicitly for all $z$, for all velocities $v_W$ and $v_C$, and for all maximum angular velocities $w_W$ and $w_C$. Nevertheless various planar collision avoidance problems have been investigated as the game of two cars or the homicidal chauffeur game over the last three decades, see, e.g., [3], [11], [24]–[28], [31], [33], [36], and [37]. Note that state-space constraints such as (10) and (11) have not been taken into account in those publications.

We first focus on the game of two cars. Given the velocities $v_W$ and $v_C$ the maximum angular velocities $w_W(v_W)$ and $w_C(v_C)$ are prescribed, too. In the re-

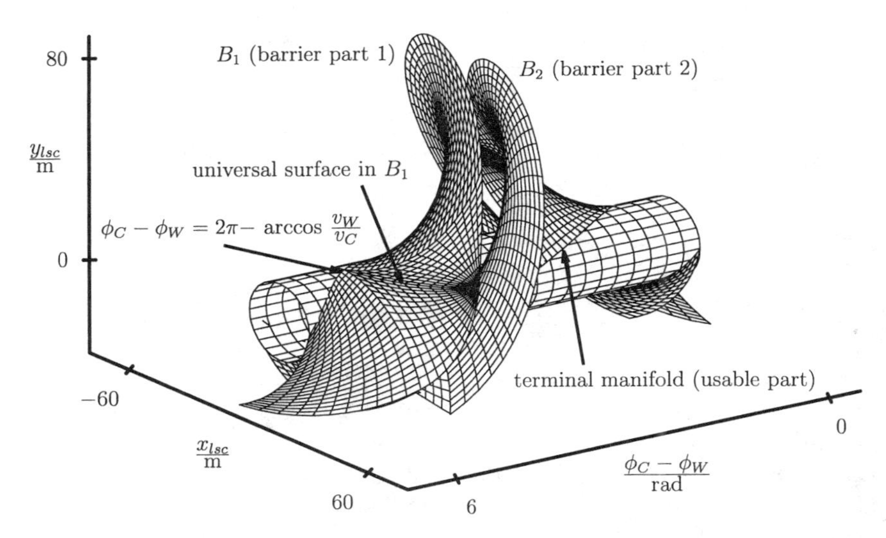


**Figure 2**: Barrier manifold for the classical game of two cars, (see [[18]], pp. 239–243). For velocities $v_W \equiv 15\text{ms}^{-1}$ and $v_C \equiv 30\text{ms}^{-1}$ and a time $t \in [t_f - 2, t_f]$ all $z(t)$ with a guaranteed minimum distance $d^* \equiv 15$m span the depicted manifold. A universal manifold, the terminal manifold (cylinder), and the boundary of the terminal manifold's usable part are marked.

duced three-dimensional state-space (here the state variables are the line of sight coordinates $x_{lsc}$ and $y_{lsc}$ of W and $\phi_C - \phi_W$), semipermeable hypermanifolds can be constructed by means of the barrier concept ([17], [18] and [29]), see Figure 2. These barriers contain all scenarios leading to the same guaranteed minimum distance

$$d^*(z) := \min_{\gamma_W \in \Gamma_W} \max_{\gamma_C \in \Gamma_C} \min_{t_0 \leq t < \infty} d(z(t)), \tag{13}$$

if both drivers use their optimal strategy $\gamma_W^*(z) = u_W^*(z)$ and $\gamma_C^*(z) = (u_C^*(z), \eta_C^*(z))^\top$. When starting the game at $z(t_0) = z_0$ the guaranteed minimum distance $d^*(z)$ is constant along the resulting trajectory. C can usually increase $d^*(z)$, if W uses a nonoptimal strategy. Vice versa, C usually has to accept a decrease of $d^*(z)$ if he uses a nonoptimal strategy. In order to obey the state constraint (11), the game of two cars is embedded into a hierarchical formulation:

1. The correct driver C stays on the freeway as long as there holds $d^*(z(t)) > 5\text{m}$ for the game of two cars. Collision is impossible in this case (best outcome for C).
2. If there holds $d^*(z(t)) = 5\text{m}$, the correct driver C has to use an optimal strategy such that $\dot{d}^*(z(t)) \equiv 0\text{ms}^{-1}$ and avoid collision in any case. In this case, it is better to leave the freeway than to risk a frontal collision (second-best outcome for C).
3. If there holds $d^*(z(t)) < 5\text{m}$, the correct driver C stays on the freeway since collision avoidance cannot be guaranteed (worst outcome for C).

The correct driver C simply slows down if a lower $v_C$ provides a larger $d^*(z)$ and vice versa accelerates if a higher $v_C$ provides a larger $d^*(z)$. Note that nonoptimal driving of W can significantly improve the outcome for C. The hierarchical formulation based on the game of two cars is discussed in detail in [9] and [34]. Again neither an optimal strategy for W nor for C can be computed in closed form. Optimal strategies are only known along optimal trajectories and therefore $\gamma_C^*(z) = (u_C^*(z), \eta_C^*(z))^\top$ has to be synthesized in the relevant parts of the state-space. For this purpose the approximation properties of neural networks are exploited; see [4], [14], [15], [23], and [38]. Since optimal strategies are generally discontinuous due to the linear appearance of the control variables in the kinematic equations, it turns out to be better to approximate the gradient $d_z^*(z)$ of the value function $d^*(z)$ instead of the optimal strategies. The approximation $\tilde{\gamma}_C^*(z)$ of an optimal strategy can then be obtained by substituting the approximated gradient $\tilde{d}_z^*(z)$ into the optimal control laws owing to the minimax principle

$$\tilde{\gamma}_C^*(z) = (\tilde{u}_C^*(z), \tilde{\eta}_C^*(z))^\top = \left(u_C^*\left(z, \tilde{d}_z^*(z)\right), \eta_C^*\left(z, \tilde{d}_z^*(z)\right)\right)^\top; \tag{14}$$

see [6], [9], and [30]. For the neural network, feedforward multilayered perceptrons are used; see [1], [2], [10], [32], [39], [40], and Figure 3. Neural network topology and weights specify the neural mapping. A priori the topology, i.e., type

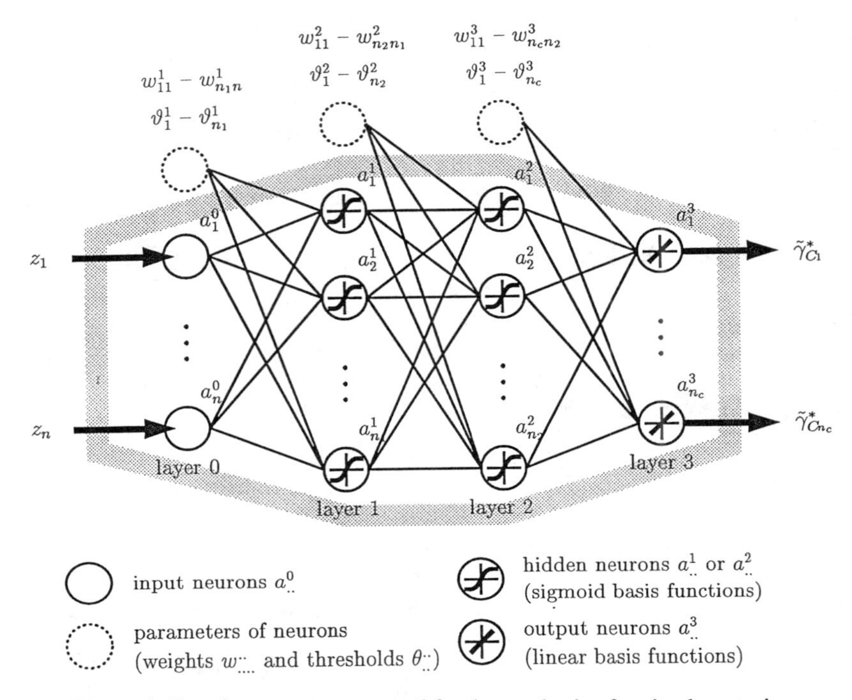


**Figure 3**: Four-layer perceptron used for the synthesis of optimal strategies.

and number of basis functions, must be prescribed suitably. We consider exemplarily the synthesis of the mapping $z \mapsto \gamma_C^*(z)$ with a four layer perceptron; see Figure 3. Dependent on the vector $\boldsymbol{w} := (w_{11}^1, \ldots, w_{n_c n_2}^3, \vartheta_1^1, \ldots, \vartheta_{n_c}^3)^\top$ the neural network $NN : z \mapsto \tilde{\gamma}_C{}^*(z; \boldsymbol{w})$ maps the state $z$ to the approximation $\tilde{\gamma}_C{}^*(z; \boldsymbol{w})$ of an optimal strategy $\gamma_C^*(z)$. The multivariate mapping $NN$ is composed of the mappings

$$a_k^0 = x_k, k = 1, 2, \ldots, n, \tag{15}$$

$$a_k^1 = \mathrm{sd}\left(\sum_{l=1}^{n} w_{kl}^1 a_l^l - \vartheta_k^1\right), \qquad k = 1, 2, \ldots, n_1, \tag{16}$$

$$a_k^2 = \mathrm{sd}\left(\sum_{l=1}^{n_1} w_{kl}^2 a_l^1 - \vartheta_k^2\right), \qquad k = 1, 2, \ldots, n_2, \tag{17}$$

$$\tilde{\gamma}_{Ck}^* = \sum_{l=1}^{n_2} w_{kl}^3 a_l^2 - \vartheta_k^3, \qquad k = 1, 2, \ldots, n_c. \tag{18}$$

The sigmoid basis function sd must be strictly monotonically increasing and must satisfy $\mathrm{sd}(\pm\infty) = \pm 1$. For sd usually hyperbolic tangent tanh or normalized inverse tangent $\frac{2}{\pi}$ arctan are used. The data set $(z_k, \gamma_C^*(z_k))$, $k = 1, 2, \ldots$, at points on many optimal paths is divided into a training data set with $k \in I_T$ and a cross-validation data set with $k \in I_V$ ($I_T \cap I_V = \emptyset$). Using a differentiable norm $\|.\|$ the approximation errors

$$\varepsilon_T(w) : = \sum_{k \in I_T} \left\| \tilde{\gamma}_C^*(z_k; w) - \gamma_C^*(z_k) \right\| , \tag{19}$$

$$\varepsilon_V(w) : = \sum_{k \in I_V} \left\| \tilde{\gamma}_C^*(z_k; w) - \gamma_C^*(z_k) \right\| \tag{20}$$

enable training and validation of the perceptron. After initializing $w$ the approximation error $\varepsilon_T(w)$ of the perceptron has to be iteratively improved. The iteration steps are usually determined by a supervised back-propagation learning method, see [1], [2], [10], [32], [39], and [40]. The training terminates if the validation error $\varepsilon_V(w)$ does not decrease any longer.

For the collision avoidance problem investigated here, the Stuttgart Neural Network Simulator (SNNS) has been used, see [40] and [41]. Three-layer perceptrons turn out to be appropriate. The inherent symmetry of the problem has been exploited. Simulations against various maneuvers of the wrong-way driver show a satisfactory performance of the method; see [9], [34], and Figure 4. This concept has been patented by the German patent office; see [34].

## 3 Advanced Formulation

The approach described in the preceding section uses a hierarchical formulation based on the game of two cars. The many optimal paths needed for the synthesis can be calculated analytically. However, for a more sophisticated modeling optimal paths can be obtained only numerically by the method of characteristics and the numerical solution of the associated multipoint boundary value problems. For an advanced approach, the state constraint (11) is obeyed explicitly in the differential game. Then, the reduction of the state-space dimension from six to three as in the classical game of two cars is no longer possible. Furthermore, the correct driver C can accelerate and decelerate, which adds the state variable $v_C$ and the control variable $\eta_C$. Hence, we end up with a modified differential game.

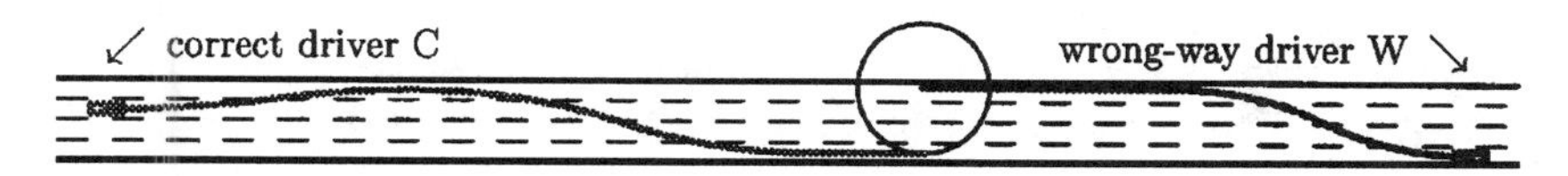


**Figure 4**: Simulation with the real-time collision avoidance method. Since the wrong-way driver W is initially on the right lane, the correct driver C moves to the left lane. Due to the change of W to the left lane, C is forced to steer to the right lane to pass W with the maximum achievable distance. Note that slowing down is not always optimal for C.

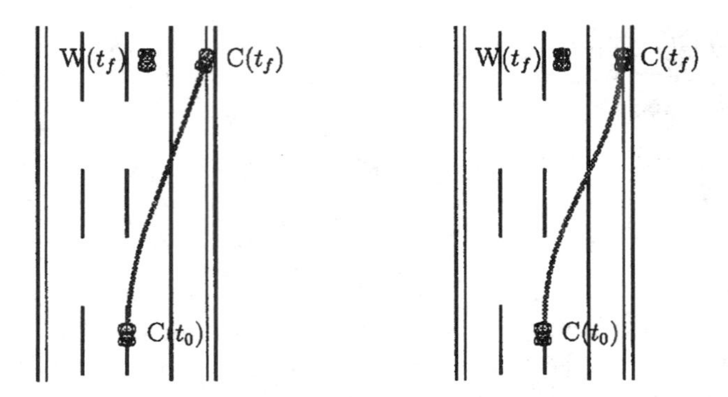


**Figure 5**: Qualitative shape of constrained paths owing to the right boundary of the freeway. Left: Using the constraint (11). Right: Using a modified constraint which also takes into account the driving direction.

### 3.1 Modification of the Constraint (11)

Using the state constraint (11), the correct driver C would always hit the constraint at terminal time and would not stay on the freeway thereafter. To avoid such unrealistic situations, optimal paths should qualitatively look like the right-hand-side figure in Figure 5. Therefore, a new first-order state constraint is introduced. This new constraint, more stringent than (11), must also take into account the driving direction $\phi_C$. If the state $z$ lies on this new constraint for an arbitrary $t_1$, the correct driver must turn sharply to the left ($u_C = -1$) and brake fully ($\eta_C = 1$) in order to stay on the freeway for all $t > t_1$ as long as $\phi_C \geq 0$. Depending on every admissible $x_C$ and $v_C$ and for $u_C = -1$ and $\eta_C = 1$ there exists a unique maximal angle $\phi_C \geq 0$, such that the resulting path touches the right boundary of the freeway. The triple $(x_C, v_C, \phi_C)$ satisfies a relation of the form $S(x_C, v_C, \phi_C) = 0$, which is obtained by an analytical retrogressive integration of the kinematic equations for $v_C$, $\phi_C$ and $x_C$, with respect to the initial values $\phi_C = 0$ and $x_C = 15$m. The new first-order state constraint has the form

$$\begin{aligned} S(x_C, v_C, \phi_C) &= h(x_C, v_C) + F(\phi_C) \\ &= (x_C - a)\frac{k_1{}^2 + k_2{}^2}{v_C{}^2} + \left[k_2\left(\cos\phi_C - \exp\left(-\frac{k_1}{k_2}\phi_C\right)\right) - k_1 \sin\phi_C\right] \leq 0 \end{aligned} \tag{21}$$

with $k_1 = 2b_C\eta_{C,\max}$ and $k_2 = \mu_0 g u_{C,\min}$. Since a rigorous treatment of higher-order state constraints is still pending for differential games, the constraint (21) is replaced by the even more stringent control constraint

$$\begin{aligned} &C(x_C, v_C, \phi_C, u_C, \eta_C) \\ &\quad = h(x_C, v_C)\left[1 + (r-1)\frac{(u_C - u_{C,\min}) - (\eta_C - \eta_{C,\max})}{(u_{C,\max} - u_{C,\min}) - (\eta_{C,\min} - \eta_{C,\max})}\right] \\ &\qquad + F(\phi_C) \leq 0. \end{aligned} \tag{22}$$

It is derived from (21) by the demand that the new control constraint should become active earlier than (21) depending on the actual values of the control variables. For more details, see [20]. Approximation accuracy and stiffness of the constraint (22) can be adjusted properly by the choice of $r \in [0, 1[$. Note that (21) and (22) represent only the right-hand side of the freeway. The left one can be treated analogously. Furthermore, the state constraint (10) for the wrong-way driver W can be omitted since W tries to hit the correct driver C and therefore will stay on the freeway.

## 3.2 Method of Characteristics

Putting together all those ingredients, one obtains a pursuit-evasion game with six state variables, one control variable for W, two control variables for C, and two symmetric control constraints. Because of the inherent symmetry of the game, only initial $x_C$-values larger than 7.5 m have to be considered. It is well known that optimal strategies can be obtained in principle by integrating the Isaacs' equation

$$V_z \dot{z}(z, u_W^*(z, V_z), u_C^*(z, V_z), \eta_C^*(z, V_z)) \equiv 0; \tag{23}$$

see [3], [6], [18], [31], and [35]. The Isaacs' equation is a nonlinear first-order partial differential equation for the value function $V(z)$. Here there holds

$$V(z) := \min_{\gamma_W \in \Gamma_W} \max_{\gamma_C \in \Gamma_C} \left\| x_W(t_f) - x_C(t_f) \right\| . \tag{24}$$

The terminal time $t_f$ is determined by $y(t) = 0$ for the first time. Unfortunately, for complex differential games an integration of the Isaacs' equation is impossible because of the high state-space dimension and the various singular surfaces. However, by the method of characteristics a system of ordinary differential equations can be obtained along characteristics of the Isaacs' equation. Assuming $V$ exists and is twice piecewise continuously differentiable, the gradient $V_z(z)$ of the value function satisfies

$$\dot{V}_z = -V_z \cdot \frac{\partial}{\partial z} \dot{z} \left(z, u_W^*(z, V_z), u_C^*(z, V_z), \eta_C^*(z, V_z)\right) \tag{25}$$

along a characteristic. Boundary conditions for $V_z(t_f)$ are obtained by parameterizing the end condition $y(t_f) = 0$ and differentiating the value function $V_z(z(t_f))$ as a function of those parameters; see [6]–[8], [18], [20], [24], and [31]. Finally a boundary value problem is obtained with the kinematic equations (1)–(6) and the differential equations (25) for the gradient $V_z(z)$, in detail,

$$\dot{V}_{x_W} = 0, \tag{26}$$

$$\dot{V}_{x_C} = 0, \tag{27}$$

$$\dot{V}_y = 0, \tag{28}$$

$$\dot{V}_{v_C} = V_y \cos\phi_C - V_{x_C} \sin\phi_C - V_{\phi_C} \frac{\partial w_C(v_C)}{\partial v_C} u_C, \tag{29}$$

$$\dot{V}_{\phi_W} = v_W(V_y \sin\phi_W - V_{x_W}\cos\phi_W), \tag{30}$$
$$\dot{V}_{\phi_C} = -v_C(V_y \sin\phi_C + V_{x_C}\cos\phi_C). \tag{31}$$

If the control constraint (22) is not active, the optimal controls $u_W^*, u_C^*$, and $\eta_C^*$ are obtained by the sign of their switching function $V_{\phi_W}$, $V_{\phi_C}$ and $V_{v_C}$, respectively. If (22) is active, $u_C^*$ and $\eta_C^*$ are determined from (22) and the sign of the switching function $V_{\phi_C} w_C(v_C) + V_{v_C} b_C$. Note that in the latter case terms containing the partial derivatives of $u_C^*$ and $\eta_C^*$ with respect to the state variables have to be added to the differential equations (25). Boundary conditions are the initial conditions $z(t_0) = z_0$ and the aforementioned final conditions, in detail,

$$V_{x_W}(t_f) = 2(x_W(t_f) - x_C(t_f)), \tag{32}$$
$$V_{x_C}(t_f) = 2(x_C(t_f) - x_W(t_f)), \tag{33}$$
$$V_y(t_f) = 2(x_C(t_f) - x_W(t_f))\frac{v_W \sin\phi_W(t_f) - v_C(t_f)\sin\phi_C(t_f)}{v_W \cos\phi_W(t_f) - v_C(t_f)\cos\phi_C(t_f)}, \tag{34}$$
$$V_{v_C}(t_f) = 0, \tag{35}$$
$$V_{\phi_W}(t_f) = 0, \tag{36}$$
$$V_{\phi_C}(t_f) = 0. \tag{37}$$

Bang bang as well as singular arcs of the controls are determined by the switching functions. Transition points from inactive to active control constraint (22) are determined by means of continuity conditions for the controls. All these interior point conditions give rise to a multipoint boundary value problem. After the numerical solution of the multipoint boundary value problems a realization of the optimal strategies is known in open-loop form along optimal paths. In order to gain a sufficient approximation of the optimal strategies, sufficiently many optimal paths have to be computed. These results are then used for a synthesis.

Obviously for many scenarios, e.g., for a large initial distance $y(t_0)$, the wrong-way driver W can enforce capture against all possible maneuvers of the correct driver C. In that case, an optimal strategy $\gamma_W(z)$ is not uniquely determined. From a numerical point of view, the multipoint boundary value problems due to $V(z_0) \leq 10^{-4}$m are extremely ill conditioned and cannot be solved at all. Therefore, the first task one is faced with is to single out all unusable initial scenarios. Afterward all usable initial scenarios $z_0$ with $V(z_0) > 10^{-4}\text{m}^2$ are considered as initial points $z_0$ for the pursuit-evasion game. The region of initial scenarios is discretized by a six-dimensional grid. Each grid point serves as an initial state for which a multipoint boundary value problem has to be solved numerically. Here the multiple shooting method ([6]–[8], [22], and [21]) has been employed. Additional necessary optimality conditions such as sign conditions must be checked a posteriori. The following table lists all the combinations of the initial values $z_0$ in the state-space:

| | | | | | | | | | | | |
|---|---|---|---|---|---|---|---|---|---|---|---|
| $x_W(t_0) \in$ | 0.75 | 4.125 | 7.5 | 10.875 | 14.25 | | | | | | m |
| $x_C(t_0) \in$ | 8.7 | 11.1 | 13.5 | | | | | | | | m |
| $y(t_0) \in$ | 0 | 15 | 30 | 45 | 60 | 75 | 90 | 105 | 120 | 135 | m |
| $v_C(t_0) \in$ | 20 | 25 | 30 | 35 | | | | | | | m/s |
| $\phi_W(t_0) \in$ | $\pi - 0.3$ | $\pi - 0.18$ | $\pi - 0.06$ | $\pi + 0.06$ | $\pi + 0.18$ | $\pi + 0.3$ | | | | | rad |
| $\phi_C(t_0) \in$ | $-0.3$ | $-0.2$ | $-0.1$ | 0 | 0.1 | 0.2 | 0.3 | | | | rad |

Even for a coarse discretization the number of initial states $z_0$ is considerably large.

### Numerical Solution of the Multipoint Boundary Value Problems

Because of the large number of grid points it would be very time-consuming and tedious to compute the optimal solutions starting at the grid points one by one manually. Therefore a semiautomatic procedure was developed by which at least all optimal paths with the same switching structure can be computed. Since the kernel of the multiple shooting method is a Newton method, a sufficiently accurate initial guess at the solution is needed. If at least one optimal path for one of the grid points is already known, this path can be used as an initial guess for a trajectory starting at a neighboring grid point. By checking all the neighbors and saving the solution in case of convergence and optimality, one can recursively proceed and cover connected regions of the state-space with the same switching structure; see Figure 6 for a flow chart of the procedure. Unfortunately, it turned out that the pursuit-evasion game under investigation possesses a variety of almost 60 switching structures. Figure 7 depicts a two-dimensional cross section through a part of the state-space showing some of the regions associated with different optimal control laws. For details, see [[20]]. Analogously to the first approach a data set $(z_k, \gamma_W^*(z_k), \gamma_C^*(z_k), V(z_k), V_z(z_k))$, $k = 1, 2, \ldots$, at suitable points along many computed optimal paths is generated. In order to achieve a homogeneous distribution of approximation points in the relevant part of the state-space, it is advisable to consider only initial points and intersection points of the trajectory with singular surfaces as approximation points. Note that at the terminal manifold $y(t_f) = 0$ optimal controls and value functions are known analytically. Hence, approximation points in the terminal manifold can be obtained without numerical integration.

### Synthesis of an Optimal Strategy with Neural Networks

We now follow the steps for the first approach described in the previous section and also in [30]. Three layer perceptrons turn out to be convenient and successful for the advanced approach, too. By means of the data set $(z_k, \gamma_W^*(z_k), \gamma_C^*(z_k), V(z_k), V_z(z_k))$, $k = 1, 2, \ldots$, a neural network is trained, representing the mapping from the state-space into the space spanned by the gradients of the value function. Evaluating the neural network and substituting the output into the control laws

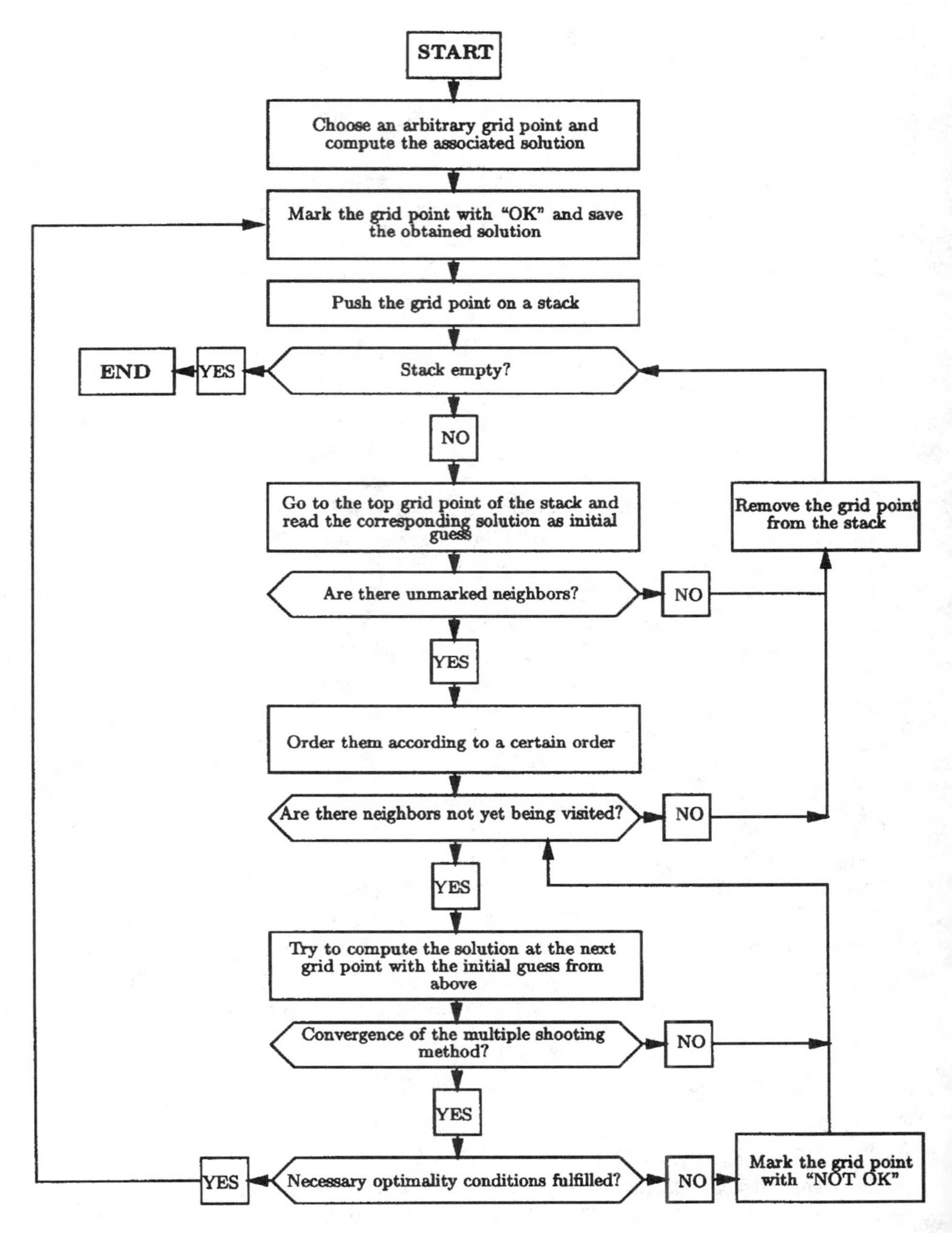


**Figure 6**: Flowchart for the recursive computation of solutions with the same switching structure starting at neighboring grid points.

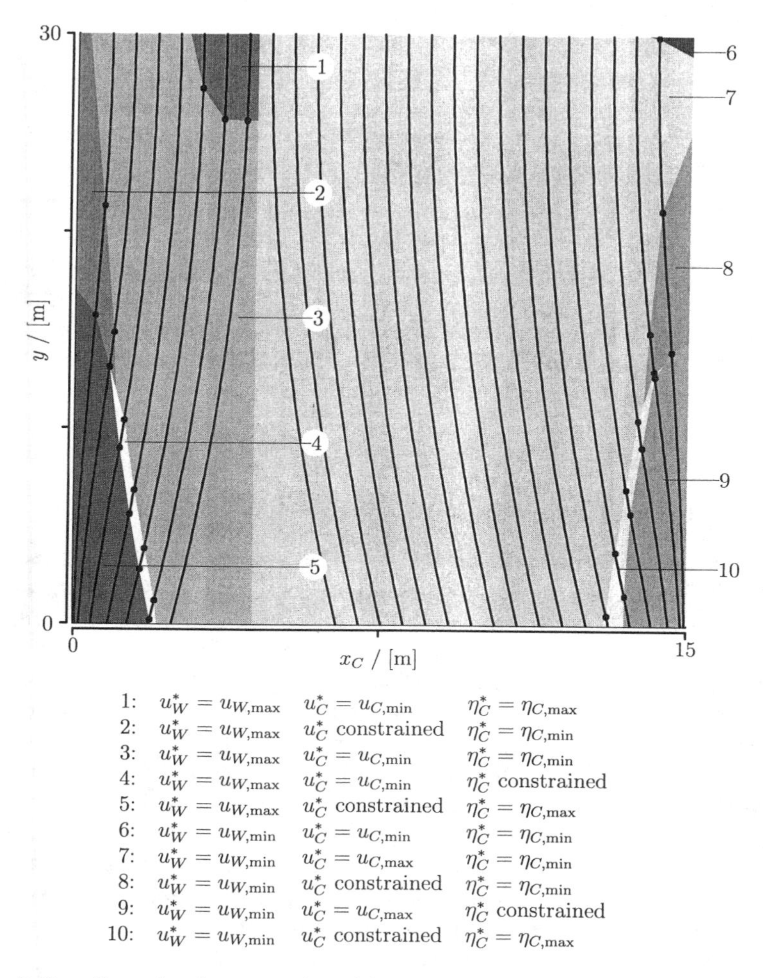


**Figure 7**: Two-dimensional cross section of the state-space for $x_W(t_0) = 3.375$ m, different $x_C(t_0)$-values, $y(t_0) = 30$ m, $v_C(t_0) = 25\,\mathrm{ms}^{-1}$, $\phi_W(t_0) = 3.08159$ rad, and $\phi_C(t_0) = 0$ rad with projections of optimal solutions (black lines). Dots on the paths mark the transition points at the singular surfaces. Shaded areas are regions with different optimal control laws.

obtained by the minimax principle (14), a real-time capable approximation $\tilde{\gamma}_C^*(z)$ of an optimal collision avoidance strategy for the correct driver C is obtained. As mentioned in Section 2 the choice of the image space of the neural mapping selected here is more advantageous than the choice of the control space as image space. The following diagram, Figure 8, sketches the loop for the approximation of optimal steering $u_C^*(z)$ and velocity control $\eta_C^*(z)$ by the neural network:

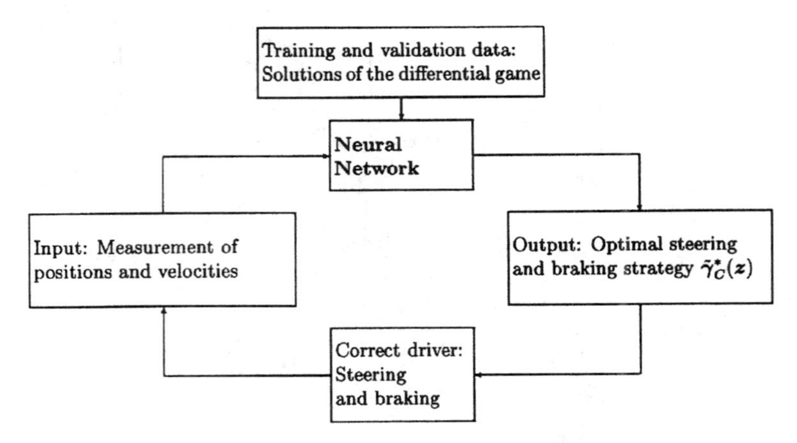


**Figure 8**: Optimal control synthesis with the neural network.

## Treatment of the Region of Unusable Initial Scenarios

The neural network can approximate an optimal strategy $\gamma_C^*(z)$ only in the region of usable initial scenarios since training data $(z_k, \gamma_W^*(z_k), \gamma_C^*(z_k), V(z_k), V_z(z_k))$, $k = 1, 2, \ldots$, exist only there. In the region of unusable initial scenarios with $V(z_0) \leq 10^{-4}$m there exists no strategy for the correct driver C, which guarantees collision avoidance. But it is unlikely that a wrong-way driver really tries to collide with another driver frontally. Therefore, in most of the cases the state enters the region of usable initial scenarios sooner or later, and the neural approximation $\tilde{\gamma}_C^*(z)$ becomes valid. In the region of unusable initial scenarios a heuristic strategy is used, e.g., keeping velocity $v_C$ constant, i.e., $\eta_C \equiv 0$, and performing a turn to $\phi_C = 0$. The on-board collision avoidance system must be able to decide reliably whether the actual scenario is situated in the region of usable or unusable initial scenarios. From the aforementioned preliminary investigations it is known that scenarios $z_0$ with large initial $y$-values are in the region of unusable initial scenarios, whereas scenarios with terminal value $y = 0$ m and $x_C \neq x_W$ are in the region of usable initial scenarios. In most cases there exists only one $y(t_0)$-value for a usable initial combination of the other five coordinates such that the resulting state $z(t_0)$ is on the separating manifold between the set of usable and unusable initial scenarios. This $y(t_0)$-value can be estimated numerically by means of the multiple shooting method if the initial condition $y(t_0) = y_0$ is replaced by the final condition $\|x_W(t_f) - x_C(t_f)\| = 10^{-4}$. Then for the separating manifold another neural network can be trained to decide whether the actual scenario is in the region of usable initial scenarios or not.

## Simulations

Many simulations have been performed to check real-time capability and reliability of the neural collision avoidance scheme. In these simulations, a network-controlled correct driver C tries to avoid a collision against a wrong-way driver following a prescribed path. Simulations against various maneuvers of the wrong-way driver show satisfactory performances of the two methods. Figures 9 and 10, for example, show initial scenarios in the region of usable initial values. Figure 9 depicts the maneuvers on the freeway. In the first simulation, the wrong-way driver W obviously behaves nonoptimally. As soon as this is noticed by the neural network, the correct driver C switches from a sharp right turn to a sharp left turn. In the last phase of the simulation the left counterpart of the constraint (22) becomes active, forcing C to turn right again in order to stay on the freeway. In the second simulation, C is faced with a wrong-way driver who behaves optimally except during the first half-second. The initial conditions enforce a sharp right turn. As soon as the state reaches a universal type singular surface, C has to move straight on for a short duration of time. This singular subarc is followed by another very short subarc of sharp right turn until the constraint (22) becomes active. C first satisfies this constraint by decelerating and then by turning to the left.

Figure 10 shows the corresponding histories of the control variables. The prescribed control of W is marked by thin dark gray lines. Owing to the linear appearance of the control variables in the dynamic system (1)–(6), the optimal controls must be either bang-bang or singular on unconstrained subarcs. With this knowledge, the synthesized controls for C (light graylines in Figure 10) are obtained by the signs of their network-approximated switching functions.

## 4 Conclusion

In the present paper two methods have been presented for the synthesis of feedback strategies for pursuit-evasion games. These methods are based both on the open-loop representations of the optimal feedback strategies along many optimal trajectories. These open-loop representations can in general be computed numerically by boundary value problem solvers, such as multiple shooting methods, even if thousands of trajectories due to various initial conditions are to be computed. However, these time-consuming computations are still difficult tasks. Parallel computing might be very promising here since the presented semiautomatic procedure of computation of the solutions of those boundary value problems requires only little communication between the different processors.

After these computations the values of the state and the open-loop control variables as well as the values of the value function of the game and the gradient of the value function w.r.t. the state variables are known. These values along the optimal trajectories can then be used to train neural networks, i.e., to determine the optimal weights occurring in the multivariate mappings represented by the networks. These networks are designed to represent here either the mapping from the state-space to the control space or from the state-space to the gradient of the value function. In the latter case, the output of the neural networks has yet to be plugged into the minimax principle to obtain the controls. After the training of

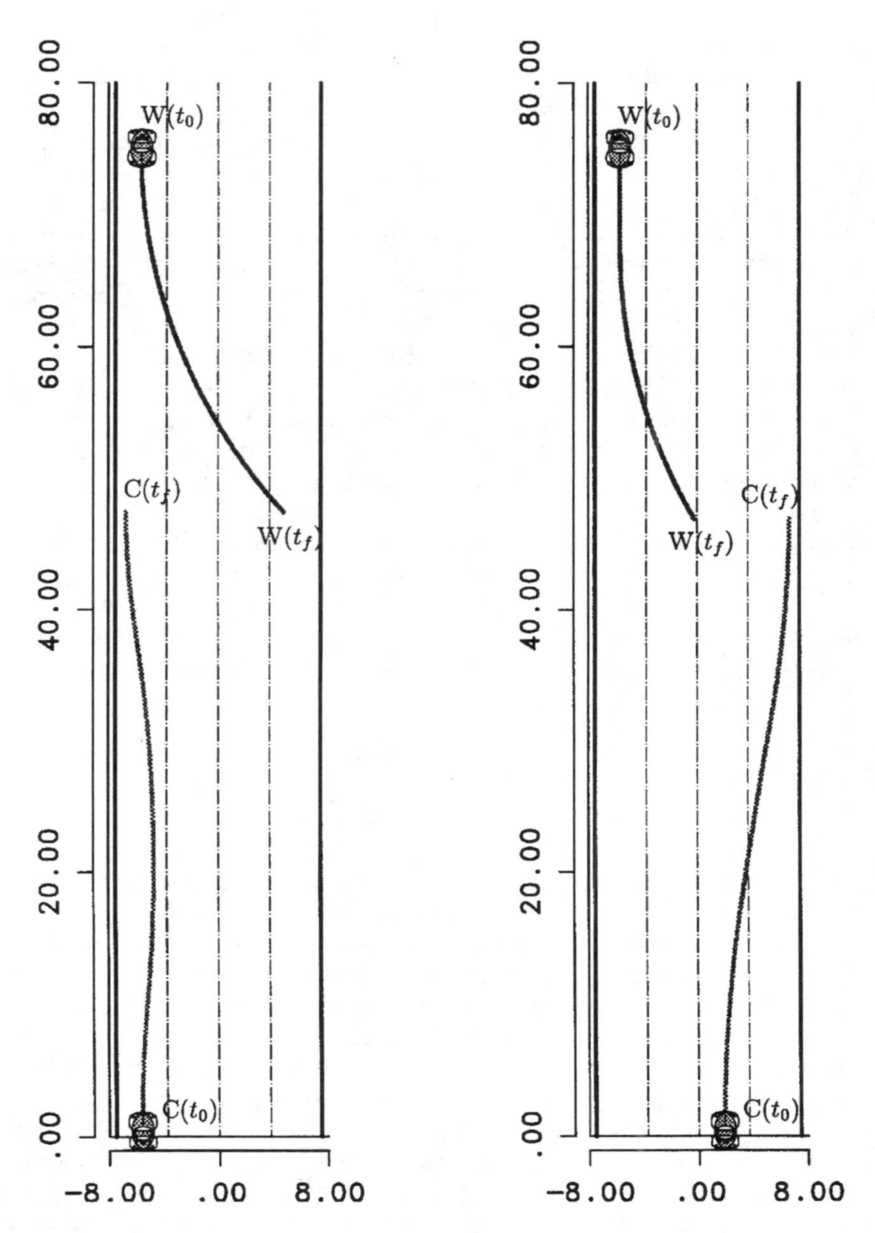


**Figure 9**: Simulations with the advanced real-time collision avoidance system (Wrong-way driver W: dark gray. Correct driver C: light gray).

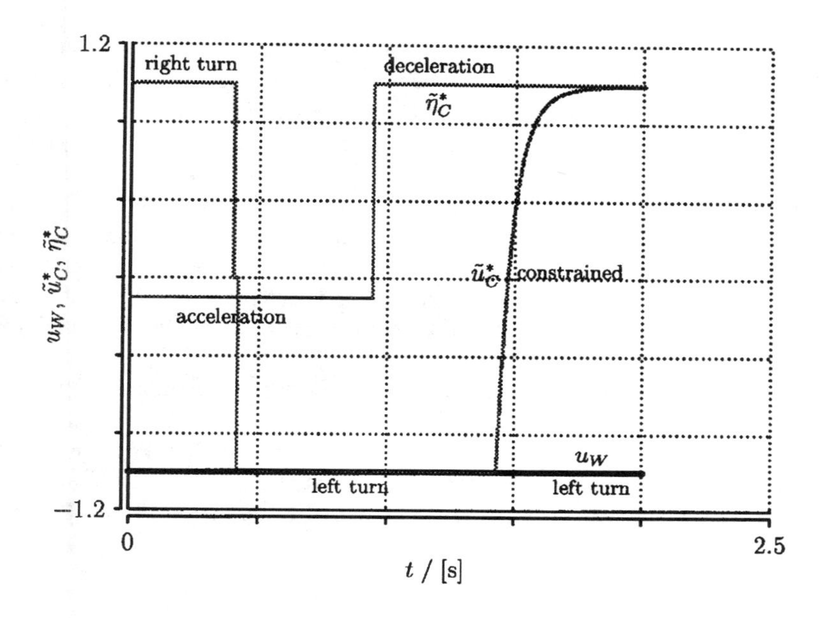


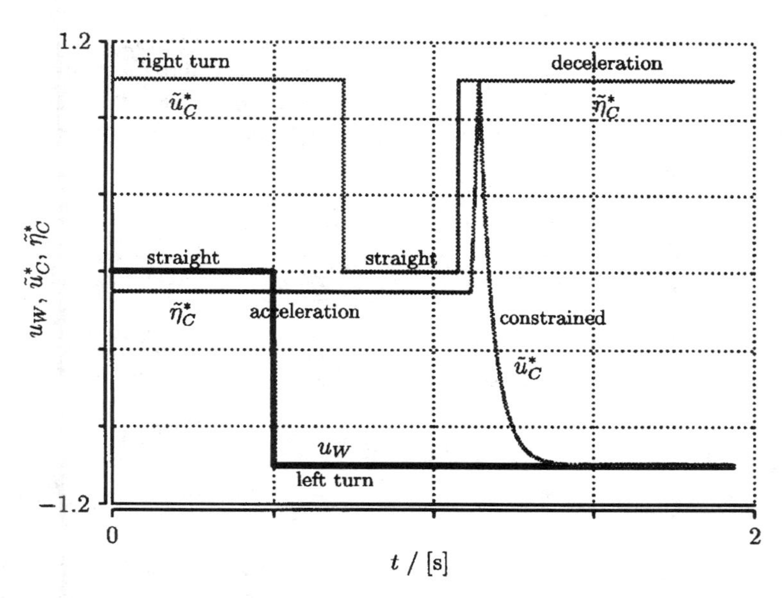


**Figure 10**: History of the control variables for the simulations of Figure 9 (Prescribed control $u_W$: dark gray. Output controls $\tilde{u}_C^*(z)$ and $\tilde{\eta}_C^*(z)$ of neural networks: light gray.

the neural networks, the neural networks are used to approximate the feedback strategies of the games in real-time. Extensive numerical results, including those of earlier papers of the authors, have shown that the approximation becomes better if more mathematical structure such as the gradient of the value function and the minimax principle is integrated.

In the present paper both methods have been used to synthesize optimal collision-avoidance strategies for situations where a car driver is faced with another driver driving on the wrong side of a freeway. Here, two modifications of the classical game of two cars, a simpler and a more realistic model, have been investigated. The latter one also includes control variable inequalities to guarantee that the two players stay on the freeway.

The successful synthesis of the optimal collision avoidance strategies gives rise to high hopes. Since six state and three control variables are involved in the more realistic collision avoidance problem, we believe that even higher-dimensional differential games can be solved following the procedures presented here. Moreover, both methods derive benefits from the fact that neural networks are easy to use.

## Acknowledgments

The authors would like to thank Dr. Stefan Miesbach, Corporate Research and Development, Siemens AG, Munich, Dipl. Math. Claudia Chini, formerly Munich University of Technology, and Dipl. Math. Matthias Gerdts, Clausthal University of Technology, for helpful comments and fruitful cooperation.

## REFERENCES

[1] Anderson, J. A. *An Introduction to Neural Networks.* MIT Press, Cambridge, MA, 1995.

[2] Annema, A. J. *Feedforward Neural Networks: Vector Decomposition Analysis, Modelling and Analog Implementation.* Kluwer, Boston, MA, 1995.

[3] Başar, T., and G. J. Olsder. *Dynamic Noncooperative Game Theory.* Academic Press, London, UK, 1982 and 1995 (2nd edition).

[4] Blum, E. K., and L. K. Li. *Approximation Theory and Feedforward Networks*, Neural Networks, **4**, 1990.

[5] Boulter, J., D. Maher, and M. F. Aburdene. *Collision Avoidance Using Ultrasonic Waves.* Report 17837, Bucknell University, Department of Electrical Engineering, Lewisburg, PA, 1994.

[6] Breitner, M. H. *Robust optimale Rückkopplungssteuerungen gegen unvorhersehbare Einflüsse: Differentialspielansatz, numerische Berechnung und Echtzeitapproximation.* Ph.D. Thesis, VDI-Verlag, VDI Fortschritt-Bericht # **596**, Reihe # **8** “Meß-, Steuerungs- und Regelungstechnik”, Düsseldorf, Germany, 1996.

[7] Breitner, M. H., and H. J. Pesch. Reentry Trajectory Optimization under Atmospheric Uncertainty as a Differential Game. *Annals of the International Society of Dynamic Games*, **1**, 1994.

[8] Breitner, M. H., H. J. Pesch, and W. Grimm. Complex Differential Games of Pursuit-Evasion Type with State Constraints, Necessary Conditions for Optimal Open-Loop Strategies (Part 1) and Numerical Computation of Optimal Open-Loop Strategies (Part 2). *Journal of Optimization Theory and Applications*, **78**, 1993.

[9] Chini, C. *Numerische Berechnung optimaler Ausweichmanöver mit einem Differentialspielansatz und Neuronalen Netzen.* Diploma Thesis, Munich University of Technology, Mathematical Institute, Munich, Germany, 1994.

[10] Gallant, S. I. *Neural Network Learning and Expert Systems.* MIT Press, Cambridge, MA, 1993.

[11] Galperin, E. A. The Cubic Algorithm for Global Games with Application to Pursuit-evasion Games. *Computers and Mathematics with Applications*, **26**, 1993.

[12] Hautzinger, H., D. Heidemann, and B. Krämer. *Fahrleistungserhebung 1990 – Inlandsfahrleistung und Kfz-Unfallrisiko in der Bundesrepublik Deutschland (alte Bundesländer).* Report of the Bundesanstalt für Straßenwesen # **M20**, Wirtschaftsverlag NW/Verlag für neue Wissenschaft, Bremerhaven, Germany, 1993.

[13] Hautzinger, H., B. Tassaux-Becker, and R. Hamacher. *Verkehrsunfallrisiko in Deutschland.* Report of the Bundesanstalt für Straßenwesen # **M58**, Wirtschaftsverlag NW/Verlag für neue Wissenschaft, Bremerhaven, Germany, 1996.

[14] Hornik, K. Approximation Capabilities of Multilayer Feedforward Networks. *Neural Networks*, **4**, 1991.

[15] Hornik, K., M. Stinchcombe, and H. White. Multi-Layer Feedforward Networks are Universal Approximators. *Neural Networks*, **2**, 1989.

[16] Isaacs, R. P. *Games of Pursuit.* Paper # **P-257**. The RAND Corporation, Santa Monica, CA, 1951.

[17] Isaacs, R. P. *Differential Games I: Introduction. Differential Games II: The Definition and Formulation. Differential Games III: The Basic Principles of the Solution Process. Differential Games IV: Mainly Examples.* Memoranda # **RM-1391**, # **RM-1399**, # **RM-1411** and # **RM-1486**. The RAND Corporation, Santa Monica, CA, 1954/55.

[18] Isaacs, R. P. *Differential Games—A Mathematical Theory with Applications to Warfare and Pursuit, Control and Optimization.* John Wiley & Sons, NY, 1965. Robert E. Krieger, Huntington, NY, 1975 (3rd edition).

[19] Kühnen, M. A., E. Brühning, and A. Schepers. *Unfallgeschehen auf Autobahnen – Strukturuntersuchung.* Report of the Bundesanstalt für Straßenwesen # **M50**, Wirtschaftsverlag NW/Verlag für neue Wissenschaft, Bremerhaven, Germany, 1995.

[20] Lachner, R. *Echtzeitsynthese optimaler Strategien für Differentialspiele schneller Dynamik mit Anwendungen bei der Kollisionsvermeidung.* Ph.D. Thesis, Clausthal University of Technology, Mathematical Institute, Clausthal-Zellerfeld, Germany, 1997.

[21] Lachner, R., M. H. Breitner, and H. J. Pesch. Optimal Strategies of a Complex Pursuit-Evasion Game. *Journal of Computing and Information*, **4**, 1994.

[22] Lachner, R., M. H. Breitner, and H. J. Pesch. Three-Dimensional Air Combat Analysis—An Example for the Numerical Solution of Complex Differential Games. *Annals of the International Society of Dynamic Games*, **3**, 1996.

[23] Leshno, M., V. Y. Lin, A. Pinkus, and S. Schocken. Multilayer Feedforward Networks with a Nonpolynomial Activation Function Can Approximate Any Function. *Neural Networks*, **6**, 1993.

[24] Lewin, J. *Differential Games: Theory and Methods for Solving Game Problems with Singular Surfaces.* Springer, London, UK, 1994.

[25] Merz, A. W. *The Homicidal Chauffeur—A Differential Game.* Ph.D. thesis, Report No. 94305, Department of Aeronautics and Astronautics, Stanford University, Stanford, CA, 1971.

[26] Miloh, T. The Game of Two Elliptical Ships. *Optimal Control Applications & Methods*, **4**, 1983.

[27] Miloh, T., and S. D. Sharma. Maritime Collision Avoidance as a Differential Game. *Schiffstechnik*, **24**, 1977.

[28] Miloh, T., and M. Pachter. Ship Collision Avoidance and Pursuit-Evasion Differential Games with Speed-Loss in a Turn. *Computers and Mathematics with Applications*, **18**, 1989.

[29] Pachter, M., and T. Miloh. The Geometric Approach to the Construction of the Barrier Surface in Differential Games. *Computers and Mathematics with Applications*, **13**, 1987.

[30] Pesch, H. J., I. Gabler, S. Miesbach, and M. H. Breitner. Synthesis of Optimal Strategies for Differential Games by Neural Networks. *Annals of the International Society of Dynamic Games*, **3**, 1996.

[31] Petrosjan, L. A. *Differential Games of Pursuit.* World Scientific, River Edge, NY, 1993.

[32] Scherer, A. *Neuronale Netze—Grundlagen und Anwendungen.* Vieweg, Wiesbaden 1996.

[33] Shinar, J., and R. Tabak. New Results in Optimal Missile Avoidance Analysis. *Journal of Guidance, Control & Dynamics*, **17**, 1994.

[34] Siemens AG, M. H. Breitner and H. J. Pesch. *Verfahren zur Kollisionsvermeidung von einem entgegenkommenden Fahrzeug und einem ausweichenden Fahrzeug mit Hilfe neuronaler Netze.* German patent # **19534942**, 1995/1998.

[35] Subbotin, A. I. *Generalized Solutions of First-Order PDEs.* Birkhäuser, Boston, MA, 1995.

[36] Vincent, T. L. Collision Avoidance at Sea. In: P. Hagedorn, H. W. Knobloch, and G. J. Olsder, (Eds.), *Differential Games and Applications, Lecture Notes in Control and Information Sciences* **3**, Springer, Berlin, Germany, 1977.

[37] Vincent, T. L., E. M. Cliff, W. J. Grantham, and W. Y. Peng. Some Aspects of Collision Avoidance. *AIAA Journal*, **12**, 1974.

[38] White, H., A. R. Gallant, K. Hornik, M. Stinchcombe, and J. Wooldridge. *Artificial Neural Networks—Approximation and Learning Theory.* Blackwell, Oxford, UK, 1992.

[39] White, D. A., and D. A. Sofge. *Handbook of Intelligent Control—Neural, Fuzzy and Adaptive Approaches.* Van Nostrand Reinhold, NY, 1992.

[40] Zell, A. *Simulation Neuronaler Netze.* Addison-Wesley, Bonn, Germany, 1996.

[41] Zell, A., et al. *SNNS—Stuttgart Neural Network Simulator (Version 4.0): User Manual.* Universität Stuttgart, Institut für paralleles und verteiltes Hochleistungsrechnen, Stuttgart, Germany, 1995.
URL: http://www.informatik.uni-stuttgart.de/ipvr/bv/projekte/snns/snns.html

# Rendezvous-Evasion as a Multistage Game with Observed Actions

Wei Shi Lim
Department of Decision Sciences
National University of Singapore
Singapore

**Abstract**

We consider a rendezvous-evasion zero-sum game that examines how two agents of team $R$ can optimally *rendezvous* while *evading* an enemy searcher $S$. We investigate the game in a discrete framework where the search region consists of $n$ identical locations along a directed cycle $D$ known to all players. The agents $R_1$, $R_2$, and the searcher $S$ start at distinct locations and, at each integer time, they can move to any one of the other locations or stay still. The game ends when some location is occupied by more than one player. If $S$ is at this location, $S$ (maximizer) wins and the payoff is 1; otherwise $R$ (minimizer) wins and the payoff is 0. The value of the game $v_n$ is the probability that $S$ wins under optimal play. We model the rendezvous-evasion problem as a multistage game where all players are obliged to announce their actions truthfully at the end of each step. We also assume that the agents $R_1$ and $R_2$ can jointly randomize their strategies. We prove that we need only consider a special class of strategies for $R$ and show that $v_3 = \frac{5}{9} \approx 0.55556$ and $v_4 = \frac{17}{32} \approx 0.53125$. We also prove that if the players have no common knowledge of $D$, the value of the game is $\frac{47}{76} \approx 0.61842$ when there are three locations.

## 1 Introduction

We analyze a rendezvous-evasion zero-sum game which models the situation faced by a team $R$, comprising of two agents $R_1$ and $R_2$, and an enemy searcher $S$. The players are randomly placed in a known search region and the game ends when at least two players meet, i.e., they come within a given detection distance. The team $R$ is a pair of "rendezvousers" whose objective is to meet each other before either of them is captured by the searcher $S$ whereas the aim of $S$ is to take captive the agents. That is, $R$ wins with payoff 0 if its agents successfully *rendezvous* while *evading* $S$. Otherwise, $S$ wins with payoff 1; the value of the game $v_n$ is the probability that $S$ wins under optimal play.

The rendezvous-evasion game is the first attempt to incorporate rendezvous search into search games with mobile hiders and to study both of them in the same context. Search games with mobile hiders, as proposed by R. Isaacs [8] have been extensively studied in the minimax environment in the last few decades [7], [8], [11]. The rendezvous search problem asks how two players randomly placed in a known region can meet together in least expected time [1], [3], [4], [5], [10]. There are essentially two versions of rendezvous search, namely, the symmetric version where players are indistinguishable and hence have to use the same mixed strategy, and the asymmetric version where players can agree to use different pure strategies. In the rendezvous-evasion game studied in this paper, the agents are involved in an asymmetric rendezvous problem with each other and at the same time, each agent is also playing the role of the hider in a search game with mobile hiders against $S$. This game has potential applications in search and rescue operations in the military where a soldier is downed in an enemy territory and has to be rescued by a fellow soldier before either of them is captured by the enemy and the movements of the soldiers can be detected by radar.

We consider the situation where the search region consists of $n$ identical locations along a directed cycle $D$ such that at every integer time, a player can either stay where he is or move to any one of the other locations. We model the rendezvous-evasion problem as multistage game with observed actions where all players are obliged to announce their actions truthfully at the end of each step. By this we mean that

(1) all players known the actions chosen at all previous steps $0, 1, 2, \cdots, k-1$ when choosing their actions at step $k$; and
(2) all players move simultaneously at each step $k$.

The class of multistage games with observed actions has found many applications in the areas of economics, political science, and biology [6]. We also assume that the agents engage in joint randomization, i.e., the agents preform a randomization experiment before the game begins and jointly decide upon a pure strategy (which describes the actions of both agents) that they will employ during the game. The case where players are not obliged to announce their actions at the end of each step is an entirely different class of games considered by the author in [9].

This paper is organized as follows. In Section 2, we give a precise formulation of the rendezvous-evasion game and show that if $R$ restricts to a special class of strategies, $S$ cannot do better than adopt the random strategy (i.e., visit each location with probability $1/n$ at each step). In particular, we show that $v_3 = \frac{5}{9} \approx 0.55556$ and $v_4 = \frac{17}{32} \approx 0.53125$. In Section 3, we consider the rendezvous-evasion game on three locations where the players do not share a common notion of the directed cycle $D$ and prove that in this instance, the value of the game is $\frac{47}{76} \approx 0.61842 > v_3$.

## 2 The Rendezvous-Evasion Game $\Gamma_n$

In this section, we formalize the rendezvous-evasion game $\Gamma_n$ where all players share a common notion of a directed cycle $D$ containing $n$ identical locations. The main results in this section are the determination of the values $v_3$ and $v_4$.

Although the same approach applies to all $n$ theoretically, implementation proves computationally tedious when $n$ becomes large. However, we provide a general upper bound for the value $v_n$.

At the start of the game $\Gamma_n$, $R_1$, $R_2$, and $S$ are randomly placed on $n$ identical locations such that no two of them occupy the same location. The players do not share a common labeling of the locations. This assumption is crucial for otherwise the game would end by the first step with the rendezvous agents agreeing to meet at each location equiprobably. At each integer time, a player can either stay where he is or move to any one of the other locations. We say that two players meet if they are at the same location at the same time. We model $\Gamma_n$ as a multistage game with observed actions where all players knew the actions chosen (by everyone) at all previous steps $0, 1, 2, \cdots, k-1$ before choosing their actions simultaneously at step $k$. That is, all previous actions are common knowledge to all parties in the game. We assume that all players adopt the following convention when labeling the locations: The location which they are initially placed is labeled as 0 and the remaining locations are labeled in an increasing order along the direction of $D$. Each player has, therefore a different refence point. We use $S(i)$, $R_1(i)$, and $R_2(i)$ to denote the respective locations which $S$, $R_1$, and $R_2$ each labels as $i$. The action of $S$ when $S$ visits location $S(i)$ is denoted by $(i)$. Similarly, a typical action of $R$ specifies a pair of locations which agents $R_1$ and $R_2$ visit; we use the pair $(j, k)$ to denote the action of $R$ where $R_1$ visits location $R_1(j)$ while $R_2$ visits location $R_2(k)$. We shall say that the *history* at the beginning of step $k$ is a record of all previous actions of all players. Here, a pure strategy for a player is a function which specifies an action for each step and each history; mixed strategies specify probability mixtures over the actions at the step. We use $S_n$ and $R_n$ to denote the set of mixed strategies of $S$ and $R$, respectively. If $R_1$ meets $R_2$ before any one of them is captured by $S$, the *payoff* is 0 and we say that $R$ (minimizer) wins. Otherwise, $S$ (maximizer) wins and the *payoff* is 1. The game ends when at least two players meet. The expected payoff $\pi_n(\sigma, \rho)$ of the game when $S$ and $R$ use strategies $\sigma$ and $\rho$, respectively, is the expected probability that $S$ wins. We say that $(\sigma^*, \rho^*)$ is an optimal strategy pair if

$$\pi_n(\sigma^*, \rho) \geq \pi_n(\sigma^*, \rho^*) \geq \pi_n(\sigma, \rho^*), \qquad \forall \sigma \in \mathcal{S}_n, \quad \rho \in \mathcal{R}_n.$$

The value $v_n$ of the game is defined to be

$$v_n = \max_{\sigma \in \mathcal{S}_n} \min_{\rho \in \mathcal{R}_n} \pi_n(\sigma, \rho) = \min_{\rho \in \mathcal{R}_n} \max_{\sigma \in S_n} (\sigma, \rho).$$

## 2.1 A Restricted Class of Strategies for $R$

The rendezvous-evasion game $\Gamma_n$ is essentially a finite problem since there is only a countable number of pure strategy pairs. However, this number can be very large. For instance, the number of (n-1)-step pure strategy pairs is $n^{3(n-1)}$ (each player has $n$ possible actions at each step). In this section, we show that we can reduce the search for an optimal strategy pair by considering a special class of strategies for $R$.

Let $\Psi_n$ denote the class of strategies of $R$ such that if $R$ uses action $(x, y)$ at step $i$ with probability $p$, then $R$ also uses each of the following pairs of actions $(x + j, y + j) \pmod{n}$, $j = 1, \ldots, n - 1$, with probability $p$; that is $R$ is equally likely to visit any pair of locations which are at the same distance apart (the random element $j$ is chosen independently at each step). Let $\sigma_n^*$ denote the random strategy for $S$, i.e., the strategy which says that $S$ visits each location with probability $1/n$ at every step independent of all previous moves. The following lemma shows that knowledge of the directed cycle $D$ allows the rendezvous agents to jointly randomize in such a way that forces $S$ to do no better than use the random strategy $\sigma_n^*$.

**Lemma 2.1.** *Suppose the rendezvous team $R$ uses strategies from $\Psi_n$, then the random strategy $\sigma_n^*$ is a best response for $S$.*

**Proof.** By virtue of the class of strategies $\Psi_n$, $R_1$ is equally likely to be at any one of the $n$ locations at every step independent of previous actions. The same can be said about $R_2$. Hence, $S$ cannot do better than visit each of the $n$ locations equiprobably at each step, which is precisely the random strategy $\sigma_n^*$. □

**Lemma 2.2.** *Suppose $S$ uses the random strategy $\sigma_n^*$, then it is not a best response for $S$ to engage in "repeated" search, that is, the agents of $R$ do not search a pair of locations which are at the same distance apart for a second time.*

**Proof.** We first note that this result is not surprising, indeed, since if the game continues after $R$'s first attempt to meet by visiting a pair of locations at distance $d$ apart, the agents will not meet in their subsequent attempts and would have wasted the steps when $S$ is using strategy $\sigma_n^*$. Let $r = (x_1, y_1, x_2, y_2, \ldots)$ denote a pure strategy of $R$ which engages the agents in "repeated" search at some step. Let $\Delta$ denote the set of integers $k(\leq n - 1)$ such that $(x_k - y_k)$ is congruent (mod $n$) to $(x_j - y_j)$ or 0 for some $j < k$. The set $\Delta$ is nonempty by the choice of $r$. Modify strategy $r$ to strategy $\tilde{r}$ by changing the actions of $R_1$ and $R_2$ at step $k$ for all $k$ in the set $\Delta$. For each $k$, choose a pair locations $(\alpha_k, \delta_k)$ so that instead of visiting locations $R_1(x_k)$ and $R_2(y_k)$ at step $k$, $R_1$ and $R_2$ visit locations $R_1(\alpha_k)$ and $R_2(\delta_k)$, respectively, and $(\alpha_k - \delta_k)$ is not congruent (mod $n$) to any of the $(x_j - y_j)$ or $(\alpha_j - \delta_j)$ for all $j < k$ or 0. Such a pair of $(\alpha_k, \delta_k)$ exists for each $k$ since $k$ is at most $n - 1$. Let $c_j^r$ and $c_j^{\tilde{r}}$ denote the events that $S$ wins at step $j$ when $R$ uses strategy $r$ and $\tilde{r}$, respectively (and $S$ uses strategy $\sigma_n^*$). We first observe that the modification of strategy $r$ to strategy $\tilde{r}$ ensures that Prob $(c_j^r) =$ Prob $(c_j^{\tilde{r}})$, for all $j$ not in $\Delta$, $j \leq n - 1$. Second, for all $k$ in $\Delta$, the agents do not meet at step $k$ when using $r$. However, with strategy, $\tilde{r}$, they meet with a positive probability, so that we have Prob $(c_k^r) >$ Prob $(c_k^{\tilde{r}})$. Third, strategy $\tilde{r}$ guarantees that the game ends by step $n - 1$. This implies that Prob $(c_j^r) \geq$ Prob $(c_j^{\tilde{r}})(= 0)$, for all $j > n - 1$. Combining these three results we have $\pi_n(\sigma_n^*, r) > \pi_n(\sigma_n^*, \tilde{r})$. Hence, strategy $r$

is not a best response to strategy $\sigma_n^*$ of $S$ and, consequently, any mixed strategy of $R$ which play $r$ with a positive probability cannot be a best response. □

Lemmas 2.1 and 2.2 together give the following result:

**Corollary 2.1.** *There is an optimal strategy pair of the form* $(\sigma_n^*, \rho_n^*)$, *where* $\rho_n^*$ *belongs to* $\Psi_n$ *and does not involve repeated search.*

It follows from the above corollary regarding the potential choices of $\rho_n^*$ that we need only consider a much reduced extensive form of the game $\Gamma_n$, which is explained in the next section.

## 2.2 Reduced Extensive Form of $\Gamma_n$

We look for optimal strategy pairs of the form $(\sigma_n^*, \rho_n^*)$. By Corollary 2.1, we can narrow our search for $\rho_n^*$ to strategies in $\Psi_n$ which do not involve repeated search, i.e., $(n-1)$-step strategies where the agents do not visit a pair of locations at the same distance apart for a second time (the initial distance between the agents is $1, 2, \ldots,$ or $n-1$ ensures that the agents meet by $(n-1)$ steps). We simplify the extensive form of $\Gamma$ by considering only actions of $R$ which are of the form $(0, d)$, $d = 1, 2, \ldots, n-1$ (all other actions of $R$ can be obtained by a rotation of one of these $(n-1)$ actions about $D$). In general, at each step $k (k = 1, 2, \ldots, n-1) = f\ R$ and $S$ simultaneously choose from $(n-k)$ and $n$ possible actions (the possible actions of $R$ depend on $R's$ previous actions). Since actions are announced at the end of each step, there are $n^{k-1}(n-1)\ldots(n-k+1)$ information sets, each being a singleton. By our restriction on $\rho_n^*$, the game ends by step $(n-1)$ so that the number of terminal nodes is $n^{n-1}(n-1)!$. However, we can further reduce the number of terminal nodes: The action of $R$ at step $(n-1)$ is determined by the teams's previous $(n-2)$ moves and since we are interested in the optimal moves of $R$ (rather than $S's$, who is using the random strategy $\sigma_n^*$) we can omit showing the actions of $S$ for the last step. In fact, it is not necessary to show the actions of $S$ at step $(n-2)$ as well, because any information learnt will not be useful since the action of $R$ at step $(n-1)$ is already determined by $R's$ previous $(n-2)$ moves. Hence we may prune the number of terminal nodes by a factor of $n^2$ to $n^{n-3}(n-1)!$ by writing the payoffs at each terminal node as $c + t/n$ where $c$ is the probability that $S$ wins by steps $(n-2)$ and $t$ is the probability that the game continues into step $(n-1)$ ($c$ and $t$ are dependent on $R's$ strategies). We can then proceed to solve $\Gamma_n$ by backward induction.

### 2.2.1 Determination of $v_3$

In this subsection, we show that $v_3$ is $\frac{5}{9}$ and describe an optimal strategy pair $(\sigma_3^*, \rho_3^*)$. We know from the previous section that we can consider the reduced extensive form of $\Gamma_3$ which consists of two terminal nodes branching from two possible step-1 actions of $R$, namely $(0, 1)$ and $(0, 2)$ (See Figure 1). Regardless of $R's$ step-1 action, we find that $c$ is $\frac{1}{2}$ and $t$ is $\frac{1}{6}$. Hence, both terminal nodes have

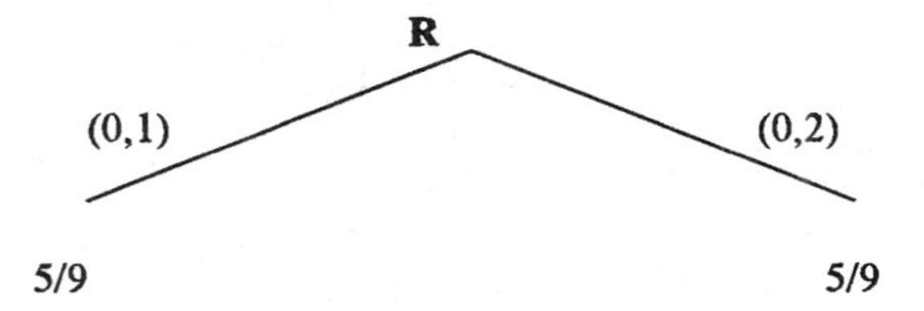


**Figure 1**: Reduced extensive form of $\Gamma_3$.

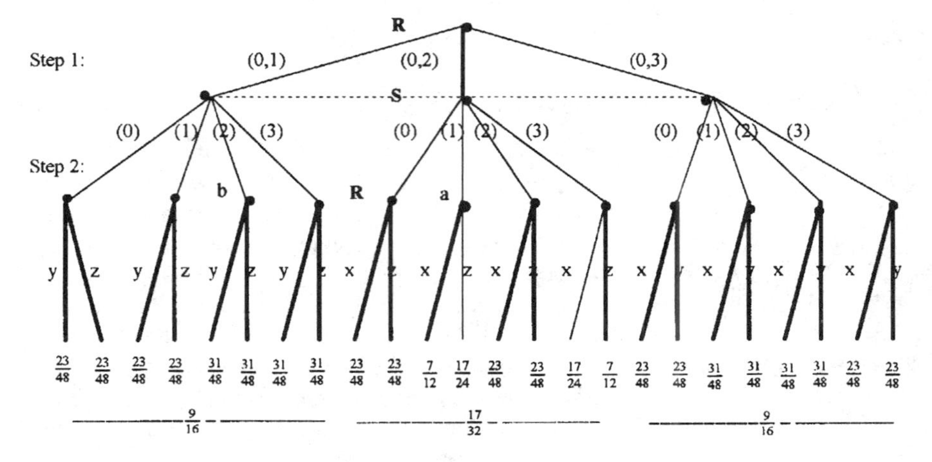


x denotes action (0,1);
y denotes action (0,2);
z denotes action (0,3)

**Figure 2**: Reduced extensive form of $\Gamma_4$.

the same payoff of $\frac{1}{2} + \frac{1}{6}\frac{1}{3} = \frac{5}{9}$, implying that $R$ is indifferent between visiting locations (0, 1) or (0, 2) at step 1. This is illustrated in Figure 2. Thus we have the following theorem:

**Theorem 2.1.** *The value of the game $\Gamma_3$ is $\frac{5}{9}$ and $(\sigma_3^*, \rho_3^*)$ is an optimal pair where $\sigma_3^*$ is the random strategy for S, and $\rho_3^*$ is given as follows:*

- *Strategy $\rho_3^*$ of R:*

  *Step* 1 : *Choose each of the actions* (0, 1), (1, 2), (2, 0) *with probability* $\frac{1}{3}$.
  *Step* 2 : *Choose each of the actions* (0, 2), (1, 0), (2, 1) *with probability* $\frac{1}{3}$.

At this point, we note that if the players are not obliged to announce their actions at the end of each step (while all other dynamics of the game remain the same), the value of the game is unchanged. This shown in [9]. This observation raises the natural question of whether the announcement of actions at the end of each step does make a difference to the value of the game. We shall show in the next subsection that it does for the game $\Gamma_4$, and the reason being that with the announcement of actions at the end of each step, some extra information may be gathered by $R$ to enhance the agents' chances of rendezvous.

### 2.2.2 Determination of $v_4$

For $\Gamma_4$, we shall see that certain actions of $R$ at step 1 allows the rendezvous team to deduce exactly their initial distance apart just after one step instead of two (the game ends latest by three steps). We also show that $v_4$ is $\frac{17}{32}$ and describe an optimal strategy pair $(\sigma_4^*, \rho_r^*)$.

Consider the reduced extensive form of $\Gamma_4$ which is shown in Figure 2. At step 1, $R$ has three different actions represented by (0, 1), (0, 2), and (0, 3) and $S$ has four actions (0), (1), (2), and (3). Upon the announcement of actions at the end of step 1, there are twelve information sets and $R$ has two choices of actions at step 2 (depending on their action at step 1) and $R's$ step 3 action (not shown) is thus uniquely determined. The respective payoffs are shown at the terminal nodes. For example, if $R$ uses action (0, 2) and $S$ uses action (1) at step 1, the probability that $S$ wins at step 1 is $\frac{1}{2}$. If $R$ then uses action (0, 1) at step 2, the probability that $S$ wins at step 2 is $\frac{1}{3}\frac{1}{4} = \frac{1}{12}$ and the game continues with probability 0. That is, $c$ is $\frac{1}{2} + \frac{1}{12} = \frac{7}{12}$ and $t$ is 0 so that the payoff is $c + t/4 = \frac{7}{12}$. Hence, in this instance, $R$ is able to conclude after step 1 that they are initially placed with $R_2$ at one position behind $R_1$ (since using (0,1) ensures meeting at step 2). If $R$ uses action (0, 3) at step 2 instead, the probability that $S$ wins at step 2 is $\frac{1}{3}\frac{2}{4} = \frac{1}{6}$ and the game continues with probability $\frac{1}{6}$ so that the payoff in this case is $\frac{1}{2} + \frac{1}{6} + \frac{1}{24} = \frac{17}{24}$. We solve for the optimal strategy $\rho_n^*$ by using the technique of backward induction on the reduced extensive form and the optimal moves for $R$ at each step is shown in thick lines. For example, at information set $a$, $R$ being a minimizer prefers action (0, 1) to (0, 3), while at information set $b$ is indifferent between the two actions (0, 2) and (0, 3). Since we are assuming that $S$ is using the random strategy $\sigma_4^*$, we write the expected payoffs at the final line, each of which is obtained by taking the average of the four payoffs corresponding to $R's$ optimal choices at step 2 (for example, $\frac{17}{32} = \frac{1}{4}(\frac{23}{48} + \frac{7}{12} + \frac{23}{48} + \frac{7}{12})$). Hence, $R's$ optimal action at step 1 is (0, 2) and the value of the game $\frac{17}{32}$.

**Theorem 2.2.** *The value of the game $\Gamma_4$ is $\frac{17}{32}$ and $(\sigma_4^*, \rho_4^*)$ is an optimal strategy pair where $\sigma_4^*$ is the random strategy for $S$ and $\rho_4^*$ is given as follows:*

- *Strategy $\rho_4^*$ of $R$:*

  *Step* 1 : *Choose each of the actions $\phi_i(0, 2)$ $(i = 0, 1, 2, 3)$ with probability $\frac{1}{4}$ where $\phi_i$ denotes rotation along $D$ by distance $i$, for example, $\phi_2(0, 2) = (2, 0)$).*

  *Step* 2 :

| Action of $R$ at step 1 | Observed action of $S$ at step 1 | Action of $R$ at step 2 |
|---|---|---|
| $\phi_i(0, 2)$ | $\phi_i(3)$ | Choose each of the actions $\phi_k(0, 3)$ $(k = 0, 1, 2, 3)$ with probability $\frac{1}{4}$* |
| $\phi_i(0, 2)$ | $\phi(0)$ (1) or (2) | Choose each of the actions $\phi_k(0, 1)$ $(k = 0, 1, 2, 3)$ with probability $\frac{1}{4}$ |

* *The game ends with probability one at step 2.*

*Step* 3 : *Choose each of the actions* $\phi_k(0, 4)$ $(k = 0, 1, 2, 3)$ *with probability* $\frac{1}{4}$.

We have shown in [9] that when players are not obliged to reveal their actions at the end of each step, the value of the game when there are four locations along $D$ is $\frac{9}{16}$ $(> v_4)$. Hence, the announcement of actions in this context helps $R$ secure a lower value.

In general, to analyze the game $\Gamma_n$, we can apply backward induction to the reduced extensive form of $\Gamma_n$. However, the structure of the reduced game tree increases in complexity as $n$ increases, which renders the same approach technically infeasible. Nonetheless, we provide a general upper bound for $v_n$. In [9], we consider a game which is similar to $\Gamma_n$ in *all* aspects except that players are not obliged to announce their actions at the end of each step. We showed in [9] that the value of the game is $((1 - 2/n)^{n-1} + 1)/2$. The optimal strategy of $R$ is a randomized version of the "one-stays-still-the-other-searches" strategy while the optimal strategy of $S$ is the random strategy $\sigma_n^*$ (as defined on page 140 here) Since the randomized version of "the one-stays-still-the-other-searches" strategy is an element in the restricted class of strategies of $R$, $\Psi_n$ (as defined on page 140), we can therefore conclude that $v_n \leq ((1 - 2/n)^{n-1} + 1)/2$. (Please refer to [9] for more details.)

## 3 The Rendezvous-Evasion Game $\tilde{\Gamma}$

We now consider a variant of the above problem where the directed cycle $D$ is not known to the players. In particular, we consider the problem $\tilde{\Gamma}$ where the search region is comprised of three identical locations. We prove in this section that the value $\tilde{v}$ of the game is $\frac{47}{76}$ and observe that $v_3 < \tilde{v}$, that is, knowledge of the directed cycle $D$ helps $R$ obtain a lower value. All terminologies used are similar to that defined in Section 2 except the way players label the locations; here we assume that the players label their initial locations as 0, the next new location that they visit as 1 and the remaining third location as 2. For example, a strategy for the searcher $S$ whereby he moves to a new location at step 1, remains still at step 2 and moves on to another new location at step 3 and stays there henceforth is denoted by $(1, 1, 2, 2, 2, \ldots)$.

We analyze $\tilde{\Gamma}$ by using the notion of backward induction. At step 1, there are three possible actions for $R$, namely $(0, 0)$, $(0, 1)$, and $(1, 1)$ (by symmetry action

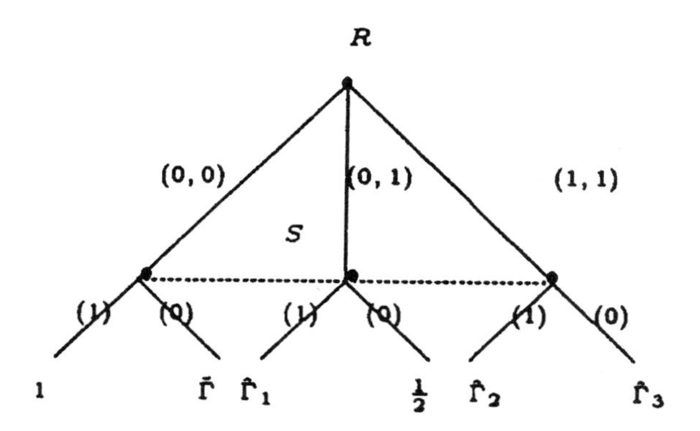


**Figure 3**: $\tilde{\Gamma}$ with all possible step 1 actions, and subsequent subgames or payoffs (if game ends by step 1).

(1, 0) is similar to action (0, 1) at step 1). For $S$, he can either stay where he is (by using action (0)) or move to a new location (by using action (1)). Since players are obliged to announce their actions at the end of each step, the subgame that they play at step 2 depends on their actions at step 1. This is given in Figure 3, where we indicate at the terminal nodes, either the payoffs (if the game has ended) or the subgames $\hat{\Gamma}_i$'s which the players will encounter at step 2 (which we will proceed to solve). For example, if $R$ uses action (0, 0) at step 1 and $S$ uses action (1), $S$ is sure to meet one of the agents so the payoff is 1 and the game ends immediately. However, if none of the players move at step 1, there is no meeting at all and no extra information is gained by anyone upon the announcement of actions. It is thus reasonable to assume that at step 2, all players play the game $\tilde{\Gamma}$ again.

By assuming optimal play from step 2 onward, we illustrate the extensive form of $\tilde{\Gamma}$ in Figure 4, where the payoffs are given at the terminal nodes and $\hat{v}$ is the value of the subgame $\hat{\Gamma}_i$ (the respective payoffs will be explained later in Sections 3.1 and 3.2).

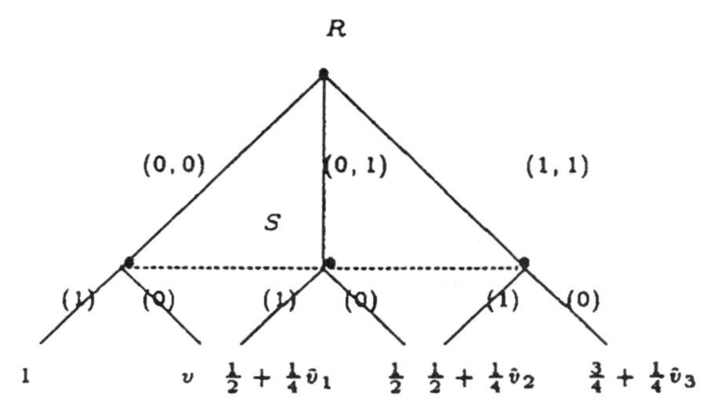


**Figure 4**: Extensive form of $\tilde{\Gamma}$ at step 1, assuming optimal play from step 2 onward.

We note that if at the beginning of any subgame $R_1$ and $R_2$ know how their labelings of the three locations are related, the value of this subgame must be $\frac{1}{3}$ with $R_1$ and $R_2$ coordinating to meet at each of the three locations equiprobably while $S$ visits each location with probability $\frac{1}{3}$. We now proceed to analyze each of the subgames $\hat{\Gamma}_i$, whose value is denoted by $\hat{v}_i$.

### 3.1 Subgame $\hat{\Gamma}_1$

If $R$ uses action $(0, 1)$ and $S$ uses action $(1)$ at step 1, $S$ wins at step 1 with probability $\frac{1}{2}$ and the game continues into step 2 (with $\hat{\Gamma}_1$ being played) with probability $\frac{1}{4}$ so that if optimal play is assumed from step 2 onward, the payoff in this case is $\frac{1}{2}+\frac{1}{4}\hat{v}_1$ as given in Figure 4. In the instance where $\hat{\Gamma}_1$ is being played, the labelings of $R_1$, $R_2$, and $S$ are equally likely to be related in the following ways:

$$\begin{cases} L_1: \\ S(0) = R_1(1) = R_2(1) \\ S(1) = R_1(2) = R_2(0) \\ S(2) = R_1(0) = R_2(2), \end{cases} \qquad \begin{cases} L_2: \\ S(0) = R_1(2) = R_2(1) \\ S(1) = R_1(1) = R_2(0) \\ S(2) = R_1(0) = R_2(2). \end{cases}$$

Since $R_1$ has stayed still at step 1, there are only two possible actions for $R_1$ at step 2 while $R_2$ having moved at step 1 has three possible actions so that as a team, $R$ has six possible actions at step 2. We compute the payoff of the game for each pair of actions of $S$ and $R$. The strategic form is given in Table 1. For instance, if all players return to their initial positions at step 2, they are faced with the same subgame at step 3 and no information is learnt about their relative labelings of the locations. Assuming optimal play from step 3 onward, the payoff is thus given by $\hat{v}_1$. There are two particular entries that are worth mentioning. When $S$ uses action $(0)$ and $R$ uses action $(1, 2)$ at step 2, $S$ wins if their labelings are related in the way described in $L_1$. No meeting occurs otherwise. This implies that if the game does not end at step 2, $R_1$ and $R_2$ would be able to deduce, upon the announcement of $S$'s action at the end of step 2 (as action $(0)$) that their labelings of the three locations are related in the manner as given in $L_2$. In this instance, we know that the value of the subgame which follows has to be $\frac{1}{3}$, which explains the payoff of $\frac{2}{3}(= \frac{1}{2} + \frac{1}{2} \times \frac{1}{3})$ given in Table 1. The other entries which have payoffs $\frac{2}{3}$ can be explained in the same fashion. Analogous argument applies to the entries with payoffs given by $\frac{1}{6}(= \frac{1}{2} \times \frac{1}{3})$ where $R$ wins with probability $\frac{1}{2}$ at step 2 and $R$ is able to deduce their relative labelings if the game continues into step 3.

Since $\hat{v}_1 \geq 0$, we observe from Table 1 that actions $(0, 0)$, $(0, 1)$, and $(1, 2)$ of $R$ are dominated by action $(0, 2)$. Thus, the payoff matrix can be reduced to that in Table 2. With respect to the payoff matrix in Table 2, it can be checked that the value $\hat{v}_1$ is $\frac{9}{19}$ and the optimal strategies of $S$ and $R$ are given by $(\frac{5}{19}, \frac{5}{19}, \frac{9}{19})$ and $(\frac{7}{19}, \frac{6}{19}, \frac{6}{19})$.

**Table 1**: Payoff matrix of $\hat{\Gamma}_1$.

| | | Actions of $R$ | | | | | |
|---|---|---|---|---|---|---|---|
| | | (0, 0) | (0, 1) | (0, 2) | (1, 0) | (1, 1) | (1, 2) |
| Actions of $S$ | (0) | $\hat{v}_1$ | 1 | 0 | $\frac{1}{2}$ | 1 | $\frac{2}{3}$ |
| | (1) | 1 | $\hat{v}_1$ | 0 | 1 | $\frac{1}{2}$ | $\frac{2}{3}$ |
| | (2) | 1 | 1 | 1 | $\frac{1}{6}$ | $\frac{1}{6}$ | 1 |

**Table 2**: Reduced payoff matrix of $\hat{\Gamma}_1$.

| | | Actions of $R$ | | |
|---|---|---|---|---|
| | | (0, 2) | (1, 0) | (1, 1) |
| Actions of $S$ | (0) | 0 | $\frac{1}{2}$ | 1 |
| | (1) | 0 | 1 | $\frac{1}{2}$ |
| | (2) | 1 | $\frac{1}{6}$ | $\frac{1}{6}$ |

### 3.2 Subgame $\hat{\Gamma}_2$

If $R$ uses action (1,1) while $S$ uses action (1) at step 1, $S$ wins with a probability of $\frac{1}{2}$ at step 1 and the games continues into step 2 with probability $\frac{1}{4}$ so that if optimal play is assumed from step 2 onward, the payoff is $\frac{1}{2} + \frac{1}{4}\hat{v}_2$ (Figure 4). It is evident that if subgame $\hat{\Gamma}_2$ is played at step 2, the players know that their labelings of the three locations must be cyclic permutations of each other, i,e., either

$$\begin{cases} L_3: \\ S(0) = R_1(1) = R_2(2) \\ S(1) = R_1(2) = R_2(0) \\ S(2) = R_1(0) = R_2(1), \end{cases} \quad \text{or} \quad \begin{cases} L_4: \\ S(0) = R_1(2) = R_2(1) \\ S(1) = R_1(1) = R_2(2) \\ S(2) = R_1(0) = R_2(0). \end{cases}$$

Since each of $R_1$ and $R_2$ has three possible actions at step 2, $R$ as a team has nine possible actions. We compute the probability that $S$ wins for each possible choice of actions by $S$ and $R$. This is summarized in Table 3. For example, when $R$ chooses action $(1, 2)$ and $S$ chooses action (0) at step 2, we know that $S$ wins if their labelings are associated in the manner of $L_3$ and the game continues otherwise. Thus the payoff must be given by $\frac{2}{3}(= \frac{1}{2} + \frac{1}{2} \times \frac{1}{3})$.

**Table 3**: Payoff matrix of $\hat{\Gamma}_2$.

| | | Actions of $R$ | | | | | | | | |
|---|---|---|---|---|---|---|---|---|---|---|
| | | (0, 0) | (0, 1) | (0, 2) | (1, 0) | (1, 1) | (1, 2) | (2, 0) | (2, 1) | (2, 2) |
| Actions of $S$ | (0) | $\hat{v}_2$ | $\frac{1}{2}$ | $\frac{1}{2}$ | $\frac{1}{2}$ | 1 | $\frac{2}{3}$ | $\frac{1}{2}$ | $\frac{2}{3}$ | 1 |
| | (1) | 1 | $\frac{1}{2}$ | $\frac{2}{3}$ | $\frac{1}{2}$ | $\hat{v}_2$ | $\frac{1}{2}$ | $\frac{2}{3}$ | $\frac{1}{2}$ | 1 |
| | (2) | 1 | $\frac{2}{3}$ | $\frac{1}{2}$ | $\frac{2}{3}$ | 1 | $\frac{1}{2}$ | $\frac{1}{2}$ | $\frac{1}{2}$ | $\hat{v}_2$ |

**Table 4**: Reduced payoff matrix of $\hat{\Gamma}_2$.

| | | Actions of $R$ | | | | | |
|---|---|---|---|---|---|---|---|
| | | (0, 1) | (0, 2) | (1, 0) | (1, 2) | (2, 0) | (2, 1) |
| | (0) | $\frac{1}{2}$ | $\frac{1}{2}$ | $\frac{1}{2}$ | $\frac{2}{3}$ | $\frac{1}{2}$ | $\frac{2}{3}$ |
| Actions of $S$ | (1) | $\frac{1}{2}$ | $\frac{2}{3}$ | $\frac{1}{2}$ | $\frac{1}{2}$ | $\frac{2}{3}$ | $\frac{1}{2}$ |
| | (2) | $\frac{2}{3}$ | $\frac{1}{2}$ | $\frac{2}{3}$ | $\frac{1}{2}$ | $\frac{1}{2}$ | $\frac{1}{2}$ |

Since all numerical entries in Table 3 are at least $\frac{1}{2}$, $\hat{v}_2$ is also at least $\frac{1}{2}$. Thus, actions (0, 0), (1, 1), and (2, 2) of $R$ are dominated by actions (0, 1), (1, 0), and (2, 1), respectively. The reduced payoff matrix of $\hat{\Gamma}_2$ is given in Table 4.

By observing the symmerty of the payoff matrix given in Table 4, one concludes that an optimal strategy for both $S$ and $R$ is to choose each action (in Table 4) equiprobably. Hence $\hat{v}_2$ is $\frac{5}{9}$.

Now if $R$ uses action (1,1) while $S$ uses action (0) at step 1, it is not possible for the agents to meet each other without meeting $S$. And the only instance where there is no meeting is when the agents interchange their positions at step 1. Hence, $S$ wins with a probability of $\frac{3}{4}$ at step 1 and the game continues into step 2 with probability $\frac{1}{4}$. This explains the term $\frac{3}{4} + \frac{1}{4}\hat{v}_3$ at the right-most terminal node of Figure 4. We shall later see that since $\hat{v}_3$ is nonnegative, its magnitude is insignificant as action (1, 1) of $R$ at step 1 is a dominated action.

In Table 5, we rewrite $\tilde{\Gamma}$ from Figure 4 into its strategic form and substitute the values of $\hat{v}_1$ and $\hat{v}_2$. Since $\tilde{v}$ is the value of the game $\tilde{\Gamma}$, it is clear that $\tilde{v}$ is at least $\frac{1}{2}$. Hence, action (0,1) of $R$ dominates actions (0, 0) and (1, 1) which reduces the choice of $S$ at step 1 to action (0) with eventual payoff of $\frac{1}{2}$ or action (1) with eventual payoff of $\frac{47}{76}$ ($> \frac{1}{2}$). The value $\tilde{v}$ is thus $\frac{47}{76}$.

Summarizing the optimal strategy pair $(\sigma^*, \rho^*)$ for $S$ and $R$ we have

**Strategy $\sigma^*$ of $S$:**

Step 1. $S$ moves to a new location $S$ (1);
Step 2. Randomize over three actions:

Action 1: Return to location $S(0)$;
Action 2: Stay at location $S(1)$;
Action 3: Go to location $S(2)$;

**Table 5**: Strategic form of $\tilde{\Gamma}_2$ when optimal play is assumed from step 2 onward.

| | | Actions of $R$ | | |
|---|---|---|---|---|
| | | (0, 0) | (0, 1) | (1, 1) |
| Actions of $S$ | (1) | 1 | $\frac{47}{76} (= \frac{1}{2} + \frac{1}{4}\hat{v}_1)$ | $\frac{23}{36} (= \frac{1}{2} + \frac{1}{4}\hat{v}_1)$ |
| | (0) | $v$ | $\frac{1}{2}$ | $\frac{3}{4} + \frac{1}{4}\hat{v}_3 (\geq \frac{3}{4})$ |

$S$ chooses these actions according to the probabilities $\frac{5}{19}$, $\frac{5}{19}$, and $\frac{9}{19}$.

Step $k (k \geq 3)$: $S$ visits each of the three locations with probability $\frac{1}{3}$.

**Strategy $\rho^*$ of $R$:**

Step 1. $R_1$ stays at location $R_1(0)$, while $R_2$ moves to location $R_2(1)$;
Step 2. There are three types of action, namely:

Type 1: $R_1$ stays at location $R_1(0)$ while $R_2$ visits location $R_2(2)$;
Type 2: $R_1$ visits location $R_1(1)$ while $R_2$ visits location $R_2(0)$;
Type 3: $R_1$ visits location $R_1(1)$ while $R_2$ visits location $R_2(1)$;
$R$ chooses Type 1 action with probability $\frac{7}{19}$, Type 2 action with probability $\frac{6}{19}$, and Type 3 action with probability $\frac{6}{19}$.

Step 3. $R_1$ and $R_2$ coordinate to meet at each of the three locations with probability $\frac{1}{3}$.

The justification for the action of $R$ at step 3 follows from our analysis $\hat{\Gamma}_1$ earlier on.

Hence, we have established the final result:

**Theorem 3.1.** *The value $\tilde{v}$ of the game $\tilde{\Gamma}$ is $\frac{47}{76}$ and the optimal strategy pair is given by $(\sigma^*, \rho^*)$ as described above.*

Although $\tilde{\Gamma}$ is solved in a fairly straightforward manner, the solution obtained generates as interesting observation—if we remove the requirement that players are obliged to announce their actions at the end of each step $(\sigma^*, \rho^*)$ is still optimal in this new game (which is dealt with in [9]). An explanation is provided in [9]. Lastly, we note that the same approach may not be feasible for the general case as the number of possible strategy pairs becomes very large as $n$ increases.

## REFERENCES

[1] Alpern, S. The Rendezvous Search Problem. *SIAM Journal of Control and Optimization*, **33**, 673–683, 1995.

[2] Alpern, S. and A. Beck. Rendezvous Search on the Line with Bounded Resources. European Journal of Operational Research, **101**, 3, 1997.

[3] Alpern, S. and S. Gal. Rendezvous Search on the Line with Distinguishable Players. *SIAM Journal of Control and Optimization*, **33**, 4, 1270–1276, 1995.

[4] Anderson, E. J. and S. Essegaier. Rendezvous Search on the Line with Indistinguishable Players. *SIAM Journal of Control and Optimization*. **33**, 1637–1642, 1995.

[5] Anderson, E. J. and R. R. Weber. The Rendezvous Problem on Discrete Locations. *Journal of Applied Probability*, **28**, 839–851, 1990.

[6] Fudenberg, D. and J. Tirole. *Game Theory*. MIT Press, Cambridge, MA, 1991.

[7] Gal, S. *Search Games*. Academic Press, New York, 1960.

[8] Isaacs, R. *Differential Games*, Wiley, New York, 1965.

[9] Lim, W. S. A Rendezvous-Evasion Game on Discrete Locations with Joint Randomization. *Advances in Applied Probability*. **29**, 1004–1017, 1997.

[10] Lim, W. S., S. Alpern, and A. Beck. Rendezvous Search on the Line with More Than Two Players. *Operations Research*. **45**, 3, 357–364, 1997.

[11] Ruckle, W. H. Pursuit on a Cyclic Graph. *International Journal of Game Theory*, **10** 91–99, 1983.

[12] Schelling, T. *The Strategy of Conflict*. Harvard University Press, Cambridge, MA, 1960.

# Identification and Construction of Singular Surfaces in Pursuit-Evasion Games

A. A. Melikyan
Institute for Problems in Mechanics
Russian Academy of Sciences
Moscow, Russia

Josef Shinar
Faculty of Aerospace Engineering
Technion-Israel Institute of Technology
Haifa, Israel

**Abstract**

The existence of singular surfaces in pursuit-evasion games is a rule rather than an exception. Identification and construction of these surfaces, which indicate some discontinuity or other singular phenomenon, is an inherent part of the game solution. Recently, a general approach, based on the method of singular characteristics, was developed for constructing singular paths and surfaces in optimal control and differential games, as well as for the generalized (viscosity) solutions of first-order PDEs. In the present paper a brief description of this technique is given and the procedure of identification and construction for several main singularities is demonstrated using two-parametric first-order PDEs. The existence of the different types of singular manifolds, which depend on the parameters, are identified, including dispersal, equivocal, and focal surfaces. The method above is also applied to a linear pursuit-evasion game with elliptical vectograms.

## 1 Introduction

The existence of singular surfaces in pursuit-evasion games [1] and [2] is a rule rather than an exception. Identification and construction of these surfaces, which indicate some discontinuity or other irregular phenomenon, are an inherent part of the game solution. The recent success in the theory of optimal control and differential games can be attributed to the development of two essentially interconnected mathematical activities, one is associated with the generalized Bellman–Isaacs equation [3], and the other with viscosity solution theory (VST) for the first-order PDE [4].

The most important contribution of the VST to the theory of optimal control and differential games is the fact that the generalized (viscosity) solution to the

Bellman–Isaacs equation coincides with the value-function. In other words, the dynamic programming approach is justified for a quite general class of problems. This allows a researcher to solve the Bellman–Isaacs equation as a separate mathematical boundary value problem for the first-order PDE without reference to the original dynamic equations, control parameters' restrictions, and cost-function.

The traditional approach to the solution of a game problem actually consists of two interconnected stages: (1) the solution of a first-order PDE; (2) the construction of the feedback controls. Thus, the VST allows us to separate this stage, as is the case in linear-quadratic problems, where the value of the game and the Hamiltonian are smooth functions. This splitting of the problem seems to be attractive since each stage is less complicated for analysis than the whole problem and involves less detail.

We restrict our attention to the first stage—to the solution of the first-order (Bellman–Isaacs) PDE. Two questions arise in this connection: (1) how should this PDE be solved and (2) what advantage does this approach have? The present paper includes two examples, the first one illustrates "how" it can be solved, and second one shows "what advantages" this approach has.

As to the first question, there exist several methods of analytical and numerical solutions. The classical method of characteristics (MC), which reduces the PDE to the ODE system, is one of them. But it does not work in the vicinity of a singular surface, corresponding to the nonsmoothness of the viscosity solution or the Hamiltonian. Not all the singular surfaces require a special construction technique. Some of them can be found using classical MC and simple results of VST (see the section Dispersal Surface).

A generalization of the classical MC was recently developed to extend the MC to those singular paths and manifolds which require special constructions [5,6]. This technique is called the method of singular characteristics (MSC). The equations for SC also have the form of the ODE system with the same structure as the classical (regular) characteristics, but they are expressed in terms of a renewed (modified) Hamiltonian. The essential difference between the MSC and the traditional approach of [1] is that the equations for SC are written only in terms of PDEs, i.e., in terms of an unknown (Bellman) function and its gradient, without reference to control parameters, etc. The MSC may also be useful for the problems with smooth (classical) solution but with nonsmooth Hamiltonian; see, e.g., [7], where a generalization of the Kelley condition is obtained using the MSC approach.

To make further comparison, recall that the construction procedure suggested in [1] is a local one and states the following: use regular characteristics until they fail for some reason, then search for some singular surface to match it with the previous constructions. The MSC approach actually suggests the same, but it is "better equipped" in two senses:

(1) equations for singular surfaces are presented in a closed and invariant form with some necessary inequality conditions in invariant form;
(2) the notion of viscosity solution.

Usually, when the regular procedure fails some edge of a possible singular surface comes up. The identification of that surface is based mainly on the verification of the appropriate initial conditions and necessary inequality conditions subsequently for each type of surface from the list of several known singularities. These two improvements of the Isaacs approach simplify the constructions, as the authors hope to demonstrate, making them more uniform.

It is important to note that originally the first-order PDEs boundary value problem does not possess any trajectories (paths), neither regular nor singular ones, which exist in control or game problems. These trajectories arise as a consequence of the method used for the solution—the method of characteristics. This phenomenon is demonstrated in the first example, which assumes no control background but possesses several well-known singular lines (singular characteristics).

The advantage of the MSC approach can be summarized as follows. All necessary conditions and equations for SC have an invariant form, independent of a specific coordinate system. The singular characteristic system is a closed ODE system of a standard form without control parameters. As a rule, it is easier to understand the nature of a singularity and derive some appropriate equations if the definition itself of a singular point is given for an arbitrary first-order PDE in abstract mathematical terms. The second example given here refers to a problem, solved in [9], where a focal surface was found. The identification of this type of singularity and the construction of the corresponding singular trajectories, using the classical methods of [1], presented, at that time, a considerable difficulty. The application of the MSC approach makes this task a straightforward operation. The equations of the trajectories in the focal surface, as well as the equations for the equivocal surfaces in a modified problem, are written in a standard closed form.

In this paper, a short description of the MSC technique and a definition of singular sets in an invariant form is given. Such is the spirit of the necessary conditions of singularity and the system of singular characteristics. The effectiveness of the approach is demonstrated first by the complete solution of a rather simple example, namely a two-dimensional first-order PDE problem with two parameters. Finally, the same method is applied to perform a new analysis of the singular surfaces of a three-dimensional linear pursuit-evasion game with elliptical vectograms, [9]. In both examples the existence of several types of singular surfaces, depending on the parameters of the problem, are identified.

## 2 Generalized Bellman–Isaacs Equation

Consider a conventional fixed-time differential game [1]:

$$\dot{x} = f(x, u, v), \qquad u \in U, \quad v \in V, \quad x = (x_0, x_1, \dots, x_n), \tag{1}$$
$$x_0 = t \in [t_0, T], \qquad J = \Phi(x(T)) \to \min_u \max_v .$$

Here $x \in R^{n+1}$ is the augmented state vector, $x_0$ being the time-variable, $\dot{x}_0 = 1$; $f = (1, f_1, \ldots, f_n) \in R^{n+1}$ is the right-hand side vector. Including the time as a component in the state vector simplifies the notations in the sequel.

Let $S = S(x)$, $x \in [t_0, T] \times R^n$, be the value (Bellman function) of the game (1). Omitting some details we give here a description of two modern generalizations of the basic, Bellman–Isaacs, equation for the game (1) [1]:

$$H(x, \partial S/\partial x) = 0, \quad x \in X = [t_0, T) \times R^n, \quad S(x) = \Phi(x), \quad x \in T \times R^n \quad (2)$$
$$H(x, p) = \min_u \max_v \langle p, f(x, u, v)\rangle, \qquad u \in U, \quad v \in V,$$
$$p = (p_0, p_1, \ldots, p_n) \in R^{n+1}.$$

One can see that $H(x, p) = p_0 + H'(x_0, \ldots, x_n, p_1, \ldots, p_n)$, where $H'$ is the conventional Hamiltonian. The gradient $\partial S/\partial x$ doesn't exist, in general, for all points $x \in X$; thus the equation (2) has to be understood in a generalized sense.

One of the generalizations, A. I. Subbotin's inequalities, [3], uses the directional derivatives $\partial S/\partial f$ with $f = f(x, u, v)$:

$$\min_u \max_v \frac{\partial S}{\partial f} \geq 0 \geq \max_v \min_u \frac{\partial S}{\partial f}, \qquad u \in U, \quad v \in V. \quad (3)$$

It is proven that the function $S(x)$ is the value of the game (1) if it satisfies the boundary condition $S(x) = \Phi(x)$, $\quad x \in T \times R^n$, and (3), which is reduced to the form (2) for the points where $S(x)$ is smooth.

The other approach, Ref. [4], introduces generalized continuous solutions to (2). A continuous function $S(x)$ is called a viscosity solution of the terminal value problem (2) if for every test-function $\phi(x) \in C^1$, such that local minimum (maximum) of the difference $S(x) - \phi(x)$ is attained at $x^* \in X$, the following inequality holds:

$$H(x^*, \phi_x(x^*)) \leq 0, \qquad (H(x^*, \phi_x(x^*)) \geq 0). \quad (4)$$

For the initial value problem the inequalities in (4) are reversed. The viscosity solution of (2) is proven to exist, to be unique, and equal to the value of the game (1). The connections between the two approaches are discussed in [4].

The approach of A. Subbotin (3) still deals with the control variables and their restrictions, as expressed by (3). The approach of the viscosity solution presented in (4) deals with the Hamiltonian and boundary condition, since the controls are eliminated by calculating the extrema in (2). The MSC relations do not use controls either. Note that in the solution procedure of a game by the MSC approach only two functions, $H(x, p)$ and $\Phi(x)$, are used. In this sense there is no difference whether a game or an abstract mathematical boundary-value problem is solved, although in a game the Hamiltonian may have some specific properties.

In this paper the procedure of solving an abstract mathematical boundary value problem, which embraces the first stage of the game solution as well, will be followed (as is already indicated in the Introduction). For the points, where the Bellman function is smooth, relations (3), (4) are reduced to the conventional equation (2). The generalizations introduced by (3), (4) become essential for the singular points defined in the sequel.

## 3 On Singular Surfaces

### 3.1 Definition of Singularities

Under certain regularity conditions optimal paths for the game (1) are known to be described by the Hamiltonian system with initial (terminal) conditions at some $t_* \in [t_0, T]$, [1]:

$$\dot{x} = H_p, \qquad \dot{p} = -H_x, \qquad x(t_*) = x^*, \quad p(t_*) = p^* \quad (p = \partial S/\partial x). \quad (5)$$

The failure of this regular procedure, following [1], is connected with the existence of some singularity. The formalization of Isaacs' approach leads to the following definitions [6].

A regular point is any inner point $x^0$ of the domain of definition $X$ of the Bellman function $S(x)$, which has a neighborhood $D$, where the function $S$ is twice differentiable and satisfies the basic equation $H(x, p) = 0$ with twice differentiable $H(x, p)$ in the neighborhood $N$ of the point $(x^0, p^0)$, $p^0 = S_x(x^0)$.

All points which are not regular are called singular points. A singular set (surface, path, manifold) consists of singular points. Usually the existence of singular sets is connected with either $S$ or $H$ (or both) not being smooth. For other closely related definitions, see [1] and [2].

It is important to distinguish between two types of singular surfaces, those which do not contain singular optimal paths (like a dispersal surface), denoted in the sequel by $\Delta$, and others that do (like universal and equivocal surfaces), to be denoted by $\Gamma$. Singular surfaces of the second category are those that require special techniques for their construction. We start the analysis with the first type of singular surfaces, without paths, which is the relatively simple case.

### 3.2 Dispersal Surface

A dispersal surface arises in the backward construction procedure of optimal trajectories, when two families of regular paths meet each other. Therefore, there is no singular path lying on $\Delta$, but the gradient $p = \partial S/\partial x$ is generally discontinuous on this surface. There is no need for any special technique to construct the dispersal surface. However, an optimality check is necessary for $x \in \Delta$ using either (3) or (4).

Let $D^+$, $D^-$ be half-neighborhoods of $\Delta$, with $D = D^+ + \Delta + D^-$, while $p^+$, $p^-$ denote the corresponding gradients of $S(x)$:

$$p^s = \partial S^s/\partial x, \qquad S^s(x) = S(x), \qquad x \in D^s, \quad s = +, -. \quad (6)$$

Here and in the sequel we suppose that, at least, $S^s(x) \in C^2(D^s)$.

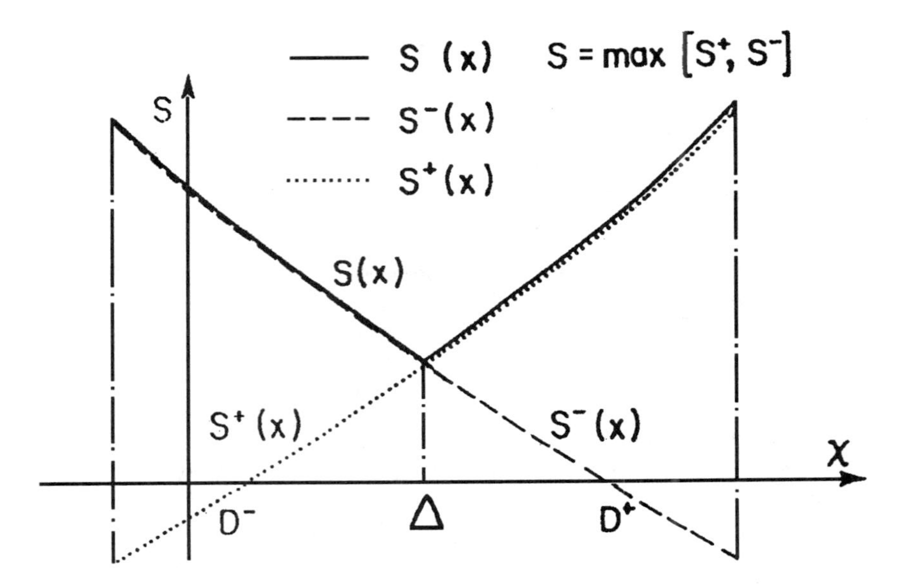


**Figure 1**: Smooth extensions of two branches of the solution.

Let the nonzero vector $p^+ - p^-$ for $x \in \Delta$ be directed from $D^+$ to $D^-$, which implies the following representation:

$$S = \min[S^+, S^-], \quad x \in D, \tag{7}$$

while $S^+ = S^-$ for $x \in \Delta$, as shown in Figure 1. We suppose here that the functions $S^{\pm}(x)$, defined in $D^{\pm}$, have extensions of the class $C^2$ to $D$. Using (4) it can be shown [6], that for $x \in \Delta$ the following optimality (viscosity) condition must hold:

$$h(\tau) = H(x, (1+\tau)/2p^+ + (1-\tau)/2p^-) \geq 0, \qquad |\tau| \leq 1, \quad x \in \Delta. \tag{8}$$

In the case of representation $S = \max[S^+, S^-]$ the inequality in (9) must be reversed.

In general, the construction of a dispersal surface consists of two steps. First, the surface $\Delta' = \{x : S^+ = S^-\}$ must be found, which is the "dispersal surface in the geometrical sense." The second step is to check all points of $\Delta'$ using (8) in order to find the optimal surface $\Delta \subset \Delta'$. In some problems not all of the points of $\Delta'$, pass the test (8) to be included in optimal solution, as will be shown in the examples presented in this paper.

### 3.3 Equations for Singular Paths

For the construction of singular surfaces containing singular paths, denoted by $\Gamma$, the procedure (5) has to be modified. An approach for the analysis of the equations for singular paths is given in [5]. In this paper the method of singular characteristics

(MSC), a technique developed in [5]–[7], is described and used for this purpose. This approach is based on the ODE system of singular characteristics, written in terms of $x$ and $p = \partial S/\partial x$ only, not using the singular controls. In addition to the Hamiltonian $H$ the MSC approach uses several functions $F_i(x, p, S)$, which may also depend upon the Bellman function $S$ and represent necessary optimality (viscosity) conditions of the type: $F_i(x, p, S) = 0, x \in \Gamma$. Such conditions can be derived by using the generalizations (3) or (4). Since the equality $H = 0$ also holds on $\Gamma$ it can be assumed that, say, $F_0 = H$.

The meaning of the additional conditions $F_i = 0$ is similar to the switching condition in optimal control. There is, however, a very important difference. The vector $p$ is a continuous function of $x$ on the switching surface, while on a singular surface $\Gamma$ it may be, in general, discontinuous. Therefore, one has to specify the value of $p$ (either $p^+$ or $p^-$, as defined in previous subsection), in the equalities $F_i(x, p, S) = 0$. This remark allows us to introduce the notions of the singular and nonsingular sides of a singular surface $\Gamma$. The singular side of $\Gamma$ (either $D^+$ or $D^-$), is the side where a nontrivial necessary condition $F_i = 0$ holds. From this definition it follows that both sides of a switching surface and no side of a dispersal surface are singular.

It is shown in [5] that for a singular surface with codimension $m, 0 < m < n+1$, the complete set of the functions $F_i$, including the Hamiltonian, should have $2m + 1$ functions. In this paper we restrict our considerations to $m = 1$, so the following three necessary optimality conditions in the form of equality must hold on the singular side of $\Gamma$:

$$F_0(x, p, S) = 0, \qquad F_1(x, p, S) = 0, \qquad F_{-1}(x, p, S) = 0, \qquad x \in \Gamma. \quad (9)$$

Using the Jacobi brackets (which become the Poisson brackets, if there is no dependence upon $S$)

$$\{RG\} = \langle R_x + pR_S, G_p\rangle - \langle G_x + pG_S, R_p\rangle$$

and the so-called singular Hamiltonian $H^\sigma$:

$$\mu \cdot H^\sigma(x, p, S) = \{F_{-1}F_0\}F_1 + \{F_1F_{-1}\}F_0 + \{F_0F_1\}F_{-1} = \mu \sum_i \lambda_i F_i \quad (10)$$

the following singular characteristic system in the $(x, p, S)$ space, restricted to the manifold defined by (9), is considered:

$$\begin{aligned} \dot{x} &= H^\sigma_p = \sum_i \lambda_i F_{ip}, \\ \dot{p} &= -H^\sigma_x - pH^\sigma_S = -\sum_i \lambda_i (F_{ix} + pF_{iS}), \\ \dot{S} &= \langle p, H^\sigma_p\rangle. \end{aligned} \quad (11)$$

Note that $\mu\lambda_i$ is the multiplier of $F_i$ in (10). The homogeneity multiplier $\mu$ is found from the condition $\dot{x}_0 = 1$ in the first equation (11). The functions $F_i$ in (9) are the first integrals of the system (11); this requirement gives the expressions

(10) for $\lambda_i$. If all the functions (9) do not depend upon $S$ then the system (11) takes the Hamiltonian form (5). Thus, in general, singular characteristics are not the solutions of a Hamiltonian system, which is the case for regular ones.

The functions $F_i$ must be found specially for each type of singularity. In [5]–[8] the complete set of $F_i$ is found for universal, equivocal, and focal surfaces. The procedure is demonstrated in the next two subsections for the equivocal and focal surfaces.

### 3.4 Equivocal Surface

An equivocal surface is a switching surface for both players, containing optimal paths. One of the players is the governing player of the equivocal surface. This player has the option to keep the trajectory on the surface, or to traverse it. If, reaching the equivocal surface, the governing player does not switch his control then the second player's optimal strategy is such that a sliding motion takes place along the surface. The Bellman function is continuous, but its gradient is discontinuous at the equivocal surface. The extrema in (3) are attained at the unique vectors $u$, $v$. There may be one exception: the extremum for the nongoverning player may be nonunique on the singular side of the surface.

Let the regular paths approach $\Gamma$ in direct (forward) time from the $D^+$ side. If the vector $(p^+ - p^-), x \in \Gamma$ is directed to $D^-$, then $S = \min[S^+, S^-]$ and $\partial S/\partial f = \min[\dot{S}^+, \dot{S}^-]$, where $\dot{S} = \langle \partial S/\partial x, f(x, u, v)\rangle$; the left side equality takes place in (3), the surface is governed by the player P with the control $u$, and the minimum with respect to $u$ in (2) is unique.

If the vector $(p^+ - p^-)$ is directed to $D^+$, then $S = \max[S^+, S^-]$ and $\partial S/\partial f = \max[\dot{S}^+, \dot{S}^-]$, right equality takes place in (3), the surface is governed by the player E with the control $v$, and the maximum with respect to $v$ in (2) is unique.

It is proved in [5] that if the extrema in (2) are unique, which means that the corresponding Hamiltonian is smooth, then the optimal paths from $D^+$ approach $\Gamma$ tangentially. The tangency condition, as well as the condition of the above-mentioned nonuniqueness, are expressed in terms of the vector $p^+$, which means that the $D^+$-side is a singular one. Thus, there may exist four types of equivocal surfaces, depending upon the governing player and tangency or nonuniqueness conditions. In the backward procedure for the equivocal surface $\Gamma$ the Bellman function can be considered to be known in advance for $x \in D^-$, since optimal regular paths approach $\Gamma$ (in inverse time) from $D^-$. In this paper only smooth Hamiltonians, $H \in C^2$, implying the tangency condition, are considered. Thus, in terms of the gradient $p = p^+$ and $S = S^+$ the following set of optimality conditions (9) is true for $x \in \Gamma$:

$$F_0 = H(x, p) = 0, \qquad F_1(x, S) = S - S^-(x) = 0, \tag{12}$$
$$F_{-1} = \{F_1 H\} = \langle H_p, p - p^-\rangle = 0.$$

Here we have the Bellman equation, the continuity condition for $S(x)$, and the tangency condition, since the vector $(p^+ - p^-)$ is normal to $\Gamma$, and $H_p$ is the velocity vector, see (5). Using the functions (12) one can calculate $\lambda_i$, $\mu$ in (10), (11):

$$\lambda_0 = 1, \qquad \lambda_{-1} = 0, \qquad \mu = \{\{F_1 H\} F_1\}, \qquad F_{1p} = 0.$$

Thus, the system (11), without the last equation, can be written as follows:

$$\dot{x} = H_p, \qquad \dot{p} = -H_x - \frac{\{\{HF_1\}H\}}{\{\{F_1 H\}F_1\}}(p - p^-(x)), \tag{13}$$
$$\{\{HF_1\}H\} \geq 0, \qquad \{\{F_1 H\}F_1\} \leq 0.$$

The inequalities here are also necessary optimality conditions [6]. For constructing the equivocal surface $\Gamma$ equations (13) can be integrated in inverse time, starting from an edge of $\Gamma$, which has to be found in advance. For a nonsmooth Hamiltonian, say, when $H = \max[H^+, H^-]$, the third function in (12) should be taken as $F_{-1} = H^-$. The corresponding singular equations can be derived from (11), see [6].

### 3.5 Focal Surface

Regular paths approach a focal surface from both sides and join the singular paths, lying on the surface, but the gradient of the Bellman function is discontinuous on the surface [1] and [2]. In the case of a smooth Hamiltonian, considered in the present paper, it is proven in [6], that regular paths approach the surface tangently from both sides, and both sides of the focal surface are singular. In the backward construction procedure of a focal surface both branches $S^+$, $S^-$ have to be constructed simultaneously. Since the gradient $p$ is a discontinuous function of $x$ on the focal surface, i.e., $(p^+ - p^-) \neq 0$, the conditions (9) or (12) have to be written for each side of $\Gamma$ as $x$ approaches (in inverse time) the focal surface from $D^+$ and $D^-$. Both functions $S^-(x)$ and $S^+(x)$ have to be considered as unknown. The method used in the previous subsection for an equivocal surface cannot be used here. Note, that Eq. (13) requires that $S^-(x)$ be twice differentiable. Since in the case of the smooth Hamiltonian the tangency condition (12) holds for each side of $\Gamma$, none of the functions $S^+$, $S^-$ has an extension of the class $C^2$ into $D = D^+ + \Gamma + D^-$ from $D^+$ and $D^-$. Thus, both functions cannot be used in (12).

To solve this case, the existence of a smooth auxiliary function $V(x)$, which is equal to $S(x)$ only on the focal surface, is assumed:

$$V(x) = S(x), \qquad x \in \Gamma, \qquad V(x) \in C^2(D).$$

The system (13) is invariant with respect to the substitution $S^-(x) \to V(x)$. Using this unknown function and its gradient $q = \partial V / \partial x$ one can write for both sides of $\Gamma$ the conditions (12) and derive the following ODE system of the order

$3(n+1)$ with the state vector $(x, p^+, p^-)$:

$$\begin{aligned} \dot{x} &= H_p(x, p^+) \qquad [= H_p(x, p^-)], \\ \dot{p}^+ &= K(x, p^+, p^-), \qquad \dot{p}^- = K(x, p^-, p^+). \end{aligned} \tag{14}$$

Here $K(x, p^+, p^-)$ is obtained from the right-hand side function of the second equation in (13) with $p = p^+$ and by substituting the term $\langle V_{xx} H_p(x, p^+), H_p(x, p^+)\rangle$ in the expression for $HF_1H$ by $-\langle H_p(x, p^-), H_x(x, p^-)\rangle$.

If for some symmetry or other reason the surface $\Gamma$ is a hyperplane, say, $x_n = 0$, then the system (14) is simplified. Assuming that $S(x) \in C^2(\Gamma)$, the auxiliary function $V$ can be taken as

$$\begin{aligned} &V(x_0, \ldots, x_n) = S^{\pm}(x_0, \ldots, x_{n-1}, 0), \qquad x \in D, \\ &q_n = 0, \qquad q_i = p_i^+ = p_i^-, \qquad i = 0, 1, \ldots, n-1, \quad x \in \Gamma. \end{aligned} \tag{15}$$

Thus, the vectors normal to $\Gamma$, like $p - p^-$ in (13), have only one non-zero component: $p^{\pm} - q = (0, \ldots, 0, p_n^{\pm})$. This allows us to reduce the system (14) to the system of the order $2n$ with respect to vectors $\bar{x} = (x_0, \ldots, x_{n-1})$, $\bar{p} = (p_0, \ldots, p_{n-1})$ and one algebraic or differential equation with respect to $p_n$ (the former having two roots $p_n^+$ and $p_n^-$):

$$\begin{aligned} \dot{x}_k &= H_k(\bar{x}, 0; \bar{p}, p_n^+), \qquad (= H_k(\bar{x}, 0; \bar{p}, p_n^-)), \\ \dot{x}_n &= H_n(\bar{x}, 0; \bar{p}, p_n^+) = 0, \quad (= H_n(\bar{x}, 0; \bar{p}, p_n^-)), \\ \dot{p}_k &= -H_{x_k}(\bar{x}, 0; \bar{p}, p_n^+), \qquad (= -H_{x_k}(\bar{x}, 0; \bar{p}, p_n^-)), \\ \dot{p}_n &= -H_{x_n}(\bar{x}, 0; \bar{p}, p_n) - \{H_n H\}/H_{nn}, \quad (p_n = p_n^{\pm}), \\ &k = 0, 1, \ldots, n-1. \end{aligned} \tag{16}$$

The equalities in parentheses in each line of (16) are the necessary conditions for the existence of a singular hyperplane $x_n = 0$.

This system can also be obtained directly by writing the equations (13) twice (once for each side of $\Gamma$) with the function $V(x)$ from (15) instead of $S^-(x)$ in (12). Here $H_k$, $H_n$, $H_{nn}$ are the first and second derivatives with respect to $p_k$, $p_n$.

The parameters $p_n^+$, $p_n^-$ can be found from the equation for $x_n$ in (16) and substituted into differential equations for $x_k$, $p_k$. Note that the resulting $2n$-order system is of a Hamiltonian type, while the full $2(n+1)$-order system is not.

If one prefers to deal only with differential equations, then the equality $H_n = 0$ should be differentiated with respect to time to obtain the last equation in (16). One can show that this equation coincides with the corresponding equation in (13):

$$\dot{p}_n = -H_{x_n} - Qp_n,$$

$Q$ being the ratio of $\{\{HF_1\}H\} = p_n\{H_n H\}$ and $\{\{F_1 H\}F_1\} = p_n^2 H_{nn}$. Note that the equality $H_n = 0$ follows from the tangency condition in (12), which in light of (15) takes the form $\{F_1 H\} = H_n p_n = 0$.

In the next two sections examples assuming the existence of a singular focal hyperplane are presented.

### 3.6 Boundary of the Indifferent Zone

Indifferent zone (IZ) is called a domain $X_z$ in the game space $X$, where the game value is a constant, and consequently its gradient is a zero-vector:

$$S(x) = C - \text{const}, \qquad p = \partial S/\partial x = 0, \qquad x \in X_z \subset X.$$

Since, generally, $p \neq 0$ for $x \in X \backslash X_z$, the boundary $\partial X_z$ is a singular surface according to the definition of Subsection 3.1. Such a zone can arise in a game or control problem with the Hamiltonian positively homogenous of the order 1, as is the case in (2):

$$H(x, \lambda p) = \lambda H(x, p), \qquad \lambda > 0.$$

Optimal controls, defined in (2), are arbitrary in IZ, since $\langle p, f(x, u, v)\rangle \equiv 0$ for $x \in X_z$. The boundary of IZ can be, generally, presented as $\partial X_z = \Gamma_0 + \Gamma_z$ with $\Gamma_0 \subset t_0 \times R^n$, i.e., $\Gamma_0$ is a part (which may be empty) of the initial plane $t = t_0$. During the solution of a game there is no need to construct something within $X_z$, one needs only to identify it. The following statement gives the sufficient conditions for the identification.

**Lemma.** *Let the continuous function $S^*(x)$ be defined in the domain $X \setminus X_z$ and satisfies there the viscosity solution's conditions for the problem (2); let $S^*(x) = C - \text{const}$ for $x \in \Gamma_z$, i.e., $S^*$ is a constant on the essential part of the boundary of IZ; let $\Gamma_z$ be a hypersuface, locally smooth everywhere, except the finite number of its sections at $t = t_i, i = 1, \ldots, m$, i.e., except the points of $\Gamma_z \cap (t_i \times R^n)$; let the game Hamiltonian be positively homogenous of order 1. Then the function $S(x)$ defined in $X$ as*

$$S(x) = S^*(x), \quad x \in X\backslash X_z; \qquad S(x) = C, \quad x \in X_z,$$

*is a viscosity solution of (2) or, in other words, the value of the game (1).*

The proof of the lemma consists in verifying the viscosity conditions on the surface $\Gamma_z$, since a constant function satisfies the Bellman equation in $X_z$. Let us consider in the vicinity of some point $x \in \Gamma_z$ the function $S^*$ as the $S^+$ from Subsection 3.2 and the constant $C$ as the $S^-$. Using the properties of the Hamiltonian one can find that the function $h(\tau)$ vanishes:

$$\begin{aligned} h(\tau) &= H(x, (1+\tau)/2p^+ + (1-\tau)/2p^-) \\ &= H(x, (1+\tau)/2p^+) = (1+\tau)/2H(x, p^+) = 0 \end{aligned}$$

and trivially satisfies the viscosity condition in (8) or the opposite one. Omitting the finite number of hyperplanes $t = t_i$ appears not to be essential for the verification of the viscosity solution.

# 4 Initial Value Example

## 4.1 Problem Formulation

Consider a two-dimensional initial value problem with respect to the function $S(x, y)$, $x, y \in R^1$, $x \geq 0$ (using componentwise notations):

$$H(x, y, p, q) = p + \sqrt{a^2 + q^2} - x\sqrt{b^2 + q^2} = 0, \qquad x > 0, \tag{17}$$
$$S(0, y) = -|y|, \qquad p = \partial S/\partial x, \quad q = \partial S/\partial y,$$

where $a$ and $b$ are positive constants.

The problem (17) is not necessarily connected with any optimal control or game problem. The objective is to find the viscosity solution to (17), which, according to [4], exists and is unique. It is easy to verify that $H \in C^2$, at least.

The regular characteristic system (5) and the corresponding initial conditions are written for problem (17) in the form

$$\dot{x} = H_p = 1, \qquad \dot{y} = H_q = q/\sqrt{a^2 + q^2} - xq/\sqrt{b^2 + q^2}, \tag{18}$$
$$\dot{p} = -H_x = \sqrt{b^2 + q^2}, \qquad \dot{q} = -H_y = 0, \qquad \dot{S} = pH_p + qH_q,$$
$$p(0, y) = -\sqrt{a^2 + 1}, \qquad q(0, y) = -\mathrm{sgn}(y), \qquad y \in R_1.$$

Since $q$ is a constant along solutions of (18), all regular characteristics for problem (17) are the following parabolas in the $(x, y)$-plane with vertical symmetry axes $x = x^+$:

$$y - C = qx/\sqrt{a^2 + q^2} - qx^2/2\sqrt{b^2 + q^2}, \qquad x^+ = \sqrt{\frac{b^2 + q^2}{a^2 + q^2}}, \qquad C - \text{const.} \tag{19}$$

Integrating system (18) with the initial values $q = \pm 1$ for the positive and negative points of the $y$-axis one can obtain the following (primary) solution:

$$S(x, y) = \min[S^+, S^-] = -|y| + \frac{1}{2}x^2\sqrt{b^2 + 1} - x\sqrt{a^2 + 1}, \tag{20}$$
$$S^{\pm}(x, y) = \mp y + \frac{1}{2}x^2\sqrt{b^2 + 1} - x\sqrt{a^2 + 1}.$$

The continuity condition $S^+ = S^-$ holds on the $x$-axis, where the characteristics, starting in $y > 0$ and $y < 0$, intersect. The last (critical) pair of characteristics, with $C = \pm\frac{1}{2}\sqrt{b^2 + 1}/(a^2 + 1)$ and $q = \pm 1$ in (19), are parabolas tangent to the $x$-axis from both sides at $x = x^+ = x_0$.

Based on the relationship between the two constant parameters of the problem the following three cases can be distinguished:

$$\begin{aligned} &(1) \quad a < b \quad (1 < x_0 < x_1 < x_2), \\ &(2) \quad a = b \quad (1 = x_0 = x_1 = x_2), \\ &(3) \quad a > b \quad (1 > x_0 > x_1 > x_2), \end{aligned} \tag{21}$$

where

$$x_0 = \sqrt{\frac{b^2+1}{a^2+1}}, \qquad x_1 = \frac{\sqrt{a^2+1}-a}{\sqrt{b^2+1}-b}, \qquad x_2 = \frac{b}{a}.$$

In the following part of this paper the terminology related to pursuit- evasion games will be used, as if (17) were an inverse-time formulation for some game problem. Thus, the above-mentioned segment $(0, x_0)$ is called a dispersal curve, despite the fact that the solution $(x, y)$ approaches this segment in the direct (forward) time $t = x$.

### 4.2 The Case 1, $a < b$

In problem (17) the viscosity condition (8) for the dispersal segment "in the geometrical sense" has the following form in terms of the primary solution (20):

$$h(\tau) = p + \sqrt{a^2+\tau^2} - x\sqrt{b^2+\tau^2} \leq 0, \qquad |\tau| \leq 1, \tag{22}$$
$$x \in (0, x_0), \quad y = 0 \qquad p = \partial S/\partial x = x\sqrt{b^2+1} - \sqrt{a^2+1}.$$

The direction of the inequality is changed because, instead of the terminal value problem of the Section 2, here an initial value problem is considered. It is easy to verify that for $a < b$ the inequality (22) holds for all $\tau$ and $x : 0 \leq x \leq x_0$. Thus, all points of the dispersal segment "in the geometrical sense" pass the optimality test.

In order to complete the solution, the domain in the right side of the tangent parabolas has to be filled by characteristics. The assumption that a segment of the $x$-axis beyond $x_0$ is a focal line appears to be useful. While $p$ is continuous, $q$ has a discontinuity on the focal segment, as in (15). This situation is illustrated in Figure 2.

From the equality $H_q = 0$, see (16), one can find the value of $q$ at a point $x = \xi, y = 0$ for both sides of the $x$-axis:

$$q^{\pm} = \mp a\sqrt{(x_2^2 - \xi^2)/(\xi^2-1)}, \qquad x_0 \leq \xi \leq x_2, \tag{23}$$
$$0 \leq |q| \leq 1, \quad x = \xi, \quad y = 0.$$

It is easy to see that problem (17) has a symmetry with respect to the $x$-axis. The superscript "+" (or "−") will be used for the upper (lower) quadrant; the quantities without the superscript should be treated as for the upper quadrant. Substituting $q$ from (23) into (19) creates the family of characteristics, corresponding to

$$C = \pm\xi/2\sqrt{(x_2^2-\xi^2)/(x_2^2-1)}, \qquad x \geq x^+ = \xi, \quad x_0 \leq \xi \leq x_2,$$

which covers the remaining part of the half-plane $x \geq 0$. Thus, the focal line ends at $x = x_2$. The value of the function $S$ at a point $(x, y)$ can be found by integrating

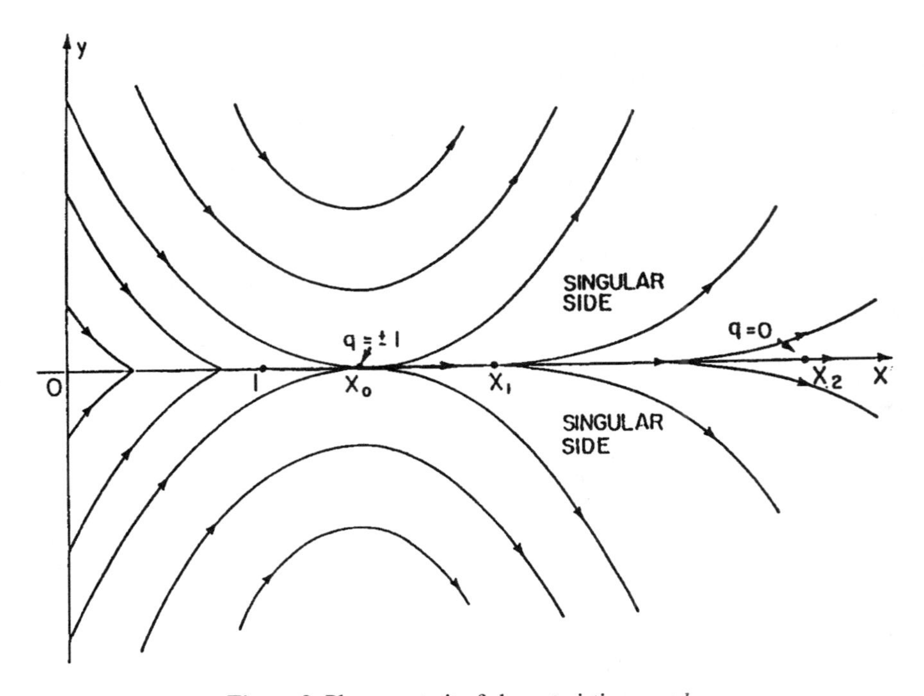


**Figure 2**: Phase portrait of characteristics, $a < b$.

the last equation in (18) along the characteristic

$$S(x, y) = \int_{\xi}^{x} (p + qH_q)\,dx. \tag{24}$$

Here $\xi = \xi(x, y)$ corresponds to the characteristic meeting the point $(x, y)$; $q(\xi)$ has to be taken from (23), and $p$ can be found from $H = 0$ in (17).

The resulting function is continuous in $X$ and smooth everywhere in $X$ except the segment $[0, x_2)$. Along this segment it satisfies the necessary condition for a viscosity solution and it satisfies (17) elsewhere. This means that the constructed function is the unique viscosity solution to the initial value problem formulated in (17). Thus, two singular curves exist in this case: dispersal and focal segments.

### 4.3 The Case 2, $a = b$

In this case the point $x_2$ coincides with $x_0$, the focal segment vanishes, and the following funnel of a one-parametric family of characteristics (with the parameter $q$), starting at the point $(x_0, 0)$, can be constructed:

$$y(x) = -q(x-1)^2/[2\sqrt{a^2+q^2}], \qquad |q| \leq 1, \tag{25}$$

as shown in Figure 3. The segment $(0, x_0)$ again is a dispersal segment. The integral (24) can be calculated along the characteristics given by (25) with $a = b$ and $\xi = 1$,

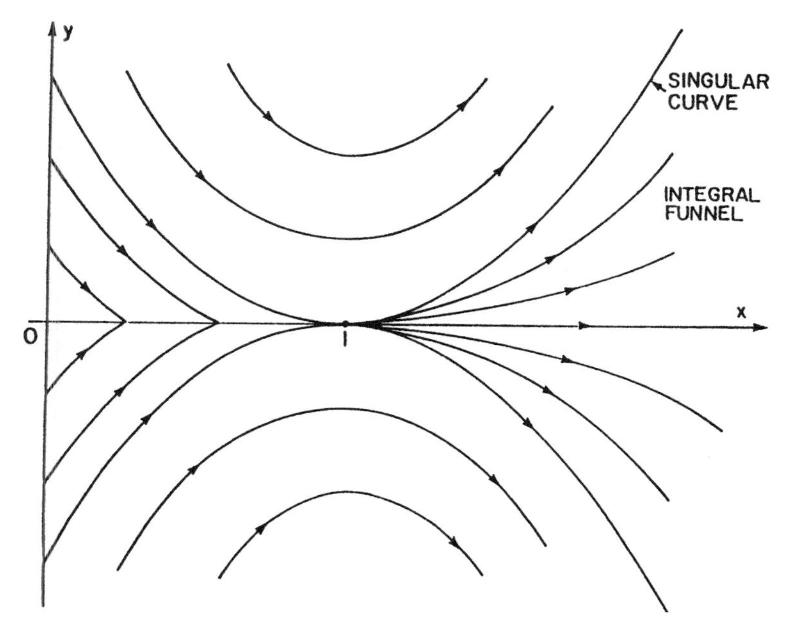


**Figure 3**: Phase portrait of characteristics, $a = b$.

generating the secondary solution, defined in the domain which is to the left of the critical parabola (with $|q| = 1$ in (25)). The secondary solution will be denoted as $V(x, y)$:

$$V(x, y) = S(x, y) = -\tfrac{1}{2}\sqrt{a^2 + 1} + a\sqrt{(x - 1)^4/4 - y^2},$$
$$|y| \le \tfrac{1}{2}(x - 1)^2/\sqrt{a^2 + 1}, \quad x \ge 1. \tag{26}$$

The functions (20) and (26) represent the unique viscosity solution for this case. It is a smooth function everywhere outside the dispersal segment $[0, x_0]$. The extremal lines of the funnel appear to be singular curves of a new type, according to the definition of Section 2. Indeed, one can calculate second derivatives of the functions (20) and (26) on this parabola equal to

$$S_{yy} = 0, \qquad V_{yy} = -2(a^2 + 1)^{3/2}/[a^2(x - 1)^2] \neq 0.$$

In the next case this singularity gives rise to the equivocal curve.

## 4.4 The Case 3, $a > b$

In this case not all the points of the segment $[0, x_0]$, $y = 0$, pass the test (22) for belonging to the dispersal surface. For $a > b$, the function $h(\tau)$ in (22) may have a unique internal maximum at $\tau = 0$, as shown in Figure 4. Thus, the condition (22) is equivalent to $h(0) \le 0$. This inequality holds only for $x \in [0, x_1]$, and $h(0) > 0$

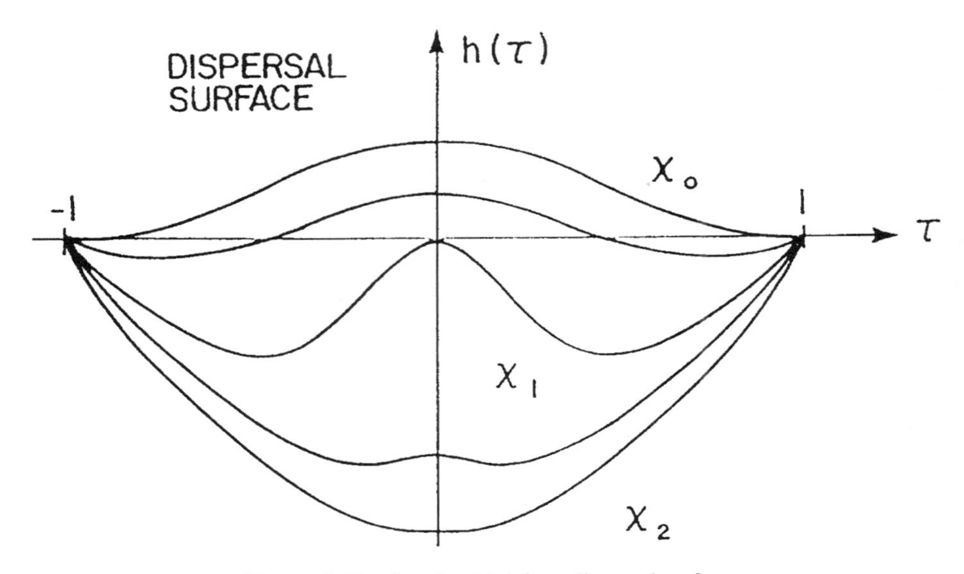


**Figure 4**: The function $h(\tau)$ for a dispersal surface.

if $x > x_1$. The value of $x_1$ is determined by solving the following equation derived from (22):

$$h(0) = x(\sqrt{b^2+1} - b) - \sqrt{a^2+1} + a = 0. \tag{27}$$

This means that the subsegment $(x_1, x_0)$ of the dispersal segment "in the geometrical sense" $[0, x_0)$ fails to be a part of the optimal dispersal surface and has to be excluded from considerations.

In order to carry out further construction of the solution it is assumed that two symmetrical equivocal curves start at the point $(x_1, 0)$, as illustrated in Figure 5. The half-plane $X$ is represented as $X = X_0 + X_1$, where $X_0$ is the (primary) domain to the left of the equivocal curves, and $X_1$ is the domain to the right of them. In $X_0$ the solution of (17) is given by (20). In the domain $X_1$ the notation $V(x, y)$ is used. Since the Hamiltonian $H$ in (17) is smooth, the tangency condition for the equivocal surface, given in (10), has to be fulfilled. Using the Hamiltonian from (17) and the function $S$ in (20) one can write the singular equations (13) for the upper part of the equivocal curve, lying in the domain $X^+ = X_0^+ + X_1^+$:

$$\begin{aligned} &\dot{y} = H_q, \qquad \dot{q} = [\sqrt{b^2+1} - (b^2 - q)/\sqrt{b^2+q^2}]/[(q+1)H_{qq}], \\ &x \geq x_1; \qquad y(x_1) = 0, \qquad q(x_1) = 0 \qquad (q = \partial V/\partial y). \end{aligned} \tag{28}$$

The equation $\dot{x} = H_p = 1$ is omitted here, as well as the equation for $p$, since after the integration of (28) it can be found from the equality $H = 0$. Note that the Hamiltonian is a first integral of the singular equations (13) and (28)). These simplifications take place because of the low dimension of the problem.

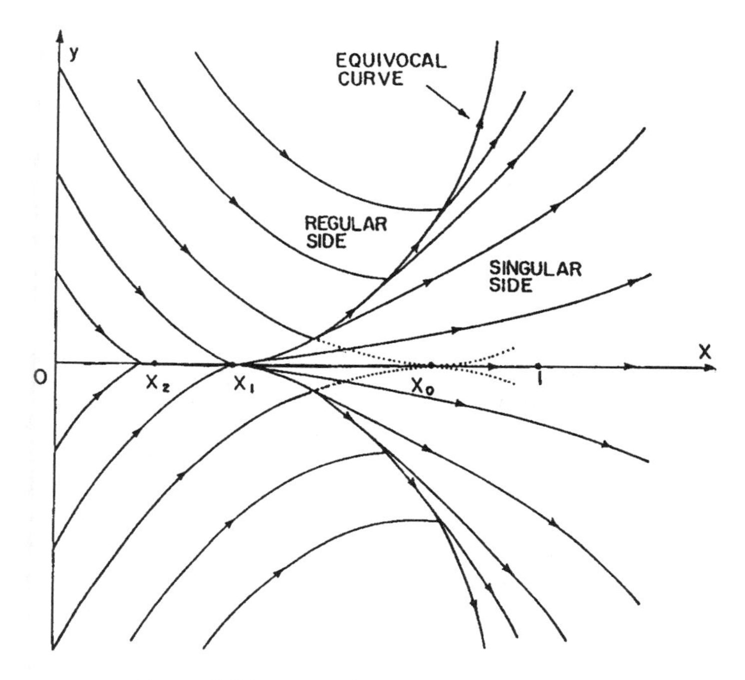


**Figure 5**: Phase portrait of characteristics, $a > b$.

The second initial condition in (28) is obtained as follows. In general, initial values for the vector $p$ should be found using equalities (12) and some of their differential corollaries [8]. This equalities should be considered both for upper and lower branches of the equivocal curve. It is easy to verify that the system (12) has the following solution, common for both branches:

$$p = \partial V/\partial x = (p^+ + p^-)/2, \qquad q = \partial V/\partial y = (q^+ + q^-)/2, \qquad (29)$$
$$p^\pm = \partial S^\pm/\partial x, \qquad q^\pm = \partial S^\pm/\partial y.$$

This means that the function $V(x, y)$ has a continuous gradient up to the point $(x_1, 0)$. Since we have $q^+ = -1$ and $q^- = 1$ in the primary domain, the condition $q(x_1) = 0$ in (28) follows from (29).

Analysis shows that the equations (28) determine a smooth curve, tangent to the $x$-axis at $x = x_1$, and $q \to -1$ with $x \to +\infty$. In order to obtain the function $V(x, y)$ in the domain $X_1$, the regular system (18) should be integrated, starting from the points of the equivocal curve with the values of $p, q$, obtained on it during its construction. The resulting viscosity solution to (17) is a smooth function everywhere, except on the dispersal segment and equivocal curves, where it satisfies the viscosity conditions (8), (12). (Note that the form of the function $h(\tau)$ on an equivocal and a focal curve is shown in the lower part of Figure 4.)

## 5 A Linear Pursuit-Evasion Game with Elliptical Vectograms

### 5.1 Problem Formulation

In this section the MSC approach is applied to a linear pursuit-evasion game with elliptical vectograms. This game is a mathematical model of a missile versus aircraft terminal engagement, assuming constant speed motion in three-dimensional Euclidean space and trajectory linearization around a nominal collision course. The trajectory linearization allows us to assume uniform relative motion along the line of sight, reducing the problem to a two-dimensional game with fixed final time. The lateral accelerations of the players, perpendicular to the respective velocity vectors, are bounded. In the line of sight coordinates, these lateral accelerations create an elliptical vectogram for each player (unless the respective velocity vectors are collinear with the line of sight, where the vectograms are circular). Moreover, it is assumed that the evader has ideal dynamics, while the pursuer's dynamics is represented by a first-order transfer function with the time constant $\tau_p$. The game has two parameters of physical significance, the speed ratio of the players $\gamma = V_E/V_P < 1$ and the ratio of their maximum lateral accelerations $\mu = (a_P)_{max}/(a_E)_{max} > 1$. In [9] this game was solved, using conventional techniques [1], [2] requiring a lot of intuition for the identification and the construction of the singular surfaces. In this paper a renewed analysis of the singular surfaces of this game is performed using the MSC approach.

By applying the method of "terminal projection" [10] the game has only two state variables, namely the components of the "zero-effort miss distance" [11] vector in the line of sight coordinates. As in [9] the game is formulated in nondimensional relative coordinates and normalized inverse time. Distances are normalized by $\tau_p^2(a_E)_{\max}$ and the inverse time by $\tau_p$. Accordingly, the equations defining the game can be summarized as follows:

$$\dot{y} = dy/d\theta = -\psi(\theta)u + \theta v; \qquad 0 \le \theta \le \theta_0; \qquad y, u, v \in R^2, \tag{30}$$

$$u \in U = \{u : \langle R_P u, u\rangle \le \mu^2\}, \qquad v \in V = \{v : \langle R_E v, v\rangle \le 1\},$$

$$\psi(\theta) = \theta + e^{-\theta} - 1, \qquad R_i = \begin{vmatrix} k_i^{-2} & 0 \\ 0 & 1 \end{vmatrix}, \qquad k_i = \cos\chi_i, \qquad i = P, E,$$

$$J = |y(0)| = \sqrt{y_1^2(0) + y_2^2(0)}.$$

In this formulation $y$ is a nondimensional vector of two components in a plane perpendicular to the initial line of sight. The independent variable is $\theta$, the normalized time-to-go (inverse time), $u$ and $v$ are control vectors of the pursuer P and the evader E, and $\chi_i (i = P, E)$ the respective angles between the velocity vector of each player and the line of sight. The objective of the player P (E) with the control $u$, ($v$) is to minimize (maximize) the cost-function $J$ in the original game formulation. The relationship between the speed ratio $\gamma$ and the parameters of the initial collision course triangle $\chi_P$ and $\chi_E$ is expressed by

$$\gamma = V_E/V_P = \sin\chi_P/\sin\chi_E = \sqrt{(1-k_P^2)/(1-k_E^2)}.$$

Note that the collision course triangle has a unique solution for any given value of $\chi_E$ only if $\gamma \leq 1$.

The augmented state-vector of the game is

$$x = (\theta, y_1, y_2), \qquad x_0 = \theta, \quad x_1 = y_1, \quad x_2 = y_2.$$

The Hamiltonian as a function of this augmented state-vector and the gradient of the Bellman function $p = \partial S/\partial x$ has the form

$$H(x,p) = p_0 + \max_u \min_v \langle \lambda, \dot{y} \rangle, \tag{31}$$
$$\lambda = (\lambda_1, \lambda_2), \qquad \lambda_1 = p_1, \quad \lambda_2 = p_2. \quad p = (p_0, p_1, p_2).$$

Using normalized inverse time is the reason that the extrema in (31) are in the opposite sense. The optimal controls of the players, obtained from (31), have the form

$$u^* = -\mu M_P^2\lambda/|M_P\lambda|, \qquad v^* = -M_E^2\lambda/|M_E\lambda|,$$
$$M_i^2 = R_i^{-1}, \qquad M_i = \begin{vmatrix} k_i & 0 \\ 0 & 1 \end{vmatrix}, \qquad i = P, E. \tag{32}$$

Substituting these results into (31) yields

$$H(x,p) = p_0 + \mu\psi\sqrt{p_1^2k_P^2 + p_2^2} - \theta\sqrt{p_1^2k_E^2 + p_2^2}. \tag{33}$$

The corresponding Bellman equation and the initial value problem of type (5) here takes the form, using (30), (31):

$$H(x,p) = 0, \quad 0 < x_0 \leq \theta_0, \qquad S(0, x_1, x_2) = \sqrt{x_1^2 + x_2^2}, \tag{34}$$

$$\begin{aligned} \dot{y}_1 &= \psi\mu k_P^2\lambda_1/|M_P\lambda| - \theta k_E^2\lambda_1/|M_E\lambda|, \qquad & y_1(0) = \alpha, \\ \dot{y}_2 &= \psi\mu\lambda_2/|M_P\lambda| - \theta\lambda_2/|M_E\lambda|, \qquad & y_2(0) = \beta, \end{aligned} \tag{35}$$

$$\begin{aligned} \dot{\lambda}_1 &= 0, \qquad & \lambda_1(0) = \alpha/\sqrt{\alpha^2+\beta^2}, \\ \dot{\lambda}_2 &= 0, \qquad & \lambda_2(0) = \beta/\sqrt{\alpha^2+\beta^2}. \end{aligned} \tag{36}$$

The equation $\dot{x}_0 = 1$ is omitted here, and $p_0(\theta)$ can be found from the first-integral condition $H = 0$ after the integration of (33), (34). From (35), (36) it also follows that

$$\dot{y}_1 = f_1(\theta, \alpha, \beta), \qquad \dot{y}_2 = f_2(\theta, \alpha, \beta) = \beta f_2^0(\theta, \alpha, \beta).$$

The integration of the system (35), (36) gives

$$y_1 = \alpha + P(\theta)\frac{k_P^2\alpha}{\sqrt{k_P^2\alpha^2 + \beta^2}} - E(\theta)\frac{k_E^2\alpha}{\sqrt{k_E^2\alpha^2+\beta^2}} = F_1(\theta, \alpha, \beta), \tag{37}$$

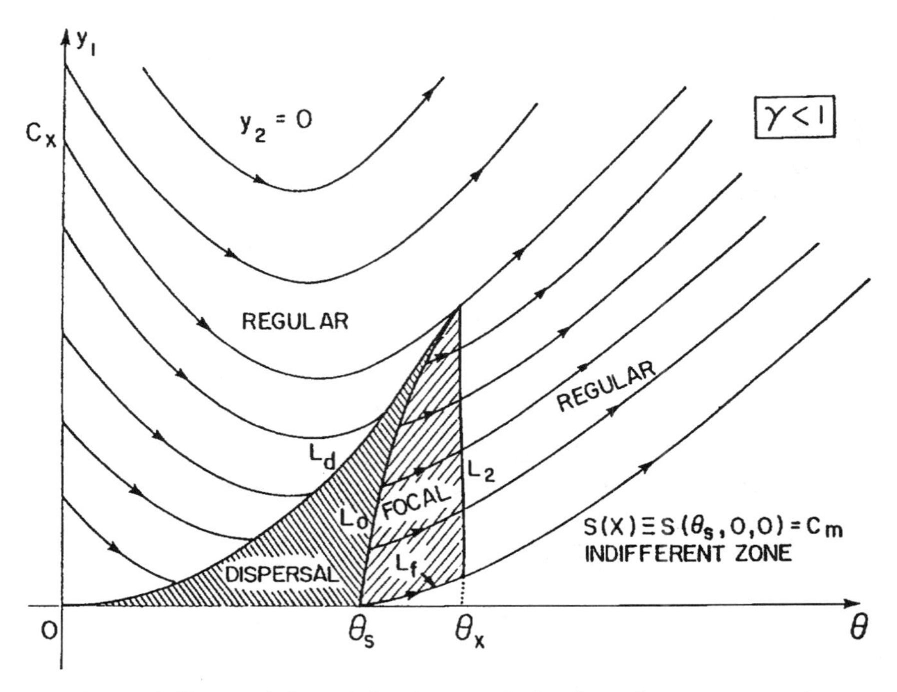


**Figure 6**: Characteristics and singular zones in the plane of symmetry, $\gamma < 1$.

$$y_2 = \beta + P(\theta)\frac{\beta}{\sqrt{k_P^2\alpha^2+\beta^2}} - E(\theta)\frac{\beta}{\sqrt{k_E^2\alpha^2+\beta^2}} = F_2(\theta,\alpha,\beta) = \beta F_2^0(\theta,\alpha,\beta),$$

$$P(\theta) = \mu(\theta^2/2 - \psi(\theta)), \qquad E(\theta) = \theta^2/2.$$

Formulas (37) define all the "primary" paths. A great majority of them do not intersect each other. However, they do not cover the entire game space.

## 5.2 Dispersal Surface

Two families of the primary paths intersect each other and create a dispersal surface in the plane of symmetry $y_2 = 0$. The boundary of this domain is a curvilinear triangle with sides $L_d^+$, $L_d^-$, $L_0$, as is shown in Figure 6. Along the line $L_0$ regular primary paths are tangent to the plane $y_2 = 0$, i.e., the equality $\dot{y}_2 = 0$ holds at the points of this line. Thus, the line $L_0$ is determined by the equality

$$L_0: \quad y_1 = F_1(\theta, \alpha^0(\theta), \beta^0(\theta)), \qquad \theta_s \le \theta \le \theta_x,$$

where $\alpha^0$, $\beta^0$ satisfy the system $y_2 = 0$, $\dot{y}_2 = 0$ with $\beta \neq 0$, or

$$F_2^0(\theta,\alpha,\beta) = 0, \quad f_2^0(\theta,\alpha,\beta) = 0.$$

The values $\theta_s, \theta_x$ are the nonvanishing roots of the equations

$$\begin{aligned} \theta_s: \quad & \theta = \mu\psi(\theta), \\ \theta_x: \quad & k_P\theta = \mu k_E \psi(\theta). \end{aligned} \tag{38}$$

These equations have finite roots only if $\mu > 1$ and $(\mu k_E / k_P) > 1$.

The dispersal zone for the problem (34) does not start immediately at the boundary $\theta = 0$, but at the points of two symmetrical lines $L_d^+, L_d^-$ in the plane $y_2 = 0$. The lines $L_d^+, L_d^-$ consist of the points, where the equation

$$y_2 = \beta F_2^0(\theta, \alpha, \beta) = 0$$

has only the single trivial root $\beta = 0$. At all other points of the dispersal surface the former equation has, in addition, two nonzero symmetric roots $\pm\beta^*$, which correspond to two paths, starting (in direct time) at the dispersal surface. This leads to the following condition for $L_d$:

$$\partial F_2(\theta, \alpha, 0)/\partial\beta = F_2^0(\theta, \alpha, 0) = 0,$$

leading to the relation

$$\alpha^*(\theta) = E(\theta)/k_E - P(\theta)/k_P$$

and to the following explicit expression for $L_d$:

$$L_d: \quad y_1 = F_1(\theta, \alpha^*(\theta), 0) = P(\theta)(k_P - 1/k_P) - E(\theta)(k_E - 1/k_E), \qquad y_2 = 0,$$
$$0 \le \theta \le \theta_s.$$

The problem (30)–(34) has two planes of symmetry: $y_1 = 0$ and $y_2 = 0$. The gradient $p$ jumps at the plane $y_2 = 0$, while the components $p_0, p_1$ are continuous. For reasons of symmetry the equality $p_2^+ = -p_2^- = p_2$ holds for $y_2 = 0$, which allows us to write the function (8) in the form

$$h(\tau) = H(x, p_0, p_1, \tau p_2) = p_0 + \mu\psi\sqrt{p_1^2 k_P^2 + \tau^2 p_2^2} - \theta\sqrt{p_1^2 k_E^2 + \tau^2 p_2^2}. \tag{39}$$

The analysis shows that $p_2^+ > 0$ at $y_2 = 0$, leading to $S = \max[S^+, S^-]$ representation of the Bellman function in the vicinity of the plane $y_2 = 0$. Thus, we have the optimality condition $h(\tau) \ge 0$, $|\tau| \le 1$, for the dispersal surface instead of (8), since instead of the terminal value problem (2) we are considering an initial value problem (32). If $\gamma \le 1$, then the function (34) satisfies this condition for all points of the dispersal surface "in the geometrical sense" and agrees with the constructions in [9]. The dispersal surface terminates along $L_0$ where the characteristics (trajectories) reach the plane of symmetry tangentially. In order to continue the solution beyond this line a new type of singularity has to be considered.

## 5.3 Focal Surface

The tangential behavior of the characteristics along the line $L_0$ suggests a type of singularity described in Subsection 3.5 as a focal surface. (Note that the Hamiltonian (33) of the present problem is a smooth function of its arguments almost everywhere.) Identification of this singular surface and the construction of the corresponding optimal trajectories presented a major difficulty in [9]. Fortunately, the phenomenon is related to the plane $y_2 = 0$, therefore the equations become simplified.

For the problem formulated in (34) the equations for focal paths (16) have the form

$$\dot{y}_1 = \partial H/\partial\lambda_1, \qquad \dot{y}_2 = 0, \qquad (40)$$
$$\dot{\lambda}_1 = 0, \qquad \dot{\lambda}_2 = -\frac{\partial^2 H}{\partial\theta\partial\lambda_2} \Big/ \frac{\partial^2 H}{\partial\lambda_2^2}.$$

The order of this system can be reduced as indicated in Subsection 3.5. From the equality $\partial H/\partial\lambda_2 = 0$ one can find

$$\left(\frac{\lambda_2}{\lambda_1}\right)^2 = \frac{(\theta k_P)^2 - (\mu\psi k_E)^2}{\mu^2\psi^2 - \theta^2} = g^2(\theta).$$

Substitution into the first equation (40) gives

$$\dot{y}_1 = \frac{\partial H}{\partial\lambda_1} = \mu\psi(\theta)\frac{k_P^2}{\sqrt{k_P^2 + g^2(\theta)}} - \theta\frac{k_E^2}{\sqrt{k_E^2 + g^2(\theta)}}. \qquad (41)$$

The last equation simplifies the construction procedure proposed in [9]. Each focal line is defined on some segment $[\theta_*, \theta_x]$, $\theta_s \le \theta_* \le \theta_x$, and naturally ends at $\theta_x$, on the line $L_2$, when $g(\theta_x) = 0$, since $g^2(\theta)$ has to be nonnegative.

Thus, the boundary of the focal region consists of the curve $L_0$, the segment $L_2$ of the line $\theta = \theta_x$, $y_2 = 0$, and of the "limiting" focal path $L_f$ defined as the solution of (37) on the interval $[\theta_s, \theta_x]$, subject to the initial condition $y_1(\theta_s) = 0$.

The optimal strategies of the players on the focal surface are also given by the relations (32) but $\lambda$ is now not a constant but the solution of the system (40). Substitution of these controls into the equations of motion (30) gives, in particular, the equation (41).

Outside the focal zone the family of regular paths, starting at the points of $L_2$, stays on the plane $y_2 = 0$. The line $L_f$, starting at $\theta_s$, is by itself a trajectory. Those trajectories, which start at $L_f$, form a surface $\Gamma_0$, which separates the region, including the part of the plane $y_2 = 0$ between the $\theta$-axis and $L_f$, from the rest of the game space. In this region the game value is the same everywhere and is equal to $S(\theta_s, 0, 0) = C_m$. This region, called in [9] *the minimal tube*, is denoted here as the "indifferent zone" (IZ). The boundary $\Gamma_0$ is a singular surface along which the viscosity conditions are satisfied (see the lemma in Subsection 3.5).

If $\gamma = 1$ ($k_E = k_P$), then $\theta_s = \theta_x$, and one can show that $L_0 = L_2$ becomes in this case a straight line segment parallel to the $y_1$-axis, as can be seen in Figure 7. As a consequence the focal zone vanishes.

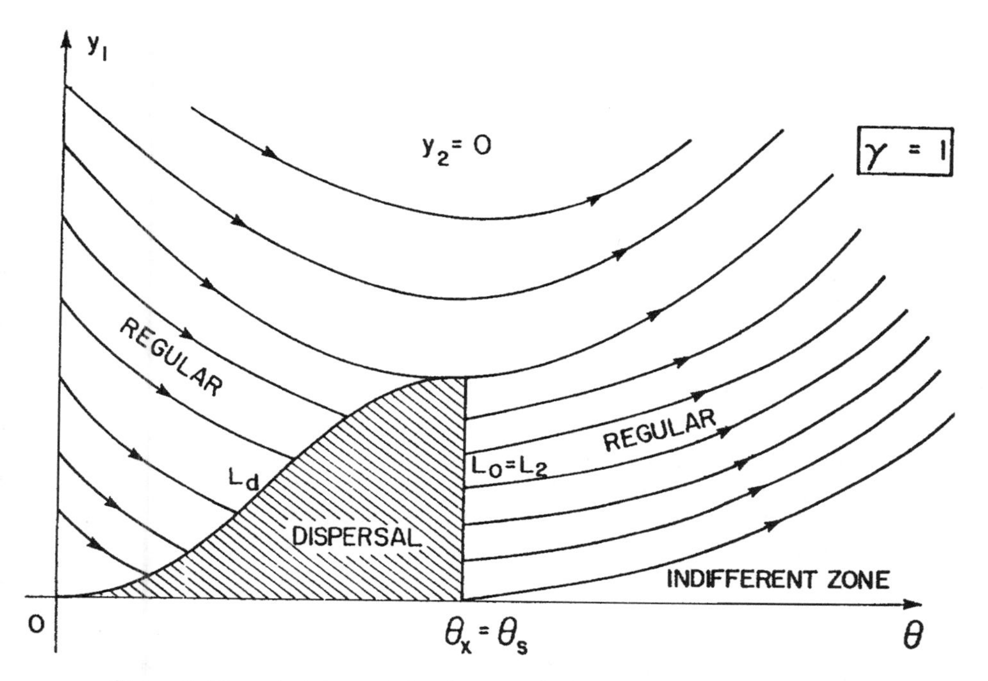


**Figure 7**: Characteristics and singular zones in the planes of symmetry, $\gamma = 1$.

## 5.4 Equivocal Surface

If $\gamma > 1$ ($k_E \geq k_P$), which is not considered in [9], it can happen that not all of the dispersal points pass the optimality test $h(\tau) \geq 0$. (Note that for $\gamma > 1$, a solution may exist only if $\sin \chi_E < 1/\gamma$, but this solution is not unique.)

If this occurs then the dispersal region on the plane $y_2 = 0$ terminates along a boundary line $L_1$ defined by one of the equivalent equations

$$h(0) = p_0 + \mu\psi|\lambda_1|k_P - \theta|\lambda_1|k_E = 0,$$

$$\mu\psi(\sqrt{\lambda_1^2 k_P^2 + \lambda_2^2} - |\lambda_1|k_P) = \theta(\sqrt{\lambda_1^2 k_E^2 + \lambda_2^2} - |\lambda_1|k_E).$$

The second equation here is the consequence of the first one and $H = 0$.

The two branches of the equivocal surface can be constructed by integrating the equations of singular characteristics (13) from the points of $L_1$:

$$\begin{aligned} &\dot{y}_1 = \partial H/\partial p_1, \qquad \dot{y}_2 = \partial H/\partial p_2, \\ &\dot{p}_0 = -H_\theta - Q(p_0 - q_0), \\ &\dot{p}_1 = -Q(p_1 - q_1), \qquad \dot{p}_2 = -Q(p_2 - q_2), \\ &Q = \{\{HF_1\}H\}/\{\{F_1H\}F_1\}; \qquad F_1(x, S) = S - S'(x), \\ &q = \partial S'/\partial x. \end{aligned} \tag{42}$$

Here $S'(x)$ is the primary solution; the Poisson bracket $\{\{HF_1\}H\}$ in the system (42) uses its second derivatives, $S'_{xx}$. The equation for $p_0$ in (42) is separated and $p_0$ can be found from $H = 0$ after the integration of the other four equations in (42). It can be shown that the initial values for $p$ at the points of $L_1$ are $p_0 = q_0, p_1 = q_1, p_2 = 0$.

The complete analysis of the equivocal surface requires a reformulation of the original pursuit-evasion game with $\gamma > 1$, as well as extensive numerical computations. This effort, beyond of the scope of the present paper, is the subject of an ongoing investigation.

## 6 Conclusions

The method of singular characteristics provides a systematic approach for the construction of singular surfaces in pursuit-evasion games. It separates the construction of these singular surfaces from the computation of the optimal feedback control strategies. Only two functions, namely the Hamiltonian and the boundary value for the Bellman function, are used in this technique. The MSC may not be convenient for the problems with relatively complicated structure of singular manifolds.

The two examples presented in this paper serve to illustrate the similarities in solving an initial value PDE problem and a differential game by the method of singular characteristics. The reconstruction of the game solution of [9] has demonstrated that applying this method is a more efficient systematic way to identify and to construct the singular game surfaces than the classical approach.

Moreover, the first (initial value problem) example indicated that by changing some parameters in the problem different types of singularities can be encountered. Symmetry in the game structure creates, in general, dispersal surfaces in the geometrical sense. If the entire dispersal surface is optimal (i.e., satisfies the necessary conditions of a viscosity solution), but the solution is still not complete, then the existence of a focal surface is most likely. When a part of the dispersal surface is shown to be nonoptimal, then the existence of an equivocal surface should be assumed and verified. This observation has motivated a new investigation related to the second example.

## REFERENCES

[1] Isaacs, R., *Differential Games*. Wiley, New York, 1965.

[2] Bernhard, P., Singular Surfaces in Differential Games: An Introduction. In: *Differential Games and Applications*. Springer Verlag, Berlin, pp. 1–33, 1977.

[3] Subbotin, A., I. A Generalization of the Basic Equation of the Theory of Differential Games. *Soviet Mathematics Doklady*, **22**, 358–362, 1980.

[4] Lions, P.-L. and P. E. Souganidis. Differential Games, Optimal Control and Directional Derivatives of Viscosity Solutions of Bellman's and Isaacs' Equations. *SIAM Journal of Control and Optimization*, **23**, No. 4, 566–583, 1985.

[5] Melikyan, A. A. *The Method of Characteristics for Constructing Singular Paths and Manifolds in Optimal Control and Differential Games*. Lecture Notes in Control and Informational Science, Vol. 156. Springer-Verlag, Berlin, pp. 81–90, 1991.

[6] Melikyan, A. A. Generalized Characteristics of First Order PDEs: Applications in Optimal Control and Differential Games. Birkhäuser Boston, 1998.

[7] Melikyan, A. A. Necessary Optimality Conditions for a Singular Surface in the Form of Synthesis. *Journal of Optimization Theory and Applications*, **82**, No. 2, 203–217, 1994

[8] Melikyan, A. A. and N. V. Ovakimyan. Differential Games of Simple Pursuit and Approach on Manifolds. Institute of Mechanics, Armenian Academy of Sciences, Preprint, Yerevan, 1993.

[9] Shinar, J., M. Medinah, and M. Biton, Singular Surfaces in a Linear Pursuit-Evasion Game with Elliptical Vectograms. *Journal of Optimization Theory and Applications*, **43**, No. 3, 431–456, 1984.

[10] Gutman, S. An Optimal Guidance for Homing Missiles. *Journal of Guidance and Control*, **2**, No. 4, 1979.

[11] Zarchan, P. *Tactical and Strategic Missile Guidance*. Progress in Astronautics and Aeronautics, Vol. 157. American Institute of Aeronautics and Astronautics, Washington, pp. 452, 1994.

# On the Numerical Solution of a Class of Pursuit-Evasion Games

Tuomas Raivio and Harri Ehtamo
Systems Analysis Laboratory
Helsinki University of Technology
Espoo, Finland

**Abstract**

This paper presents a new computational approach for a class of pursuit-evasion games of degree. The saddle-point problem is decomposed into two subproblems that are solved by turns iteratively. The subproblems are ordinary optimal control problems that can be solved efficiently using discretization and nonlinear programming techniques. Hence it is not necessary to supply an initial guess of the adjoint variables or a hypothesis on the switching structure of the solution. Furthermore, in the presented approach the numerical differentiation of the payoff is avoided. We test the algorithm with different numerical examples and compare the results with solutions obtained by an indirect method. In the test examples the method converges rapidly from a rough initial guess, and the results coincide well with the reference solutions.

## 1 Introduction

Optimal feedback strategies of a pursuit-evasion game of degree are, in principle, obtained by solving the Isaacs partial differential equation in the state space split by singular surfaces. For games with simple dynamics and low-dimensional state vector, the strategies can be determined using the theory of viscosity solutions, see [1] and references cited therein, or the "Tenet of Transition" [8] and integration in retrograde time, see also [17].

For more complex problems, indirect methods can be used to solve the multipoint boundary value problem arising from the necessary conditions of a saddle point. The methods include finite differences, quasilinearization, and the well-known multiple shooting [4], [10]. The convergence domain of indirect methods is small, and continuation and homotopy techniques are frequently required. A drawback of these techniques is the unpredictable changes in the switching structure of the solution in the course of continuation.

In this paper we provide an alternative way to solve the necessary conditions for games in which the players control their own state equations. In such games, the necessary conditions and the optimal saddle-point control histories are coupled only via the terminal payoff and the capture set. Consequently, the saddle-point problem can be decomposed into two optimal control problems that are solved by turns iteratively using either indirect methods or, as is done here, discretization and nonlinear programming [7], [16]. In particular, it is shown that if the solutions of the subproblems converge, the limit solution satisfies the necessary conditions of an open-loop representation of a feedback saddle-point trajectory. In all the numerical examples the iteration converges well; see Section 4, where the solutions of the numerical examples are compared with reference trajectories obtained by solving the necessary conditions of a saddle point using an indirect method. The examples also suggest that the method works in the presence of certain singular surfaces and separable state-dependent control constraints.

The subproblems of the method are the min and the linearized max problems for the pursuer and the evader, respectively. In [9], Krasovskii and Subbotin give sufficient conditions for the lower (maxmin) value function to coincide with the upper (minmax) value function of the game and suggest that the maxmin problem be solved as a Stackelberg game. Nevertheless, the numerical solution of this rather involved problem is not considered.

Moritz et al. [12] have suggested direct discretization of a pursuit-evasion game by control parametrization. The solution method is based on alternating minimization and maximization steps against a fixed trajectory of the opponent. The convergence of the method essentially depends on the number of optimization steps that the method is allowed to take at each iteration. A large fraction of total computation time is needed to evaluate the gradient of the payoff numerically, whereas in this paper basic sensitivity results are used to express the gradient analytically. Furthermore, it is not clear whether the solutions obtained in [12] satisfy the necessary conditions of a saddle point.

## 2 Pursuit-Evasion Game

We consider two-player, perfect information zero-sum differential games with unspecified terminal time. We assume that the dynamics of the individual players can be divided into two distinct sets of state equations independent of the states and controls of the other player,

$$\dot{x}(t) = \begin{pmatrix} \dot{x}_1(t) \\ \dot{x}_2(t) \end{pmatrix} = f(x(t), u_1(t), u_2(t), t) = \begin{pmatrix} f_1(x_1(t), u_1(t), t) \\ f_2(x_2(t), u_2(t), t) \end{pmatrix}, \tag{1}$$

$$x(0) = x_0, \qquad t \in [0, \infty).$$

The state vector $x(t)$ consists of the vectors $x_1(t) \in R^{n_1}$ and $x_2(t) \in R^{n_2}$ that describe the states of the players. Subscripts 1 and 2 refer to the pursuer and the evader, respectively. The admissible controls $u_i(t)$ of the players belong to sets $S_i \subset R^{p_i}$, $i = 1, 2$, for all $t$. The admissible control functions $u_1$ and $u_2$ are assumed to be piecewise continuous functions on the interval $[0, \tilde{T}]$ with $\tilde{T}$

sufficiently large. The terminal payoff of the game is

$$J[u_1, u_2] = q(x(T), T).$$

The game ends when the vector $(x(t), t)$ enters the target set $\Lambda \subset R^{n_1+n_2} \times R^+$, which, together with the terminal cost, are the only connecting factors between the players. The unprescribed final time of the game is defined as

$$T = \inf\{t \mid (x(t), t) \in \Lambda\}.$$

The target set $\Lambda$ is closed. The boundary $\partial\Lambda$ of $\Lambda$ is an $(n_1 + n_2)$-dimensional manifold in the space $R^{n_1+n_2} \times R^+$ and is characterized by the scalar equation

$$l(x(t), t) = 0.$$

The function $l$ as well as the functions $f$ and $q$ are assumed to be continuously differentiable in $x$ and $t$. We assume that the initial state $x_0$ of the game is selected so that the pursuer can enforce a capture against any action of the evader. Consequently, $T < \infty$ always holds.

Suppose that a pair $(\gamma_1^*, \gamma_2^*) \in \Gamma_1 \times \Gamma_2$ is a saddle-point solution in feedback strategies for the game and let $x^*(t)$ be the corresponding trajectory. The value function of the game when the players start from $(x, t)$ and apply their feedback saddle-point strategies is defined by

$$V(x, t) = \min_{\gamma_1 \in \Gamma_1} \max_{\gamma_2 \in \Gamma_2} q(x(T), T) = \max_{\gamma_2 \in \Gamma_2} \min_{\gamma_1 \in \Gamma_1} q(x(T), T).$$

In this paper we are interested in the open-loop representations $u_i^*(t) := \gamma_i^*(x^*(t), t)$, $i = 1, 2$, of the feedback strategies. The following necessary conditions hold (see [2, Theorem 8.2, p. 433]): There is an adjoint vector $p(t) \in R^{n_1+n_2}$, $t \in [0, T]$, and a Lagrange multiplier $\alpha \in R$ such that

$$\dot{x}^*(t) = f(x^*(t), u_1^*(t), u_2^*(t), t), \tag{2}$$

$$x^*(0) = x_0,$$

$$H(x^*(t), u_1^*(t), u_2(t), p(t), t) \leq H(x^*(t), u_1^*(t), u_2^*(t), p(t), t) \leq$$
$$H(x^*(t), u_1(t), u_2^*(t), p(t), t) \quad \forall t \in [0, T],\ u_i(t) \in S_i,\ i = 1, 2, \tag{3}$$

$$\dot{p}(t) = -\frac{\partial}{\partial x} H(x^*(t), u_1^*(t), u_2^*(t), p(t), t), \tag{4}$$

$$p(T) = \frac{\partial}{\partial x} q(x^*(T), T) + \alpha \frac{\partial}{\partial x} l(x^*(T), T), \tag{5}$$

$$l(x^*(T), T) = 0, \tag{6}$$

$$H(x^*(T), u_1^*(T), u_2^*(T), p(T), T) = -\frac{\partial}{\partial t} q(x^*(T), T) - \alpha \frac{\partial}{\partial t} l(x^*(T), T), \tag{7}$$

where

$$H(x(t), u_1(t), u_2(t), p(t), t) = p(t)^T f(x(t), u_1(t), u_2(t), t).$$

## 3 A Feasible Direction Method for the Game

In this section we provide a numerical method to solve the necessary conditions (2)–(7). The solution method is based on the min and the linearized max subproblems that are solved by turns iteratively, either by using indirect methods or discretization and nonlinear programming. If the iteration converges, the solution satisfies the necessary conditions; see Theorem 3.1 below.

The maxmin problem is defined by

$$\begin{aligned} &\max_{u_2} \min_{u_1} \quad q(x(T), T) \\ &\dot{x}(t) = f(x(t), u_1(t), u_2(t), t), \qquad x(0) = x_0, \\ &l(x(T), T) = 0. \end{aligned}$$

To solve it we first consider the minimization problem

$$\begin{aligned} &\mathrm{P}: \min_{u_1, T} \quad q(x_1(T), x_2^0(T), T) \\ &\dot{x}_1(t) = f_1(x_1(t), u_1(t), t), \qquad x_1(0) = x_{10}, \\ &l(x_1(T), x_2^0(T), T) = 0, \end{aligned}$$

where $x_2^0(t)$ is some fixed function of time that represents an arbitrary trajectory of the evader. Let the solution of $P$ be $\bar{x}_1(\cdot)$ with the capture time $T^0$, and let $e^0 := x_2^0(T^0)$ denote the corresponding capture point. Now, $\bar{x}_1(\cdot)$ also solves the problem where the given trajectory $x_2^0(\cdot)$ in problem P is replaced by the fixed point $e^0 \in R^{n_2}$, and the final time is fixed to $T^0$.

Next consider all the points $(e, T) \in R^{n_2+1}$ in the neighborhood of $(e^0, T^0)$, and define the value function of the pursuer's problem, corresponding to the initial state $x_0$, as a function of the capture point $(e, T)$ by

$$\begin{aligned} \tilde{V}(e, T) = \min_{u_1} \{ q(x_1(T), e, T) \mid \dot{x}_1(t) = f_1(x_1(t), u_1(t), t),\ t \in [0, T], \\ x_1(0) = x_{10};\ l(x_1(T), e, T) = 0 \}. \end{aligned} \tag{8}$$

It then holds that

$$\tilde{V}(e^0, T^0) = q(\bar{x}_1(T^0), e^0, T^0).$$

The evader's problem is to maximize the pursuer's value function (8). Therefore the original maxmin problem can equivalently be written as

$$\begin{aligned} &\mathrm{E}' : \max_{u_2, T} \tilde{V}(x_2(T), T) \\ &\dot{x}_2(t) = f_2(x_2(t), u_2(t), t), \qquad x_2(0) = x_{20}. \end{aligned}$$

Problem E′ is difficult to solve directly since $\tilde{V}(e, T)$ cannot be expressed analytically. Therefore we proceed as follows: we first approximate the solution of E′ by the solution of problem E (see below), where the final time is fixed to $T^0$ and $\tilde{V}(e, T^0)$ is linearized in the neighborhood of $e^0$. To approximate the solution of E′ for $t > T^0$ we extend the solution of E into the interval $[T^0, T^0 + \Delta T]$, $\Delta T > 0$, using a linear approximation.

The linear approximation of $\tilde{V}(e, T^0)$ in the neighborhood of $e^0$ is given by

$$\tilde{V}(e^0, T^0) + \frac{\partial}{\partial e}\tilde{V}^T(e^0, T^0)(e - e^0).$$

Basic sensitivity results (see [6, Chap. 3.4]; see also [5], [15] for the derivation of the result) imply that the gradient of the value function is given by

$$\frac{\partial}{\partial e}\tilde{V}(e^0, T^0) = \frac{\partial}{\partial e}q(\bar{x}_1(T^0), e^0, T^0) + \alpha^0\frac{\partial}{\partial e}l(\bar{x}_1(T^0), e^0, T^0), \tag{9}$$

where $\alpha^0$ is the Lagrange multiplier associated with the capture condition in the solution of P. Note that this is an analytical expression. Numerical differentiation of the payoff is avoided.

Neglecting the constant $\tilde{V}(e^0, T^0)$, the fixed final time, free final state problem E can be written as

$$\begin{aligned} \text{E:} \max_{u_2} \quad & c^T(x_2(T^0) - e^0) \\ \dot{x}_2(t) = & f_2(x_2(t), u_2(t), t), \qquad x_2(0) = x_{20}, \end{aligned}$$

where

$$c := \frac{\partial}{\partial e}q(\bar{x}_1(T^0), e^0, T^0) + \alpha^0\frac{\partial}{\partial e}l(\bar{x}_1(T^0), e^0, T^0).$$

Let the solution of E be $x_2^1(t)$, $t \in [0, T^0]$. The extension to the interval $[T^0, T^0 + \Delta T]$, $\Delta T > 0$, is done using the linear approximation

$$x_2^1(T^0 + h) = x_2^1(T^0) + \dot{x}_2^1(T^0)h, \qquad h \in [0, \Delta T]. \tag{10}$$

The extended solution of E is now inserted back into P, which is solved anew to locate the new capture point and to evaluate the linear approximation of the value function. Problems P and E are solved and updated until a constrained maximum of $\tilde{V}(e, T)$, and thus the solution of problem E′, is achieved. To summarize, the iteration proceeds as follows:

1. Fix an initial trajectory of the evader and solve P. Obtain $e^0$, $T^0$, and $\alpha^0$. Set $k := 0$.
2. Solve E using $e^k$, $T^k$ and $\alpha^k$.
3. Insert the extended solution $x_2^{k+1}$ of E into P and solve P to obtain $e^{k+1}$, $T^{k+1}$ and $\alpha^{k+1}$. If $||e^{k+1} - e^k|| < \epsilon$, where $\epsilon$ is the desired accuracy, terminate. Otherwise, set $k := k + 1$ and go to step 2.

The following theorem shows that the limit solution satisfies the necessary conditions of a saddle point; the proof is given in the Appendix:

**Theorem 3.1.** *Suppose that the iteration converges. Denote the limit solution of P and E by $u_1^*(t)$ and $u_2^*(t)$, respectively, the corresponding trajectories by $x_1^*(t)$ and $x_2^*(t)$, the capture time by $T^*$, and the Lagrange multiplier by $\alpha^*$. Then the* *solution satisfies the necessary conditions* (2)–(7) *for an open-loop saddle point or for an open-loop representation of a feedback saddle-point strategy.*

Thus the solution of the necessary conditions can be decomposed to the solution of the necessary conditions for subproblems P and E involving one player. The subproblems can be solved, e.g., using multiple shooting. Nevertheless, as the subproblems are optimal control problems, discretization and nonlinear programming can also be applied. The solutions produced by these methods approximate the optimal state and control trajectories up to an accuracy that depends on the discretization interval and the scheme being used. In addition, the Lagrange multipliers corresponding to the constraints that enforce pointwise satisfaction of the state equations approximate the adjoint trajectories (for Euler discretization and direct collocation, see [18]).

Discretizing P results in a nonlinear programming problem PD of the form

$$\begin{aligned} \text{PD: min} \quad & q(x_1^{N_1}, x_2^0(T), T) \\ g_1(\bar{x}_1, \bar{u}_1, T) & \leq \bar{0}, \\ h_1(\bar{x}_1, \bar{u}_1, T) & = \bar{0}, \\ l(x_1^{N_1}, x_2^0(T), T) & \leq 0, \\ T & > 0. \end{aligned}$$

Here $\bar{x}_1 \in R^{N_1 \times n_1}$ and $\bar{u}_1 \in R^{N_1 \times p_1}$ refer to the discretized trajectory and controls of the pursuer. The number of discretization points is $N_1$, the superscript referring to the last discretization point. Note that $x_2^0$ is a given function of $T$ and need not be discretized.

The expression for the multiplier $\alpha^0$ to be used in (9) can be obtained analytically using the necessary conditions (15) and (17) of problem P, and the definition of $H_1$; see the Appendix:

$$\begin{aligned} \alpha^0 = & -[\partial q/\partial x_1^{N_1}(x_1^{N_1}, x_2^0(T^0), T^0) f_1(x_1^{N_1}, u_1^{N_1}, T^0) \\ & + \partial q/\partial x_2(x_1^{N_1}, x_2^0(T^0), T^0) f_2(x_2^0(T^0), u_2^0(T^0), T^0) \\ & + \partial q/\partial T(x_1^{N_1}, x_2^0(T^0), T^0)]/ \\ & [\partial l/\partial x_1^{N_1}(x_1, x_2^0(T^0), T^0) f_1(x_1^{N_1}, u_1^{N_1}, T^0) \\ & + \partial l/\partial x_2(x_1^{N_1}, x_2^0(T^0), T^0) f_2(x_2^0(T^0), u_2^0(T^0), T^0) \\ & + \partial l/\partial T(x_1^{N_1}, x_2^0(T^0), T^0)]. \end{aligned}$$

Problem ED, the finite-dimensional version of problem E, is of the form

$$\begin{aligned} \text{ED: max} \quad & \eta^T(x_2^{N_2} - e^0) \\ g_2(\bar{x}_2, \bar{u}_2, T^0) & \leq \bar{0}, \\ h_2(\bar{x}_2, \bar{u}_2, T^0) & = \bar{0}, \end{aligned}$$

where

$$\eta = \frac{\partial}{\partial e} q(x_1^{N_1}, e^0, T^0) + \alpha^0 \frac{\partial}{\partial e} l(x_1^{N_1}, e^0, T^0).$$

## 4 Numerical Examples

In this section we assess the approach by solving some numerical examples that can be solved also analytically or indirectly. First the method is applied to the "Homicidal Chauffeur" game [8]. To check that the necessary conditions indeed become satisfied we solve a simple aerial dogfight in the horizontal plane. Finally, we investigate the method in a head-on encounter by solving a collision avoidance problem proposed by Lachner et al. [11]. All the subproblems are discretized using direct collocation and solved by Sequential Quadratic Programming; see [7].

**Example 4.1.** In the first example the pursuer has a positive minimum turning radius whereas the evader may redirect his velocity vector instantaneously. Both players move with constant velocities $v_1$ and $v_2$, $v_1 > v_2$, in the horizontal plane. The objective of the pursuer is to catch the evader in minimum time and the evader tries to avoid capture as long as possible. The payoff $q$ equals the terminal time $T$. The players' equations of motion are

$$\begin{aligned} \dot{x}_1 &= v_1 \cos \phi_1, & \dot{x}_2 &= v_2 \cos u_2, \\ \dot{y}_1 &= v_1 \sin \phi_1, & \dot{y}_2 &= v_2 \sin u_2, \\ \dot{\phi}_1 &= u_1. \end{aligned}$$

Here and in the other examples $x_i$ and $y_i$, $i = 1, 2$, refer to the $x$- and $y$-coordinates of the players and $\phi_1$ refers to the heading of the pursuer. The absolute angular velocity $|u_1|$ of the pursuer is restricted to be less than some given maximum

$$|u_1| \leq u_{\max}.$$

The target set $\Lambda$ can be expressed by the inequality

$$(x_1(T) - x_2(T))^2 + (y_1(T) - y_2(T))^2 \leq d^2,$$

where $d$ is the capture radius. The boundary of $\Lambda$ is the circle

$$l(x(T)) = (x_1(T) - x_2(T))^2 + (y_1(T) - y_2(T))^2 - d^2 = 0.$$

The vector $x = [x_1, y_1, \phi_1, x_2, y_2]^T$ denotes here the state of the game. Despite its simplicity the game captures salient features of pursuit-evasion games. For example, most saddle-point solutions involve singular surfaces.

The parameters of the game are $v_1 = 3$, $v_2 = 1$, $d = 1$, and $u_{1,\max} = 1$. Four cases are considered. The evader's initial position is varied according to Table 1, and the initial conditions of the pursuer are

$$x_{10} = y_{10} = \phi_{10} = 0.$$

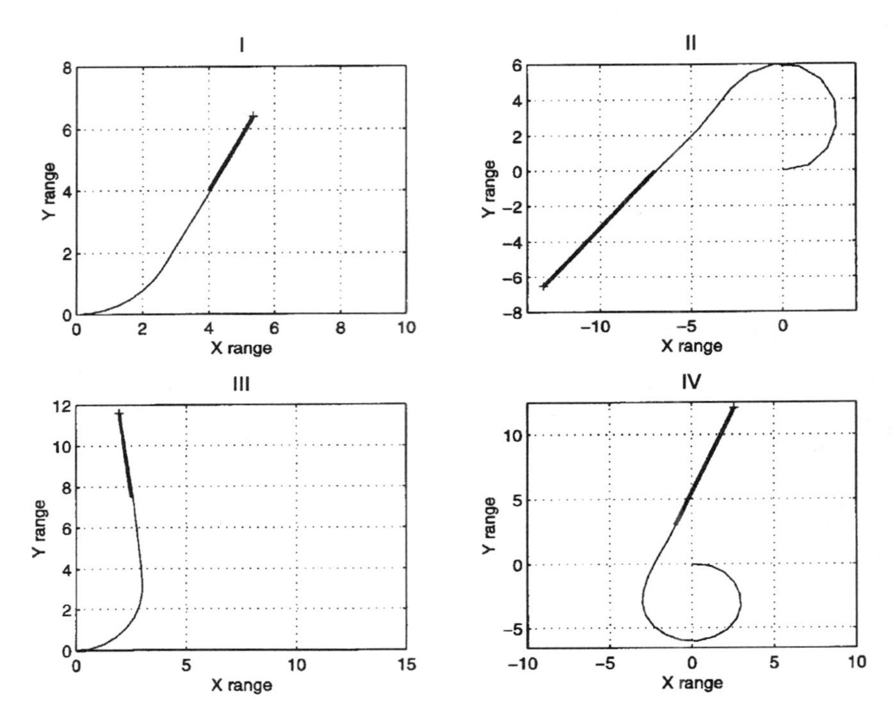


**Figure 1**: The solutions of the homicidal chauffeur game corresponding to the initial conditions given in Table 1. The thick line represents the evader's trajectory.

In the first case the evader starts in front of the pursuer. A dispersal surface is encountered in the second case, as the initial state of the evader is directly behind the pursuer. The third initial state lies near the barrier, and the fourth one is on the other side of it. The solution trajectories are presented in Figure 1, and information on the results and the computational procedure in Table 1.

All the solutions end up in a tail chase, corresponding to a universal surface. Only games where the evader is initially close to the pursuer in front of him or on the barrier would end otherwise. The treatment of a dispersal surface in case II would require instantaneous mixed strategies for the pursuer [8]. From the computational point of view, the initial guess of the pursuer's trajectory determines the side to which the play evolves. The vicinity of the barrier does not seem to affect the convergence of case III.

The last case converges to a maxmin solution, i.e., the evader plays nonoptimally. The evader should first pursue the pursuer to gain time. The method cannot identify the equivocal surface. Its tributaries are not solutions of the maxmin problem.

**Example 4.2.** We consider two aircraft moving in a horizontal plane. The pursuing aircraft attempts to capture the evader in minimum time. The evading aircraft tries

**Table 1**: Initial conditions of the evader and summary of the results of Example 1. For cases I and III the initial guess of the evader's trajectory is a straight path along the positive $x$-axis and for cases II and IV along the negative $x$-axis. The iteration was executed until the relative change of the capture point was less than $5 \times 10^{-3}$. The reference value is computed analytically. Case IV is a maxmin solution.

| Case | I | II | III | IV |
|---|---|---|---|---|
| $(x_{20}, y_{20})$ | (4,4) | (−7,0) | (2.5,7.5) | (−1,3) |
| No. of iterations | 8 | 10 | 11 | 11 |
| Computed value, sec. | 2.506 | 8.956 | 4.124 | 9.763 |
| Ref. value, sec. | 2.504 | 8.927 | 4.120 | * |
| Relative error, % | 0.08 | 0.3 | 0.08 | * |

to avoid this situation as long as possible. Assuming point mass dynamics, the players' equations of motion are

$$\begin{aligned}
\dot{x}_i &= v_i \cos \phi_i, \\
\dot{y}_i &= v_i \sin \phi_i, \\
\dot{\phi}_i &= \omega_i, \\
\dot{v}_i &= g(A_i(v_i) - B_i(v_i)v_i^2 - C_i(v_i)((\omega_i/g)^2 + 1/v_i^2)), \qquad i = 1, 2,
\end{aligned}$$

where $x_i$, $y_i$, $\phi_i$, and $v_i$ stand for the $x$- and $y$-coordinates, heading angle, and velocity of aircraft $i$. The gravitational acceleration is denoted by $g$. The coefficients $A_i$, $B_i$, and $C_i$, $i = 1, 2$, describe the thrust and the drag forces of the aircraft, see [14]. The angular velocities $\omega_{1,2}$ are constrained by

$$|\omega_i| \leq \omega_{\max,i}, \qquad i = 1, 2.$$

The pursuing aircraft is assumed slightly faster and more agile than the evading one. The capture condition is similar to the one in previous examples, the capture radius being now 100 [m].

The scenario takes place at the altitude of $h_0 = 3000$ m from the initial conditions

$$\begin{aligned}
x_{10} &= 0 \text{ m}, & x_{20} &= 3000 \text{ m}, \\
y_{10} &= 0 \text{ m}, & y_{20} &= 5000 \text{ m}, \\
\phi_{10} &= 0 \text{ rad}, & \phi_{20} &= -1 \text{ rad}, \\
v_{10} &= 200 \text{ m/s}, & v_{20} &= 100 \text{ m/s}.
\end{aligned}$$

The saddle-point trajectories produced by this method are shown in Figure 2 together with the reference solution that was obtained via solving the necessary conditions of the saddle point for this problem. The boundary value problem was solved using the BNDSCO software package [13]. The solution trajectory, as well as the value, are almost identical with the reference solution. It is easy to numerically check that the solution satisfies the necessary conditions (2)–(7) for

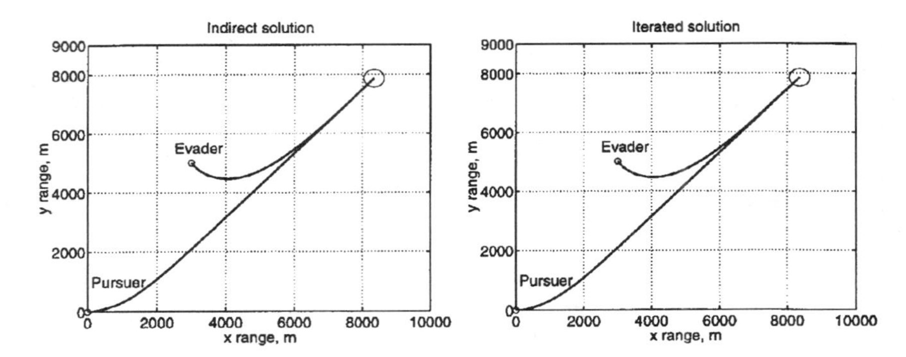


**Figure 2**: On the left, the reference solution of the aerial dogfight problem. On the right, the solution obtained by the proposed method. The circle indicates the projection of the terminal manifold (drawn larger than actual). The computed value of the game is 34.647, whereas the reference value is 34.649 [sec].

the saddle point when the Lagrange multipliers of the collocation constraints are used in approximating the adjoint variables.

**Example 4.3.** The last example, a collision avoidance problem, is a variant of the game of two cars [8]. The pursuer is a car driving in the wrong direction on a freeway defined by

$$-7.5 < x < 7.5, \qquad y \text{ free.}$$

The evader, driving in the correct direction, tries to avoid collision with the pursuer, who is assumed to aim at colliding with the evader. The players obey the state equations

$$\begin{aligned} \dot{x}_1 &= v_1 \sin\phi_1, & \dot{x}_2 &= v_2 \sin\phi_2, \\ \dot{y}_1 &= v_1 \cos\phi_1, & \dot{y}_2 &= v_2 \cos\phi_2, \\ \dot{\phi}_1 &= u_1, & \dot{\phi}_2 &= u_2, \\ & & \dot{v}_2 &= b\eta_2, \end{aligned}$$

where $\phi_i$, $i = 1, 2$, are measured clockwise from the positive $y$-axis. The velocity of the pursuer, $v_1$, is assumed constant, but the evader can accelerate and decelerate by selecting $\eta_2 \in [-0.1, 1]$ appropriately. The maximum deceleration rate is described by the constant $b$. The angular velocities of both players are constrained by

$$|u_i| \leq u_{i,\max} = \frac{\mu_0 g}{v_i},$$

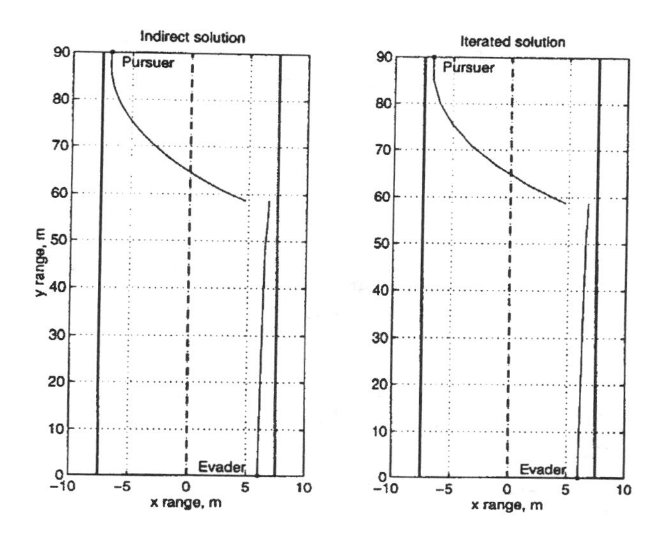


**Figure 3**: The solution of the collision avoidance problem. On the left, the solution obtained by an indirect method and on the right, the solution obtained by the proposed approach. The value is computed to be 1.991, whereas the value obtained by the indirect approach is 2.012 [m]. The final times are computed to be 2.2876 and 2.2870 seconds, respectively.

where $\mu_0$ is the friction coefficient between tires and asphalt and $g$ is the acceleration due to gravity. To stay on the road the evader has to satisfy the following additional constraint, which is here given for the right edge of the freeway:

$$\frac{x_2 - a}{v_2^2}(K_1^2 + K_2^2) + K_2(\cos\phi_2 - e^{-\phi_2(K_1/K_2)})$$
$$-K_1 \sin\phi_2 + \frac{u_2/u_{2,\max} + 1}{K_3} - \frac{\eta_2 - \eta_{2,\max}}{K_4} \leq 0. \tag{11}$$

Here $a = 6.75$, $K_1 = 2b\eta_{2,\max}$, $K_2 = \mu_0 g u_{2,\min}$, and $K_3 = K_4 = 120$. The constraint assures that the evader will stay on the freeway even after the avoidance maneuver. For details, see [11]. Note that both constraints above are mixed state and control constraints. The target set $\Lambda$ is the set of states where

$$y_1(T) \leq y_2(T).$$

The boundary of the target set is characterized by

$$l(x(T)) = y_1(T) - y_2(T) = 0,$$

that is, the game ends when the cars pass each other. The payoff is the distance at the moment of termination,

$$q(x(T)) = |x_1(T) - x_2(T)|.$$

**Table 2**: Summary of the last two examples. The reference values are computed by an indirect method. For the collision avoidance problem the number of discretization points was smaller for the pursuer.

| Example | Aerial Dogfight | Wrong Driver |
|---|---|---|
| No. of nodes | 15 | 8,15 |
| No. of iterations | 5 | 2 |
| Computed value | 34.647 s | 1.991 m |
| Ref. value | 34.649 s | 2.012 m |
| Relative error, % | 0.005 | 1.0 |

The values for the parameters are $\mu_0 = 0.55$, $b = -6$ [m/s$^2$], and $v_1 = 15$ [m/s]. The initial conditions are

$$\begin{aligned} x_{10} &= -6.75 \text{ m}, & x_{20} &= 6.0 \text{ m}, \\ y_{10} &= 90.0 \text{ m}, & y_{20} &= 0 \text{ m}, \\ \phi_{10} &= 3.202 \text{ rad}, & \phi_{20} &= 0 \text{ rad}, \\ v_{20} &= 25 \text{ m/s}, \end{aligned}$$

i.e., the game is a head-on encounter. The solution, together with the reference solution obtained by an indirect method, is given in Figure 3. The evader's discretized problem cannot fully capture the resulting rapid changes of controls. The value of the game is computed rather accurately. The results of the last two examples are summarized in Table 2.

## 5 Conclusion

In this paper, the computation of a solution for certain pursuit-evasion games of degree is decomposed into a solution of a sequence of optimal control problems involving a single player. These optimal control problems can be solved either by indirect methods or by discretization and nonlinear programming. The converged solution is shown to satisfy the regular necessary conditions of a saddle point.

In the numerical examples the iteration converges well. The method as such is a feasible direction method (see [3, Chap. 10]) where a new direction is produced by linearizing the objective function, see problem E′ and the linearized problem E. In order to ensure convergence for such methods a line search should be conducted along the produced direction. In our method the line search could, in principle, be performed by computing the value of $\tilde{V}(e, T)$ in discrete points along the line from $(e^k, T^k)$ to $(e^{k+1}, T^{k+1})$.

Numerical examples suggest that in addition to regular solution arcs the method is able to locate at least a universal surface and a switching line and can handle a dispersal surface. In Example 4.3, the switching structure of the solution, induced by state-dependent control constraints, is approximately identified. Case

IV of Example 4.1 demonstrates that an equivocal surface cannot be detected by the method. Nevertheless, the method still provides a maxmin solution, which also might be useful in practice.

The method can possibly offer a robust way to extend the recent automated solution framework of optimal control problems [19] into pursuit-evasion games. Automatically computed open-loop representations of feedback strategies would allow designers, engineers, and decision makers to easily study the worst-case performance of different systems. On the other hand, synthesis of optimal feedback strategies for complex models requires that massive amounts of open-loop representations corresponding to different initial states are computed. The flexibility of the method makes it easy to automate the computational procedure, since the convergence domain is large and the switching structure of the solution need not be specified in advance.

In the present approach we first decompose the continuous-time saddle-point problem into subproblems and then discretize them. Another possibility, to be addressed in the future, would be to first discretize the saddle-point problem and then solve this finite-dimensional setting using a suitable decomposition method. For a preliminary work along this direction, see [5].

## Appendix: Proof of Theorem 1

First, the optimal solutions $u_1^*$, $x_1^*$, $p_1$, and $\alpha^*$ satisfy the necessary conditions of optimality for P, that is,

$$\dot{x}_1^*(t) = f_1(x_1^*(t), u_1^*(t), t), \tag{12}$$

$$\forall t \in [0, T^*] : u_1^*(t) = \arg \min_{u_1(t)\in S_1} H_1(x_1^*(t), u_1(t), p_1(t), t), \tag{13}$$

$$\dot{p}_1(t) = -\frac{\partial}{\partial x_1} H_1(x_1^*(t), u_1^*(t), p_1(t), t), \tag{14}$$

$$p_1(T^*) = \frac{\partial}{\partial x_1} q(x_1^*(T^*), x_2^*(T^*), T^*) + \alpha^* \frac{\partial}{\partial x_1} l(x_1^*(T^*), x_2^*(T^*), T^*), \tag{15}$$

$$l(x_1^*(T^*), x_2^*(T^*), T^*) = 0, \tag{16}$$

$$H_1(x_1^*(T^*), u_1^*(T^*), p_1(T^*), T^*) = -\frac{\partial}{\partial t} q(x_1^*(T^*), x_2^*(T^*), T^*) - \alpha^* \frac{\partial}{\partial t} l(x_1^*(T^*), x_2^*(T^*), T^*), \tag{17}$$

where

$$H_1(x_1(t), u_1(t), p_1(t), t) = p_1(t)^T f_1(x_1(t), u_1(t), t).$$

Note that in (17), the functions $q(\cdot)$ and $l(\cdot)$ explicitly depend on time also through $x_2^*(t)$, since in problem P it is assumed that $x_2^0(t)$, which now equals $x_2^*(t)$, is a given, time-dependent function.

Second, the optimal control $u_2^*$ and the corresponding state and adjoint trajectories $x_2^*$ and $p_2$ satisfy the necessary conditions of optimality for E:

$$\dot{x}_2^*(t) = f_2(x_2^*(t), u_2^*(t), t), \tag{18}$$

$$\forall t \in [0, T^*]: \; u_2^*(t) = \arg \max_{u_2(t)\in S_2} H_2(x_2^*(t), u_2(t), p_2(t), t), \tag{19}$$

$$\dot{p}_2(t) = -\frac{\partial}{\partial x_2} H_2(x_2^*(t), u_2^*(t), p_2(t), t), \tag{20}$$

$$p_2(T^*) = \frac{\partial}{\partial x_2} c^T (x_2^*(T^*) - e^*) = c, \tag{21}$$

where

$$H_2(x_2(t), u_2(t), p_2(t), t) = p_2(t)^T f_2(x_2(t), u_2(t), t),$$

and

$$c = \frac{\partial q}{\partial x_2}(x_1^*(T^*), x_2^*(T^*), T^*) + \alpha^* \frac{\partial l}{\partial x_2}(x_1^*(T^*), x_2^*(T^*), T^*).$$

Combining (12) and (18) gives (2); combining (14) and (20) gives (4). The terminal conditions of the adjoint variables, (15) and (21), coincide with (5) and the capture condition (16) with (6). Summing $H_1$ and $H_2$ yields the Hamiltonian of the game and conditions (13) and (19) correspond to the saddle-point condition (3). Furthermore, at time $T^*$,

$$\begin{aligned} &H_1(x_1^*(T^*), u_1^*(T^*), p_1(T^*), T^*) + H_2(x_2^*(T^*), u_2^*(T^*), p_2(T^*), T^*) \\ &\quad = -\frac{\partial}{\partial t}\{q(x_1^*(T^*), x_2^*(T^*), T^*) + \alpha^* l(x_1^*(T^*), x_2^*(T^*), T^*)\} \\ &\quad\quad + p_2(T^*)^T f_2(x_2^*(T^*), u_2^*(T^*), T^*). \end{aligned} \tag{22}$$

The time derivative in the right-hand side of (22) can be written as (recall that $q(\cdot)$ and $l(\cdot)$ here explicitly depend on time also through $x_2^*$):

$$\begin{aligned} &-\frac{\partial}{\partial x_2}\{q(x_1^*(T^*), x_2^*(T^*), T^*) + \alpha^* l(x_1^*(T^*), x_2^*(T^*), T^*)\}\dot{x}_2^*(T^*) \\ &\quad -\frac{\partial}{\partial t}\{q(x_1^*(T^*), x_2^*(T^*), T^*) + \alpha^* l(x_1^*(T^*), x_2^*(T^*), T^*)\} \\ &= -p_2(T^*)^T f_2(x_2^*(T^*), u_2^*(T^*), T^*) \\ &\quad -\frac{\partial}{\partial t}\{q(x_1^*(T^*), x_2^*(T^*), T^*) + \alpha^* l(x_1^*(T^*), x_2^*(T^*), T^*)\}; \end{aligned} \tag{23}$$

above the time derivative $\partial/\partial t$ is taken only with respect to the explicit $t$-dependence.

Substituting (23) into (22) yields

$$H_1(x_1^*(T^*), u_1^*(T^*), p_1(T^*), T^*) + H_2(x_2^*(T^*), u_2^*(T^*), p_2(T^*), T^*)$$
$$= -\frac{\partial}{\partial t} q(x_1^*(T^*), x_2^*(T^*), T^*) - \alpha^* \frac{\partial}{\partial t} l(x_1^*(T^*), x_2(T^*), T^*), \quad (24)$$

which gives (7). □

**Remark.** The proof can be easily extended to include separable mixed state and control constraints, too. Suppose constraints of the form $g_1(x_1, u_1) \leq 0$, $g_2(x_2, u_2) \leq 0$ arise for the pursuer and the evader, respectively. Let $\mu_1(t)$ and $\mu_2(t)$ be the Lagrange multipliers corresponding to these constraints at optimum. The only effect the additional constraints have on the proof above is to add the terms $\mu_1^T g_1$, $\mu_2^T g_2$ to the definitions of $H_1$ and $H_2$ above. The terminal condition (22) and hence the derived equation (24) remain unaffected, since $\mu_2^T(T^*) g_2(x_2^*(T^*), u_2(T^*)) = 0$.

## Acknowledgments

The authors wish to thank Dr. Rainer Lachner, TU-Clausthal, Germany, for providing the collision avoidance problem and the reference solution.

## REFERENCES

[1] Bardi, M., S. Bottacin, and M. Falcone. "Convergence of Discrete Schemes for Discontinuous Value Functions of Pursuit-Evasion Games", In: G. J. Olsder, ed. *New Trends in Dynamic Games and Applications*, pp. 273–304. Birkhäuser, Boston, 1996.

[2] Basar, T. and G. Olsder. *Dynamic Noncooperative Game Theory*, 2nd ed. Academic Press, London, 1995.

[3] Bazaraa, M., H. Sherali, and C. Shetty. *Nonlinear Programming, Theory and Algorithms*. 2nd ed., Wiley, New York, 1993.

[4] Breitner, M., H. Pesch, and W. Grimm. "Complex Differential Games of Pursuit-Evasion Type with State Constraints, Part 2: Multiple Shooting and Homotopy", *Journal of Optimization Theory and Applications*, **78**, No. 3, 442–464, 1993.

[5] Ehtamo, H. and T. Raivio. A Feasible Direction Method for Saddle Point Problems. *Proceedings of the 8th International Symposium on Dynamic Games and Applications*, Maastricht, The Netherlands, pp. 184–188, 1998.

[6] Fiacco, A. *Introduction to Sensitivity and Stability Analysis in Nonlinear Programming*. Academic Press, New York, 1983.

[7] Hargraves, C. and S. Paris. Direct Trajectory Optimization Using Nonlinear Programming and Collocation. *Journal of Guidance, Control and Dynamics*, **10**, No. 4, 338–342, 1987.

[8] Isaacs, R. *Differential Games*. Reprint. Krieger, New York, 1975.

[9] Krasovskii, N. and A. Subbotin. *Game Theoretical Control Problems*. Springer-Verlag, New York, 1988.

[10] Lachner, R., M. Breitner, and H. Pesch. Three-Dimensional Air Combat: Numerical Solution of Complex Differential Games. In: G. Olsder, ed. *New Trends in Dynamic Games and Applications*, pp. 165–190. Birkhäuser, Boston, 1996.

[11] Lachner, R., M. Breitner, and H. Pesch. Real-Time Collision Avoidance Against Wrong Drivers: Differential Game Approach, Numerical Solution and Synthesis of Strategies with Neural Networks. *Proceedings of the 7th International Symposium on Dynamic Games and Applications*, Kanagawa, Japan, 564–572, 1996.

[12] Moritz, K., R. Polis, and K. Well. Pursuit-Evasion in Medium-Range Air-Combat Scenarios. *Computers and Mathematics with Applications*, **13**, No. 1–3, 167–180, 1987.

[13] Oberle, H. and W. Grimm. BNDSCO—A Program for the Numerical Solution of Optimal Control Problems. Program manual, 1989.

[14] Raivio, T., H. Ehtamo, and R. Hämäläinen. Simulating the Worst Possible Prediction Error of System Trajectories. *Proceedings of the European Simulation Multiconference*, Budapest, Hungary, pp. 299–303, 1996.

[15] Raivio, T., H. Ehtamo, and R. Hämäläinen. A Decomposition Method for a Class of Pursuit-Evasion Games. *Proceedings of the 7th International Symposium on Dynamic Games and Applications*, Kanagawa, Japan, pp. 784–795, 1996.

[16] Seywald, H. Trajectory Optimization Based on Differential Inclusion. *Journal of Guidance, Control and Dynamics*, **17**, No. 3, 480–487, 1994.

[17] Shinar, J., M. Guelman, and A. Green. An Optimal Guidance Law for a Planar Pursuit-Evasion Game of Kind, *Computers and Mathematics with Applications*, **18**, No. 1–3, 35–44, 1989.

[18] Stryk von, O. Numerische Lösung optimaler Steuerungsprobleme: Diskretisierung, Parameteroptimierung und Berechnung der adjungierten Variabeln. Doctoral Dissertation, Munich University of Technology, 1994 (In German).

[19] Virtanen, K., H. Ehtamo, T. Raivio, and R. Hämäläinen. VIATO—Visual interactive Aircraft Trajectory Optimization. *IEEE Transactions on Systems, Man and Cybernetics*, Part C, *Applications and Reviews*, **29**, 3, 409–421, 1999.

# PART III
# Coupled Dynamic and Stochastic Games

# Infinite Horizon Dynamic Games with Coupled State Constraints

Dean A. Carlson
The University of Toledo
Department of Mathematics
Toledo, Ohio, USA

Alain B. Haurie
University of Geneva
Department of Management Studies
Geneva, Switzerland
and
GERAD-École des HEC
Montréal, Canada

**Abstract**

In this paper we investigate the existence of equilibrium solutions to a class of infinite-horizon differential games with a coupled state constraint. These differential games typically represent dynamic oligopoly models where each firm controls its own dynamics (decoupled controls). In the absence of a coupled state constraint, the interactions among players take place essentially in the objective function. The introduction of a coupled state constraint in such models can be motivated, for example, by the consideration of a global environmental constraint imposed on competing firms. Our approach for the characterization of asymptotic equilibria under such a constraint is an adaptation of Rosen's *normalized equilibrium* concept [10] to this infinite-horizon dynamic game framework. We utilize the *Turnpike property* and the notion of an attracting normalized steady-state equilibrium for the definition of a time invariant multiplier to provide conditions for the existence of dynamic equilibria. We consider discounted and undiscounted dynamic game models in our investigation.

## 1 Introduction

In this paper we investigate the existence of equilibrium solutions to a class of infinite-horizon differential games with a coupled state constraint. These differential game typically represent dynamic oligopoly models where each firm controls its own dynamics (decoupled controls) and so, in the absence of a coupled state constraint, the interactions among players take place essentially in the objective function. The introduction of a coupled state constraint in a dynamic oligopoly model can be motivated, for example, by the consideration of a global environmental constraint imposed on competing firms. Our approach for the characterization of asymptotic equilibria under such a constraint will be an adaptation of Rosen's *normalized equilibrium* concept [10] to this infinite-horizon dynamic game framework. Recently, in Carlson and Haurie [2],[3], a comprehensive investigation of a class of dynamic competitive models was initiated. In these works conditions for the existence of an overtaking Nash equilibrium, conditions under which the equilibrium trajectories converge to an equilibrium steady state (the turnpike property), and conditions under which there is a unique overtaking Nash equilibrium are given. Both cases of undiscounted and discounted cost were considered. These results extended similar results for the case of optimal asymptotic control (i.e., a single player game) collected in [1] to the case of a convex infinite horizon dynamic game. While pointwise state constraints generally pose no particular difficulties in establishing the existence of an equilibrium (or optimal solution), they do create problems when the relevant necessary conditions are utilized to determine the optimal solution. Indeed, in these situations the adjoint variable will possibly have a discontinuity wherever the optimal trajectory hits the boundary of the constraint (see the recent survey paper [5]). Our approach circumvents this difficulty by introducing a relaxed asymptotic formulation of the coupled state constraint that is well adapted to the context of environment management. More specifically, we utilize the *Turnpike property* and the notion of an attracting normalized steady-state (implicit) equilibrium for the definition of a time-invariant multiplier. Indeed, this relaxation of the constraint and the use of a steady-state game for the definition of a time-invariant multiplier associated with the constraint could a priori lead to a dynamic solution which would not exhibit an equilibrium property. In the work that follows we provide conditions for the existence of dynamic equilibria under these relaxed asymptotic constraints. When the cost functionals are not discounted we show that an *averaging Nash equilibrium* exists, in which the coupled state constraint is satisfied asymptotically. When the cost functional are discounted we have to rely on the concept of an *implicit equilibrium* both for the definition of the attracting steady state and for the definition of the transient equilibrium. The interesting feature of these equilibria is that, in both cases, the solution is obtained by finding an overtaking Nash equilibrium for an associated infinite-horizon dynamic game, with decoupled control (i.e., the same sort as discussed in Carlson and Haurie [2], [3]) but where a time-invariant tax scheme takes care of the asymptotic state constraints. In an economic parlance, these equilibria can be easily induced by a time-invariant tax scheme based on the multiplier associated with the normalized steady-state equilibrium. As indicated above, the motivation to study such a problem arises in environmental economics when we have $p$ firms competing on an oligopolistic market and where the production processes generate some pollution. In its simplest form, a dynamic economic model of oligopolistic competition under a global environmental constraint can be stated as follows:

- the state variable $x_j(t)$ denotes the production capacity of firm $j$ at time $t$ which is supposed to depreciate at a rate $\mu_j$;
- $D(\sum_{j=1,\dots,p} x_j(t))$ is the inverse demand function that defines a clearing market price corresponding to the total supply $\sum_{j=1,\dots,p} x_j(t)$;
- $c_j$ is the unit production cost for firm $j$ and $\gamma_j(\cdot)$ is the investment cost function (production adjustment cost);
- $L_j(\mathbf{x}(t), \dot{x}_j(t)) = -[(D(\sum_{j=1,\dots,p} x_j(t)) - c_j)x_j(t) - \gamma_j(\dot{x}_j(t) + \mu_j x_j(t))]$ is the net cost (the negative of net profit) at time $t$ of firm $j$; and
- $e_j(x_j)$ denotes the pollutant emission rate of the $j$th firm when its production capacity is $x_j$.

The dynamic oligopoly model is then formulated as a differential game with a cost for player $j$ defined over any time interval $[0, T]$ by

$$J_j^T(\mathbf{x}(\cdot)) = \int_0^T L_j(\mathbf{x}(t), \dot{x}_j(t))\, dt,$$

and with decoupled constraints $x_j(t) \geq 0$. A global environmental constraint can be formulated as the coupled pointwise constraint

$$\sum_{j=1}^{p} e_j(x_j(t)) \leq \bar{e}, \qquad t \geq 0,$$

where $\bar{e}$ denotes some global limit on the pollutant emission level. In practice, the environment constraint need not be observed strictly at every point in time, provided that, in the long run, the global emission levels remain satisfactory. This calls for a weaker definition of the constraint where a notion of asymptotic feasibility should be introduced. The consideration of a noncooperative game under a coupled constraint may look a priori as a contradiction in terms, since the satisfaction of the coupled constraint seems to call for some sort of cooperation. Actually the concept becomes meaningful if one assumes that an extra player, e.g., a government agency like EPA,[1] is in charge of designing an emission taxation scheme which would induce the oligopolistic firms to satisfy globally the environmental constraint. Then the equilibrium concept, with coupled constraint, serves as the economic rational for designing the taxing scheme. Furthermore, the concept of normalized equilibrium, introduced by Rosen [10] is the natural approach for relating a common dual price vector (or multiplier) to an equilibrium under a coupled constraint. The interplay between a dynamic oligopoly model, the turnpike property, the design of an asymptotic emission tax through the solution of a steady-state static game with coupled emission constraint has been studied, for the model sketched above, by Haurie [6] and for a more general class of differential game models by Haurie and Zaccour [9]. However these two previous papers left unclear the precise definition of the dynamic game with asymptotic coupled constraint and the proof that

[1]Environmental Protection Agency.

a sort of dynamic equilibrium was then obtained. The results we establish in the present paper indicate that the method proposed in [6] can be used to define, in a precise and operational way (that is, permitting a relatively easy computation), a dynamic equilibrium with coupled asymptotic state constraint. The plan of this paper is as follows. In Section 2 we introduce the basic model, define terminology, and give basic hypotheses to make the dynamic game well defined. In Section 3 we investigate the corresponding steady-state game and we review the notion of a normalized equilibrium and the concept of strict diagonal convexity. These ideas will allow us to provide conditions under which a unique steady-state equilibrium exists. In Section 4 we establish the existence of an averaging optimal solution for the coupled infinite-horizon dynamic game. In Section 5 the case of a differential game with discounted cost is considered. We show that the proposed approach satisfies an *implicit equilibrium* property. Finally, in Section 6 we summarize the utility of our results by returning to the environmental management example briefly described above and discuss other possible formulations of global environmental constraints in oligopolistic markets.

## 2 The Model and Basic Hypotheses

We consider a $p$-player dynamic game in which the state trajectory of the $j$th player, $j = 1, 2, \dots p$, is a vector-valued function $x_j(\cdot) : [0, +\infty) \to \mathbb{R}^{n_j}$. The cost accumulated by each player up to time $T$ is described by an integral functional.

$$J_j^T(\mathbf{x}(\cdot)) = \int_0^T L_j(x(t), \dot{x}_j(t))\, dt, \tag{1}$$

in which $x = (x_1, x_2, \dots, x_p) \in \mathbb{R}^n = \mathbb{R}^{n_1} \times \mathbb{R}^{n_2} \times \cdots \times \mathbb{R}^{n_p}$ and the notation $\dot{x}_j(t)$ denotes the time derivative of $x_j(\cdot)$ at time $t > 0$. The functions $L_j : \mathbb{R}^n \times \mathbb{R}^{n_j} \to \mathbb{R}$, $j = 1, 2, \dots, p$, are assumed to be continuous functions in which for each $j = 1, 2, \cdots$ the function $(\mathbf{x}, z_j) \to L_j(\mathbf{x}, z_j)$ is convex and continuously differentiable in the $(x_j, z_j)$ coordinates. In addition, we impose the constraints

$$x(0) = \mathbf{x}_0 \in \mathbb{R}^n, \tag{2}$$

$$(x_j(t), \dot{x}_j(t)) \in X_j \subset \mathbb{R}^{2n_j} \qquad \text{a.e.} \quad t \geq 0, \quad j = 1, 2, \dots, p, \tag{3}$$

$$h_l(\mathbf{x}(t)) \leq 0 \qquad \text{for} \quad t \geq 0, \quad l = 1, 2, \dots k, \tag{4}$$

in which the sets $X_j$,$j = 1, 2, \dots, p$, are convex and compact and the functions $h_l : \mathbb{R}^n \to \mathbb{R}$ are continuously differentiable and convex, $l = 1, 2, \dots, k$. We will refer to the problem described by (1), (2), and (3) as the uncoupled constraint problem (UCP) and to be problem described by (1), (2), (3), and (4) as the coupled constraint problem (CCP). The consideration of a pointwise state constraint (4) complicates significantly the model and the characterization of an optimal (equilibrium) solution. However, in the realm of environmental management problems, the satisfaction of a global pollution constraint is often a "long term" objective of the regulating agency instead of strict compliance to a standard at all points of time. This observation motivates the following definition of a relaxed version of the coupled constraint for an infinite horizon game (we use the notation meas$[\cdot]$ to denote Lebesgue measure):

**Definition 2.1.** A trajectory $\mathbf{x}(\cdot)$ (see Definition 2.2 below) is said to asymptotically satisfy the coupled state constraint (2.4) if the following holds: For each $\varepsilon > 0$ there exists a number $B(\varepsilon) > 0)$ such that

$$\text{meas}\left[\{t \geq 0 : h_l(\mathbf{x}(t)) > \varepsilon\}\right] < B(\varepsilon) \qquad \text{for} \quad l = 1, 2, \ldots, k. \tag{5}$$

**Remark 2.1.** Clearly this definition is inspired from the turnpike theory in optimal economic growth (see, e.g., [1]). As in the case of a "turnpike," for any $\varepsilon > 0$ the asymptotically admissible trajectory will spend most of its journey in the $\varepsilon$-vicinity of the admissible set.

Additionally we require the following definitions.

**Definition 2.2.** We say that a function $\mathbf{x} : [0, +\infty) \to \mathbb{R}^n$ is:

1. An admissible trajectory for UCP if $\mathbf{x}(\cdot)$ is locally absolutely continuous, $J_j^T(\mathbf{x}(\cdot))$ is finite for all $T \geq 0$, and the constraints (2)–(3) hold.
2. Asymptotically admissible for CCP if it is admissible for UCP and asymptotically satisfies the coupled constraint (4).
3. A pointwise admissible trajectory for CCP if $\mathbf{x}(\cdot)$ is locally continuous, $J_j^T(\mathbf{x}(\cdot))$ is finite for all $T \geq 0$, and the constraints (2)–(4) hold.

**Remark 2.2.** When no confusion arises we shall refer to only admissible trajectories with no explicit reference to the uncoupled constraint or coupled constraint problem.

**Definition 2.3.** Let $\mathbf{x}^*(\cdot) : [0, +\infty) \to \mathbb{R}^n$ be an admissible trajectory. For $j = 1, 2, \ldots, p$ we define the set of admissible responses

$$A_j(\mathbf{x}^*(\cdot)) = \left\{y_j : [0, +\infty) \to \mathbb{R}^{n_j} : \left[\mathbf{x}^{*(j)}, y_j\right](\cdot) \text{ is admissible} \right\},$$

where we use the notation

$$\left[\mathbf{x}^{*(j)}, y_j\right](\cdot) = \left(x_1^*(\cdot), x_2^*(\cdot), \ldots, x_{j-1}^*(\cdot), y_j(\cdot), x_{j+1}^*(\cdot), \ldots, x_p^*(\cdot)\right).$$

**Remark 2.3.** The notation

$$\left[\mathbf{x}^{*(j)}, y_j\right](\cdot) = \left(x_1^*(\cdot), x_2^*(\cdot), \ldots, x_{j-1}^*(\cdot), y_j(\cdot), x_{j+1}^*(\cdot), \ldots, x_p^*(\cdot)\right)$$

represents a unilateral change for the $j$th player from the trajectory $x_j(\cdot)$ to $y_j(\cdot)$.

Finally we introduce the various types of equilibria we consider in our work.

**Definition 2.4.** An admissible trajectory $\mathbf{x}^*(\cdot)$ is called:

1. A strong Nash equilibrium if for $j = 1, 2, \ldots, p$ we have

$$J_j^\infty(\mathbf{x}^*(\cdot)) = \lim_{\mathbf{T}\to+\infty} \mathbf{J}_\mathbf{j}^\mathbf{T}(\mathbf{x}^*(\cdot))$$

is finite and if for all $y_j \in A_j(\mathbf{x}^*(\cdot))$ we have

$$\liminf_{T\to\infty} J_j^T\left(\left[\mathbf{x}^{*(j)}, y_j\right](\cdot)\right) \geq J_j^\infty(\mathbf{x}^*(\cdot)). \tag{6}$$

2. An overtaking Nash equilibrium if for every $\epsilon > 0$, $j = 1, 2, \ldots, p$, and $y_j \in A_j(\mathbf{x}^*(\cdot))$ there exist $T_j = T(j, \epsilon, y) > 0$ such that

$$J_j^T(\mathbf{x}^*(\cdot)) < J_j^T\left(\left[\mathbf{x}^{*(j)}, y_j\right](\cdot)\right) + \epsilon \tag{7}$$

for all $T \geq T_j$ or, equivalently,

$$\liminf_{T\to\infty}\left[J_j^T\left(\left[\mathbf{x}^{*(j)}, y_j\right](\cdot)\right) - J_j^T(\mathbf{x}^*(\cdot))\right] \geq 0 \tag{8}$$

for all $y_j \in A_j(\mathbf{x}^*(\cdot))$.

3. An average cost optimal Nash equilibrium if for $j = 1, 2, \ldots, p$ we have

$$\liminf_{T\to\infty}\frac{1}{T}\left[J_j^T\left(\left[\mathbf{x}^{*(j)}, y_j\right](\cdot)\right) - J_j^T\left(\mathbf{x}^*(\cdot)\right)\right] \geq 0 \tag{9}$$

for all $y_j \in A_j(\mathbf{x}^*(\cdot))$.

**Remark 2.4.** If $\mathbf{x}^*(\cdot)$ is an overtaking Nash equilibrium, then it is easy to see that it is an average cost Nash equilibrium as well. To see this we note that by (7) for all $T \geq T_j$ we have

$$\frac{1}{T}\left[J_j^T\left(\left[\mathbf{x}^{*(j)}, y_j\right](\cdot)\right) - J_j^T\left(\mathbf{x}^*(\cdot)\right)\right] \geq -\frac{\epsilon}{T},$$

which, upon taking the limit inferior, immediately implies (9) holds.

## 3 The Steady-State Equilibrium Problem and Normalized Equilibrium

We now define the associated steady-state game. To do this we define the $p$ scalar functions $\Phi_j : \mathbb{R}^n \to \mathbb{R}$ by the formulas

$$\Phi_j(\mathbf{x}) = L_j(\mathbf{x}, 0), \tag{10}$$

where, as usual, $\mathbf{x} = (x_1, x_2, \ldots, x_p)$. We now consider the convex game of finding a Nash equilibrium for the cost functions $\Phi_j$ subject to the constraints

$$(x_j 0) \in X_j, \quad j = 1, 2, \ldots, p, \tag{11}$$

$$h_l(\mathbf{x}) \leq 0. \quad l = 1, 2, \ldots, k. \tag{12}$$

Games of this sort have been studied by researches too numerous to list. For our treatment we refer exclusively to the semial paper of Rosen [10]. In Rosen's work, two concepts are introduced which are particularly important for our results. The first of these is the notion of a *normalized Nash equilibrium*. The second idea is that of *strict diagonal convexity* which provides sufficient conditions for a unique Nash equilibrium. To discuss these ideas we let $\mathbf{r} = (r_1, r_2, \ldots, r_p) \in \mathbb{R}^p$ be a fixed constant vector of positive real numbers (i.e., $r_j > 0$), define $\sigma : \mathbb{R}^n \times \mathbb{R}^p$ by

$$\sigma(\mathbf{x}, \mathbf{r}) = \sum_{j=1}^{p} r_j \Phi_j(\mathbf{x}), \tag{13}$$

and let

$$\mathcal{R} = \{\mathbf{x} \in \mathbb{R}^n : (x_j, 0) \in X_j, \; h_l(\mathbf{x}) \leq 0, \; 1 = 1, 2, \ldots, k\}.$$

We assume that the set $\mathcal{R}$ is closed, convex, and bounded. We further assume that the Slater constraint qualification holds (we could also use a less restrictive constraint qualification condition). That is, there exists $\bar{\mathbf{x}} \in$ int $\mathcal{R}$. Under these conditions, if $\hat{\mathbf{x}}$ is a Nash equilibrium, then for each $j = 1, 2, \ldots, p$ there exists (by the Karush–Kuhn–Tucker theorem) multipliers $\mu_j = (\mu_{j1}, \mu_{j2}, \ldots \mu_{jk}) \in \mathbb{R}^k$ such that

$$\begin{aligned} \mu_{jl} \geq 0 \quad & \text{for} \quad l = 1, 2, \ldots, k, \\ \mu_{jl} h_l(\bar{\mathbf{x}}) = 0 \quad & \text{for} \quad l = 1, 2, \ldots, k, \\ \nabla_j \Phi_j(\bar{\mathbf{x}}) + \sum_{l=1}^{k} \mu_{jl} \nabla_j h_l(\bar{\mathbf{x}}) = 0, & \end{aligned} \tag{14}$$

where $\nabla_j$ denotes the gradient with respect to the variables $x_j$. With this notation we can now define what is meant by a normalized equilibrium.

**Definition 3.1.** We say that a Nash equilibrium for the convex game described by (10) and (12), say $\bar{\mathbf{x}}$, is a normalized equilibrium with respect to the weights $\mathbf{r} = (r_1, r_2, \ldots, r_p)$ if there exists a vector $\mu_0 \in \mathbb{R}^k$ so that the vectors

$$\mu_j = \frac{1}{r_j} \mu_0, \qquad j = 1, 2, \ldots, p,$$

form a set of Lagrange multipliers satisfying (14).

With this definition we state the following theorem given in Rosen [10]:

**Theorem 3.1.** *Under the Slater constraint qualification, for each* $\mathbf{r} \in \mathbb{R}^p$, *there exists a normalized Nash equilibrium.*

**Proof.** See Rosen [10, Theorem 3]. □

We now discuss conditions for uniqueness beginning with the following definition:

**Definition 3.2.** We say that the function $\sigma(\cdot, \cdot) : \mathbb{R}^n \times \mathbb{R}^p \to \mathbb{R}$ is diagonally strictly convex in $\mathbf{x}$ for $\mathbf{r} = (r_1, r_2, \ldots, r_p) \in \mathbb{R}^p, r_j > 0$, if for all $\mathbf{x}^0, \mathbf{x}^1 \in \mathbb{R}^n$ we have

$$\sum_{j=1}^{p} r_j \left\{ \left(x_j^1 - x_j^0\right)' \left(\nabla_j \Phi_j(\mathbf{x}^0) - \nabla_j \Phi_j(\mathbf{x}^1)\right) \right\} < 0.$$

This condition was introduced by Rosen [10] who used it to establish the following result:

**Theorem 3.2.** *If the Slater constraint qualification holds and if $\sigma(\cdot,\cdot)$ is diagonally strictly convex for every $\mathbf{r} \in Q$, where $Q$ is a convex subset of the positive orthant in $\mathbb{R}^p$, then for each $\mathbf{r} \in Q$, there is a unique normalized equilibrium point.*

**Proof.** See Rosen [10, Theorem 4]. □

**Remark 3.1.** Under additional differentiability conditions one can show that $\sigma(\cdot,\cdot)$ is diagonal strictly convex if the symmetric matrix

$$G(\mathbf{x},\mathbf{r}) + G(\mathbf{x},\mathbf{r})'$$

is positive definite, where $G(\mathbf{x},\mathbf{r})$ is the Jacobian of the vector valued function

$$g(\mathbf{x},\mathbf{r}) = [r_1\nabla_1\Phi_1(\mathbf{x}), r_2, \nabla_2\Phi_2(\mathbf{x}), \ldots, r_p\nabla_p\Phi_p(\mathbf{x})]'$$

with respect to $\mathbf{x}$. Consequently, the conditions given above are verifiable in these situations.

In this section we have seen that the steady-state game is amenable to a treatment according to the theory developed by Rosen. In the remainder of the paper we will show that the solution of the steady-state game, in particular the vector of Kuhn–Tucker multipliers, can be used to define an equilibrium solution, in a weaker sense, to the dynamic game.

## 4 Equilibria Solutions for the Undiscounted Dynamic Game

We begin this section by summarizing the results appearing in Carlson and Haurie [2], beginning with the following additional assumptions:

**Assumption 4.1.** There exists a fixed $\mathbf{r} = (r_1, r_2, \ldots, r_p) \in \mathbb{R}^n$, with $r_j > 0$ so that the associated steady-state game has a unique normalized Nash equilibrium, denoted by $\hat{\mathbf{x}}$, with multipliers $\mu_j = \mu_0/r_j$.

**Remark 4.1.** Conditions under which this assumption holds were given in the previous section. An equivalent form for writing this condition is that there exists a unique normalized steady-state equilibrium, say $\hat{\mathbf{x}}$, and an associated multiplier, say $\hat{\mathbf{p}} \in \mathbb{R}^n$, satisfying

$$0 \in \partial_{pj}\tilde{H}_j(\hat{\mathbf{x}}, \hat{p}_j; \mu_0), \tag{15}$$

$$0 \in \partial_{xj}\tilde{H}_j(\hat{\mathbf{x}}, \hat{p}_j; \mu_0), \tag{16}$$

$$\mu_0 h(\hat{\mathbf{x}}) = 0, \tag{17}$$

$$h(\hat{\mathbf{x}}) \leq 0, \tag{18}$$

$$\mu_0 \geq 0, \tag{19}$$

in which

$$\begin{aligned}\tilde{H}_j(\mathbf{x}, p_j, \mu_0) &\doteq \sup_z \left\{ p_j' z - L_j(\mathbf{x}, z) - \frac{1}{r_j} \sum_{l=1}^{k} \mu_{0l} h_l(\mathbf{x}) \right\} \\ &= \sum_z \{p_j' z - L_j(\mathbf{x}, z)\} - \frac{1}{r_j} \sum_{l=1}^{k} \mu_{0l} h_l(\mathbf{x}) \\ &= H_j(\mathbf{x}, p_j) - \frac{1}{r_j} \sum_{l=1}^{k} \mu_{0l} h_l(\mathbf{x}),\end{aligned} \tag{20}$$

where the supreme above is over all $z \in \mathbb{R}^{n_j}$ for which $(x_j, z_j) \in X_j$.

**Assumption 4.2.** There exists $\varepsilon_0 > 0$ and $S > 0$ such that for any $\bar{\mathbf{x}} \in \mathbb{R}^n$ satisfying $||\bar{\mathbf{x}} - \hat{\mathbf{x}}|| < \varepsilon_0$ there exists a trajectory $\mathbf{w}(\bar{\mathbf{x}}, \cdot)$ defined on $[0, S]$ such that $\mathbf{w}(\bar{\mathbf{x}}, 0) = \hat{\mathbf{x}}$, $\mathbf{w}(\bar{\mathbf{x}}, S) = \bar{\mathbf{x}}$, and $(w_j(\bar{\mathbf{x}}, t), \dot{w}_j(\bar{\mathbf{x}}, t)) \in X_j$ a.e. $0 \le t \le S$.

**Remark 4.2.** This assumption is a controllability assumption which says that in a neighborhood of the steady-state equilibrium, $\hat{\mathbf{x}}$, we can drive the system to the steady state. (i.e., turnpike) in a uniform length of time (in this case $S$) and still maintain the uncoupled constraints. This condition is required to utilize the results of Carlson and Haurie [2] to conclude the existence of an overtaking Nash equilibrium for an associated uncoupled dynamic game which we describe below.

With these assumptions in place we consider an associated uncoupled dynamic game that is obtained by adding to the accumulated costs (1) a penalty term which imposes a cost to the player for violating the coupled constraint. To do this we use the multipliers defining the normalized steady-state Nash equilibrium. That is, we consider the new accumulated costs

$$\tilde{J}_j^T(\mathbf{x}(\cdot)) = \int_0^T \left[ L_j(\mathbf{x}(t), \dot{x}_j(t)) + \frac{1}{r_j} \sum_{l=1}^{k} \mu_{0l} h_l(\mathbf{x}(t)) \right] dt, \tag{21}$$

where we assume that the initial condition (2) and the uncoupled constraints (3) are included. Note here that we are not using the standard method of Lagrange since we have introduced a constant multiplier instead of a continuous function as is usually done for variational problems. As we shall see, the fact that we have an autonomous infinite-horizon problem and the fact that the asymptotic turnpike property holds allow us to use as a constant multiplier the one resulting from the solution of the steady-state game and then obtain an asymptotically admissible dynamic equilibrum. Under these above assumptions we have the following result that is a direct consequence of Theorem 3.1 of [2]:

**Theorem 4.1.** *The associated uncoupled dynamic game with costs given by* (21) *and constraints* (2) *and* (3) *has an overtaking Nash equilibrium, say* $\mathbf{x}^*(\cdot)$, *which additionally satisfies*

$$\lim_{t \to \infty} \mathbf{x}^*(t) = \hat{\mathbf{x}}. \tag{22}$$

**Proof.** See Theorem 3.1 of [2]. □

**Remark 4.3.** Under an additional strict diagonal convexity assumption the above theorem insures the existence of a unique overtaking Nash equilibrium. The specific hypothesis required is that the combined cost function $\sum_{j=1}^{p} r_j L_j(t, \mathbf{x}, z_j)$ is diagonally strictly convex in $(x, z)$, i.e., verifies

$$\sum_{j=1}^{p} r_j \left[ (z_j^1 - z_j^0)' \left( \frac{\partial}{\partial_{z_j}} L_j \left(t, \mathbf{x}^1, z_j^1\right) - \frac{\partial}{\partial_{z_j}} L_j \left(t, \mathbf{x}^0, z_j^0\right) \right) \right.$$
$$\left. + (x_j^1 - x_j^0)' \left( \frac{\partial}{\partial_{x_j}} L_j \left(t, \mathbf{x}^1, z_j^1\right) - \frac{\partial}{\partial_{x_j}} L_j \left(t, \mathbf{x}^0, z_j^0\right) \right) \right] > 0, \tag{23}$$

for all

$$\mathbf{x}^1, \mathbf{z}^1 \quad \text{and} \quad \mathbf{x}^0, \mathbf{z}^0.$$

**Remark 4.4.** We observe that the equilibrium trajectory guaranteed in the above result is not generally pointwise admissible for the dynamic game with coupled constraints. On the other hand, since it enjoys the turnpike property (22) it is an easy matter to see that it is asymptotically admissible for the original coupled dynamic game. Indeed, since the functions $h_l(\cdot)$ are continuous and since (22) holds, for every $\varepsilon > 0$ there exists $T(\varepsilon) > 0$ such that for all $t > T(\varepsilon)$, we have

$$h_l(\hat{\mathbf{x}}) - \varepsilon < h_l(\mathbf{x}^*(t)) < h_l(\bar{\mathbf{x}}) + \varepsilon \leq \varepsilon$$

for all $l = 1, 2, \ldots, k$ since $h_l(\bar{\mathbf{x}}) \leq 0$. Thus we can take $B(\varepsilon) = T(\varepsilon)$ in (5). As a consequence of this fact one is led directly to investigate whether this trajectory is some sort of equilibrium for the original problem. This in fact is the following theorem:

**Theorem 4.2.** *Under the above assumptions the overtaking Nash Equilibrium for the UCP* (2), (3), (21) *is also an averaging Nash equilibrium for the coupled dynamic game when the coupled constraint* (4) *are interpreted in the asymptotic sense as described in Definition* 2.1.

**Proof.** As a result of Theorem 4.1 there exist an overtaking Nash equilibrium for the associated uncoupled dynamic game, say $\mathbf{x}^*(\cdot)$. We demonstrate that this asymptotically admissible trajectory is an averaging Nash equilibrium for the coupled dynamic game over the class of all asymptotically admissible trajectories. To see this for each $j = 1, 2, \ldots, p$ we let $y_j(\cdot) : [0, +\infty) \to \mathbb{R}^{n_j}$ be such that the trajectory, $[\mathbf{x}^{*(j)}, y_j](\cdot)$ is asymptotically admissible. Then for $\varepsilon > 0$ there exists $\hat{T}(\varepsilon) > 0$ such that for all $T > \hat{T}(\varepsilon)$ we have

$$\int_0^T L_j(\mathbf{x}^*(t), \dot{x}_j^*(t)) + \frac{1}{r_j} \sum_{l=1}^{k} \mu_{0l} h_l(\mathbf{x}^*(t))\, dt$$
$$< \int_0^T L_j([\mathbf{x}^{*(j)}, y_j](t), \dot{y}_j(t))$$

$$+\frac{1}{r_j}\sum_{l=1}^{k}\mu_{0l}h_l([\mathbf{x}^{*(j)},y_j](t))\,dt+\varepsilon,$$

or, equivalently,

$$-\varepsilon+\frac{1}{r_j}\sum_{l=1}^{k}\mu_{0l}\int_0^T h_l(\mathbf{x}^*(t))-h_l([\mathbf{x}^{*(j)},y_j](t))\,dt$$
$$<\int_0^T L_j([\mathbf{x}^{*(j)},y_j](t),y_j(t))-L_j(\mathbf{x}^*(t),\dot{x}_j^*(t))\,dt.$$

Since $\mathbf{x}^*(\cdot)$ enjoys the turnpike property (22) we can assume that $\hat{T}(\varepsilon)$ has been chosen sufficiently large to insure that for $l=1,2,\ldots,k$ we have

$$h_l(\hat{\mathbf{x}})-\varepsilon<h_l(\mathbf{x}^*(t))<h_l(\hat{\mathbf{x}})+\varepsilon$$

for all $t\geq\hat{T}(\varepsilon)$. Further, since $\left[\mathbf{x}^{*(j)},y_j\right](\cdot)$ is asymptotically admissible we have that there exists a constant $B(\varepsilon)>0$ such that for all $T>0$ we have

$$\text{meas}\left[S_\varepsilon^T\right]=\text{meas}\left[\left\{t\in[0,T]:\,h_l\left(\left[\mathbf{x}^{*(j)},y_j\right](t)\right)\geq\varepsilon\right\}\right]<B(\varepsilon).$$

From these facts, when combined with the complementary slackness condition

$$\frac{\mu_{0l}}{r_j}h_l(\hat{\mathbf{x}})=0\qquad\text{for}\quad l=1,2,\ldots,k,$$

we may write

$$\frac{1}{T}\int_0^T\left\{L_j(\left[\mathbf{x}^{*(j)},y_j\right](t),\dot{y}_j(t))-L_j(\mathbf{x}^*(t),\dot{y}_j(t))\right\}\,dt$$
$$>-\frac{1}{T}\varepsilon+\frac{1}{r_j}\sum_{l=1}^{k}\mu_{0l}\left[\frac{1}{T}\int_0^{\hat{T}(\varepsilon)}h_l(\mathbf{x}^*(t))\,dt-\frac{1}{T}\int_{S_\varepsilon^T}h_l([\mathbf{x}^{*(j)},y_j](t))\,dt\right]$$
$$-\left(\frac{1}{r_j}\sum_{l=1}^{k}\mu_{0l}\right)\frac{T-\hat{T}(\varepsilon)}{T}\varepsilon-\frac{1}{r_j}\sum_{l=1}^{k}\mu_{0l}\frac{1}{T}\int_{[0,T]\backslash S_\varepsilon^T}h_l([\mathbf{x}^{*(j)},y_i](t))\,dt,$$
$$>-\frac{\epsilon}{T}\left[1+\left(\frac{1}{r_j}\sum_{l=1}^{k}\mu_{0l}\right)(T-\hat{T}(\varepsilon))+\text{meas}\,\left[[0,T]\backslash S_\varepsilon^T]\right]\right]$$
$$+\frac{1}{r_j}\sum_{l=1}^{k}\mu_{0l}\left[\frac{1}{T}\int_0^{\hat{T}(\varepsilon)}h_l(\mathbf{x}^*(t))-\frac{1}{T}\int_{S_\varepsilon^T}h_l([\mathbf{x}^{*(j)},y_j](t))\,dt\right].$$

Since each of the trajectories $\mathbf{x}^*(\cdot)$ and $[\mathbf{x}^{*(j)},y_j](\cdot)$ are bounded we have that the second term tends to zero as $T\to+\infty$. Consequently,

$$\liminf_{T\to+\infty}\frac{1}{T}\int_0^T L_j\left([\mathbf{x}^{*(j)},y_j](t)\right)-L_j(\dot{\mathbf{x}}_j^*(t))\,dt>-\varepsilon\left[2+\left(\frac{1}{r_j}\sum_{l=1}^{k}\mu_{0l}\right)\right].$$

for every $\varepsilon > 0$ which implies

$$\liminf_{T\to+\infty} \frac{1}{T} \int_0^T L_j([\mathbf{x}^{*(j)}, y_j](t), \dot{y}_j(t)) - L_j(\mathbf{x}^*(t), \dot{\mathbf{x}}_j^*(t))\, dt \geq 0$$

as desired. □

We conclude this section by demonstrating that the long-term average cost for each player for the Nash equilibrium given in the above theorem is precisely the steady-state cost obtained from the solution of the steady-state equilibrium problem. That is, we have the following result:

**Theorem 4.3.** *For the averaging Nash equilibrium given by Theorem 4.2 we have*

$$\lim_{T\to+\infty} \frac{1}{T} \int_0^T L_j(\mathbf{x}^*(t), \dot{\mathbf{x}}_j^*(t))\, dt = L_j(\hat{\mathbf{x}}, 0). \tag{24}$$

**Proof.** Let $\varepsilon > 0$ be given. Since $\mathbf{x}^*(\cdot)$ satisfies the turnpike property there exists $T(\varepsilon) > 0$ such that.

$$|L_j([\mathbf{x}^{*(j)}, \hat{x}_j](t), 0)\, dt - L_j(\hat{\mathbf{x}}, 0)| < \frac{\varepsilon}{3}.$$ □

Further, as a consequence of the Controllability Assumption 4.2, we can choose $T(\varepsilon)$ sufficiently large to insure that $||\mathbf{x}^*(T(\varepsilon)) - \hat{\mathbf{x}}|| < \varepsilon_0$. Consequently, there exists a function $\mathbf{w}(\cdot) : [0, S] \to \mathbb{R}^n$ so that

$$\mathbf{y}(t) \doteq \begin{cases} \mathbf{x}^*(t), & 0 \leq t \leq T(\varepsilon), \\ \mathbf{w}(t), & T(\varepsilon) \leq t \leq T(\varepsilon) + S, \\ \hat{\mathbf{x}}, & T(\varepsilon) + S \leq t, \end{cases}$$

is an asymptotically admissible trajectory. Further, since the trajectories $[\mathbf{x}^{*(j)}, y_j(\cdot)](\cdot)$ enjoy the turnpike property (and hence are asymptotically admissible) we can find $\hat{T}(\varepsilon) > T(\varepsilon) + S$ so that for all $T > \hat{T}(\varepsilon)$ and $j = 1, 2, \ldots, p$ we have

$$\begin{aligned} \frac{1}{T} \int_0^T L_j(\mathbf{x}^*(t), \dot{x}_j^*(t))\, dt &\leq \frac{1}{T} \int_0^T L_j([\mathbf{x}^{*(j)}, y_j](t), \dot{y}_j(t))\, dt + \frac{\varepsilon}{3} \\ &= \frac{1}{T} \int_0^{T(\varepsilon)+S} L_j([\mathbf{x}^{*(j)}, y_j](t), \dot{y}_j(t))\, dt \\ &\quad + \frac{1}{T} \int_{T(\varepsilon)}^{T(\varepsilon)+S} L_j([\mathbf{x}^{*(j)}, \hat{x}_j], 0)\, dt + \frac{\varepsilon}{3}. \end{aligned}$$

Thus, we may write

$$\begin{aligned} &\frac{1}{T} \int_0^T L_j(\mathbf{x}(t), \dot{x}_j^*(t))\, dt - L_j(\hat{\mathbf{x}}, 0) \\ &\quad \leq \frac{1}{T} \int_0^{T(\varepsilon)+S} \left[L_j([\mathbf{x}^{*(j)}, y_j](t), \dot{y}_j(t)) - L_j(\hat{\mathbf{x}}, 0)\right] dt \end{aligned}$$

$$+\frac{1}{T}\int_{T(\varepsilon)+S}^{T}\left[L_j(\left[\mathbf{x}^{*(j)},\hat{x}_j\right](t),0)-L_j(\hat{\mathbf{x}},0)\right]dt+\frac{\varepsilon}{3}$$

$$<\frac{1}{T}\int_{0}^{T(\varepsilon)+S}\left[L_j(\left[\mathbf{x}^{*(j)},y_j\right](t),\dot{y}_j(t))-L_j(\hat{\mathbf{x}},0)\right]dt$$

$$+\left[\frac{T-T(\varepsilon)-S}{T}+1\right]\frac{\varepsilon}{3}.$$

As $T(\varepsilon)+S$ is fixed there exists $T_1 > \hat{T}(\varepsilon)$ so that for all $T \leq T_1$

$$\frac{1}{T}\int_0^T L_j(\mathbf{x}^*(t),\dot{x}_j^*(t))\,dt - L_j(\hat{\mathbf{x}},0) < \varepsilon.$$

On the other hand, the convexity and smoothness conditions placed on $L_j$ gives us

$$\frac{1}{T}\int_0^T L_j(\mathbf{x}^*(t),x_j^*(t)) - L_j(\hat{\mathbf{x}},0)\,dt \geq \frac{1}{T}\frac{\partial}{\partial \dot{x}_j}L_j(\hat{\mathbf{x}},0)'(x_j^*(T)-x_j^*(0)),$$

and since $(x_j^*(t),\dot{x}_j(t)) \in X_j$, a compact set, it follows that there exists $T_2 > 0$ such that for all $T > T_2$ we have

$$\frac{1}{T}\int_0^T L_j(\mathbf{x}^*(t),\dot{x}_j^*(t))\,dt - L_j(\hat{\mathbf{x}},0) > -\varepsilon.$$

Therefore, for all $T \geq \max\{T_1,T_2\}$, we have

$$-\varepsilon < \frac{1}{T}\int_0^T L_j(\mathbf{x}^*(t),\dot{x}_j^*(t))\,dt - L_j(\hat{\mathbf{x}},0), < \varepsilon,$$

implying the desired result.

## 5 The Discounted Case

In this section we investigate the questions considered above when the cost functionals are discounted by a uniform discount rate $\rho > 0$. That is, we modify our costs to become

$$J_j^T(\mathbf{x}(\cdot)=\int_0^T e^{-pt}L_j(\mathbf{x}(t),\dot{x}_j(t))\,dt. \tag{25}$$

The introduction of the discount factor modifies the approach to the above arguments in several ways.

- We must replace the steady-state equilibrium game with an *implicit steady-state game*.
- Since the discount rate introduces a tendency to delay tough decisions to the future we must modify the means in which we require the pointwise state constraint to be satisfied.

- We need not considert the weaker notions of overtaking and averaging Nash equilibria since the discount factor forces all of the imporoper integrals to converge.

We begin our investigations by discussing the first two points indicated above.

### 5.1 The Implicit Steady-State Equilibrium Problem

Let us introduce the steady-state dynamic equilibrium conditions with positive discounting as follows: we assume there exists a steady state $\hat{\mathbf{x}}$, and an associated costate $\hat{\mathbf{p}} \in \mathbb{R}^n$ satisfying

$$0 \in \partial_{pj} \tilde{H}_j(\hat{\mathbf{x}}, \hat{p}_j; \mu_0), \tag{26}$$

$$\rho \hat{p} \in \partial_{xj} \tilde{H}_j(\hat{\mathbf{x}}, \hat{p}_j; \mu_0), \tag{27}$$

$$\mu_0 h(\hat{\mathbf{x}}) = 0, \tag{28}$$

$$h(\hat{\mathbf{x}}) \leq 0, \tag{29}$$

$$\mu_0 \geq 0, \tag{30}$$

where $\tilde{H}_j(\hat{\mathbf{x}}, \hat{p}_j; \mu_0)$ is defined as in (20). As indicted in Feinstein and Luenberger [4] for the control case and by Haurie and Roche [8] for the differential game case, these steady-state conditions with a positive discount rate can be interpreted as an *implicit equilibrium solution*. More precisely, the vector $\hat{\mathbf{x}}$ is a normalized equilibrium solution for the static game with payoffs.

$$L_j(\mathbf{x}, \rho(x_j - \hat{x}_j)), \qquad \text{where} \quad (x_j, \rho(x_j - \hat{x}_j)) \in X_j, \; j = 1, \ldots, p, \tag{31}$$

and coupled constraint $h(\mathbf{x}) \leq 0$. The conditions insuring the uniqueness of such a steady-state implicit equilibrium are not easy to obtain. Indeed the fixed-point argument that is inherent in the definition is at the origin of this difficulty. We shall, however, assume the following:

**Assumption 5.1.** There exists a fixed $\mathbf{r} = (r_1, r_2, \ldots, r_p) \in \mathbb{R}^n$, with $r_j > 0$ so that there exists a unique implicit normalized Nash equilibrium, denoted by $\hat{\mathbf{x}}$, with multipliers $\mu_j = \mu_0 / r_j$ for the game defined in (31).

With this assumption, as in the undiscounted case, we now introduce an associated discounted game with uncoupled canstraints by considering the perturbed cost functionals

$$\tilde{J}_j^T(\mathbf{x}(\cdot)) = \int_0^T e^{-\rho t} \left[ L_j(\mathbf{x}(t), \dot{x}_j(t)) + \frac{1}{r_j} \sum_{l=1}^{k} \mu_{0l} h_l(\mathbf{x}(t)) \right] dt, \tag{32}$$

with uncoupled constraints

$$(x_j(t), \dot{x}_j(t)) \in X_j, \qquad j = 1, 2, \ldots, p. \tag{33}$$

**Theorem 5.1.** *5.1 Under Assumption* 5.1 *there exists a Nash equilibrium, say* $\mathbf{x}^*(\cdot)$, *for the associated discounted uncoupled game. In addition, if the combined*

*Hamiltonian*

$$\sum_{j=1}^{m} \tilde{H}_j(\mathbf{x}, p_j; \mu_o)$$

*is sufficiently stricly diagonally concave in* $\mathbf{x}$ *and convex* $\mathbf{p}$ *relative to the discount rate* $\rho$*, then this Nash equilibrium,* $\mathbf{x}^*(\cdot)$ *enjoys the turnpike property. That is,*

$$\lim_{t\to+\infty} \mathbf{x}^*(t) = \hat{\mathbf{x}},$$

*in which* $\hat{\mathbf{x}}$ *is the unique solution of the implicit steady-state game.*

**Proof.** This result follows directly from Theorems 2.5 and 3.2 of [2]. □

**Remark 5.1.** The precise conditions under which the turnpike property in the above result is valid are given in Carlson and Haurie [2], and for brevity we refer the reader there for specific details.

The natural question to ask now, as before, is what type of optimality is implied by the above theorem for the original dynamic discounted game. The perturbed cost functions (5.8) seem to indicate that, with the introduction of the discount rate, it is appropriate to consider the coupled isoperimetric constraint

$$\int_0^{+\infty} e^{-\rho t} h_l(\mathbf{x}(t))\, dt \le 0, \qquad l = 1, 2, \ldots, k. \tag{34}$$

However, the multipliers $\mu_{0l}$ are not those associated with the coupled isoperimetric constraints (34) but those defined by the normalized steady-state implicit equilibrium problem. As we shall see in the next section the solution of the auxiliary game with decoupled constrols and cost functionals (32) enjoys a weaker dynamic equilibrium propterty which we call *Implicit Nash equilibrium*.

## 5.2 Existence of an Implicit Nash Equilibrium

The introduction of a positively discounted isoperimetric constraint (34) would enable each of the players to delay costly decisions to the future. This of course would not lead the players to meet the coupled-state constraint for most of the time in the future. To address this issue we introduce the following definition of admissibility relative to a trajectory that asymptotically satisfies the constraints as in Definition 2.1:

**Definition 5.1.** 5.1 Let $\bar{\mathbf{x}}(\cdot)$ be a fixed admissible trajectory for the discounted uncoupled dynamic game and let $\bar{\mathbf{x}}^\infty \in \mathbb{R}^n$ be a constant vector such that

$$\begin{aligned} \lim_{t\to+\infty} \bar{\mathbf{x}}(t) &= \bar{\mathbf{x}}^\infty, \\ h_l(\bar{\mathbf{x}}^\infty) &\le 0 \qquad \text{for} \quad l = 1, 2, \ldots, k, \\ (\bar{x}_j^\infty) &\in X_j \qquad \text{for} \quad j = 1, 2, \ldots, p. \end{aligned} \tag{35}$$

We say a trajectory $\mathbf{x}(\cdot)$ asymptotically satisfies the coupled state constraint (4) for the discounted dynamic game relative to $\bar{\mathbf{x}}(\cdot)$ if

$$\int_0^{+\infty} e^{\rho t} h_l(\mathbf{x}(t))\, dt \le \int_0^{+\infty} e^{-\rho t} h_l(\bar{\mathbf{x}}(t))\, dt. \tag{36}$$

**Remark 5.2.** In the above the definition we view the coupled constraint as an isoperimetric-type constraint where the expression

$$\rho \int_0^{+\infty} e^{-\rho t} h_l(\mathbf{x}(t))\, dt$$

may be viewed as a "discounted average value" of the pointwise constraint (4) over the infinite time interval $[0. + \infty)$. with these terms, a unilateral change from a given asymptotically admissible strategy by any of the players is admissible if it does not exceed the "discounted average value" of the constraint.

**Definition 5.2.** 5.2 An admissible trajectory $\bar{\mathbf{x}}(\cdot)$ is an implicit Nash equilibrium for the discounted dynamic game if there exists a constant vector $\bar{\mathbf{x}}^{\infty}$ so that (35) holds and if for all $y_j(\cdot)$, for which the trajectories $\left[\bar{\mathbf{x}}^{(j)}, y_j\right](\cdot)$ asymptotically satisfy the coupled-state constraint (4) relative to $\bar{\mathbf{x}}(\cdot)$, the following holds

$$\int_0^{+\infty} e^{-\rho t} L_j(\bar{\mathbf{x}}(t), \dot{\bar{x}}_j(t))\, dt \le \int_0^{+\infty} e^{-\rho t} L_j(\left[\bar{\mathbf{x}}^{(j)}, y_j\right](t), \dot{y}_j(t))\, dt.$$

With this definition in hand we are now able to obtain our discounted version of Theorem 4.2.

**Theorem 5.2.** *Under the assumptions given above there exists an implicit Nash equilibrium for the discounted coupled dynamic game.*

**Proof.** From Theorem 5.1 there exists a Nash equilibrium for the associated discounted uncoupled game, say $\mathbf{x}^*(\cdot)$. Let $y_j(\cdot)$ be any unilateral deviation by player $j$ that asymptotically satisfies the coupled-state constraint (4) relative to $\mathbf{x}^*(\cdot)$. Then from the equilibrium property for $\mathbf{x}(\cdot)$ and the fact that $\left[\mathbf{x}^{*(j)}, y_j\right](\cdot)$ satifies (36) we have

$$\begin{aligned} 0 &\ge \int_0^{+\infty} e^{-\rho t} \frac{1}{r_j} \sum_{l=1}^{k} \mu_{0l} h_l(\left[\mathbf{x}^{*(j)}, y_j\right](t))\, dt \\ &\quad - \int_0^{+\infty} e^{-\rho t} \frac{1}{r_j} \sum_{l=1}^{k} \mu_{0l} h_l(\mathbf{x}^*(t))\, dt \\ &\ge \int_0^{+\infty} e^{-\rho t} L_j(\mathbf{x}^*(t), \dot{x}_j^*(t))\, dt \\ &\quad - \int_0^{+infty} e^{-\rho t} L_j(\left[\mathbf{x}^{*(x}), y_j\right](t), \dot{y}_j(t))\, dt, \end{aligned}$$

giving us the desired result. □

## 6 Environmental Management Model Revisited

The turnpike property of dynamic economic models provides an interesting analytic tool for environmental economics. Indeed, one can easily relate the extremal steady state satisfying an environmental constraint with the notion of *sustainable development*. Furthermore, as we have demonstrated in the present paper, the extremal steady state can be used for the design of a tax that would eventually drive the economic system toward the long-term sustainable economic state. The economic model sketched in the Introduction is one of the simplest forms of a dynamic oligopoly game with a global emission constraint. The pollution constraint may in some cases be represented as a limit on a stock of nuisances instead of a limit on a flow. Our approach would easily accommodate such models. In a typical production system the state variables $x_j$ would be the installed capacities in a variety of production equipment and the control variables would include the use of these equipments to produce a variety of end-products. An additional state variable could represent the net accumulation of pollutant due to emissions associated with production activities and subject to the natural elimination process. This pollution state variable will then bear the environment constraint. Haurie and Krawczyk [7] propose such a model where the pollution process is associated with the flow dynamics of a river basin. The development of such models would however call for an extension of our theory to the case of differential games with coupled controls (the payoffs involving the controls of different players) which should not be too difficult to achieve.

## 7 Conclusion

In this paper we have extended the theory of the theory of infinite-horizon dynamic competitive processes to the case where the payers have to satisfy jointly, in their equilibrium, a coupled-state constraint. The consideration of an infinite horizon and the motivating example of an environmental constraint sugguested the introduction of a relaxed asymptotic form of the coupled constraint. Under such a reformulation of the game we have been able to show that the use of the asymptotic steady-state game for the definition of a time-invariant multiplier (to be interpreted in the environmental management context as a constant emission tax) permits the construction of a auxiliary decoupled differential game whose Nash equilibrium is also an equilibrium (in a weaker sense) of the original game under the asymptotic coupled constraint.

## Acknowledgments

Research supported by FNRS-Switzerland, FCAR-Québec, NSERC-Canada, National Science Foundation Grant No. INT-9500782, and a University of Toledo Faculty Research Award.

# REFERENCES


[1] Carlson, D. A., A. Haurie, and A. Lezarowitz. *Infinite Horizon Optimal Control: Deterministic and Stochastic Systems.* 2nd ed. Springer-Verlag, New York, 1991.

[2] Carlson, D. A and A. Haurie. A Turnpike Theory for Infinite Horizon Open-Loop Differential Games with Decoupled Contols. In: G. J. Olsder, ed., *New Trends in Dynamic Games and Applications.* Annals of the International Society of Dynamic Games, pp. 353–376. Birkhäuser, Boston, 1995.

[3] Carlson, D. A. and A. Haurie. A Turnpike Theory for Infinite Horizon Open-Loop Competitive Processes. *SIAM Journal on Control and Optimization*, **34** (4), 1405–1419, July 1996.

[4] Feinstein, C. D. and D. G. Luenberger. Analysis of the Asymptotic Behavior of Optimal Control Trajectories. *SIAM Journal on Control and Optimization*, **19**, 561–585, 1981.

[5] Hartl, R. F., S. P. Sethi, and R. G. Vickson. A Survey of the Maximum Principles for Optimal Control Problems with State Constraints. *SIAM Review*, **37**, 181–218, 1995.

[6] Haurie, A. Environmental Coordination in Dynamic Oligopolistic Markets. *Group Decision and Negotiation*, **4**, 49–67, 1995.

[7] Haurie, A. and J. Krawczyk. Optimal Taxation of Agents in a River Basin with Lumped and Distributed Emissions. *Environmental Modelling and Assessment.* To appear.

[8] Haurie, A. and M. Roche. Turnpikes and Computation of Piecewise Open-Loop Equilibria in Stochastic Differential Games. *Journal on Economic Dynamics and Control*, **18**, 317–344, 1994.

[9] Haurie, A. and G. Zaccour. Differential Game Models of Global Environmental Management. In: C. Carraro and J. A. Filar, ed., *Control and Game-Theoretic Models of the Environment*, Annals of the International Society of Dynamic Games, pp. 3–23. Birkhäuser, Boston, 1995.

[10] Rosen, J. B. Existence and Uniqueness of Equilibrium Points for Concave $n$-Person Games. *Econometrica*, **33** (3), 520–534, July 1965.

# Constrained Markov Games: Nash Equilibria

Eitan Altman
INRIA B.P. 93
Sophia-Antipolis Cedex, France

Adam Shwartz
Technion-Israel Institute of Technology
Department of Electrical Engineering
Haifa, Israel

**Abstract**

In this paper we develop the theory of constrained Markov games. We consider the expected average cost as well as discounted cost. We allow different players to have different types of costs. We present sufficient conditions for the existence of stationary Nash equilibrium. Our results are based on the theory of sensitivity analysis of mathematical programs developed by Dantzig, Folkman, and Shapiro [9], which was applied to Markov Decision Processes (MDPs) in [3]. We further characterize all stationary Nash equilibria as fixed points of some coupled Linear Programs.

## 1 Introduction

Constrained Markov decision processes arise in situations when a controller has more than one objective. A typical situation is when one wants to minimize one type of cost while keeping other costs lower than some given bounds. Such problems arise frequently in computer networks and data communications, see Lazar [20], Hordijk and Spieksma [16], Nain and Ross [23], Ross and Chen [26], Altman and Shwartz [1] and Feinberg and Reiman [12]. The theory of constrained MDPs goes back to Derman and Klein [10], and was developed by Hordijk and Kallenberg [15], Kallenberg [18], Beutler and Ross [6], [7], Ross and Varadarajan [27] Altman and Shwartz [2], [3], Altman [4], [5], Spieksma [32], Sennott [28], [29], Borkar [8], Feinberg [11], Feinberg and Shwartz [13], [14], and others.

In all these papers, a single controller was considered. A natural question is whether this theory extends to Markov games with several (say $N$) players, where player $i$ wishes to minimize $C_i^0$, subject to some bounds $V_i^j, i = 1, ..., N$, on the costs $C_i^j$, $j = 1, \ldots, B_i$. Although a general theory does not exist, several applications to telecommunications have been analyzed (see [17],[19]). The problem studied in these references is related to dynamic decentralized flow control by several (selfish) users, each of which seeks to maximize its own throughput. Since voice and video traffic typically require the end-to-end delays to be bounded, this problem was posed as a constrained Markov game. For (static) games with constraints, see, e.g., Rosen [24]. Some other theoretical results on zero-sum constrained Markov games were obtained by Shimkin [30]. A related theory of approachability for stochastic games was developed by Shimkin and Shwartz [31], and extended to semi-Markov games, with applications in telecommunications, by Levi and Shwartz [21], [22].

In this paper we present sufficient conditions for the existence of stationary Nash equilibria. Our results are based on the theory of sensitivity analysis of mathematical programs developed by Dantzig, Folkman, and Shapiro [9]. This theory was applied to Markov Decision Processes in [3], and we rely on this application here. We further characterize all stationary Nash equilibria as fixed points of some coupled Linear Programs.

## 2 The Model and Main Result

We consider a game with $N$ players, labeled $1, \ldots, N$. Define the tuple $\{\mathbf{X}, \mathbf{A}, \mathcal{P}, c, V\}$ where:

- $\mathbf{X}$ is a finite state space. Generic notation for states will be $x, y$.
- $\mathbf{A} = \{\mathbf{A}_i\}, i = 1, ..., N$, is a finite set of actions. We denote by $\mathbf{A}(x) = \{\mathbf{A}_i(x)\}_i$ the set of actions available at state $x$. A generic notation for a vector of actions will be $\mathbf{a} = (a_1, ..., a_N)$ where $a_i$ stands for the action chosen by player $i$. Denote $\mathcal{K} = \{(x, \mathbf{a}) : x \in \mathbf{X},\ \mathbf{a} \in \mathbf{A}(x)\}$ and set $\mathcal{K}_i = \{(x, a_i) : x \in \mathbf{X},\ a_i \in \mathbf{A}_i(x)\}$.
- $\mathcal{P}$ are the transition probabilities; thus $\mathcal{P}_{x\mathbf{a}y}$ is the probability to move from state $x$ to $y$ if the vector $\mathbf{a}$ of actions is chosen by the players.
- $c = \{c_i^j\}, i = 1, ..., N,\ j = 0, 1, ..., B_i$, is a set of immediate costs, where $c_i^j : \mathcal{K} \to \mathbb{R}$. Thus player $i$ has a set of $B_i + 1$ immediate costs; $c_i^0$ will correspond to the cost function that is to be minimized by that player, and $c_i^j$, $j > 0$, will correspond to cost functions on which some constraints are imposed.
- $V = \{V_i^j\}, i = 1, ..., N,\ j = 1, ..., B_i$, are bounds defining the constraints (see (3) below).

Let $M_1(G)$ denote the set of probability measures over a set $G$. Define a history at time $t$ to be a sequence of previous states and actions, as well as the current state: $h_t = (x_1, \mathbf{a}_1, ..., x_{t-1}, \mathbf{a}_{t-1}, x_t)$. Let $\mathbf{H}_t$ be the set of all possible histories of length $t$. A policy $u^i$ for player $i$ is a sequence $u^i = (u_1^i, u_2^i, ...)$ where $u_t : \mathbf{H}_t \to M_1(\mathbf{A}_i)$ is a function that assigns to any history of length $t$, a probability measure over the set of actions of player $i$. At time $t$, the controllers choose independently of each other actions $\mathbf{a} = (a_1, ..., a_N)$, where action $a_i$ is chosen by player $i$ with probability $u_t(a_i|h_t)$ if the history $h_t$ was observed. The class of all policies defined as above for player $i$ is denoted by $U^i$. The collection $U = \times_{i=1}^{N} U^i$ is called the class of multipolicies ($\times$ stands for the product space).

A stationary policy for player $i$ is a function $u^i : \mathbf{X} \to M_1(\mathbf{A}_i)$ so that $u^i(\cdot|x) \in M_1(\mathbf{A}_i(x))$. We denote the class of stationary policies of player $i$ by $U_S^i$. The set $U_S = \times_{i=1}^N U_S^i$ is called the class of stationary multipolicies. Under any stationary multipolicy $u$ (where the $u^i$ are stationary for all the players), at time $t$, the controllers, independently of each other, choose actions $\mathbf{a} = (a_1, ..., a_N)$, where action $a_i$ is chosen by player $i$ with probability $u^i(a_i|x_t)$ if state $x_t$ was observed at time $t$. Under a stationary multipolicy the state process becomes a Markov chain with transition probabilities $P_{xy}^w = \sum_{\mathbf{a}} \mathcal{P}_{x\mathbf{a}y} w(\mathbf{a}|x)$.

For $u \in U$ we use the standard notation $u^{-i}$ to denote the vector of policies $u^k$, $k \neq i$; moreover, for $v^i \in U^i$, we define $[u^{-i}|v^i]$ to be the multipolicy where, for $k \neq i$, player $k$ uses $u^k$, while player $i$ uses $v^i$. Define $U^{-i} := \bigcup_{u \in U} \{u^{-i}\}$. For $\mathbf{a} \in \mathbf{A}(x)$ and $a \in \mathbf{A}_i(x)$, we use the obvious notation $\mathbf{a}^{-i}$, $[\mathbf{a}^{-i}|a]$, and the set $\mathbf{A}^{-i}(x)$, $i = 1, ..., N$, $x \in \mathbf{X}$.

A distribution $\beta$ for the initial state (at time 1) and a multipolicy $u$ together define a probability measure $P_\beta^u$ which determines the distribution of the stochastic process $\{X_t, A_t\}$ of states and actions. The corresponding expectation is denoted as $E_\beta^u$.

Next, we define the cost criteria that will appear in the constrained control problem. For any policy $u$ and initial distribution $\beta$, define the $i, j$-discounted cost by

$$C_\alpha^{i,j}(\beta, u) = (1 - \alpha) \sum_{t=1}^{\infty} \alpha^{t-1} E_\beta^u c_i^j(X_t, A_t), \tag{1}$$

where $0 < \alpha < 1$ is fixed. The $i, j$-expected average cost is defined as

$$C_{ea}^{i,j}(\beta, u) = \overline{\lim_{T \to \infty}} \frac{1}{T} \sum_{t=1}^{T} E_\beta^u c_i^j(X_t, A_t). \tag{2}$$

We assume that all the costs $C^{i,j}$, $j = 0, ..., B_i$, corresponding to any given player $i$ are either discounted costs with *the same discount factor* $\alpha = \alpha_i$ or they are all expected average costs. However, the discount factors $\alpha_i$ may vary between players, and some player may have discounted costs while others may have expected average costs.

A multipolicy $u$ is called $i$-feasible if it satisfies

$$C^{i,j}(\beta, u) \leq V_i^j \qquad \text{for all} \quad j = 1, ..., B_i. \tag{3}$$

It is called feasible if it is $i$-feasible for all the players $i = 1, ..., N$. Let $U_V$ be the set of feasible policies. A policy $u \in U_V$ is called constrained Nash equilibrium if for each player $i = 1, ..., N$ and for any $v^i$ such that $[u^{-i}|v^i]$ is $i$-feasible,

$$C^{i,0}(\beta, u) \leq C^{i,0}(\beta, [u^{-i}|v^i]). \tag{4}$$

Thus, any deviation of any player $i$ will either violate the constraints of the $i$th player, or if it does not, it will result in a cost $C^{i,0}$ for that player that is not lower than the one achieved by the feasible multipolicy $u$.

For any multipolicy $u$, $u^i$ is called an optimal response for player $i$ against $u^{-i}$ if $u$ is $i$-feasible, and if for any $v^i$ such that $[u^{-i}|v^i]$ is $i$-feasible, (4) holds. A multipolicy $v$ is called an optimal response against $u$ if for every $i = 1, ..., N$, $v^i$ is an optimal response for player $i$ against $u^{-i}$.

We introduce the following assumptions:

- ($\Pi_1$) Ergodicity: If there is at least one player that uses the expected average cost criteria, then the unichain ergodic structure holds, i.e., under any stationary multipolicy $u$, the state process is an irreducible Markov chain with one ergodic class (and possibly some transient states).
- ($\Pi_2$) Strong Slater condition: For any stationary multipolicy $u$, and for any player $i$, there exists some $v^i$ such that

$$C^{i,j}(\beta, [u^{-i}|v^i]) < V_i^j \qquad \text{for all} \quad j = 1, ..., B_i. \tag{5}$$

We are now ready to introduce the main result.

**Theorem 2.1.** *Assume that $\Pi_1$ and $\Pi_2$ hold. Then there exists a stationary multipolicy $u$ which is constrained-Nash equilibrium.*

## 3 Proof of Main Result

We begin by describing the way an optimal stationary response for player $i$ is computed for a given stationary multipolicy $u$. Fix a stationary multipolicy $u$. Denote the transition probabilities induced by players other than $i$, when player $i$ uses action $a_i$, by $\mathcal{P}^{u,i} = \{\mathcal{P}^{u,i}_{xa_iy}\}$ where

$$\mathcal{P}^{u,i}_{xa_iy} := \sum_{\mathbf{a}^{-i} \in \mathbf{A}^{-i}} \prod_{l \neq i} u^l_x(a_l|x) \mathcal{P}_{x\mathbf{a}y}, \qquad \mathbf{a} = [\mathbf{a}^{-i}|a_i].$$

That is, we consider new transition probabilities to go from $x$ to $y$ as a function of the action $a_i$ of player $i$, for fixed stationary policies of players $l$, $l \neq i$. Similarly, define

$$c_i^{j,u}(x, a_i) := \sum_{\mathbf{a}^{-i} \in \mathbf{A}^{-i}} \prod_{l \neq i} u^l_x(a_l|x) c_i^j(x, \mathbf{a}), \qquad \mathbf{a} = [\mathbf{a}^{-i}|a_i].$$

Next we present a Linear Program (LP) for computing the set of all optimal responses for player $i$ against a stationary policy $u^{-i}$. The following LP will be related to both the discounted and average case; if the $i$th player uses discount costs, then $\alpha_i$ below will stand for its discount factor. If it uses the expected average costs then $\alpha_i$ below is set to 1. Recall that $\beta$ is the initial distribution.

**LP(i, u)**:
Find $z^* := \{z^*(y, a)\}_{y,a}$ that minimizes $\mathcal{C}^{i,0}(z) := \sum_{y \in \mathbf{X}} \sum_{a \in \mathbf{A}_i(y)} c_i^{0,u}(y, a) z(y, a)$ subject to:

$$\sum_{y \in \mathbf{X}} \sum_{a \in \mathbf{A}_i(y)} z(y, a) \left[\delta_r(y) - \alpha_i \mathcal{P}^{u,i}_{yar}\right] = [1 - \alpha_i]\beta(r), \qquad r \in \mathbf{X}, \tag{6}$$

$$C^{i,j}(z) := \sum_{y\in\mathbf{X}} \sum_{a\in\mathbf{A}_i(y)} c_i^{j,u}(y,a)z(y,a) \le V_i^j, \qquad 1 \le j \le B_i, \tag{7}$$

$$z(y,a) \ge 0, \qquad \sum_{y\in\mathbf{X}} \sum_{a\in\mathbf{A}_i(y)} z(y,a) = 1. \tag{8}$$

Define $\Gamma(i,u)$ to be the set of optimal solutions of $\mathbf{LP}(i,u)$.

Given a set of nonnegative real numbers $z = \{z(y,a), y \in \mathbf{X}, a \in \mathbf{A}_i(y)\}$, define the point to set mapping $\gamma(z)$ as follows: If $\sum_a z(y,a) \neq 0$ then $\gamma_y^a(z) := \{z(y,a)[\sum_a z(y,a)]^{-1}\}$ is a singleton: for each $y$, we have that $\gamma_y(z) = \{\gamma_y^a(z) : a \in \mathbf{A}_i(y)\}$ is a point in $M_1(\mathbf{A}_i(y))$. Otherwise, $\gamma_y(z) := M_1(\mathbf{A}_i(y))$, i.e., the (convex and compact) set of all probability measures over $\mathbf{A}_i(y)$. Define $g^i(z)$ to be the set of stationary policies for player $i$ that choose, at state $y$, action $a$ with probability in $\gamma_y^a(z)$.

For any stationary multipolicy $v$ define the occupation measures

$$f(\beta, v) := \{f^i(\beta, v; y, a) : y \in \mathbf{X},\ a \in \mathbf{A}_i(y),\ i = 1, ..., N\}$$

as follows. Let $P(v)$ be the transition probabilities of the Markov chain representing the state process when the players use the stationary multipolicy $v$. If player $i$ uses the discounted cost $\alpha_i < 1$, then

$$f^i(\beta, v; y, a) := (1-\alpha_i)\sum_{x\in\mathbf{X}} \beta(x)\left(\sum_{s=1}^{\infty} \alpha_i^{s-1}[P(v)^s]_{xy}\right) v^i(a|y). \tag{9}$$

If player $i$ uses the expected average cost, then

$$f^i(\beta, v; y, a) := \pi^v(y) v^i(a|y),$$

where $\pi^v$ is the steady-state (invariant) probability of the Markov chain describing the state process, when policy $v$ is used (which exists and is unique by Assumption $\Pi_1$).

**Proposition 3.1.** *Assume $\Pi_1$–$\Pi_2$. Fix any stationary multipolicy $u$.*

(i) *If $z^*$ is an optimal solution for $\mathbf{LP}(i,u)$ then any element $w$ in $g^i(z^*)$ is an optimal stationary response of player $i$ against the stationary policy $u^{-i}$. Moreover, the multipolicy $v = [u^{-i}|w]$ satisfies $f^i(\beta, v) = z^*$.*

(ii) *Assume that $w$ is an optimal stationary response of player $i$ against the stationary policy $u^{-i}$, and let $v := [u^{-i}|w]$. Then $f^i(\beta, v)$ is optimal for $\mathbf{LP}(i,u)$.*

(iii) *The optimal sets $\Gamma(i,u)$, $i = 1, ..., N$, are convex, compact, and upper semi-continuous in $u^{-i}$, where $u$ is identified with points in $\times_{i=1}^{N} \times_{x\in\mathbf{X}} M_1(\mathbf{A}_i(x))$.*

(iv) *For each $i$, $g^i(z)$ is upper semicontinuous in $z$ over the set of points which are feasible for $\mathbf{LP}(i,u)$ (i.e., the points that satisfy constraints (6)–(8)).*

**Proof.** When all players other than $i$ use $u^{-i}$, then player $i$ is faced with a constrained Markov Decision Process (with a single controller). The proof of (i) and (ii) then follows from [3, Theorems 2.6]. The first part of (iii) follows from standard properties of Linear Programs, whereas the second part follows from an application of the theory of sensitivity analysis of Linear Programs by Dantzig, Folkman, and Shapiro [9] and in [3, Theorem 3.6] to $\mathbf{LP}(i, u)$. Finally, (iv) follows from the definition of $g^i(z)$. □

Define the point to set map

$$\Psi : \bigtimes_{i=1}^{N} M_1(\mathcal{K}_i) \to 2^{\left\{\bigtimes_{i=1}^{N} M_1(\mathcal{K}_i)\right\}}$$

by

$$\Psi(\mathbf{z}) = \bigtimes_{i=1}^{N} \Gamma(i, g(z)),$$

where $\mathbf{z} = (z_1, \ldots, z_N)$, each $z_i$ is interpreted as a point in $M_1(\mathcal{K}_i)$ and $g(z) = (g^1(z_1), \ldots, g^N(z_N))$.

**Proposition 3.2.** *Assume $\Pi_1$–$\Pi_2$. In the case that no player uses the expected average cost criterion, assume further that, for some $\eta > 0$, the initial distribution satisfies $\beta(x) > \eta$ for all $x$. Then there exists a fixed point $\mathbf{z}^* \in \Psi(\mathbf{z}^*)$.*

**Proof.** By Proposition 3.1(i) and (9), it follows that under the stated conditions, all solutions $z^*$ of $\mathbf{LP}(i, u)$ satisfy $\sum_a z^*(x, a) > \eta'$ for some $\eta' > 0$. But under this restriction the set $\gamma_y(z)$ is a singleton, and hence the range of the function $g^i(z)$ is also single points. The proof now follows directly from Kakutani's fixed point theorem applied to $\Psi$. Indeed, by Proposition 3.1(iii) and (iv), $\Gamma(i, g(z))$ is a composition of two upper semicontinuous functions: $g(\cdot)$ and $\Gamma$, which have convex compact ranges. Hence $\Psi$ is upper semicontinuous in $\mathbf{z}$, and has a compact range. Since $g(\cdot)$ can be considered as a regular (i.e., not set valued) function, the composition also has a convex range. Therefore the conditions of Kakutani's theorem hold. □

**Proof of Theorem 2.1.** Under the conditions of Proposition 3.2, the proof is obtained by combining Proposition 3.1(i) with Proposition 3.2. Indeed, Proposition 3.1(i) implies that for any fixed point $\mathbf{z}$ of $\Psi$, the stationary multipolicy $g = \{g^i(z^i); i = 1, \ldots, N\}$ is constrained Nash equilibrium. It therefore remains to treat the case where all players use a discounted criterion and, moreover, the initial distribution $\beta$ satisfies $\beta(x) = 0$ for some $x$.

Given such $\beta$, let $\beta_n$ be a sequence of initial distributions satisfying $\beta_n(x) > \eta_n > 0$ and $\beta_n(x) \to \beta(x)$ for each $x$ (so that $\eta_n \to 0$). Let $u_n$ be a constrained Nash equilibrium multipolicy for the problem with initial distribution $\beta_n$: this was just shown to exist. If we identify the multipolicies $u_n$ with points in $\bigtimes_{i=1}^{N} \bigtimes_{x \in \mathbf{X}} M_1(\mathbf{A}_i(x))$, then they lie in a compact set. Let $u$ be any limit point. We claim that $u$ is a constrained Nash equilibrium multipolicy for the problem with initial distribution $\beta$. To show this we need to establish that:

(i) $u$ satisfies (3) for each $i$; and
(ii) if $[u^{-i}|v^i]$ is $i$-feasible, then (4) holds.

From (1) it follows that all costs are linear (hence continuous) functions of the frequencies $f^i(\beta, v)$ which are, in turn, continuous functions of $(\beta, v)$ (see (9)). In fact, the costs are linear in $\beta$. Therefore the costs $C^{i,j}(\beta, u)$ are continuous in $(\beta, u)$. Since (3) holds for $(\beta_n, u_n)$, the continuity implies that it holds for $(\beta, u)$ and (i) is established.

Now fix some $i$ and suppose that $v^i$ is such that $[u^{-i}|v^i]$ is $i$-feasible. Note that it is possible that, for some $i$, $[u_n^{-i}|v^i]$ is not $i$-feasible for any $n$ large. However, by assumption $\Pi_2$, we can find some $\tilde{v}^i$ so that (5) holds. Fix an arbitrary $\epsilon$ and note that (5) holds also if we replace $\tilde{v}^i$ by $v^i_\epsilon = \epsilon\tilde{v}^i + (1-\epsilon)v^i$. By the continuity and the linearity we established in proving (i) above, this implies that $[u_n^{-i}|v^i_\epsilon]$ is $i$-feasible for all $n$ large enough. Therefore,

$$\begin{aligned} C^{i,0}(\beta, u) &= \lim_{n\to\infty} C^{i,0}(\beta_n, u_n) \\ &\leq \lim_{n\to\infty} C^{i,0}(\beta_n, [u_n^{-i}|v^i_\epsilon]) \\ &= C^{i,0}(\beta, [u^{-i}|v^i_\epsilon]). \end{aligned}$$

Using the continuity again, we have

$$C^{i,0}(\beta, [u^{-i}|v^i]) = \lim_{\epsilon\to 0} C^{i,0}(\beta, [u^{-i}|v^i_\epsilon])$$

and (ii) follows. □

**Remark 3.1.** (i) The Linear Program formulation **LP**$(i, u)$ is not only a tool for proving the existence of a constrained Nash equilibrium; in fact, due to Proposition 3.1(ii), it can be shown that any stationary constrained Nash equilibrium $w$ has the form $w = \{g^i(z^i); i = 1, ..., N\}$ for some $\mathbf{z}$ which is a fixed point of $\Psi$. Indeed, if $w$ is a constrained Nash equilibrium, then it follows from Proposition 3.2 that $f(\beta, w)$ is a fixed point of $\Psi$.

(ii) It follows from [3, Theorems 2.4 and 2.5] that if $\mathbf{z} = (z^1, ..., z^N)$ is a fixed point of $\Psi$, then any stationary multipolicy $g$ in $\times_{i=1}^N g^i(z^i)$ satisfies $C^{i,j}(\beta, g) = C^{i,j}(z)$, $i = 1, ..., N$, $j = 0, ..., B_i$. Conversely, if $w$ is a constrained Nash equilibrium, then

$$C^{i,j}(\beta, w) = \sum_{y\in\mathbf{X}} \sum_{a\in\mathbf{A}_i(y)} f^i(\beta, w; y, a) c_i^{j,w}(y, a)$$

(and $f(\beta, w)$ is a fixed point of $\Psi$).

## Acknowledgments

Research of the second author was supported in part by the Israel Science Foundation, administered by the Israel Academy of Sciences and Humanities.

## REFERENCES

[1] Altman, E. and A. Shwartz. Optimal Priority Assignment: A Time Sharing Approach. *IEEE Transactions on Automatic Control*, **AC-34**, No. 10, 1089–1102, 1989.

[2] Altman, E. and A. Shwartz. Markov Decision Problems and State-Action Frequencies. *SIAM Journal of Control and Optimization*, **29**, No. 4, 786–809, 1991.

[3] Altman, E. and A. Shwartz. Sensitivity of Constrained Markov Decision Problems. *Annals of Operations Research*, **32**, 1–22, 1991.

[4] Altman, E. Denumerable Constrained Markov Decision Processes and Finite Approximations. *Mathematics of Operations Research*, **19**, No. 1, 169–191, 1994.

[5] Altman, E. Asymptotic Properties of Constrained Markov Decision Processes. *ZOR—Methods and Models in Operations Research*, **37**, No. 2, 151–170, 1993.

[6] Beutler, F. J. and K. W. Ross. Optimal Policies for Controlled Markov Chains with a Constraint. *Journal of Mathematical Analysis and Applications*, **112**, 236–252, 1985.

[7] Beutler, F. J. and K. W. Ross. Time-Average Optimal Constrained Semi-Markov Decision Processes. *Advances of Applied Probability*, **18**, No. 2, 341–359, 1986.

[8] Borkar, V. S. Ergodic Control of Markov Chains with Constraints—The General Case. *SIAM Journal of Control and Optimization*, **32**, No. 1, 176–186, 1994.

[9] Dantzig, G. B., J. Folkman, and N. Shapiro. On the Continuity of the Minimum Set of a Continuous Function. *Journal of Mathematical Analysis and Applications*, **17**, 519–548, 1967.

[10] Derman, C and M. Klein. Some Remarks on Finite Horizon Markovian Decision Models. *Operations Research*, **13**, 272–278, 1965.

[11] Feinberg, E. A. Constrained Semi-Markov Decision Processes with Average Rewards. *ZOR*, **39**, 257–288, 1993.

[12] Feinberg, E. A. and M. I. Reiman. Optimality of Randomized Trunk Reservation. *Probability in the Engineering and Informational Sciences*, **8**, 463–489, 1994.

[13] Feinberg, E. A. and A. Shwartz. Constrained Markov Decision Models with Weighted Discounted Rewards. *Mathematics of Operations Research*, **20**, 302–320, 1995.

[14] Feinberg, E. A. and A. Shwartz. Constrained Discounted Dynamic Programming. *Mathematics of Operations Research*, **21**, 922–945, 1996.

[15] Hordijk, A. and L. C. M. Kallenberg. Constrained Undiscounted Stochastic Dynamic Programming. *Mathematics of Operations Research*, **9**, No. 2, 276–289, 1984.

[16] Hordijk, A. and F. Spieksma. Constrained Admission Control to a Queuing System. *Advances of Applied Probability*, **21**, 409–431, 1989.

[17] Hsiao, M. T. and A. A. Lazar. Optimal Decentralized Flow Control of Markovian queueing Networks with Multiple Controllers. *Performance Evaluation*, **13**, 181–204, 1991.

[18] Kallenberg, L. C. M. *Linear Programming and Finite Markovian Control Problems*. Mathematical Centre Tracts 148, Amsterdam, 1983.

[19] Korilis, Y. A. and A. Lazar. On the Existence of Equilibria in Noncooperative Optimal Flow Control. *Journal of the Association for Computing Machinery*, **42**, No. 3, 584–613, 1995.

[20] Lazar, A. Optimal Flow Control of a Class of Queuing Networks in Equilibrium. *IEEE Transactions on Automatic Control*, **28**, No. 11, 1001–1007, 1983.

[21] Levi, R. and A. Shwartz. A Theory of Approachability and Throughput-Cost Tradeoff in a Queue with Impatient Customers. EE Pub. 936, Technion, 1994.

[22] Levi, R. and A. Shwartz. Throughput-Delay Tradeoff with Impatient Arrivals. *Proceedings of the 23rd Allerton Conference on Communications, Control and Computing*, Allerton, IL, 1994.

[23] Nain, P. and K. W. Ross. Optimal Priority Assignment with hard Constraint. *Transactions on Automatic Control*, **31**, No. 10, 883–888, IEEE 1986.

[24] Rosen, J. B. Existence and Uniqueness of Equilibrium Points for Concave $n$-Person Games. *Econometrica*, **33**, 520–534, 1965.

[25] Ross, K. W. Randomized and Past-Dependent Policies for Markov Decision Processes with Multiple Constraints. *Operations Research*, **37**, No. 3, 474–477, 1989.

[26] Ross K. W. and B. Chen. Optimal Scheduling of Interactive and NonInteractive Traffic in Telecommunication Systems. *IEEE Transactions on Automatic Control*, **33**, No. 3, 261–267, 1988.

[27] Ross, K. W. and R. Varadarajan. Markov Decision Processes with Sample Path Constraints: The Communicating Case. *Operations Research*, **37**, No. 5, 780–790, 1989.

[28] Sennott, L. I. Constrained Discounted Markov Decision Chains. *Probability in the Engineering and Informational Sciences*, **5**, 463–475, 1991.

[29] Sennott, L. I. Constrained Average Cost Markov Decision Chains, *Probability in the Engineering and Informational Sciences* 7, 69–83, 1993.

[30] Shimkin, N. Stochastic Games with Average Cost Constraints. Annals of the International Society of Dynamic Games, Vol. 1: *Advances in Dynamic Games and Applications* (T. Basar and A. Haurie, eds.) Birkhauser, Boston, 1994.

[31] Shimkin, N. and A. Shwartz. Guaranteed Performance Regions for Markovian Systems with Competing Decision Makers, *IEEE Transactions on Automatic Control*, **38**, 84–95, 1993.

[32] Spieksma, F. M. Geometrically Ergodic Markov Chains and the Optimal Control of Queues. Ph.D. thesis, University of Leiden, 1990.

# Piecewise-Deterministic Differential Games and Dynamic Teams with Hybrid Controls

Eitan Altman
INRIA B.P. 93
Sophia-Antipolis Cedex, France

Tamer Başar
University of Illinois
Coordinated Science Laboratory
Urbana, Illinois, USA

Zigang Pan
Polytechnic University
EE Department
Brooklyn, New York, USA

### Abstract

We consider a class of linear-quadratic differential games with piecewise-deterministic dynamics, where the changes from one structure (for the dynamics) to another are governed by a finite-state Markov process. Player 1 controls the continuous dynamics, whereas Player 2 controls the rate of transition for the finite-state Markov process; both have access to the states of both processes. Player 1 wishes to minimize a given quadratic performance index, while Player 2 wishes to maximize the same quantity. For this class of differential games, we present results on the existence and uniqueness of a saddle point, and develop computational algorithms for solving it. Analytical (closed-form) solutions are obtained for the case when the continuous state is of dimension 1. The paper also considers the modified problem where Player 2 is also a minimizer, that is, when the underlying problem is a *team*, and illustrates the theoretical results in this context on a problem that arises in high-speed telecommunication networks. This application involves combined admission and rate-based flow control, where the former corresponds to control of the finite-state Markov process, and the latter to control of the continuous linear system.

## 1 Introduction

We consider in this paper the optimal control of linear jump parameter systems under a quadratic cost criterion, when the transition probabilities associated with the underlying Markov chain are also controlled. We study two scenarios:

The transition probabilities are controlled cooperatively (which leads to a team problem), and they are controlled antagonistically (which leads to a zero-sum differential game). In both cases, the control governing the switches takes values in a finite set, and is allowed to depend on the present and past values of the hybrid state—the same information that the "continuous" controller is endowed with. We treat both the finite- and infinite-horizon cases.

One of the difficulties associated with these classes of problems is that the associated dynamic programming equations cannot be solved explicitly, and may not even admit continuously differentiable value functions. It turns out that they can be solved in the scalar case (for which we provide a complete set of results), and for some specific classes of controlled rate matrices in the higher-dimensional case. For the cases when they do not admit closed-form solutions, we provide schemes whereby they can be approximated, both from below and above, and suboptimal policies can be constructed. We illustrate the results obtained for the scalar case on a telecommunications problem that features admission and rate-based controls, where the former controls the probabilities of accepting different types of traffic to the network, and the latter controls the flow of the traffic through the network.

Research on control systems with Markovian switching structure was initiated more than 30 years ago by Krassovskii and Lidskii [8], [9] and Florentin [10], with follow-up work in the late 1960s and early 1970s addressing the stochastic maximum principle [11], [12], dynamic programming [12], and linear-quadratic control [13], [14] in this context. The late 1980s and early 1990s have witnessed renewed interest on the topic, with concentrated research on theoretical issues like controllability and stabilizability [15], [6] in linear-quadratic systems. Perhaps the first theoretical development in a differential games context was reported in [16], where a general model was adopted that allows in a multiple player environment the Markov jump process (also controlled by the players) to determine the mode of play, in addition to affecting the structures of the system and the cost. Results in [16] also apply, as a special case, to zero-sum differential games with Markov jump parameters and state feedback information for both players. More recently, this framework has also been used in worst-case ($H^\infty$) control design problems where the underlying system changes structure at random points of time according to a finite-state Markov process with a constant rate matrix, and with the system dynamics affected by an additional (continuous-valued) disturbance process; some selected publications on this topic are [17], [3], [7], and [18].

The paper is organized as follows. Section 2 introduces the general model studied, and the solution concepts used. Section 3 provides the associated dynamic programming equations, presents some relevant results on the existence and uniqueness of viscosity solutions from the recent paper [19], and discusses closed-form solutions in some special cases. Sections 4 and 5 study respectively computational algorithms, and bounding and approximation techniques. The illustrative example involving a high speed communication network is presented in Section 6, and the paper ends with the concluding remarks of Section 7.

## 2 General Model

The general model studied in this paper fits the framework of controlled piecewise-deterministic Markov processes, for which a precise mathematical formulation can be found in [5, Chap. III]. Here, we simply identify some salient elements of this formulation that is relevant to our context, and refer the reader to [5] for the requisite mathematical background.

Let $\theta(t)$, $t \geq 0$, be a controlled continuous-time Markov process, taking values in a finite state space $\mathcal{S}$, of cardinality $s$. Transitions from state $i \in \mathcal{S}$ to $j \in \mathcal{S}$ occur at a rate controlled by Player 2, who chooses at time $t$ an action $u^2(t)$ from among a finite set $\mathbf{U}_2(i)$ of actions available at state $i$. Let $\mathbf{U}_2 := \bigcup_{i \in \mathcal{S}} \mathbf{U}_2(i)$. The controlled rate matrix (of transitions within $\mathcal{S}$) is

$$\Lambda = \{\lambda_{i,a,j}\}, \qquad i, j \in S, \quad a \in \mathbf{U}_2(i),$$

where henceforth we drop the "commas" in the subscripts of $\lambda$. The $\lambda_{iaj}$'s are real numbers such that for any $i \neq j$, and $a \in \mathbf{U}_2(i)$, $\lambda_{iaj} \geq 0$, and for all $a \in \mathbf{U}_2(i)$ and $i \in \mathcal{S}$, $\lambda_{iai} = -\sum_{j \neq i} \lambda_{iaj}$. Now introduce a second dynamics:

$$\frac{dx}{dt} = A(\theta(t))x + B(\theta(t))u^1(t), \qquad x(0) = x_0, \tag{1}$$

where $x(t) \in \mathbb{R}^n$ for each $t \geq 0$, $x_0$ is a fixed (known) initial state, $u^1$ is another control, taking values in $\mathbf{U}_1 = \mathbb{R}^r$, which is applied by another player, Player 1. Fix some initial state $i_0$ of the controlled Markov chain $\mathcal{S}$, and the final time $t_f$ (which may be infinite). Consider the class of policies $\mu^k \in \mathcal{U}_k$ for Player $k$ ($k = 1, 2$), whose elements (taking values in $\mathbf{U}_k$) are of the form of Borel-measurable maps

$$u^k(t) = \mu^k(t, x_{[0,t]}; \theta_{[0,t]}), \qquad t \in [0, t_f). \tag{2}$$

Here, a further restriction on the $\mu^k$'s is that they are piecewise continuous in the first argument, and piecewise Lipschitz continuous in the second argument.

Define $\mathcal{X} = \mathbb{R}^n \times \mathcal{S}$ to be the global state space of the system and $\mathcal{U} := \mathcal{U}_1 \times \mathcal{U}_2$ to be the class of multistrategies $\mu := (\mu^1, \mu^2)$. Define the running (immediate) cost $L : \mathcal{X} \times \mathbf{U}_1 \to \mathbb{R}$ as

$$L(x, i, u^1) = |x|^2_{Q(i)} + |u^1|^2_{R(i)}, \tag{3}$$

where $Q(\cdot) \geq 0$ and $R(\cdot) > 0$, and $|x|_Q$ denotes the Euclidean (semi-) norm. To any fixed initial state $(x_0, i_0)$ and a multistrategy $\mu \in \mathcal{U}$, there corresponds a unique probability measure $P^{\mu}_{x_0,i_0}$ on the canonical probability space $\Omega$ of the states and actions of the players, equipped with the standard Borel $\sigma$-algebra, $\mathcal{B}$. Denote by $\mathbf{E}^{\mu}_{x_0,i_0}$ the expectation operator corresponding to $P^{\mu}_{x_0,i_0}$. By a possible abuse of notation, denote by $(x(t), \theta(t))$, $u(t)$, $t \in [0, t_f)$, the stochastic processes corresponding to the states and actions, respectively. It is important to note here that when the multistrategy $\mu \in \mathcal{U}$ is memoryless (which will turn out to be the case for the optimum one), the pair $\big(x(t), \theta(t)\big)$, $t \geq 0$, is a Markov process on $(\Omega, \mathcal{B}, P^{\mu}_{x_0,i_0})$. In terms of this notation and convention, and for each fixed initial state $(x_0, i_0)$, multistrategy $\mu \in \mathcal{U}$, and a horizon of duration $t_f$, we introduce the discounted (expected) cost function:

$$J_\beta(x_0, i_0, \mu; t_f) := \mathbf{E}^{\mu}_{x_0,i_0}\Big\{|x(t_f)|^2_{Q_f(\theta(t_f))}e^{-\beta t_f} + \int_0^{t_f} e^{-\beta t} L(x(t), \theta(t), u^1(t))\, dt\Big\}, \tag{4}$$

where $\beta \geq 0$ is a discount factor (allowed also to be *zero*), and $Q_f(\cdot) \geq 0$ is a terminal state weighting matrix. We further denote the *cost-to-go* from any time-state pair $(t; x, i)$ by

$$J_\beta(t; x, i, \mu; t_f) := \mathbf{E}^{\mu}_{x,i}\Big\{|x(t_f)|^2_{Q_f(\theta(t_f))}e^{-\beta(t_f - t)} + \int_t^{t_f} e^{-\beta(\sigma - t)} L(x(\sigma), \theta(\sigma), u^1(\sigma))\, d\sigma\Big\}. \tag{5}$$

For $t_f = \infty$, we shall simply drop $t_f$ as an argument of $J_\beta$ in both cases, and take $Q_f \equiv 0$. For this infinite-horizon problem, we have to ensure that the cost is finite for at least one stationary policy of Player 1, and for all possible realizations of the rate matrix. A sufficient condition for this is the following, which we invoke throughout our analysis, when dealing with infinite-horizon team and game problems:

**Condition 2.1.** *The pair $(A(\theta(t)), B(\theta(t)))$ is stochastically stabilizable for all $x$-independent stationary policies of Player 2, and $(A(i), Q(i))$ is observable for each $i \in \mathcal{S}$.* [1]

The problem addressed in this paper is the derivation of a solution to

$$\hat{J}_\beta(t; x, i; t_f) := \operatorname*{val}_{\mu \in \mathcal{U}} J_\beta(t; x, i, \mu; t_f), \tag{6}$$

where $\mathrm{val}_{\mu \in \mathcal{U}}$ stands for

$$\inf_{\mu^1 \in \mathcal{U}_1} \sup_{\mu^2 \in \mathcal{U}_2} \quad \text{and} \quad \inf_{\mu^1 \in \mathcal{U}_1} \inf_{\mu^2 \in \mathcal{U}_2}$$

operations for the game and team cases, respectively, with again $t_f$ dropped as an argument in the infinite-horizon problem. For the team problem, $\hat{J}_\beta$ stands clearly for the global minimum of $J_\beta$, and in this case we also seek a multistrategy $\mu^* \in \mathcal{U}$, such that

$$J_\beta(t; x, i, \mu^*; t_f) = \hat{J}_\beta(t; x, i; t_f).$$

[1] Stochastic stabilizability means that there exists a quadratic Lyapunov function for the linear jump parameter system (1) for each fixed achievable rate matrix. Together with observability, this implies that there exists a stabilizing controller that renders the quadratic terminal cost exponentially go to zero; see [6] and [7].

For the game problem, $\hat{J}_\beta$ stands for the upper value, and in this case we seek a $\mu^{1*} \in \mathcal{U}_1$ such that

$$\sup_{\mu^2 \in \mathcal{U}_2} J_\beta(t; x, i, \mu^{1*}, \mu^2; t_f) = \hat{J}_\beta(t; x, i; t_f). \tag{7}$$

If the inf and sup operations can be interchanged, that is,

$$\hat{J}_\beta(t; x, i; t_f) = \sup_{\mu^2 \in \mathcal{U}_2} \inf_{\mu^1 \in \mathcal{U}_1} J_\beta(t; x, i, \mu^1, \mu^2; t_f),$$

then $\hat{J}_\beta$ is the *value* of the game, and if furthermore (in addition to (7)),

$$\inf_{\mu^1 \in \mathcal{U}_1} J_\beta(t; x, i, \mu^1, \mu^{2*}; t_f) = \hat{J}_\beta(t; x, i; t_f), \tag{8}$$

for some $\mu^{2*} \in \mathcal{U}_2$, then the pair $(\mu^{1*}, \mu^{2*})$ constitutes a *saddle-point* solution for the game.

## 3 The Solution Process

Introduce a backward controlled Markov operator $\mathcal{A}^u$ associated with the system (1)–(4) as follows. For each $\psi(t, \cdot, i)$ such that $\psi(\cdot, \cdot, i) \in \mathcal{C}^1$ for all $i \in \mathcal{S}$ (where $\mathcal{C}^1$ is the vector space of all functions with continuous partial derivatives), and for each $u = (u^1, a) \equiv (u^1, u^2) \in \mathbf{U}(i)$,

$$\mathcal{A}^u \psi(t, x, i) := \frac{\partial \psi(t, x, i)}{\partial t} + [D_x \psi(t, x, i)]\, f(x, u^1, i) + \sum_{j \in \mathcal{S}} \lambda_{iaj}\, \psi(t, x, j),$$

where $D_x$ stands for the gradient operator, and $f(x, u^1, i) := A(i)x + B(i)u^1$. Further introduce, for each function $\psi(x, i)$ for which $\psi(\cdot, i) \in \mathcal{C}^1$ for all $i \in \mathcal{S}$, and for each $u \in \mathbf{U}_1 \times \mathbf{U}_2(i)$, $i \in \mathcal{S}$,

$$G^u \psi(x, i) := [D_x \psi(x, i)]\, f(x, u^1, i) + \sum_{j \in \mathcal{S}} \lambda_{iaj}\, \psi(x, j).$$

Next we introduce the dynamic programming equations associated with the team and game problems above, where the continuous-differentiability of the value functions will be replaced simply with continuity, and their solutions interpreted in the generalized (viscosity) sense if differentiability fails—a notion to be introduced shortly. The "val" operator below will have the same interpretation as before, but now applied to static game or team problems, over the product action space $\mathbf{U}_1 \times \mathbf{U}_2(i)$.

For finite $t_f$, consider, subject to the given boundary condition, the Hamilton–Jacobi–Bellman (HJB, in short) (or Isaacs) equation:

$$\beta\psi(t, x, i) = \operatorname*{val}_{u^1, u^2} \left[\mathcal{A}^{(u^1, u^2)} \psi(t, x, i) + L(x, i, u^1)\right], \qquad \psi(t_f, x, i) = |x|^2_{Q_f(i)}. \tag{9}$$

For infinite $t_f$, consider its infinite-horizon counterpart:

$$\beta\psi(x,i) = \underset{u^1,u^2}{\text{val}} \left[ G^{(u^1,u^2)}\psi(x,i) + L(x,i,u^1) \right]. \tag{10}$$

Associate with (9) and (10) the corresponding sets, $\mathcal{C}_f$ and $\mathcal{C}$, respectively, of functions $\psi$ continuous in $(t,x)$ and $x$. We can view these functions as $s$-dimensional vector-valued functions, with the $i$th component being $\psi(t,x,i)$ or $\psi(x,i)$. We first introduce the notion of a viscosity solution for (9) for the team case.

**Definition 3.1.**

(i) $V \in \mathcal{C}_f$ is a viscosity subsolution of (9) if, for all $i \in \mathcal{S}$,

$$\beta\,\psi(t_0,x_0,i) + \sup_{u^1,u^2} \left[ -\mathcal{A}^{(u^1,u^2)}\psi(t_0,x_0,i) - L(x_0,i,u^1) \right] \le 0,$$

$$\psi(t_f,x,i) \le |x|^2_{Q_f(i)}, \tag{11}$$

whenever $\psi(\cdot,\cdot,i)$ is such that $V(t,x,i) - \psi(t,x,i)$ attains a local maximum at $(t_0,x_0)$, with $\psi(t_0,x_0,j) = V(t_0,x_0,j)$ for all $j \in \mathcal{S}$.

(ii) $V \in \mathcal{C}_f$ is a *viscosity supersolution* of (9) if, for all $i \in \mathcal{S}$,

$$\beta\,\psi(t_0,x_0,i) + \sup_{u^1,u^2} \left[ -\mathcal{A}^{(u^1,u^2)}\psi(t_0,x_0,i) - L(x_0,i,u^1) \right] \ge 0,$$

$$\psi(t_f,x,i) \ge |x|^2_{Q_f(i)}, \tag{12}$$

whenever $\psi(\cdot,\cdot,i)$ is such that $V(t,x,i) - \psi(t,x,i)$ attains a local minimum at $(t_0,x_0)$, with $\psi(t_0,x_0,j) = V(t_0,x_0,j)$ for all $j \in \mathcal{S}$.

(iii) $V \in \mathcal{C}_f$ is a *viscosity solution* of (9) if it is both a viscosity subsolution and a viscosity supersolution.

The counterpart of this definition in the infinite-horizon case is similar, with only operator $\mathcal{A}$ replaced by $G$, dependence on $t$ dropped, and the terminal condition inequality also dropped. In the zero-sum differential game case, again the same definition applies, with $\sup_{u^1,u^2}$ replaced by $\sup_{u^1} \min_{u^2}$, where the minimization is over a finite set.

The following theorems are now from [19], establishing the existence and uniqueness of viscosity solutions to (9) and (10):

**Theorem 3.1** (Team Problem).

(i) *Let $t_f$ be finite and $\beta > 0$. Then, the HJB equation* (9) *has a unique viscosity solution $\psi$ in $\mathcal{C}_f$, and the value of* (5) *(i.e.,* (6)*) equals $\psi$. Moreover, any Markov policy that chooses at time $t$, for all $t \in [0,t_f]$, actions that achieve the argmin in* (9)*, given that the state at that time is $(x,i)$, is optimal.*

(ii) *Let Condition* 2.1 *be satisfied, and $\beta > 0$. Then, there exists a unique viscosity solution $\psi$ in $\mathcal{C}$ to* (10)*, with the further properties that:*

(a) $\psi(x,i) \leq J_\beta(x,i,\mu)$ *for every* $\mu \in \mathcal{U}$ *that satisfies*

$$\varliminf_{t_1\to\infty} e^{-\beta t_1}\mathbf{E}^{\mu}_{x,i}\psi(x(t_1),\theta(t_1)) \leq 0. \tag{13}$$

(b) *Any stationary policy* $g$ *that chooses at state* $(x,i)$ *actions that achieve the argmin in* (10) *satisfies* $\psi(x,i) \geq J_\beta(x,i,g)$*, provided that*

$$\varlimsup_{t_1\to\infty} e^{-\beta t_1}\mathbf{E}^{g}_{x,i}\psi(x(t_1),\theta(t_1)) \geq 0. \tag{14}$$

**Proof.** A proof of the first statement of (i) is given in [19]. The second statement, that is the optimality of the Markov policy that minimizes the right-hand side of (9), follows from the results of [5, Chap III.8–III.10].

Proof of part (ii) is also given in [19], with Condition 2.1 guaranteeing the boundedness of the infinite-horizon cost. The two properties given follow from the results given in Chapters III.8–III.10 of [5]. □

**Remark 3.1.** The condition $\beta > 0$ seems to be necessary to guarantee the uniqueness of the viscosity solution; it is not needed, however, for its existence. It will be shown shortly that for the scalar case (i.e., $n = 1$), and with $t_f$ finite, the solution of the HJB equation is in fact differentiable and unique for all $\beta \geq 0$. Again in the scalar case, but with $t_f = \infty$, there exists a unique nonnegative solution when $\beta = 0$.

We now state the counterpart of Theorem 3.1 for the game setting, separately for finite and infinite horizons.

**Theorem 3.2** (Zero-Sum Game: Finite Horizon). *Let* $t_f$ *be finite, and* $\beta > 0$. *Then,* (9) *admits a unique viscosity solution* $\psi$ *in* $C_f$*, with "val" interpreted simultaneously as* "inf sup" *and* "sup inf." *Furthermore:*

(i) *The dynamic game has a (saddle-point) value* $\hat{J}_\beta \equiv \psi$.

(ii) *Let* $\overline{\mu}^2$ *be a policy that chooses at time* $t$*, for all* $t \in [0, t_f)$*, an action that achieves the maximum in the static game:*

$$\sup_{u^2\in \mathrm{U}_2(i)} \inf_{u^1\in \mathrm{U}_1} \left[\mathcal{A}^{(u^1,u^2)}\psi(t,x,i) + L(x,i,u^1)\right] = \beta\psi(t,x,i), \tag{15}$$

*given that the state at that time is* $(x,i)$.

*If* $\overline{\mu}^2$ *is in* $\mathcal{U}_2$ *(i.e., if the continuity and Lipschitz conditions are satisfied), then it is an optimal policy for Player* 2.

(iii) *Any Markov policy* $\overline{\mu}^1 \in \mathcal{U}_1$ *for Player* 1 *that chooses at time* $t$*, for all* $t \in [0, t_f)$*, actions that achieve the minimum in the static game:*

$$\inf_{u^1\in \mathrm{U}_1} \sup_{u^2\in \mathrm{U}_2(i)} \left[\mathcal{A}^{(u^1,u^2)}\psi(t,x,i) + L(x,i,u^1)\right] = \beta\psi(t,x,i), \tag{16}$$

*given that the state at that time is* $(x,i)$*, is optimal for Player* 1.

**Proof.** The first statement of the theorem on the existence of a unique viscosity solution is from [19]. We prove here the remaining properties (i)–(iii).

(ii) Consider the Markov policy $\overline{\mu}^1$ given in part (iii). Let $\mu^2 \in \mathcal{U}_2$ be an arbitrary policy for Player 2. Let $u(\cdot) = (u^1(\cdot), u^2(\cdot))$ be the actions used by the players when $\mu = (\overline{\mu}^1, \mu^2)$ is adopted. Since the value of the game in (9) is $\beta\psi(t, x, i)$, and $\mu^2$ is not necessarily maximizing, we have

$$\mathcal{A}^{(u(t))}\psi(t, x, i) + L(x, i, u^1(t)) \leq \beta\psi(t, x, i).$$

It then follows from Dynkin's formula of integration that

$$\begin{aligned}
\psi(t, x, i) &= E^{\mu}_{x,i}\left[\int_t^{t_f} -e^{-\beta(s-t)}\mathcal{A}^{(u(s))}\psi(s, x(s), \theta(s))\, ds \right.\\
&\qquad \left. + e^{-\beta(t_f-t)}\psi(t_f, x(t_f), \theta(t_f))\right]\\
&\geq E^{\mu}_{x,i}\left[\int_t^{t_f} e^{-\beta(s-t)}L(x(s), \theta(s), u^1(s))\, ds \right.\\
&\qquad \left. + e^{-\beta(t_f-t)}\psi(t_f, x(t_f), \theta(t_f))\right]\\
&\equiv J_\beta(t; x, i, \mu; t_f).
\end{aligned}$$

Hence, $\mu^2$ cannot be a maximizing policy unless (15) is satisfied.

(iii) Let $\overline{\mu}^2$ be an optimal policy for Player 2, as described in (ii). Let $\mu^1 \in \mathcal{U}_1$ be an arbitrary policy for Player 1. Let $u(s) = (u^1(s), u^2(s))$ be the actions used by the players dictated by $\mu = (\mu^1, \overline{\mu}^2)$. Then it follows from a reasoning similar to that in part (ii) that

$$\mathcal{A}^{(u(t))}\psi(t, x, i) + L(x, i, u^1(s)) \geq \beta\psi(t, x, i).$$

Again using Dynkin's formula, we obtain the following equality and inequality:

$$\begin{aligned}
\psi(t, x, i) &= E^{\mu}_{x,i}\left[\int_t^{t_f} -e^{-\beta(s-t)}\mathcal{A}^{(u(s))}\psi(s, x(s), \theta(s))\, ds \right.\\
&\qquad \left. + e^{-\beta(t_f-t)}\psi(t_f, x(t_f), \theta(t_f))\right]\\
&\leq E^{\mu}_{x,i}\left[\int_t^{t_f} e^{-\beta(s-t)}L(x(s), \theta(s), u^1(s))\, ds \right.\\
&\qquad \left. + e^{-\beta(t_f-t)}\psi(t_f, x(t_f), \theta(t_f))\right]\\
&\equiv J_\beta(t; x, i, \mu; t_f).
\end{aligned}$$

This again shows that $\mu^1$ cannot be a minimizing policy for Player 1 unless (16) is satisfied.

Part (i) is obtained by combining (ii) and (iii). □

**Theorem 3.3** (Zero-Sum Game: Infinite Horizon).

*Let $t_f = \infty$, $\beta > 0$, and Condition 2.1 be satisfied. Then, (10) admits a unique viscosity solution $\psi$ in $\mathcal{C}$, with "val" interpreted simultaneously as "inf sup" and "sup inf." Furthermore, if $\overline{\mu}^1 \in \mathcal{U}_1$ is a stationary policy for Player 1 that chooses at state $(x, i)$ actions that achieve the minimum in*

$$\inf_{u^1 \in \mathbf{U}_1} \sup_{u^2 \in \mathbf{U}_2(i)} \left[ G^{(u^1,u^2)} \psi(x, i) + L(x, i, u^1) \right], \tag{17}$$

*and $\overline{\mu}^2 \in \mathcal{U}_2$ is a stationary policy for Player 2 that chooses at state $(x, i)$ actions that achieve the maximum in*

$$\sup_{u^2 \in \mathbf{U}_2(i)} \inf_{u^1 \in \mathbf{U}_1} \left[ G^{(u^1,u^2)} \psi(x, i) + L(x, i, u^1) \right], \tag{18}$$

*then we have:*

(i) *$\psi(x, i) \leq J_\beta(x, i, \mu)$ for every $\mu \in \mathcal{U}$ that satisfies $\mu^2 = \overline{\mu}^2$ and*

$$\underline{\lim}_{t_1 \to \infty} e^{-\beta t_1} \mathbf{E}^{\mu}_{x,i} \psi(x(t_1), \theta(t_1)) \leq 0. \tag{19}$$

(ii) *$\psi(x, i) \geq J_\beta(x, i, \mu)$, provided that $\mu^1 = \overline{\mu}^1$ and*

$$\overline{\lim}_{t_1 \to \infty} e^{-\beta t_1} \mathbf{E}^{\mu}_{x,i} \psi(x(t_1), \theta(t_1)) \geq 0. \tag{20}$$

**Proof.** The first statement of the theorem on the existence of a unique viscosity solution is again from [19]; we prove here only the remaining properties (i)–(ii).

Choose an arbitrary $\mu^1 \in \mathcal{U}_1$, and pick $\mu^2 = \overline{\mu}^2$. With $u(t)$ denoting the actions of both players taken at time $t$, we have by (10):

$$G^{u(t)} \psi(x(t), \theta(t)) - \beta \psi(x(t), \theta(t)) + L(x(t), \theta(t), u^1(t)) \geq \beta \psi(x(t), \theta(t)). \tag{21}$$

By applying Dynkin's formula as in the proof of the previous theorem, we get (as in [5, p. 146]),

$$\psi(x, i) \geq E^{\mu}_{x,i} \int_0^{t_1} e^{-\beta t} L(x(t), \theta(t), u^1(t)) dt + e^{-\beta t_1} E^{\mu}_{x,i} \psi(x(t_1), \theta(t_1)).$$

Part (i) is obtained by taking the limit as $t_1 \to \infty$ along a sequence for which the last term tends to a nonpositive value. Part (ii) is established by symmetrical arguments. □

Equations (9) and (10) do not admit closed-form solutions in all cases, but they do in some special cases. To investigate these cases, let us first consider the finite-horizon case, and stipulate a structure for $\psi(t, x, i)$ that is quadratic in $x$:

$$\psi(t, x, i) := x^T P(i, t) x, \qquad t \in [0, t_f], \quad i \in \mathcal{S}, \tag{22}$$

where $P(i, t)$ is an $n \times n$ symmetric matrix for each $i \in \mathcal{S}, t \in [0, t_f]$. Substituting this structural form into (9), we obtain

$$\beta x^T P(i,t)x = x^T P_t(i,t)x + \min_{u^1}\left[2\left(A(i)x + B(i)u^1\right)^T P(i,t)x + |u^1|^2_{R(i)}\right] + \text{opt}_a \sum_{j\in\mathcal{S}} \lambda_{iaj}\, x^T P(j,t)x, \tag{23}$$

where $P_t$ denotes the partial derivative of $P$ with respect to $t$, and "opt" stands for minimization in the team problem and maximization in the game problem. Note that for the game problem there is a complete separation of the min and max operations, and hence a value exists in (9), provided of course that this quadratic structure is valid. We now introduce a useful definition.

**Definition 3.2.** We shall say that for the team problem (respectively, game problem) a square symmetric matrix $P_1$ dominates another square symmetric matrix $P_2$ of the same dimension if the difference matrix $P_1 - P_2$ is nonpositive definite (respectively, nonnegative definite).

Now continuing with the derivation, the unique minimizing control $u^1$ in (23) is

$$\mu^1_{\text{opt}}(x, i, t) = -R^{-1}(i)B^T(i)P(i,t)x, \tag{24}$$

whose substitution into (23) leads to

$$0 = x^T\left(P_t(i,t) - \beta P(i,t) + Q(i) + A^T(i)P(i,t) + P(i,t)A(i) - P(i,t)B(i)R^{-1}(i)B^T(i)P(i,t)\right)x + \underset{a\in\mathbf{U}_2(i)}{\text{opt}} \sum_{j\in\mathcal{S}} \lambda_{iaj} x^T P(j,t)x.$$

Hence, the quadratic structure is the right one provided that the optimization over $u^2 = a$ is independent of $x$, i.e.,

$$\underset{a\in\mathbf{U}_2(i)}{\text{opt}} \sum_{j\in\mathcal{S}} \lambda_{iaj}\, x^T P(j,t)x = x^T\left(\underset{a\in\mathbf{U}_2(i)}{\text{opt}} \sum_{j\in\mathcal{S}} \lambda_{iaj}\, P(j,t)\right)x, \tag{25}$$

where the second optimization is in the sense of *domination of matrices* (cf. Definition 3.2), and $P(j,t)$'s are obtained as nonnegative-definite matrices satisfying the coupled matrix Riccati differential equations:

$$P_t(i,t) = A^T(i)P(i,t) + P(i,t)A(i) - \beta P(i,t) + Q(i) - P(i,t)B(i)R^{-1}(i)B^T(i)P(i,t) + \underset{a\in\mathbf{U}_2(i)}{\text{opt}} \sum_{j\in\mathcal{S}} \lambda_{iaj}\, P(j,t); \tag{26}$$

$$P(i, t_f) = Q_f(i), \quad i = 1, \ldots, s.$$

Of course, for the scalar case (i.e., $n = 1$), condition (25) is automatically satisfied. Furthermore, even in the vector case $\beta$ can be allowed to be *zero*. This now brings us to the following theorem.

**Theorem 3.4.**

(i) *Let $t_f$ be finite, $\beta \geq 0$, and assume that there exist nonnegative-definite matrix-valued functions $P(i,t)$, $i \in \mathcal{S}$, $t \in [0, t_f)$, satisfying* (25) *and* (26). *Then,*

$$\psi(t, x, i) = x^T P(i, t)x \,. \tag{27}$$

*Furthermore, a Markov policy $\mu^{2*}$ that uses at time t an action (depending on i and t, but not on x) that achieves the optimum in* (25) *(minimum in the team case, and maximum in the game case) is overall optimal. A corresponding optimal Markov policy for Player* 1 *is*

$$\mu^{1*}(x, i, t) = -R^{-1}(i)B^T(i)P(i,t)x. \tag{28}$$

(ii) *Let $t_f = \infty$, $\beta \geq 0$, and assume that there exist nonnegative-definite matrices $P(i)$, $i \in \mathcal{S}$, satisfying*

$$\begin{aligned}\beta P(i) = {} & A^T(i)P(i) + P(i)A(i) - P(i)B(i)R^{-1}(i)B^T(i)P(i) \\ & + Q(i) + \operatorname*{opt}_a \sum_{j \in \mathcal{S}} \lambda_{iaj} P(j), \qquad i \in \mathcal{S},\end{aligned} \tag{29}$$

*with the properties that there exist $f_i \in \mathbf{U}_2(i)$, $i \in \mathcal{S}$, such that for all $a \in \mathbf{U}_2(i)$ and all $i \in \mathcal{S}$,*

$$\sum_{j \in \mathcal{S}} \left(\lambda_{if_ij} - \lambda_{iaj}\right) P(j) \geq 0 \qquad \text{if "opt" = "max"}, \tag{30}$$

$$\sum_{j \in \mathcal{S}} \left(\lambda_{if_ij} - \lambda_{iaj}\right) P(j) \leq 0 \qquad \text{if "opt" = "min"}, \tag{31}$$

*where $\geq 0$ (respectively, $\leq 0$) stands for nonnegative-definite (respectively, nonpositive-definite),* **and** *for all $i \in \mathcal{S}$, $x \in \mathbb{R}^n$,*

$$\lim_{t_1 \to \infty} e^{-\beta t_1} \mathbf{E}^{\mu^*}_{x,i} x^T(t_1) P(i) x(t_1) = 0, \tag{32}$$

*where $\mu^* = (\mu^{1*}, \mu^{2*})$, with*

$$\mu^{1*}(x, i) = -R^{-1}(i)B^T(i)P(i)x, \qquad \mu^{2*}(i) = f_i \,. \tag{33}$$

*Under these conditions,* (33) *constitutes a pair of optimal (team-optimal or saddle-point) policies.*

**Proof.**

Part (i) follows directly from Theorem 3.1(i) for the team problem, and from Theorem 3.2 for the game problem.

Part (ii) follows from Theorem 3.1(ii) for the team case, and Theorem 3.3 for the game case, by following the derivation given prior to the statement of the theorem. □

The statement of Theorem 3.4 can be strengthened for the scalar case, which we now discuss for the infinite-horizon problem. Let $F$ denote the set of stationary policies for Player 2 that depend only on the current state of the Markov chain. With $a$ fixed as $f \in F$, let us denote the solution of (29) by $P^f(i)$, $i \in \mathcal{S}$. For each such fixed $f \in F$, (29) is in fact precisely the set of coupled algebraic Riccati equations that arises in the quadratic optimal control of standard linear jump parameter systems with a constant rate matrix [6], [7]. Using that standard theory, we can readily conclude that (29) for $n = 1$, i.e., the linearly coupled set of quadratic equations:

$$\beta P(i) = 2A(i)P(i) - [P(i)]^2 N(i) + Q(i) + \sum_{j \in \mathcal{S}} \lambda_{if(i)j}\, P(j), \qquad i \in \mathcal{S}, \quad (34)$$

where

$$N(i) := [B(i)]^2 / R(i), \qquad i \in \mathcal{S}, \quad (35)$$

admits a unique positive solution set $P^f(i)$, $i \in \mathcal{S}$, provided that $N(i) \neq 0$ (i.e., $B(i) \neq 0$) and $Q(i) > 0$ for all $i \in \mathcal{S}$—conditions we henceforth assume to hold. We now describe a policy iteration algorithm that will lead to a solution of (29) for $n = 1$. For a reason which will become clear shortly, we first rewrite (29) for $n = 1$ in the following equivalent form:

$$(\beta + \bar{\lambda}_i)P(i) = 2A(i)P(i) - [P(i)]^2 N(i) + Q(i) + \operatorname*{opt}_{a \in \mathbf{U}_2(i)} \sum_{j \in \mathcal{S}} \lambda_{iaj}\, P(j) + \bar{\lambda}_i P(i), \quad (36)$$

where

$$\bar{\lambda}_i := \max_{a \in \mathbf{U}_2(i)} \sum_{j \neq i} \lambda_{iaj} \equiv \max_{a \in \mathbf{U}_2(i)} |\lambda_{iai}|, \qquad i \in \mathcal{S}. \quad (37)$$

Note that this reorganization of (29) makes the last two terms of (36) nonnegative for all $P(j)$, $j \in \mathcal{S}$. Now, let $f \in F$ and $P^f > 0$, identified earlier, and to be relabeled as $(f^0, P^{f^0})$, be the starting choices for the policy iteration algorithm (to be described next). Given such a pair $(f^k, P^{f^k})$ at the $k$th step, a new (improved) pair at the $(k+1)$st step can be generated as follows:

$$f^{k+1}(i) = \arg \operatorname*{opt}_{a \in \mathbf{U}_2(i)} \sum_{j \in \mathcal{S}} \lambda_{iaj}\, P^{f^{(k)}}(j), \qquad i \in \mathcal{S}, \quad (38)$$

$$P^{f^{(k+1)}}(j) > 0,\ i \in \mathcal{S},\ \text{ solves (36) with } a \text{ fixed at } f^{(k+1)}. \quad (39)$$

Note that $f^{(k+1)}$ exists (but may not be unique) since $\mathbf{U}_2(i)$, $i \in \mathcal{S}$, is a finite set, and $P^{f^{(k+1)}}$ exists, and is unique in the class of nonnegative solutions since $N(i) \neq 0$ and $Q(i) > 0$, as pointed out earlier. $P^{f^{(k+1)}}$ will have the further property that

$$\text{if } \text{“opt”} = \text{“min”} : P^{f^{(k+1)}} \leq P^{f^{(k)}}, \qquad \forall i \in \mathcal{S}, \quad (40)$$

$$\text{if } \text{“opt”} = \text{“max”} : P^{f^{(k+1)}} \geq P^{f^{(k)}}, \qquad \forall i \in \mathcal{S}. \quad (41)$$

To see this for the team problem (“opt” = “min”), let us introduce an *inner-loop* iteration, using (36). With $a$ fixed at $f \in F$, denote the last three terms of (36) by

$\tilde{Q}(f, P; i)$, and note that

$$P'(i) \geq P(i), \quad \forall i \in \mathcal{S} \quad \Rightarrow \quad \tilde{Q}(f, P'; i) \geq \tilde{Q}(f, P; i), \quad \forall i \in \mathcal{S}. \tag{42}$$

Consider the iteration (*inner-loop*):

$$\begin{aligned}\left(\beta + \bar{\lambda}_i - 2A(i)\right) P^k_{(\ell+1)}(i) &= -[P^k_{(\ell+1)}(i)]^2 N(i) + \tilde{Q}(f, P^k_{(\ell)}; i),\\ \ell = 0, 1, \ldots; \qquad & P^k_{(0)}(i) = P^{f^{(k)}}(i),\end{aligned} \tag{43}$$

where $P^k_{(\ell+1)}(i) > 0$ exists, and is unique, for each $i \in \mathcal{S}$ (note that $\tilde{Q} > 0$). Now, since $P^{f(k)} > 0$ solves (36) with $a = f^{(k)}$, and since, from the definition of $f^{(k+1)}$,

$$\tilde{Q}(f^{(k+1)}, P^k_{(0)}; i) \leq \tilde{Q}(f^{(k)}, P^k_{(0)}; i), \tag{44}$$

we have

$$P^k_{(1)}(i) \leq P^k_{(0)}(i). \tag{45}$$

Hence, the inner-loop iteration generates a monotonically nondecreasing positive sequence, which has to converge to the solution of (36) with $a$ fixed at $a = f^{(k+1)}$, which is $P^{f^{(k+1)}}$. This then verifies (40), which says that the outer-loop iteration generates a monotonically nonincreasing positive sequence, which therefore should converge:

$$\lim_{k\to\infty} P^{f^{(k)}}(i) = P^{f^*}(i), \qquad i \in \mathcal{S}. \tag{46}$$

Since $\mathbf{U}_2(i)$'s are finite, this convergence is in fact achieved in only a finite number of steps. This also says that there should exist a minimizing stationary policy for Player 2:

$$\mu^{2^*} = f^*. \tag{47}$$

Identical arguments as above hold for the game case ("opt" = "max"), where now we have a monotonically nondecreasing sequence of $P^{f^{(k)}}$'s, which is bounded above because (36) admits a positive solution for every $a = f, \; f \in F$.

We now provide a precise statement of this result in the following theorem:

**Theorem 3.5.** *Consider the framework of Theorem* 3.4($ii$), *but for the scalar case, with* $B(i) \neq 0, \; Q(i) > 0, \forall i \in \mathcal{S}$. *Then:*

(i) *The coupled set of equations* (29) *admits a unique positive solution,* $P^*$, *which can be generated through the outer-loop algorithm* (38).

(ii) *There exists a unique stationary optimal policy for Player* 1, *given by*

$$\mu^{1^*}(x, i) = -\left[B(i)P^*(i)/R(i)\right] x, \qquad i \in \mathcal{S}.$$

(iii) *There is an optimal (minimizing in the team case, and maximizing in the game case) stationary policy for Player* 2, *generated by the outer-loop algorithm* (38).

**Proof.** The result follows from Theorem 3.4(ii) and the derivation given prior to the statement of the theorem. Perhaps the only additional point that should be touched upon is that condition (32) is necessarily satisfied here since for $\mu^2 = f^*$ the problem involves a standard quadratic optimization for jump linear systems with a constant rate matrix, and the positivity of $Q(i)$'s and the nonsingularity of $B(i)$'s guarantee the cited property [6]. □

## 4 Additional Algorithms and Perspectives

### 4.1 Mathematical Programming

We show here that the solution of (29) (whenever it exists and is smooth—which is certainly guaranteed for the scalar case) can be obtained as the value of a (finite) quadratic program. As a byproduct, we obtain an alternative proof of existence and uniqueness for the scalar case. First we introduce some useful definitions:

**Definition 4.1.** A function $\phi(\cdot, i) : \mathbb{R}^n \to \mathbb{R}, \phi(\cdot, i) \in \mathcal{C}$, is a *superharmonic function* if it satisfies the following inequality in the viscosity sense (cf. Definition 3.1), for all $x \in \mathbb{R}^n$, $i \in \mathcal{S}$, and $u^1 \in \mathbf{U}_1$ and $a \in \mathbf{U}_2(i)$:

$$\beta\phi(x, i) \leq [D_x\phi(x, i)]\, f(x, i, u^1) + \sum_{j \in \mathcal{S}} \lambda_{iaj}\, \phi(x, j) + L(x, i, u^1). \tag{48}$$

**Definition 4.2.** A function $\phi(\cdot, i) : \mathbb{R}^n \to \mathbb{R}$, $\phi(\cdot, i) \in \mathcal{C}$, is a *subharmonic function* if it satisfies the following inequality in the viscosity sense, for all $x \in \mathbb{R}^n$, $i \in \mathcal{S}$, $a \in \mathbf{U}_2(i)$, and for some stationary control policy $\mu^1 \in \mathcal{U}_1$ (for Player 1):

$$\beta\phi(x, i) \geq [D_x\phi(x, i)]\, f(x, i, u^1) + \sum_{j \in \mathcal{S}} \lambda_{iaj}\, \phi(x, j) + L(x, i, \mu^1(x, i)). \tag{49}$$

**Theorem 4.1.** *Let Condition* 2.1 *be satisfied.*

(i) *Consider the infinite-horizon team problem, and assume that it admits an optimal stationary team policy $\mu^*$. Then, its value function $\hat{J}_\beta(x, i)$ is the largest superharmonic function (componentwise) among functions $\phi \in \mathcal{C}$ that satisfy the additional condition:*

$$\lim_{t_1 \to \infty} e^{-\beta t_1}\, \mathbf{E}_{x,i}^{\mu^*} \phi(x(t_1), \theta(t_1)) = 0. \tag{50}$$

(ii) *If the value function $\hat{J}_\beta(x, i)$ of the zero-sum game exists and is continuous in $x$, then it is the smallest nonnegative subharmonic function among those that correspond to finite-cost policies of Player* 1.[2]

**Proof.** (i) Consider an arbitrary superharmonic function $\phi$ satisfying the given condition (50), and integrate both sides of (48) from $t = 0$ to $t = t_1 > 0$. Take the

[2]A policy $\mu^1 \in \mathcal{U}_1$ is a *finite-cost* policy for Player 1, if under it the infinite-horizon cost is finite for each $\mu^2 \in \mathcal{U}_2$.

expected value of both sides under $\mu^*$, and conditioned on $(x(0) = x, \theta(0) = i)$, and then apply Dynkin's formula of integration (see [5, p. 146, eq. (9.7)]) to get

$$\phi(x,i) \le \mathbf{E}_{x,i}^{\mu^*} \int_0^{t_1} e^{-\beta t} L(x(t),\theta(t),u^1(t))\,dt + e^{-\beta t_1}\mathbf{E}_{x,i}^{\mu^*}\phi(x(t_1),\theta(t_1)).$$

Since $\mu^*$ is optimal, in view of condition (50) imposed on $\phi$ the second term above goes to zero as $t_1 \to \infty$, and hence we arrive at the inequality:

$$\phi(x,i) \le \mathbf{E}_{x,i}^{\mu^*} \int_0^{\infty} e^{-\beta t} L(x(t),\theta(t),u^1(t))\,dt = \hat{J}_\beta(x,i).$$

Since $\hat{J}_\beta(x,i)$ also satisfies (48), as well as (50) under the given stabilizability and observability conditions, it follows that it is the largest superharmonic function among those satisfying (50).

(ii) Let $\bar{\mu}^1$ be a fixed stationary finite-cost control policy for Player 1, and consider a corresponding arbitrary subharmonic nonnegative function $\phi$. Let $\bar{\mu} := (\bar{\mu}^1, \mu^2)$, where $\mu^2 \in \mathcal{U}_2$ is an arbitrary stationary policy for Player 2. Then, following the first step in the proof of part (i) above, we have the inequality:

$$\begin{aligned}\phi(x,i) &\ge E_{x,i}^{\overline{\mu}} \int_0^{t_1} e^{-\beta t} L(x(t),\theta(t),u^1(t))\,dt + e^{-\beta t_1}\mathbf{E}_{x,i}^{\overline{\mu}}\,\phi(x(t_1),\theta(t_1)) \\ &\ge E_{x,i}^{\overline{\mu}} \int_0^{t_1} e^{-\beta t} L(x(t),\theta(t),u^1(t))\,dt,\end{aligned}$$

where the second line has followed because $\phi$ is nonnegative. Taking the limit as $t_1$ goes to infinity, we get

$$\phi(x,i) \ge J_\beta(x,i;\bar{\mu}) \ge \hat{J}_\beta(x,i),$$

where the second inequality has followed by picking $\mu^1$ and $\mu^2$ as saddle-point policies. Since $\hat{J}_\beta(x,i)$ is a subharmonic function, the result follows. □

Theorem 4.1 enables us to formalize a mathematical program to compute the solution $P(\cdot)$ of (29) in the general dimensional case, provided that it exists:[3]

**Theorem 4.2.** *Assume that Condition* 2.1 *holds. Consider equation* (29)*, and assume that it admits a nonnegative-definite solution $P(i)$, $i \in \mathcal{S}$, for both the team and game cases. Fix an initial state $(x_0, k)$.*

(i) *For the team problem, $\hat{J}_\beta(x_0,k)$ is given by the value of the following quadratic program:*

**QP1**$(k)$: *Find $P(i) > 0$, $i \in \mathcal{S}$, so as to maximize $x_0^T P(k) x_0$ subject to*

$$0 \le -\beta P(i) + A^T(i)P(i) + P(i)A(i)$$

[3] We should note that a unique nonnegative-definite solution exists for each fixed rate matrix, and under the given stochastic stabilizability and observability conditions, as has already been demonstrated in the literature; see [6] and [7].

$$- P^T(i)B(i)R^{-1}(i)B^T(i)P(i) + Q(i)$$
$$+ \sum_{j \in \mathcal{S}} \lambda_{iaj} P(j), \qquad \forall i \in \mathcal{S}, \quad a \in \mathbf{U}_2(i). \tag{51}$$

(ii) *For the game problem, $\hat{J}_\beta(x_0, k)$ is obtained as the value of the following quadratic program:*
**QP2($k$):** *Find $P(i) > 0$, $i \in \mathcal{S}$, so as to minimize $x_0^T P(k) x_0$ subject to*

$$0 \geq -\beta P(i) + 2A(i)P(i) - P^T(i)B(i)R^{-1}(i)B^T(i)P(i) + Q(i)$$
$$+ \sum_{j \in \mathcal{S}} \lambda_{iaj} P(j), \qquad \forall i \in \mathcal{S}, \quad a \in \mathbf{U}_2(i). \tag{52}$$

**Proof.** This is a direct consequence of Theorem 4.1, obtained by specializing it to functions $\phi$ of the form $\phi(x,i) = x^T P(i)x$. More precisely, consider the team problem. Theorem 3.4(ii) implies that $\hat{J}_\beta$ has the form $x^T P(i)x$.[4] Theorem 4.1(i) implies that it is the largest superharmonic function among those which have this form. By the definition of superharmonic functions, this means that $\hat{J}_\beta$ is the largest function among those that satisfy the constraints

$$\beta x^T P(i)x \leq 2(A(i)x + B(i)u^1)^T P(i)x$$
$$+ \sum_{j \in \mathcal{S}} \lambda_{iu^2j} x^T P(j)x + x^T Q(i)x + u^{1^T} R(i)u^1,$$

for all $x, i, u^1, u^2$. The above constraint is equivalent to

$$\beta x^T P(i)x \leq \inf_{u^1} \Big[ 2(A(i)x + B(i)u^1)P(i)$$
$$+ \sum_{j \in \mathcal{S}} \lambda_{iu^2j} x^T P(j)x + x^T Q(i)x + u^{1^T} R(i)u^1 \Big],$$

for all $x, i$, and $u^2$, which is further equivalent to (by carrying out the minimization in $u^1$):

$$\beta x^T P(i)x \leq x^T \left(A^T(i)P(i) + P(i)A(i) - P(i)B(i)R^{-1}(i)B^T(i)P(i) + Q(i)\right) x$$
$$+ \sum_{j \in \mathcal{S}} \lambda_{iu^2j} x^T P(j)x,$$

for all $x, i$, and $u^2$. This is equivalent to the constraint (51).

(ii) Consider the class of functions $F_1$ of the form $\phi(x,i) = x^T P(i)x$. Since the value function is the smallest subharmonic function, it is, in particular, the smallest among $F_1$. It is moreover, the smallest among the following class of policies $F_2 \subset F_1$, defined as follows: $\phi \in F_2$ if (49) holds for all $x \in \mathbb{R}^n$, $i \in \mathcal{S}$, $a \in \mathbf{U}_2(i)$, and for $u^1(x,i)$ given in (33). This is equivalent to the following. The value function is the smallest function of the form $x^T P(i)x$ satisfying, for all $x$, $i$,

[4]This is of course true, provided that conditions (30)–(31) hold.

and $u^2$,

$$\beta x^T P(i)x \geq x^T \left(A^T(i)P(i) + P(i)A(i) - P(i)B(i)R^{-1}(i)B^T(i)P(i) + Q(i)\right) x + \sum_{j\in\mathcal{S}} \lambda_{iu^2j} x^T P(j)x.$$

This is, finally, equivalent to the constraint (52). □

It is appropriate to list here some useful properties of the mathematical program **QP1**($k$). First note that the feasible region satisfying the constraints (51) is nonempty; this follows from the existence hypothesis of the theorem, and also directly from the fact that $P(i) = 0, i \in \mathcal{S}$, is feasible. Moreover, it is a closed region. Second, if $P^*(i)$ is the optimal solution of **QP1**($i$), for $i \in \mathcal{S}$, then $P^*(i)$, $i \in \mathcal{S}$, are feasible for **QP1**($k$) for any $k \in \mathcal{S}$ (which follows from Theorem 4.1). Consequently, if optimal solutions $P^*(i)$ have already been computed for $i \in \mathcal{S}' \subset \mathcal{S}$, then one can substitute these for $P(i)$ in (51), when computing $P^*(j)$ for $j \notin \mathcal{S}'$.

Similar properties can be observed in connection with the mathematical program (52). The feasible region satisfying the constraints in (52) is nonempty by the existence hypothesis of the theorem. Let $P^*(i)$ be optimal solutions of **QP2**($i$), $i \in \mathcal{S}$. Then $P^*(i)$, $i \in \mathcal{S}$, are feasible for **QP2**($k$) for any $k \in \mathcal{S}$ (this follows from Theorem 4.1). Consequently, if optimal solutions $P^*(i)$'s have been already computed for $i \in \mathcal{S}' \subset \mathcal{S}$, then one can readily substitute these for $P(i)$, $i \in \mathcal{S}'$, in (52), when computing $P^*(j)$ for $j \notin \mathcal{S}'$.

## 4.2 Value Iteration

We now restrict our discussion to the scalar case ($n = 1$), for which Theorem 3.5 already presented the complete solution and also provided an iterative procedure for numerical computation. Here, we present another iterative algorithm to compute the solution, which is based on value iteration:

**(A1)** Set $P_0(j) := 0$, $j \in \mathcal{S}$. Introduce constants $\overline{\lambda}_i$, $i \in \mathcal{S}$, as in (37).
**(A2)** Compute $P_\ell(i) > 0$, $i \in \mathcal{S}$, $\ell \geq 1$, iteratively by solving the following set of quadratic equations:

$$0 = -(\overline{\lambda}_i + \beta)P_{\ell+1}(i) + 2A(i)P_{\ell+1}(i) - [P_{\ell+1}(i)]^2 N(i) + Q(i) + \operatorname*{opt}_{a\in\mathbf{U}_2(i)} \sum_{j\in\mathcal{S}} \lambda_{iaj} P_\ell(j) + \overline{\lambda}_i P_\ell(i), \qquad i \in \mathcal{S}. \tag{53}$$

**Theorem 4.3.** *Consider either the team or the game problem, for the scalar case, and under the conditions $B(i) \neq 0$, $Q(i) > 0$, $i \in \mathcal{S}$. Positive solutions $P(j)$, $j \in \mathcal{S}$, of* (29) *can be obtained as the limit of the set of nondecreasing sequences $\{P_\ell(j)\}_{\ell\geq 0}$.*

**Proof.** The proof uses ideas similar to those employed in the result that led to Theorem 3.5, but some of the details are different. To see that the $P_{(\ell)}(j)$'s are nondecreasing, write (53) as

$$0 = -(\overline{\lambda}_i + \beta)P_{\ell+1}(i) + 2A(i)P_{\ell+1}(i) - P_{\ell+1}(i)^2 N(i) + Q_\ell(i), \tag{54}$$

where

$$Q_\ell(i) := Q(i) + \operatorname*{opt}_{a \in \mathbf{U}_2(i)} \sum_{j \in \mathcal{S}} \lambda_{iaj} P_\ell(j) + \overline{\lambda}_i P_\ell(i). \tag{55}$$

Set $Q_0(i) = Q(i)$. It follows from (55) that for any $i$, $\ell \geq 0$ (compare this with (42)):

$$\text{if} \quad \forall\, j \in \mathcal{S}, \;\; P_{\ell+1}(j) \geq P_\ell(j), \quad \text{then} \quad Q_{\ell+1}(i) \geq Q_\ell(i). \tag{56}$$

This is true because under **(A1)** coefficients of the $P_\ell(j)$'s in

$$m(a; P_\ell) := \sum_{j \in \mathcal{S}} \lambda_{iaj} P_\ell(j) + \overline{\lambda}_i P_\ell(i) \tag{57}$$

are each nonnegative, for every $a \in \mathbf{U}_2(i)$. Hence, under the hypothesis of (56),

$$m(a; P_{\ell+1}) \geq m(a; P_\ell) \qquad \text{for each} \quad a \in \mathbf{U}_2(i), \tag{58}$$

from which (56) follows. Now, (54) is a standard Riccati equation that corresponds to a system that always remains in state $i$, and where the weighting on the quadratic cost for the state is $Q_\ell(i)$. Therefore, its solution $P_{\ell+1}(i)$ will be increasing in $Q_\ell(i)$. It then follows that

$$\text{if} \quad Q_{\ell+1}(i) \geq Q_\ell(i), \quad \text{then} \quad P_{\ell+2}(i) \geq P_{\ell+1}(i). \tag{59}$$

In view of the fact $P_1(i) \geq P_0(i) = 0$, (56)–(59) establish by induction the desired result that the sequences $\{P_\ell(i)\}$ and $\{Q_\ell(i)\}$ are nondecreasing for each $i \in \mathcal{S}$, and therefore have respective limits, with the former satisfying (29). □

Next, we present a monotone nonincreasing iterative scheme to compute $P$ for the game problem, based on a feasible solution of the mathematical program. This is an alternative way to establish in the scalar case that the solution $P$ of (29) exists and is finite.

**Lemma 4.1.** *Fix an arbitrary $k \in \mathcal{S}$. Let $P_0$ be an arbitrary vector that is feasible in the quadratic program* (52) *for that $k$. Define $P_\ell$, $\ell = 1, 2, \ldots$, to be generated by the iterations* (53). *Then $P(j)$, $j \in \mathcal{S}$, are obtained as the limit of the nonincreasing sequences $\{P_\ell(j)\}$.*

**Proof.** We write (53) as

$$0 = -(\overline{\lambda}_i + \beta)P_{\ell+1}(i) + 2A(i)P_{\ell+1}(i) - [P_{\ell+1}(i)]^2 N(i) + Q_\ell(i), \tag{60}$$

where

$$Q_\ell(i) := Q(i) + \max_{a \in \mathbf{U}_2(i)} \sum_{j \in \mathcal{S}} \lambda_{iaj} P_\ell(j) + \overline{\lambda}_i P_\ell(i). \tag{61}$$

Introduce

$$\Delta(i) := -\beta P_0(i) + 2A(i)P_0(i) - [P_0(i)]^2 N(i) + Q(i) + \max_{v\in \mathbf{U}_2(i)} \sum_{j\in \mathcal{S}} \lambda_{ivj} P_0(j),$$

and let $\tilde{Q}(i) = Q_0(i) + \Delta(i)$. It follows from (52) that $\Delta(i) \leq 0$, and hence $\tilde{Q}(i) \leq Q_0(i)$, and that $P_0(i)$ is the solution of the standard Riccati equation

$$0 = -(\overline{\lambda}_i + \beta)P_0(i) + 2A(i)P_0(i) - [P_0(i)]^2 N(i) + \tilde{Q}(i)$$

corresponding to a system that always remains in state $i$, and where the weighting on the quadratic cost for the state is $\tilde{Q}(i)$.

Since $\tilde{Q}(i) \leq Q_0(i)$, the above discussion leads to the conclusion that $P_1(i) \leq P_0(i)$. This further implies that $Q_1(i) \leq Q_0(i)$, which by the same argument leads to $P_2(i) \leq P_1(i)$. We may thus continue iteratively and conclude that the sequence $P_n(i)$ is monotonically nonincreasing. Since $P_\ell(i)$'s are positive, they converge to a finite limit $P$. It is now easy to check that the limit indeed satisfies (29). □

In the above, we made use of the monotonicity to establish the convergence of the iterative scheme. It turns out that the iterative scheme would converge (though not necessarily monotonically) when starting from any initial condition $P_0$. The reason is that any initial condition $P_0$ can be lower bounded by $\underline{P}_0 = 0$ and upper bounded by some $\overline{P}_0$ that is a feasible vector in the quadratic program (52). It then follows that for any $\ell$,

$$\underline{P}_\ell \leq P_\ell \leq \overline{P}_\ell.$$

The convergence of $P_\ell$ to the solution $P$ of (29) (for $n = 1$) now follows from the fact that both $\underline{P}_\ell$ and $\overline{P}_\ell$ converge to $P$, as established above.

## 5 Nonquadratic Value Functions and Quadratic Bounds

When the continuous part of the state is not of dimension 1 (i.e., $n > 1$), conditions (30) and (31) will, in general, not hold. We discuss in this section some computable bounds on the value function and some heuristics for obtaining suboptimal policies that are easy to compute and to implement—all for the infinite-horizon case. We assume throughout that the stochastic stabilizability and observability conditions of Theorem 4.1 hold.

We first show that the value is bounded by quadratic functions—both upper and lower bounds for the team problem as well as for the game. Then, we devise a scheme for obtaining an approximation to the nonquadratic value, and a corresponding sequence of suboptimal policies.

### 5.1 Quadratic Bounds

For the team problem, a simple *upper bound* for the value is obtained by restricting Player 2 to policies that are *not functions of the $x$ component of the state*, but are possibly dependent on the initial state $x_0$, and are furthermore stationary with respect to the state of the Markov process. Denote the class of such policies (for Player 2) by $M_2^0$, and note that this set is finite. Hence, for any initial state pair $(x_0, i_0)$,

$$\hat{J}_\beta(x_0, i_0) \leq \min_{\mu^2 \in M_2^0} \inf_{\mu^1 \in \mathcal{U}_1} J_\beta(x_0, i_0, \mu) =: \bar{J}_\beta(x_0, i_0).$$

When Player 2 uses such a policy $\mu^2 \in M_2^0$, Player 1 is faced with a linear-quadratic control problem with a constant (control-independent) rate matrix, for which we have (as has already been discussed in previous sections):

$$\min_{\mu^1 \in \mathcal{U}_1} J_\beta(x_0, i_0, \mu) = x_0^T P(i, \mu^2) x_0, \tag{62}$$

where $P(i, \mu^2)$, $i \in \mathcal{S}$, is the set of unique nonnegative-definite solutions of the following linearly coupled Riccati equations:

$$\begin{aligned}\beta P(i) = A^T(i)P(i) + P(i)A(i) - P(i)B(i)R^{-1}(i)B^T(i)P(i) \\ + Q(i) + \sum_{j \in \mathcal{S}} \lambda_{i\mu^2(i)j} P(j), \qquad i \in \mathcal{S}.\end{aligned} \tag{63}$$

Let $\hat{\mu}^2 = \hat{\mu}^2(i, x_0)$ be a policy for Player 2 that minimizes (62) subject to (63), and define

$$\hat{\mu}^1(x, i; x_0) = -R^{-1}(i)B^T(i)P(i, \hat{\mu}^2)x,$$

which is the control that achieves the unique minimum in (62) with $\mu^2 = \hat{\mu}^2$. Let $\hat{\mu} := (\hat{\mu}^1, \hat{\mu}^2)$ denote the resulting composite policy. Clearly, this policy achieves the bound $\bar{J}_\beta(x_0, i_0)$.

For the game case, by following a similar procedure, this time a *lower bound* for the value can be obtained, by again restricting Player 2 to policies in $M_2^0$. In this case, $\hat{\mu}^2$ will be obtained by maximizing (62) subject to (63).

Next, we obtain a **lower bound** on the value of the team. From Theorem 3.1, it follows that the value $\hat{J}_\beta(x, i)$ is given by $\psi(x, i)$, where the latter satisfies

$$\begin{aligned}\beta\psi(x, i) = \min_{u^1} \left[ \left(A(i)x + B(i)u^1\right)^T D_x\psi(x, i) + |u^1|^2_{R(i)} \right] \\ + |x|^2_{Q(i)} + \min_a \sum_{j \in \mathcal{S}} \lambda_{iaj}\psi(x, j).\end{aligned} \tag{64}$$

Since $\lambda_{iai} < 0$ and $\lambda_{iaj} \geq 0$, for $j \neq 0$, the following inequality readily follows from (64):

$$\begin{aligned}(\beta + \bar{\lambda}_i)\psi(x, i) \geq \min_{u^1} \left[ \left(A(i)x + B(i)u^1\right)^T D_x\psi(x, i) + |u^1|^2_{R(i)} \right] \\ + |x|^2_{Q(i)} + \sum_{j \in \mathcal{S}, j \neq i} (\min_a \lambda_{iaj})\psi(x, j),\end{aligned} \tag{65}$$

where

$$\bar{\lambda}_i := \max_{a \in \mathbf{U}_2} |\lambda_{iai}|, \qquad i \in \mathcal{S},$$

as already defined earlier by (37). Consider now the following equality version of (65):

$$\begin{aligned}\beta\tilde{\psi}(x,i) = \min_{u^1}\left[\left(A(i)x + B(i)u^1\right)^T D_x\tilde{\psi}(x,i) + |u^1|^2_{R(i)}\right] + |x|^2_{Q(i)} \\ + \sum_{j\in\mathcal{S}, j\neq i} (\min_a \lambda_{iaj})\tilde{\psi}(x,j) - \bar{\lambda}_i\tilde{\psi}(x,i). \qquad (66)\end{aligned}$$

This admits a nonnegative quadratic solution, which lower-bounds $\hat{J}_\beta(x,i)$—as we next show. First note that (66) is similar to the HJB equation of a jump linear-quadratic control problem with a constant rate matrix, with the only difference being that the rates do not necessarily add up to *zero*, i.e.,

$$\sum_{j\in\mathcal{S}, j\neq i} \min_a \lambda_{iaj} - \bar{\lambda}_i \neq 0.$$

But this can be fixed by adding to both sides of (66) the quantity

$$\left(\bar{\lambda}_i - \sum_{j\in\mathcal{S}, j\neq i} \min_a \lambda_{iaj}\right)\tilde{\psi}(x,i).$$

This then makes (66) precisely the HJB equation associated with a legitimate jump linear-quadratic control problem, with a new discount factor

$$\beta + \bar{\lambda}_i - \sum_{j\in\mathcal{S}, j\neq i} \min_a \lambda_{iaj},$$

which is indeed nonnegative. Hence, using the standard theory (under the given stochastic stabilizability and observability conditions), one can conclude that (66) admits a quadratic solution of the form $\psi'(x,i) = x^T P'(i)x$, where $P'$ is the unique nonnegative-definite solution set to the following linearly coupled Riccati equations:

$$\begin{aligned}\beta P'(i) = A^T(i)P'(i) + P'(i)A(i) - P'(i)B(i)R^{-1}(i)B^T(i)P'(i) \\ + Q(i) + \sum_{j\in\mathcal{S}, j\neq i}\left(\min_a \lambda_{iaj}\right)P'(j) - \bar{\lambda}_i P'(i), \qquad i \in \mathcal{S}. \quad (67)\end{aligned}$$

We now show that this quadratic function $\psi'(x,i)$ lower-bounds $\hat{J}_\beta(x,i)$ for all $x$ and $i$. Toward this end we first observe that, for each fixed $i \in \mathcal{S}$, (66) can also be viewed as the HJB equation for a deterministic linear-quadratic control problem whose system dynamics are given by (1) permanently locked to state $i$, whose immediate cost is $L_{\tilde{\psi}}(x,i,u)$, where

$$L_{\tilde{\psi}}(x,i,u) := |u^1|^2_{R(i)} + |x|^2_{Q(i)} + \sum_{j\in\mathcal{S}, j\neq i}(\min_a \lambda_{iaj})\,\tilde{\psi}(x,j), \qquad (68)$$

and for which the discount factor is $\beta + \bar{\lambda}_i$. Let $\psi^0 := \hat{J}_\beta$, and define $\psi^\ell \in \mathcal{C}$, $\ell = 1, 2, \ldots$, recursively to be the unique nonnegative viscosity solutions of the following sequence of HJB equations:

$$(\beta + \bar{\lambda}_i)\psi^\ell(x, i) = \min_{u^1}\left[\left(A(i)x + B(i)u^1\right)^T D_x\psi^\ell(x, i) + |u^1|^2_{R(i)}\right] + |x|^2_{Q(i)} + \sum_{j\in\mathcal{S}, j\neq i}(\min_a \lambda_{iaj})\psi^{\ell-1}(x, j), \quad i \in \mathcal{S}. \tag{69}$$

We note that an inductive argument would show that such solutions necessarily exist, since the instantaneous cost is continuous in each case.[5] We further observe that (64), of which $\hat{J}_\beta$ is the solution, can be viewed as the HJB equation of the same system as above, i.e., a system that is always in state $i$, with the same discount factor, but with an immediate cost of

$$L(x, i, u) = |u^1|^2_{R(i)} + |x|^2_{Q(i)} + \min_a \sum_{j\in\mathcal{S}} \lambda_{iaj}\hat{J}_\beta(x, j) + \bar{\lambda}_i\hat{J}_\beta(x, j).$$

Since $0 \leq L_{\psi^0}(x, i, u) \leq L(x, i, u)$, it follows that the corresponding values are ordered accordingly, i.e., $\psi^1 \leq \psi^0 = \hat{J}_\beta$. This implies that $L_{\psi^1}(x, i, u) \leq L_{\psi^0}(x, i, u)$, and thus $\psi^2 \leq \psi^1$. Proceeding in this fashion, it follows by induction that $\{\psi^\ell\}$ is indeed a nonincreasing sequence, which therefore has a limit for each $x \in \mathbb{R}^n$ and $i \in \mathcal{S}$. Finally, $\psi' = \lim_{\ell\to\infty}\psi^\ell$ is easily seen to be the (classical) solution of (66).

**Remark 5.1.**

(i) An alternative way to show that $\psi' \leq \hat{J}_\beta$ is to note that $\psi'$ is a superharmonic function, and then apply Theorem 4.1(i).

(ii) The bounds that we obtained above are based on transforming an LQ (linear-quadratic) problem with controlled jumps into one with control-free jumps. In this transformation we did not make use of the LQ properties, so that the bounds are valid for other dynamics and cost structures as well (provided that both $f$ and $L$ are continuous, $L$ is also nonnegative, and the infinite-horizon discounted cost is finite for some stationary controller).

For the game case we can obtain an **upper bound** $\psi'$ on the value by employing the above approach, where $\psi'$ is now the solution of

$$(\beta + \min_a|\lambda_{iai}|)\psi(x, i) = \min_{u^1}\left[\left(A(i)x + B(i)u^1\right)^T D_x\psi(x, i) + |u^1|^2_{R(i)}\right] + |x|^2_{Q(i)} + \sum_{j\in\mathcal{S}, j\neq i}(\max_a \lambda_{iaj})\psi(x, j). \tag{70}$$

---

[5]Note that the term that starts the recursion, $\hat{J}_\beta$, is continuous in $x$, as per Theorem 3.1.

## 5.2 A Dynamic Approximation and Suboptimal Team Policies

We now describe a procedure for obtaining an approximation to the value function of the $n$-dimensional team problem, which involves solutions to a sequence of team problems of the type that were used in arriving at the upper bound $\bar{J}_\beta$ in the previous subsection. Consider a partition of the time interval into finite subintervals, determined by the increasing sequence of natural numbers $0 < t_1 < t_2 < t_3 < \cdots$. First solve the original team problem by restricting Player 2 to policies that are open-loop in $x$ and stationary closed-loop in $\theta$. Such policies were already introduced in the previous subsection, in connection with the derivation of an upper bound for the team problem. The class of such policies were denoted by $M_2^0$, and the team optimal controllers for the two players (which necessarily exist) were denoted by $\hat{\mu}$. Here we add a subscript to this notation, i.e., $\hat{\mu}_0$, where the subscript "0" signifies that these policies were derived by fixing the initial value of $x$ at time $t = 0$ at $x_0$. Let us recall that the derivation of these policies followed a fairly simple construction, as it involved the solution of a finite optimization problem, subject to constraints imposed by $s$ linearly coupled Riccati equations. For each $j = 1, 2, \ldots$, let us now introduce the notation $\hat{\mu}_j$ to denote a similar team-optimal policy pair, obtained over the interval $[t_j, \infty)$, with the initial state now being $x = x(t_j) =: x_j$. We denote the class to which the corresponding policy for Player 2 (i.e., $\hat{\mu}_j^2$) belongs by $M_2^j$. Note that elements of $M_2^j$ are restricted to the interval $[t_j, \infty)$.

We now construct a sequence of policies (with improving performance) by way of concatenation as follows:

Consider the policy $\mu_j^*$ that uses $\hat{\mu}_0$ on $[0, t_1)$, $\hat{\mu}_1$ on $[t_1, t_2)$, ..., $\hat{\mu}_{j-1}$ on $[t_{j-1}, t_j)$, and $\hat{\mu}_j$ on $[t_j, \infty)$. Then, we have the following:

**Lemma 5.1.** *The sequence of corresponding costs,*

$$\{\hat{J}_\beta(x, i; \mu_j^*)\}_{j\geq 0},$$

*is nonincreasing.*

**Proof.** For any $j = 1, 2, \ldots$,

$$\begin{aligned}
\hat{J}_\beta(x, i; \mu_j^*) &= E_{x,i}^{\mu_j^*} \int_0^{t_{j-1}} e^{-\beta t} L(x(t), \theta(t), u^1(t)\, dt \\
&\quad + E_{x,i}^{\mu_j^*} \int_{t_{j-1}}^{t_j} e^{-\beta t} L(x(t), \theta(t), u^1(t)\, dt \\
&\quad + E_{x,i}^{\mu_j^*} \int_{t_j}^{\infty} e^{-\beta t} L(x(t), \theta(t), u^1(t)\, dt \\
&= E_{x,i}^{\mu_{j-1}^*} \int_0^{t_{j-1}} e^{-\beta t} L(x(t), \theta(t), u^1(t)\, dt \\
&\quad + E_{x,i}^{\mu_j^*} \int_{t_{j-1}}^{t_j} e^{-\beta t} L(x(t), \theta(t), u^1(t)\, dt
\end{aligned}$$

$$
\begin{aligned}
&+ \min_{\mu | x = x_j} E^{\mu}_{x,i} \int_{t_j}^{\infty} e^{-\beta t} L(x(t), \theta(t), u^1(t) \, dt \\
&\le E^{\mu^*_{j-1}}_{x,i} \int_0^{t_{j-1}} e^{-\beta t} L(x(t), \theta(t), u^1(t) \, dt \\
&+ E^{\mu^*_{j-1}}_{x,i} \int_{t_{j-1}}^{\infty} e^{-\beta t} L(x(t), \theta(t), u^1(t) \, dt \\
&= \hat{J}_\beta(x, i; \mu^*_{j-1}).
\end{aligned}
$$

A few words are now in order for the steps used above, and one notation introduced: The breaking of the interval $[0, \infty)$ into three subintervals in step 1 is consistent with the construction of policy $\mu^*_j$. The first two expressions have been carried intact to the second step, with the expectation now taken under the policy $\mu^*_{j-1}$, since it agrees with $\mu^*_j$ on the subinterval $[0, t_j)$ as per the construction. The last expression in the second step stands for minimization of the cost over the interval $[t_j, \infty)$, over $\mu_1 \in \mathcal{U}_1$ restricted to the same interval and $\mu^2 \in M^j_2$—with the conditioning with respect to the observed value $x_j$ of $x(t_j)$ as well as that of the Markov chain, and then the entire expression averaged with respect to the probability measure corresponding to $\mu^*_j$ over the subinterval $[0, t_j)$. This expression clearly bounds from below the same for an arbitrary control picked over the interval $[t_j, \infty)$, and in particular $\mu^*_{j-1}$—which leads to the inequality and the second expression of the third step. The fourth step simply follows from the definition of $\mu^*_j$. □

The significance of the result given above, and the accompanying recursive construction, is that one can obtain a constantly improving sequence of approximations to a possibly nonquadratic value function, which is also nondifferentiable, in terms of piecewise quadratic value functions, whose construction is fairly easy. As a byproduct, we also obtain a sequence of easily computable policies for the team, which improve with increasing $j$. The limiting properties of this sequence of policies, and ditto for the value functions, still need to be studied.

## 6 An Illustrative Example

To motivate the class of hybrid problems studied in this paper, and to illustrate the theory for the scalar case, consider the following communication network, which can be viewed as a modified version of the models studied recently in [1] and [2], with a new component. It is assumed that the network has linearized dynamics (for the control of queue length), and all performance measures (such as throughput, delays, loss probabilities, etc.) are determined essentially by a bottleneck node. Both these assumptions have theoretical as well as experimental justifications, see [1].

Let $q_t$ denote the queue length at a bottleneck link, and let $r_t$ denote the effective service rate available for traffic of the given source at that link at time $t$.

We let $r_t$ be arbitrary, but assume that the controllers have perfect measurements of it. Let $\xi_t$ denote the (controlled) source rate at time $t$, whose shifted version is defined as $u_t^1 := \xi_t - r_t$, which is determined by controller 1, called the flow controller. Consider the following linearized dynamics for the queue length:

$$\frac{dq}{dt} = u^1, \tag{71}$$

which is called *linearized* because the end-point effects have been ignored. The objectives of the flow controller are

(i) to ensure that the bottleneck queue size stays around some desired level $\bar{Q}$; and

(ii) to achieve good tracking between input and output rates.

In particular, the choice of $\bar{Q}$ and the variability around it have direct impact on loss probabilities and throughput. We therefore define a shifted version of $q$:

$$x_t := q_t - \bar{Q},$$

in view of which (71) now becomes

$$\frac{dx}{dt} = u^1. \tag{72}$$

An appropriate local cost function that is compatible with the objectives stated above would be the one that penalizes variations in $x_t$ and $u_t^1$ around *zero*—a candidate for which is the weighted quadratic cost function.

Suppose that there are several types (say $s$) of possible traffic, with different kinds of requirements on the performance measures. Associated with type $i$ traffic are the positive constants $Q(i)$ and $R(i)$ appearing in the immediate cost: $L(x, i, u) = |x|^2_{Q(i)} + |u^1|^2_{R(i)}$. Typically, traffic requiring higher quality of service (QoS) might have a larger $Q(i)$, which reflects the fact that it might require lower loss probabilities and higher throughput. It could be receiving a higher priority from the network in the sense that larger variations in $u^1$ will be tolerated so as to achieve the required QoS; thus the corresponding $R(i)$ might be smaller. The occurrence of these different types of traffic will be governed by a continuous-time Markov jump process, with transitions controlled by a second controller with a finite action set.

A typical admission control problem is the following: by default the system always accepts traffic of some given type, say 1. Traffic of type 1 is transmitted until a session consisting of another type of traffic of higher priority, say type $i$ ($i = 2, 3, \dots, s$), is accepted. When it is accepted, the session cannot be interrupted until it ends. Thus at states $\theta = i > 1$, there are no (admission) control actions available.

Controller 2 is thus effective only at state 1, at which the rates of arrival of sessions of higher priority traffic are to be determined. To each traffic type there corresponds two admission decisions (that are part of the action to be chosen by controller 2): 0—corresponding to low admission rate $\underline{\lambda}(j)$, and 1—corresponding to a high admission rate $\bar{\lambda}(j)$. The control action at state 1 is thus of the form $u^2 = (u_2^2, \ldots, u_s^2)$, $u_j^2 \in \{0, 1\}$. The controlled transition rates have the form

$$\lambda_{1,u^2,j} = \underline{\lambda}(j)1_{\{u_j^2=0\}} + \bar{\lambda}(j)1_{\{u_j^2=1\}},$$

where $1_{\{\cdot\}}$ denotes the set indicator function. Note that, if at state 1 the control action $u^2$ is fixed, then the next type of session to be accepted will be $j$ $(j > 1)$ with probability

$$[\underline{\lambda}(j)1_{\{u_j^2=0\}} + \bar{\lambda}(j)1_{\{u_j^2=1\}}] \Big/ \sum_{k=2}^{s} [\underline{\lambda}(k)1_{\{u_k^2=0\}} + \bar{\lambda}(k)1_{\{u_k^2=1\}}].$$

The problem then is to minimize the expected discounted or total expected cost (with instantaneous cost being $L$ above) with respect to the multistrategy $\mu := (\mu^1, \mu^2)$, where $\mu^1$ is the flow controller and $\mu^2$ is the admission controller, both having as arguments the current and past values of $\xi$, $r$, and $\theta$.

We now apply the results presented in the previous sections to the admission-flow control problem just formulated. In terms of the notation of Section 2, we have $A(i) = 0$ and $B(i) = 1$, for all $i \in \mathcal{S}$. Furthermore, to obtain some explicit results, and in accordance with the earlier discussion in this section, we take $\beta = 0$ (i.e., no discount factor), $s = 3$ (i.e., only three states for the Markov chain; $\mathcal{S} = \{1, 2, 3\}$), $U_2(1) = \{00, 01, 11, 10\}$, $U_2(2) = \{0\}$, $U_2(3) = \{0\}$; $\lambda_{1002} = \lambda_{1012} = 1$, $\lambda_{1003} = \lambda_{1103} = 2$, $\lambda_{1102} = \lambda_{1112} = 10$, $\lambda_{1013} = \lambda_{1113} = 20$; $Q(1) = 1$, $Q(2) = 10$, $Q(3) = 20$; $R(1) = 10$, $R(2) = 2$, and $R(3) = 1$.[6] Under this setup, for each possible choice of $\mu^2$ we computed the corresponding optimal policy for controller 1 (i.e., $\mu^1$), and the associated optimal cost function, using (24), (27), and (34). By comparing the optimal value functions obtained for all possible (four) admission policies, we found the unique optimal controller 2 to be $\mu^{2*}(1) = 00$, and the solution to (29) to be $P^*(1) = 4.141$, $P^*(2) = 4.439$, $P^*(3) = 4.352$. The unique optimal flow controller is then (from the scalar version of (24)):

$$\mu^{1*}(q, i) = \begin{cases} -0.4141\, q, & i = 1, \\ -2.220\, q, & i = 2, \\ -4.352\, q, & i = 3. \end{cases}$$

Typical system responses under all four admission control policies are depicted in Figures 1–4. Each figure consists of a set of four plots, depicting (on the vertical axis) the time history of the queue size, traffic type, the flow control, and the integral cost incurred—with appropriate subcaptions identifying each plot. For illustration purposes, we have taken the initial queue size to be 30 units, which is 10 units larger than the desired queue length of 20 units. Since the high priority traffic types (2 and 3) have stringent QoS specifications, the corresponding flow control is more aggressive to maintain the queue size around its desired value. On the other hand,

[6] $\lambda_{1ij2}$ stands for the rate of switch from state 1 to state 2 when $u^2 = (i, j)$; a similar interpretation applies to $\lambda_{1ij3}$.

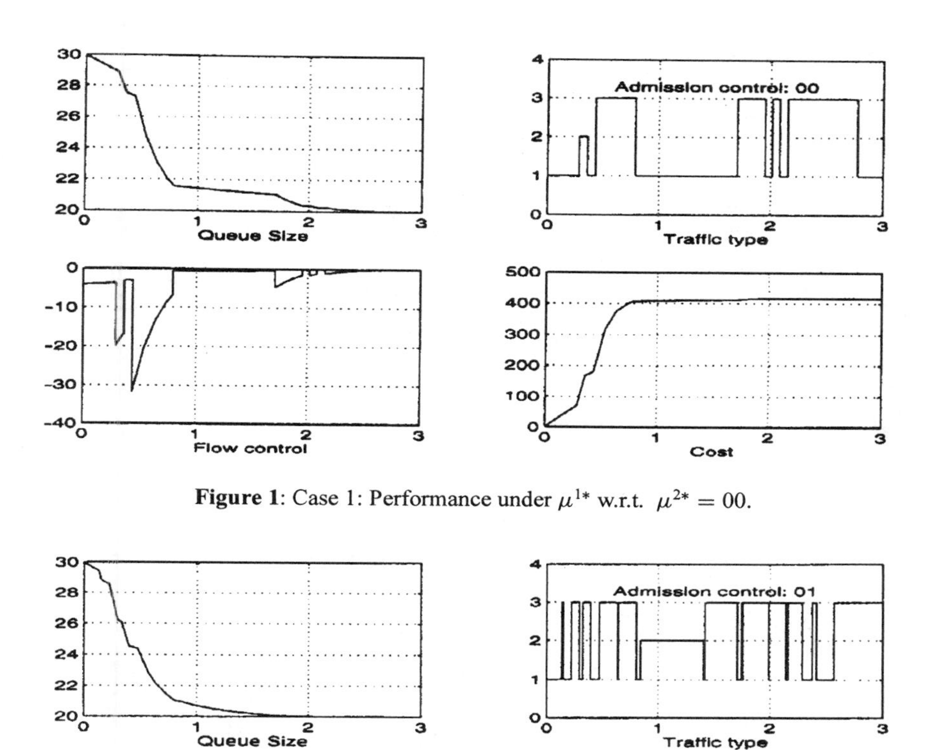


**Figure 1**: Case 1: Performance under $\mu^{1*}$ w.r.t. $\mu^{2*} = 00$.

Queue Size
Admission control: 01
Traffic type
Flow control
Cost

**Figure 2**: Case 1: Performance under $\mu^{1*}$ w.r.t. $\mu^{2*} = 01$.

for the low priority type (1), the control is smoother in order to minimize the jitter (i.e., variability) in the network. These observations are consistent with what we would have expected from the design. We further observe that by admitting the high priority traffic at a higher rate, the queue size reaches its desired value more quickly, at the expense of a larger control jitter. Although they only represent one sample path of the stochastic system in each of the four cases, the simulations corroborate the theory very well. The smallest integral cost is achieved under $\mu^2(1) = 00$, which is actually the optimal admission rule for the total (expected) value of the cost function.

We next considered the case when $R(1)$ was increased to $R(1) = 100$, with all parameter values remaining the same as above. In this case, the optimal admission

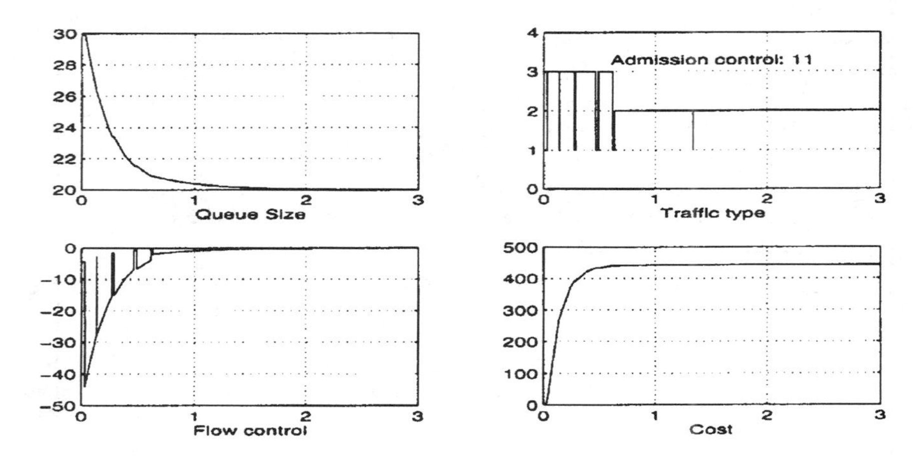


**Figure 3**: Case 1: Performance under $\mu^{1*}$ w.r.t. $\mu^{2*} = 11$.

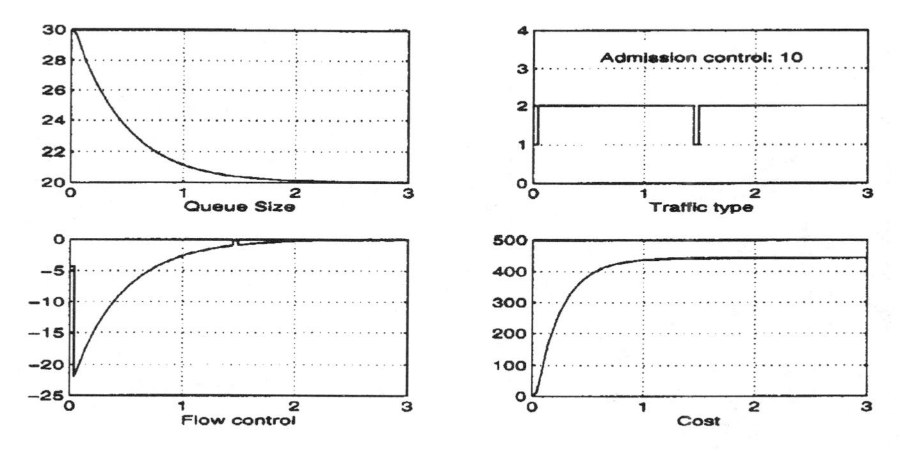


**Figure 4**: Case 1: Performance under $\mu^{1*}$ w.r.t. $\mu^{2*} = 10$.

controller turned out to be $\mu^{2*}(1) = 11$, and the solution to (29) was $P^*(1) = 4.508$, $P^*(2) = 4.476$, $P^*(3) = 4.485$. The unique optimal flow controller was then

$$\mu^{1*}(q, i) = \begin{cases} -0.0451\, q, & i = 1, \\ -2.238\, q, & i = 2, \\ -4.485\, q, & i = 3. \end{cases}$$

Typical system responses for this case are depicted in Figures 5–8, with each one again consisting of four plots, appropriately subcaptioned. Qualitative behaviors similar to those in the previous case are also observed here. Because of the increased weighting of $R(1)$, the flow control magnitude for type 1 traffic is reduced

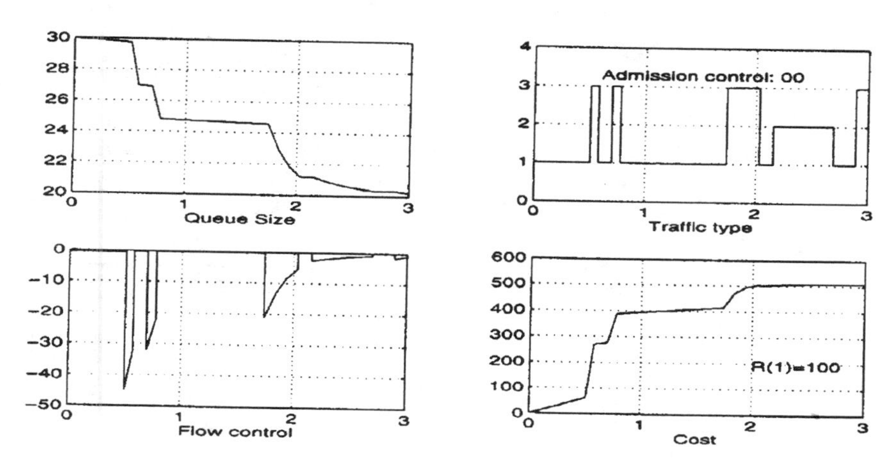


**Figure 5**: Case 2: Performance under $\mu^{1*}$ w.r.t. $\mu^{2*} = 00$.

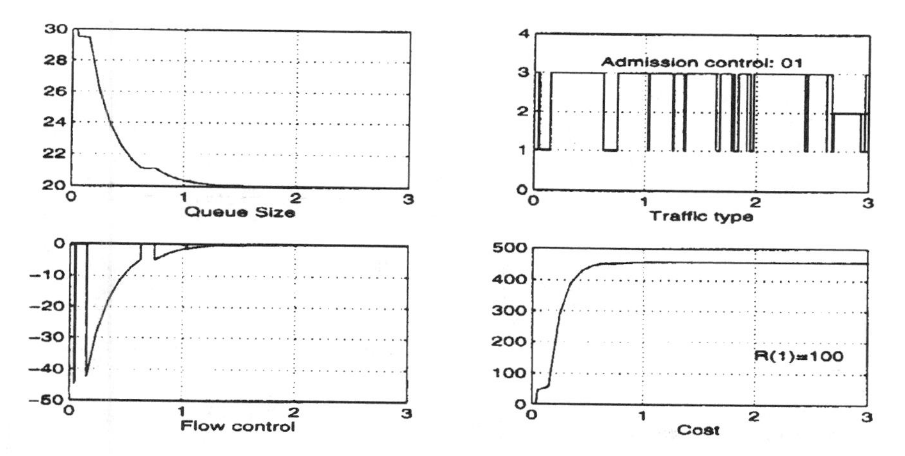


**Figure 6**: Case 2: Performance under $\mu^{1*}$ w.r.t. $\mu^{2*} = 01$.

significantly. The integral costs incurred under the four admission control laws are ranked differently from the previous case. Coincidentally, the cost incurred under the optimal admission control $\mu^{2*}(1) = 11$ is the smallest among the four particular sample paths simulated.

Finally, while holding $R(1)$ at 100, we increased $Q(3)$, from 20 to $Q(3) = 30$. With this change, the optimal controller 2 turned out to be $\mu^{2*}(1) = 10$, and the solution to (29) was $P^*(1) = 4.680$, $P^*(2) = 4.493$, $P^*(3) = 5.223$. The unique

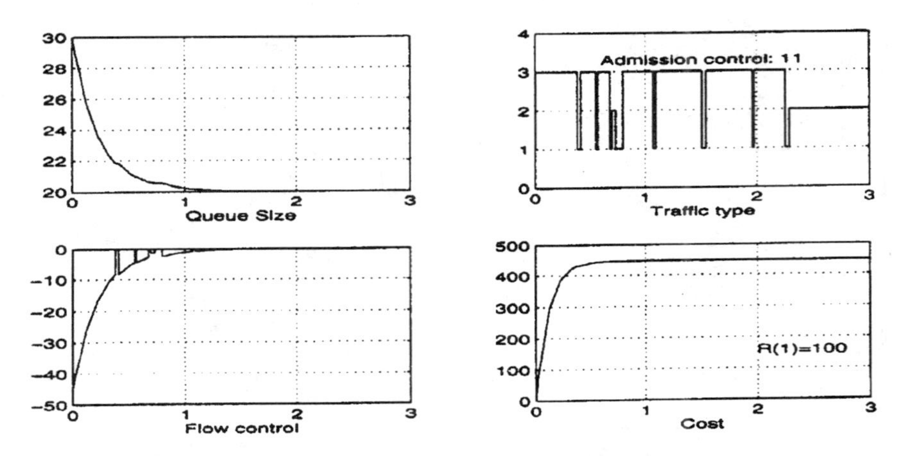


**Figure 7**: Case 2: Performance under $\mu^{1*}$ w.r.t. $\mu^{2*} = 11$.

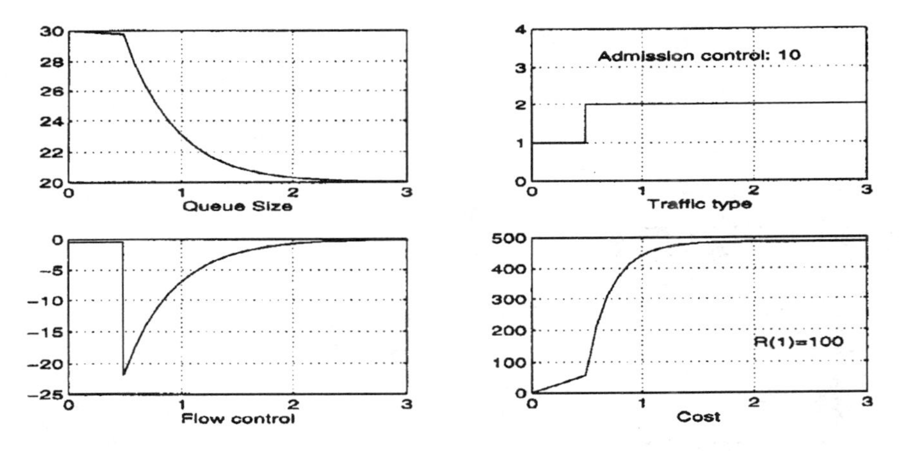


**Figure 8**: Case 2: Performance under $\mu^{1*}$ w.r.t. $\mu^{2*} = 10$.

optimal flow controller was then

$$\mu^{1*}(q,i) = \begin{cases} -0.0468\,q, & i = 1, \\ -2.246\,q, & i = 2, \\ -5.223\,q, & i = 3. \end{cases}$$

Typical system responses are again depicted in Figures 9–12, each one again consisting of four plots, appropriately subcaptioned. In all four simulations we have observed qualitative behavior similar to the previous two cases. We have seen a significant increase in the flow control magnitude for type 3 traffic; this is due to the increased weighting of $Q(3)$, which corresponds to a more stringent QoS for type 3 traffic than in the previous cases. Again coincidentally, the cost

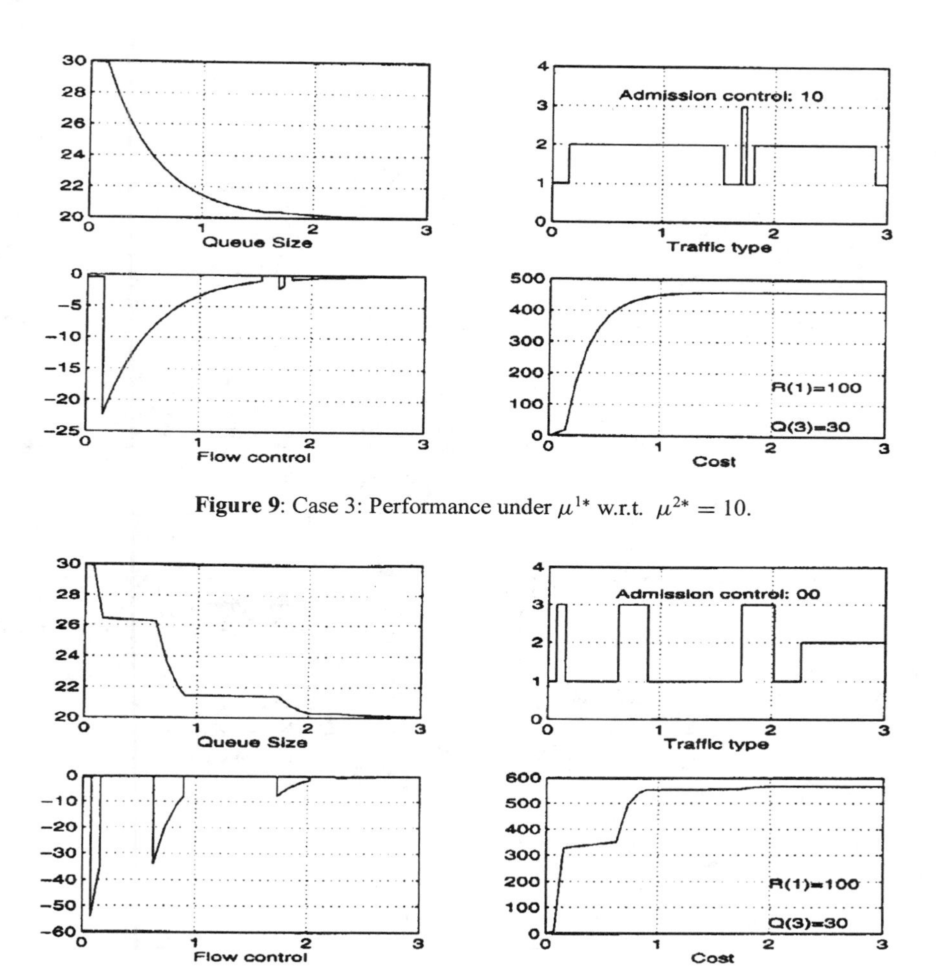


**Figure 9**: Case 3: Performance under $\mu^{1*}$ w.r.t. $\mu^{2*} = 10$.



**Figure 10**: Case 3: Performance under $\mu^{1*}$ w.r.t. $\mu^{2*} = 00$.

incurred under the optimal admission control $\mu^{2*}(1) = 10$ turned out to be the smallest among the four particular sample paths simulated.

## 7 Conclusions

Several extensions of the results of this paper can be envisioned, both for the general theoretical model and for the special telecommunication network application. For the former, one question that this paper has left unanswered is the structure of the solution to the HJB equation (9), or its infinite-horizon version (10), when the minimization over $u^2$ in (23) depends on the state $x$. There is also the issue of developing computational tools for the solutions of (9) and (10) when

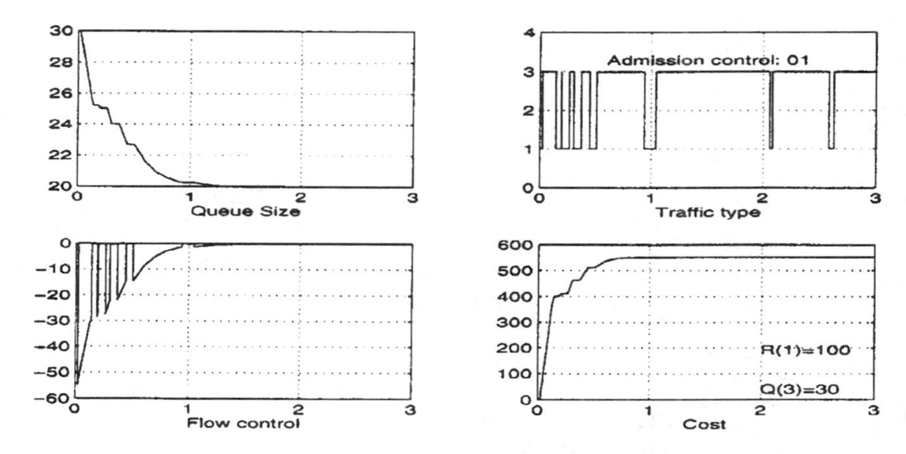


**Figure 11**: Case 3: Performance under $\mu^{1*}$ w.r.t. $\mu^{2*} = 01$.

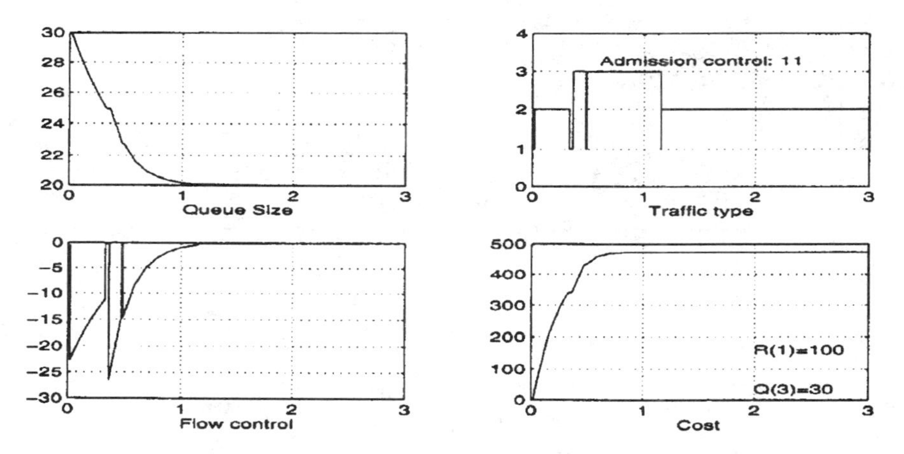


**Figure 12**: Case 3: Performance under $\mu^{1*}$ w.r.t. $\mu^{2*} = 11$.

general structural information is lacking. One extension of the general model of this paper would be the inclusion of an additional additive term in (1), which would represent an unknown disturbance—modeled either as a stochastic process with known statistics (such as a Brownian motion process) or as a completely unknown deterministic process (as in $H^\infty$ control [4]). In the latter case, one chooses as performance index the ratio of $J_\beta$ introduced here, to the energy of the unknown deterministic process, whose maximum over the unknown input will now have to be minimized with respect to the multistrategy $\mu$, in the team problem (with respect to $\mu_1$ in the game problem). For discussions on the solution to this problem for the special case when the transition rate matrix is not controlled, but under various types of different measurements for the controller (including the noise-perturbed measurement scheme not covered in this paper), see [7].

As far as the specific telecommunication network model of Section 6 is concerned, there is the potential to extend it to the more general case where the effective service rate $r_t$ is not known, but is measured in some additive noise. Furthermore, $r_t$ could be generated by a stochastic ARMA process, or by a deterministic linear model driven by an unknown deterministic process with finite energy. Such models have been considered before in [2], but for a single type traffic (i.e., with $s = 1$), and the extensions to the cases where there are multiple types of traffic (as in this paper) with fixed or controlled transition rates remain today as interesting but challenging research topics to pursue.

## Acknowledgments

The research of the second author was supported by Grants NSF ANI 98–13710 and NSF INT 98–04950 from the National Science Foundation.

The work of the third author was carried out at the ECE Department of the University of California at Santa Barbara, Santa Barbara, CA, 93106.

## REFERENCES

[1] Altman, E., F. Baccelli, and J.-C. Bolot. Discrete-Time Analysis of Adaptive Rate Control Mechanisms. In: H. G. Perros and Y. Viniotis, eds., *High Speed Networks and their Performance*, pp. 121–140. North-Holland, Amsterdam, 1994.

[2] Altman, E. and T. Başar. Optimal Rate Control for High Speed Telecommunication Networks. *Proceedings of the 34th IEEE Conference on Decision and Control*, pp. 1389–1394. New Orleans, LA, December 13–15, 1995.

[3] Başar, T. Minimax Control of Switching Systems under Sampling. *Systems & Control Letters*, **25**(5), 315–325, August 1995.

[4] Başar, T. and P. Bernhard. *$H^\infty$-Optimal Control and Related Minimax Design Problems: A Dynamic Game Approach*, 2nd ed. Birkhäuser, Boston, 1995

[5] Fleming, W. H. and H. M. Soner. *Controlled Markov Processes and Viscosity Solutions*, vol. 25. Springer-Verlag, New York, 1993.

[6] Ji, Y. and H. J. Chizeck. Controllability, Stabilizability, and Continuous-Time Markovian Jump Linear Quadratic Control. *IEEE Transactions on Automatic Control*, **AC-35**, 777–788, July 1990.

[7] Pan, Z. and T. Başar. $H^\infty$-Control of Markovian Jump Systems and Solutions to Associated Piecewise-Deterministic Differential Games. In: *Annals of Dynamic Games* (G. J. Olsder, ed.). vol. 3, pp. 61–94, Birkhäuser, Boston, 1995.

[8] Krassovskii, N. N. and E. A. Lidskii. Analytical Design of Controllers in Systems with Random Attributes I, II, III. *Automation Remote Control*, **22**, 1021–1025, 1141–1146, 1289–1294, 1961.

[9] Lidskii, E. A. Optimal Control of Systems with Random Properties. *Applied Mathematics and Mechanics*, **27**, 33–45, 1963.

[10] Florentin, J. J. Optimal Control of Continuous-Time Markov, Stochastic Systems. *Journal of Electronic Control*, **10**, 1961.

[11] Kushner, H. J. On the Stochastic Maximum Principle: Fixed Time of Control. *Journal of Mathematical Analysis and Applications*, **11**, 78–92, 1965.

[12] Rishel, R. Dynamic Programming and Minimum Principles for Systems with Jump Markov Disturbances. *SIAM Journal on Control and Optimization*, **13**, 338–371, February 1975.

[13] Sworder, D. D. Feedback Control of a Class of Linear Systems with Jump Parameters. *IEEE Transactions on Automatic Control*, **14**, 9–14, 1969.

[14] Wonham, W. M. Random Differential Equations in Control Theory. In: *Probabilistic Methods in Applied Mathematics* (A. T. Bharucha-Reid, ed.), pp. 191–199. Academic Press, New York, 1970.

[15] Mariton, M. On Controllability of Linear Systems with Stochastic Jump Parameters. *IEEE Transactions on Automatic Control*, **AC-31**, 680–683, 1986.

[16] Başar, T. and A. Haurie. Feedback Equilibria in Differential Games with Structural and Modal Uncertainties. *Advances in Large Scale Systems* (J. B. Cruz, Jr., ed.), vol. 1, pp. 163–201. JAI Press Inc., Connecticut, May 1984.

[17] de Souza, C. E. and M. Fragoso. $H^\infty$ Control of Linear Systems with Markovian Jumping Parameters. *Control Theory and Technology*, **9**, no. 2, 457–466, 1993.

[18] Pan, Z. and T. Başar. $H^\infty$ Control of Large Scale Jump Linear Systems via Averaging and Aggregation. *Proceedings of the 34th IEEE Conference on Decision and Control*, New Orleans, Louisiana, pp. 2574–2579, December 13–15, 1995.

[19] Xiao, M. and T. Başar. Optimal control of piecewise deterministic nonlinear systems with controlled transitions: viscosity solutions, their existence and uniqueness. *Proceedings of the 38th IEEE Conference on Decision and Control*, Phoenix, Arizona, December 7–10, 1999.

# A Game Variant of the Stopping Problem on Jump Processes with a Monotone Rule

Jun-ichi Nakagami, Masami Kurano, and Masami Yasuda
Department of Mathematics and Informatics
Chiba University
Chiba, Japan

**Abstract**

A continuous-time version of the multivariate stopping problem is considered. Associated with vector-valued jump stochastic processes, stopping problems with a monotone logical rule are defined under the notion of the Nash equilibrium point. The existence of an equilibrium strategy and its characterization by integral equations are obtained. Illustrative examples are provided.

## 1 Introduction

In social life or in business, group decision making is often alleviated by taking into account each individual's opinion in the whole group. How can we impose each opinion on the group decision? As one abstraction of such a situation, we shall try to propose a multivalued stopping game by introducing a monotone logical function to sum up each individual's opinion. The discrete-time case has already been discussed [4], [10]. Here we consider the continuous-time case, which is formulated as a multiobjective extension of Karlin's model [3] and as a rule's extension of Sakaguchi's model [7]. As a related result, Presman and Sonin [6] have obtained the multiperson best choice problem on the Poisson stream but their rule of a decision to stop is different from ours. Szajowski and Yasuda [8] treat the case when the process is a Markov Chain.

The situation of our problem is as follows. A group of $p$ players observes a $p$-dimensional stochastic process. Each player can decide to stop or to continue the process at any time when the $p$-dimensional successive offers will have happened, and the individual declarations are summed to make the group decision for the process by using a monotone logical rule. When the process is stopped by the group of $p$ players, components of the stochastic process are given to each player as a reward, so that the player can make his expected gain as large as possible.

First, we introduce some definitions and notations to formulate our stopping problem in Section 2. Then, by preparing several lemmas, we show the existence of an equilibrium stopping strategy and obtain its characterization by an integral equation in Section 3. In Section 4, examples of the underlying model are given.

## 2 Formulation

We consider a $p$-dimensional vector valued stochastic process $\{X_t; t \geq 0\}$ with $i$th component $X_t^i$, adapted to $\mathcal{F}_t$ on a probability space $(\Omega, \mathcal{F}, P)$ where $\mathcal{F}_t$ is the $\sigma$-algebra generated by $\{X_s; 0 \leq s \leq t\}$. Let us assume that the process $\{X_t; t \geq 0\}$ is an independent jump process (see, e.g., Feller [2]), that is, there are two independent stochastic sequences $Z_n = (Z_n^1, Z_n^2, \ldots, Z_n^p)$ and $\tau_n, n = 0, 1, 2, \ldots$, which satisfy $X_t = Z_n$ if $\tau_n \leq t < \tau_{n+1}$ for any $t, t \geq 0$, under the following assumption:

**Assumption 2.1.**

(a) $p$-Dimensional random vectors $Z_n = (Z_n^1, Z_n^2, \ldots, Z_n^p), n = 0, 1, 2, \ldots$, are i.i.d. with a common distribution $F$ on $R^p$, where $R = (-\infty, \infty)$.

(b) $\tau_0 = 0$ a.s. and $\tau_n - \tau_{n-1}, n = 1, 2, \ldots$ are i.i.d. with a common distribution $G$ on $R_+$, where $R_+ = [0, \infty)$ and $G(0) = 0$.

(c) $\int_{R^p} |z| \, F(dz) < \infty$ and $\mu_G = \int_{R_+} t \, G(dt) < \infty$, where $|\cdot|$ is a norm on $R^p$.

In order to denote the declaration for each player $i (i = 1, \ldots, p)$, when the process is $\{X_t; t \geq 0\}$, let $\sigma^i(t, x)$ be a $\{0, 1\}$-valued Borel measurable function on $R_+ \times R^p$ with $\sigma^i(0, x) = 1$. We call $\sigma^i = \sigma^i(\cdot, \cdot)$ an *individual strategy* for player $i$, and $\sigma = (\sigma^1, \ldots, \sigma^p)$ a *strategy*. The individual strategy $\sigma^i(t, x)$ may be interpreted as follows; when the amount $x$ of the offer has happened and the time interval remaining until termination is $t$, $\sigma^i(t, x) = 1(0)$ means player $i$ declares to stop (continue). In particular, $\sigma^i(0, x) = 1$ means that any player $i$ must declare to stop when the time remaining until termination is 0.

The individual declarations are summed up by a logical rule. A logical rule is a map $\pi : \{0, 1\}^p \to \{0, 1\}$ and is called *monotone* if $\pi(1, \ldots, 1) = 1$ and $\pi(\sigma^1, \ldots, \sigma^p) \leq \pi(\tilde{\sigma}^1, \ldots, \tilde{\sigma}^p)$ for $\sigma^i \leq \tilde{\sigma}^i$ $(1 \leq i \leq p)$. A monotone logical rule includes a wide variety in a choice system such as a unanimity rule, an equal/unequal majority rule, and a hierarchical rule, some of which are given in the last section. For example, if no less than $r$ $(\leq p)$ members in a group of $p$ players declare to stop, the group decision is to stop the process (equal majority rule). That is, $\pi(\sigma^1, \ldots, \sigma^p) = 1(0)$ if $\sum_{i=1}^p \sigma^i \geq (<) r$. Refer also to our previous papers [4] and [10].

For a strategy $\sigma$, a monotone logical rule $\pi$ and a planning horizon $T$, a *stopping time* $t(T, \sigma, \pi)$ for the group of $p$ players is defined by

$$t(T, \sigma, \pi) = \min\{\tau(\sigma, \pi), T\},$$

where $\tau(\sigma, \pi) =$ the first $\tau_k$ such that $\pi(\sigma(T - \tau_k, X_{\tau_k})) = 1$ for $k \geq 0$. Note that $t(T, \sigma, \pi)$ means the first time that the declaration $\sigma^i$ of each player $i$ is summed up for the group of $p$ players, to stop the process by the rule $\pi$. Since the monotone logical rule $\pi$ is fixed, $\pi$ is suppressed in $t(T, \sigma, \pi)$ hereafter.

An *expected reward* of player $i$ for a strategy $\sigma$ is defined by

$$u^i(T,\sigma) = E[X^i_{t(T,\sigma)}], \qquad T \geq 0.$$

Since the problem is fundamentally formulated as a noncooperative game, a notion of Nash equilibrium point (see Nash [5] and Vorobév [9] ) can be utilized. A strategy ${}^*\sigma = ({}^*\sigma^1, \ldots, {}^*\sigma^p)$ is *equilibrium* if, for each $i$,

$$u^i(T, {}^*\sigma) \geq u^i(T, {}^*\sigma^{-i}\|_{\sigma^i})$$

for any individual strategy $\sigma^i$ and any $T \geq 0$, where

$${}^*\sigma^{-i}\|_{\sigma^i} = ({}^*\sigma^1, \ldots, {}^*\sigma^{i-1}, \sigma^i, {}^*\sigma^{i+1}, \ldots, {}^*\sigma^p).$$

In this paper, we will find an equilibrium strategy ${}^*\sigma$ and the corresponding stopping time $t(T, {}^*\sigma, \pi)$ given a monotone rule $\pi$.

## 3 Lemmas and Theorems

In this section, the existence of an equilibrium strategy and its characterization are obtained. First, we will derive the integral equation of $u(T,\sigma) = (u^1(T,\sigma), \ldots, u^p(T,\sigma))$, $T \geq 0$ for given a strategy $\sigma$. Let $\mathcal{G}_n$ be the $\sigma$-algebra generated by $(Z_k, \tau_k)$, $k = 0, 1, \ldots, n-1$ and $\tau_n$ for each $n$.

**Lemma 3.1.**

$$E[X^i_{t(T,\sigma)} I_{\{t(T,\sigma)\geq\tau_n\}} | \mathcal{G}_n] = u^i(T-\tau_n, \sigma) I_{\{t(T,\sigma)\geq\tau_n\}} \qquad a.e.,$$

*where $I_A$ is the indicator for a set $A$.*

**Proof.** By Assumption 2.1, it holds that

$$t(T,\sigma) = t(T-\tau_n,\sigma) + \tau_n \qquad \text{on} \qquad \{t(T,\sigma) \geq \tau_n\} \tag{1}$$

for some $n$ and that

$$E[X^i_{t(T-\tau_n,\sigma)+\tau_n} | \mathcal{G}_n] = u^i(T-\tau_n, \sigma) \quad \text{a.e.} \tag{2}$$

So, the proof is completed by noting $\{t(T,\sigma) \geq \tau_n\} \in \mathcal{G}_n$. □

Let $Z = (Z^1, \ldots, Z^p)$ be a $p$-dimensional random variable whose distribution is $F$ and let $\mathcal{F}$ be the $\sigma$-algebra generated by $Z$. For any set $A \in \mathcal{F}$ and any $\alpha \in R$, define the operators $L^i$, $i = 1, \ldots, p$, by

$$L^i(A;\alpha) = E[Z^i I_A] + \alpha P(A^c).$$

**Lemma 3.2.** *For each $i (i = 1, \ldots, p)$, $u^i(T) = u^i(T,\sigma)$ satisfies the following integral equation:*

$$u^i(T) = L^i(\{\pi(\sigma_{(T,Z)}) = 1\}; G \circ u^i(T)), \tag{3}$$

*where $\sigma_{(T,Z)} = (\sigma^1_{(T,Z)}, \ldots, \sigma^p_{(T,Z)})$ and $G \circ u^i(T) = \int_0^T u^i(T-s)G(ds)$.*

**Proof.** By Assumption 2.1, we have

$$u^i(T) = E\left[Z_0^i I_{\{t(T,\sigma)=0\}}\right] + E\left[X^i_{t(T,\sigma)} I_{\{t(T,\sigma)\geq\tau_1\}}\right]. \quad (4)$$

From Lemma 3.1, it holds that

$$E[X^i_{t(T,\sigma)} I_{\{t(T,\sigma)\geq\tau_1\}}|\mathcal{G}_1] = u^i(T-\tau_1) I_{\{t(T,\sigma)\geq\tau_1\}} \quad \text{a.e.}$$

Thus, noting $\{t(T,\sigma) = 0\} = \{\pi(\sigma_{(T,Z_0)}) = 1\}$ and $\{t(T,\sigma) \geq \tau_1\} = \{\pi(\sigma_{(T,Z_0)}) = 1\}^c$, (3) follows from (4), replacing $Z_0$ by $Z$. □

To show the existence of an equilibrium strategy, we need several further lemmas.

Let $S$ be the set of all $\{0,1\}$-valued Borel measurable functions on $R_+ \times R^p$. For any number $\alpha \in R$ and $i$ $(i = 1, \ldots, p)$, define $\sigma^i[\alpha] \in S$ by $\sigma^i[\alpha] = 1$ if $Z^i \geq \alpha$, $= 0$ otherwise, which is called an individual strategy of a control-limit-type.

For any $(\sigma^1, \ldots, \sigma^p) \in S^p$, let us denote $\pi(\sigma) = \pi(\sigma^1_{(T,Z)}, \ldots, \sigma^p_{(T,Z)})$ for simplicity.

**Lemma 3.3.** *For any $\alpha \in R$ and $(\sigma^1, \ldots, \sigma^p) \in S^p$,*

$$L^i(\{\pi(\sigma) = 1\}; \alpha) \leq L^i(\{\pi(\sigma^{-i}\|_{\sigma^i[\alpha]}) = 1\}; \alpha).$$

**Proof.** Since $\pi$ is monotone, we have $\pi(\sigma^1, \ldots, \sigma^{i-1}, 1, \sigma^{i+1}, \ldots, \sigma^p) \geq \pi(\sigma^1, \ldots, \sigma^p) \geq \pi(\sigma^1, \ldots, \sigma^{i-1}, 0, \sigma^{i+1}, \ldots, \sigma^p)$ for all $\sigma \in S^p$. Thus, from the definition of $\sigma^{-i}\|_{\sigma^i[\alpha]}$, it follows that

$$\{Z^i - \alpha \geq 0, \pi(\sigma^{-i}\|_{\sigma^i[\alpha]}) = 1\} \supset \{Z^i - \alpha \geq 0, \pi(\sigma) = 1\}$$

and

$$\{Z^i - \alpha < 0, \pi(\sigma^{-i}\|_{\sigma^i[\alpha]}) = 1\} \subset \{Z^i - \alpha < 0, \pi(\sigma) = 1\},$$

which implies

$$E[(Z^i - \alpha) I_{\{\pi(\alpha)=1\}}] \leq E[(Z^i - \alpha) I_{\{\pi(\sigma^{-i}\|_{\sigma^i[\alpha]})=1\}}].$$

So, the proof is completed by noting $L^i(A; \alpha) = E([Z^i - \alpha) I_A] + \alpha$ for all $A \in \mathcal{F}$. □

**Lemma 3.4.** *For any $\alpha, \beta \in R$, with $\alpha \geq \beta$ and $(\sigma^1, \ldots, \sigma^p) \in S^p$,*

$$L^i(\{\pi(\sigma^{-i}\|_{\sigma^i[\alpha]}) = 1\}; \alpha) \geq L^i(\{\pi(\sigma^{-i}\|_{\sigma^i[\beta]}) = 1\}; \beta). \quad (5)$$

**Proof.** By Lemma 3.3, it holds that

$$L^i(\{\pi(\sigma^{-i}\|_{\sigma^i[\alpha]}) = 1\}; \alpha) \geq L^i(\{\pi(\sigma^{-i}\|_{\sigma^i[\beta]}) = 1\}; \alpha). \quad (6)$$

Since $\alpha \geq \beta$, then

$$L^i(\{\pi(\sigma^{-i}\|_{\sigma^i[\beta]}) = 1\}; \alpha) \geq L^i(\{\pi(\sigma^{-i}\|_{\sigma^i[\beta]}) = 1\}; \beta),$$

so that (5) follows from (6). □

**Lemma 3.5.** *For any fixed i and any strategies $\sigma = (\sigma^1, \ldots, \sigma^p) \in S^p$, let us consider the following integral equation with respect to $v(T) := v^i(T)$, for simplicity:*

$$v(T) = L^i(\{\pi(\sigma^{-i}\|_{\sigma^i[G\circ v(T)]}) = 1\}; G \circ v(T)) \tag{7}$$

*for $T \geq 0$. Then, we have:*

(i) *The solution $v(T)$ exists uniquely in $L^1([0,\infty), dG)$; and*
(ii) *$v(T) \geq u^i(T;\sigma)$ for $T \geq 0$.*

**Proof.** First we shall show the uniqueness of the solution of (7). Let $\alpha = G \circ v(T)$ and $\alpha' = G \circ v'(T)$, where $v(T)$ and $v'(T)$ are two solutions of (7) in $L^1([0,\infty), dG)$. We generally assume $\alpha \geq \alpha'$. Then, since

$$\pi(\sigma^{-i}\|_{\sigma^i[\alpha]}) = \pi(\sigma^{-i}\|_{\sigma^i[\alpha']}) \quad \text{on} \quad \{Z^i \leq \alpha'\} \cup \{Z^i > \alpha\},$$

we have

$$\begin{aligned}
&E\left[Z^i I_{\{\pi(\sigma^{-i}\|\sigma^i[\alpha])=1\}}\right] - E\left[Z^i I_{\{\pi(\sigma^{-i}\|\sigma^i[\alpha'])=1\}}\right] \\
&= E\left[Z^i I_{\{\alpha'<Z^i\leq\alpha,\pi(\sigma^{-i}\|\sigma^i[\alpha])=1\}}\right] - E\left[Z^i I_{\{\alpha'<Z^i\leq\alpha,\pi(\sigma^{-i}\|\sigma^i[\alpha'])=1\}}\right] \\
&\leq \alpha\, P\{\alpha' < Z^i \leq \alpha, \pi(\sigma^{-i}\|_{\sigma^i[\alpha]}) = 1\} - \alpha' P\{\alpha' < Z^i \leq \alpha, \pi(\sigma^{-i}\|_{\sigma^i[\alpha']}) = 1\} \\
&\leq \alpha\, P\{\pi(\sigma^{-i}\|_{\sigma^i[\alpha]}) = 1\} - \alpha' P\{\pi(\sigma^{-i}\|_{\sigma^i[\alpha']}) = 1\}.
\end{aligned}$$

It follows from (7) that $v(T) - v'(T) \leq \alpha - \alpha'$. Thus, we have $0 \leq v(T) - v'(T) \leq G \circ v(T) - G \circ v'(T)$ from Lemma 4.4, which implies

$$|v(T) - v'(T)| \leq \int_0^T |v(T-s) - v'(T-s)|\, G(ds) \qquad \text{for all} \quad T \geq 0.$$

By the well-known Gronwall–Bellman's theorem (see, e.g., Bellman [1]), we obtain the result

$$v(T) = v'(T) \qquad \text{for all} \quad T \geq 0$$

in $L^1([0,\infty), dG)$.

Next, we shall show the existence of the solution of (7). For any strategy $\sigma$ it holds from Lemma 3.2 that

$$u^i(T;\sigma) = L^i(\{\pi(\sigma_{(T,Z)}) = 1\}; G \circ u^i(T;\sigma)). \tag{8}$$

Now putting $\alpha_1^i = G \circ u^i(T,\sigma)$ and $\sigma_1^i = \sigma^i[\alpha_1^i]$, we define

$$u_1^i(T) = L^i(\{\pi(\sigma^{-i}\|_{\sigma_1^i}) = 1\}; \alpha_1^i).$$

Then we observe from Lemma 3.3 that

$$u_1^i(T) \geq u_1^i(T;\sigma). \tag{9}$$

If we define, recursively, for each $n \geq 2$,

$$u_n^i(T) = L^i(\{\pi(\sigma^{-i}\|_{\sigma_n^i}) = 1\}; \alpha_n^i), \tag{10}$$

$$\alpha_n^i = G \circ u_{n-1}^i(T) \quad \text{and} \quad \sigma_n^i = \sigma^i[\alpha_n^i],$$

we see that

$$u_n^i(T) \geq u_{n-1}^i(T) \tag{11}$$

by applying Lemma 3.4. Hence by the monotone convergence theorem, when $n \to \infty$ in (10), it holds that the limit $u^i(T) := \lim_{n\to\infty} u_n^i(T)$ equals a solution $v(T) := v^i(T)$ of (7) in $L^1([0,\infty), dG)$. Clearly, (ii) holds from (9) and (11). □

**Condition 3.1.** There are $v^i(T) \in L^1([0,\infty), dG)$, $i = 1, \ldots, p$, which satisfy the following $p$ simultaneous integral equations:

$$v^i(T) = L^i(\{\pi(^*\sigma) = 1\}; G \circ v^i(T)), \qquad i = 1, \ldots, p, \quad T \geq 0, \tag{12}$$

where $^*\sigma = (^*\sigma^1, \ldots, {}^*\sigma^p)$ and $^*\sigma^i = {}^*\sigma^i_{(T,Z)} = \sigma^i[G \circ v^i(T)]$.

We are now ready to prove the main theorem.

**Theorem 3.1.** *Under Condition 3.1, it holds that:*

(i) $u^i(T, {}^*\sigma) = v^i(T)$, $i = 1, \ldots, p$ *for* $T \geq 0$.
(ii) $^*\sigma$ *is an equilibrium strategy.*

**Proof.** (i) By Lemma 3.2, $u^i(T; {}^*\sigma)$ satisfies (12). Thus, from (i) of Lemma 3.5, the uniqueness of the solution of (12) implies (i) of Theorem 3.1. Also, (ii) follows from (ii) of Lemma 3.5. □

**Remark 3.1.** Theorem 3.1 says that under Condition 3.1 there exists an equilibrium strategy of control-limit-type, whose threshold for each player $i$ is $\alpha^i = G \circ v^i(t)$ when the remaining time interval until termination is $t$.

**Remark 3.2.** In most cases the direct verification of Condition 3.1 seems to be difficult. However, if $G(T) < 1$ for all $T > 0$ as the case that $G(ds) = \lambda e^{-\lambda s} ds$, $\lambda > 0$ (an exponential distribution), (12) has a unique solution $v(T) = (v^1(T), \ldots, v^p(T))$ in $L^\infty[0,\infty)^p$, where $L^\infty[0,\infty)$ denotes the set of all bounded Borel measurable functions on $[0,\infty)$. This result is used as the example in the next section. In fact, we define the map $U : L^\infty[0,\infty)^p \to L^\infty[0,\infty)^p$ by

$$Uu(T) = \left(L^i(\pi\{(\sigma_{u(T)})) = 1\}; G \circ u^i(T)\right) \quad i = 1, \ldots, P$$

where $u(T) \in L^\infty[0,\infty)^p$ and $\sigma_{u(T)} = (\sigma^1[G \circ u^1(T)], \ldots, \sigma^p[G \circ u^p(T)])$. Then, by the same way as in the proof of Lemma 3.5, we get

$$\|Uu - Uu'\|_T \leq G(T)\|u - u'\|_T \quad \text{for any} \quad u, u' \in L^\infty[0,\infty)^p \quad \text{and} \quad T \geq 0,$$

where $\|u\|_T = \max_{1\leq i\leq p} \sup_{0\leq t\leq T} |u^i(t)|$ for $u(T) \in L^\infty[0,\infty)^p$.

The above discussion shows that $U$ is a contraction w.r.t. the norm $\|\cdot\|_T$, so that $U$ has a unique fixed point $v_T \in L^\infty[0,\infty)^p$. Since $T$ is arbitrary, $v := \lim_{T\to\infty} v_T$ satisfies (12). Also, the uniqueness of the solution of (12) follows from Lemma 3.5.

**Remark 3.3.** If the observation cost is incurred at each arrival time of offers, a $p$-dimensional random vector (net profit) $Z_n = (Z_n^1, \dots, Z_n^p)$ is defined by

$$Z_n = Y_n - (n+1)c,$$

where $Y_n = (Y_n^1, \dots, Y_n^p)$ are i.i.d. with a common distribution $F$ on $R^p$ and $c = (c^1, \dots, c^p)$ is a constant observation cost.

The corresponding $p$-simultaneous integral equations for (12) reduces, for $T \geq 0$,

$$v^i(T) + c^i = L^i\left(\{\pi(^*\sigma) = 1\}; G \circ v^i(T)\right), \qquad i = 1, \dots, p. \tag{13}$$

Then we can prove in identical fashion that for a solution $v^i(T) \in L^1([0,\infty), dG)$, $i = 1, \dots, p$, of (13), the same theorem as Theorem 3.1 holds.

**Remark 3.4.** When $G$ is a degenerate distribution with total mass at unity, the integral equation (13) becomes

$$v^i(T) + c^i = L^i(\{\pi(^*\sigma) = 1\}; v^i(T-1)), \qquad i = 1, \dots, p, \tag{14}$$

where $^*\sigma = (^*\sigma^1, \dots, {}^*\sigma^p)$ and $^*\sigma^i_{(T,Z)} = {}^*\sigma^i[v^i(T-1)]$. Thus, if we define a sequence $\{v_n^i\}_{n=0,1,\dots}$ for $i = 1, \dots, p$, by

$$\begin{aligned} v_0^i &= E[Z^i], \\ v_n^i &= L^i(\{\pi(^*\sigma) = 1\}; v_{n-1}^i), \; n = 1, 2, \dots, \end{aligned} \tag{15}$$

recursively, then we observe that

$$v^i(T) = v_n^i \qquad \text{if} \quad n \leq T < n+1 \quad \text{for some } n.$$

Assuming that $c^i > 0$ for all $i$, it follows that $v^i := \lim_{T\to\infty} v^i(T)$ exists. Now, as $T \to \infty$ in (14), we obtain

$$\begin{aligned} &E[(Z^i - v^i)^+ P(\{\pi(^*\sigma^{-i}\|_1) = 1\} | Z^i)] \\ &\quad - E[(Z^i - v^i)^- P(\{\pi(^*\sigma^{-i}\|_0) = 1\} | Z^i)] = c^i, \end{aligned} \tag{16}$$

where $^*\sigma^{-i}\|_k = (^*\sigma^1, \dots, {}^*\sigma^{i-1}, k, {}^*\sigma^{i+1}, \dots, {}^*\sigma^p)$ for each $k = 0, 1$. We note that (16) corresponds to (3.1) of [10].

## 4 Examples

In this section, we provide some examples involving the two-person stopping problem ($p = 2$) with the unanimity and simple majority rule as the typical ones of a monotone logical rule.

**Example 4.1** (The Unanimity Rule). Let us consider a unanimity rule, that is, decide to stop only if both of players' opinion is stop. Let us define $\pi(\sigma^1, \sigma^2), \sigma^1, \sigma^2 \in \{0, 1\}$ by $\pi(\sigma^1, \sigma^2) = 1$ if $\sigma^1 = \sigma^2 = 1, = 0$ otherwise.

Then the integral equation (12) of $v^i(T)$, $T \geq 0$ becomes

$$v^i(T) = \overline{v}_T^i + \iint_D (z^i - \overline{v}_T^i) F(dz^1, dz^2), \quad i = 1, 2 \tag{17}$$

where $D = \{(z^1, z^2); z^1 \geq \overline{v}_T^1 \text{ and } z^2 \geq \overline{v}_T^2\}$. The equilibrium strategy is of control-limit-type and the threshold of player $i$ is $\overline{v}_T^i = G \circ v^i(T) = \int_{(0,T]} v^i(T-s)\, G(ds)$, $i = 1, 2$.

If $Z^1$ and $Z^2$ are i.i.d. with a common distribution $F(z)$, then $v(T) := v^1(T) = v^2(T)$ for all $T \geq 0$, (17) becomes

$$v(T) = \overline{v}_T + (1 - F(\overline{v}_T)) \int_{\overline{v}_T}^{\infty} (1 - F(z))\, dz, \tag{18}$$

where $\overline{v}_T = \int_{(0,T]} v(T-s)\, G(ds)$.

Now suppose $G(ds) = \lambda e^{-\lambda s}\, ds$, $\lambda > 0$, that is, the time interval between successive offers is exponentially distributed. Then since $d\overline{v}_T/dT = \lambda(v(T) - \overline{v}_T)$, we have the following differential equation from (18):

$$\frac{d\overline{v}_T}{dT} = \lambda(1 - F(\overline{v}_T)) \int_{\overline{v}_T}^{\infty} (1 - F(z))\, dz, \tag{19}$$

which corresponds to (10) of Sakaguchi [7].

**Example 4.2** (The Simple Majority Rule). A simple majority rule $\pi(\sigma^1, \sigma^2)$, $\sigma^1, \sigma^2 \in \{0, 1\}$ for $p = 2$ is defined by $\pi(\sigma^1, \sigma^2) = 1$ if $\sigma^1 + \sigma^2 \geq 1, = 0$ otherwise. If $Z^1$ and $Z^2$ are nonnegative and i.i.d. with a common distribution $F(z)$, then $v(T) := v^1(T) = v^2(T)$, and (13) becomes

$$v(T) = \overline{v}_T + \iint_D (z^1 - \overline{v}_T) F(dz^1) F(dz^2), \tag{20}$$

where $D = \{(z^1, z^2); z^1 \geq \overline{v}_T \text{ or } z^2 \geq \overline{v}_T\}$. The equilibrium strategy is of control-limit-type and the threshold of player $i$ is

$$\overline{v}_T = G \circ v(T) = \int_{(0,T]} v(T-s)\, G(ds).$$

Then, we have from Assumption 2.1 that $\mu_F = \int_R x F(dx) < \infty$,

$$v(T) = \mu_F - F(\overline{v}_T) \int_{[0,\overline{v}_T]} z\, F(dz) + \overline{v}_T \{F(\overline{v}_T)\}^2.$$

Since

$$\int_{[0,y]} z\, F(dz) = -y[1 - F(y)] + \int_0^y (1 - F(z))\, dz \quad \text{for all } y > 0,$$

we obtain that

$$v(T) = \mu_F + F(\overline{v}_T) \int_0^{\overline{v}_T} F(z)\, dz,$$

so that

$$\overline{v}_T = \int_{(0,T]} G(ds) \left\{ \mu_F + F(\overline{v}_{T-s}) \int_0^{\overline{v}_{T-s}} F(z)\,dz \right\}. \tag{21}$$

Next, suppose $G(ds) = \lambda e^{-\lambda s}\,ds$. Then, by elementary calculus, (21) becomes

$$\overline{v}_T \cdot \exp\{\lambda T\} = \lambda \int_0^T e^{\lambda s}\,ds \left\{ \mu_F + F(\overline{v}_s) \int_0^{\overline{v}_s} F(z)\,dz \right\}. \tag{22}$$

By taking the derivative of both sides of (22) with respect to $T$, we have the following differential equation:

$$\frac{d\overline{v}_T}{dT} = \lambda \left\{ \mu_F - \overline{v}_T + F(\overline{v}_T) \int_0^{\overline{v}_T} F(z)\,dz \right\}. \tag{23}$$

Thus, by rewriting (23),

$$d\overline{v} \Big/ \left\{ \mu_F - \overline{v} + F(\overline{v}) \int_0^{\overline{v}} F(z)dz \right\} = \lambda\,dt,$$

from which we obtain the inverse function of $\overline{v}_T$, $T(\overline{v})$, given by

$$T(\overline{v}) = \lambda^{-1} \int_0^{\overline{v}} M^{-1}(\xi)\,d\xi, \tag{24}$$

where $M(\xi) = \mu_F - \xi + F(\xi) \int_0^{\xi} F(z)\,dz$.

For a numerical example, supposing $F(z) = z,\ 0 \le z \le 1$, from (24) we get

$$T(\overline{v}) = 2\lambda^{-1} \int_0^{\overline{v}} 1/(\xi^3 - 2\xi + 1)\,d\xi,$$

so that

$$\lim_{T\to\infty} \overline{v}_T = (\sqrt{5}-1)/2 (\approx 0.6180) \qquad \text{and} \qquad \lim_{T\to\infty} v(T) = (\sqrt{5}-1)/2,$$

which are the threshold of the control-limit strategy and the expected reward for the infinite horizon problem (refer to Table 3.1 of [10] for the discrete time case).

## Acknowledgments

The authors wish to express their thanks to the referees for helpful comments and suggestions to improve this paper.

## REFERENCES

[1] Bellman, R. *Stability Theory of Differential Equations*. McGraw-Hill, New York, 1953.

[2] Feller, W. *An Introduction to Probability Theory and its Applications* II. Wiley, New York, 1966.

[3] Karlin, S. Stochastic Models and Optimal Policy for Selling an Asset, Chapter 9 in *Studies in Applied Probability and Management Sciences*. Stanford University Press, Stanford, CA, 1962.

[4] Kurano, M., M. Yasuda, and J. Nakagami. Multi-Variate Stopping Problem with a Majority Rule. *Journal of the Operations Research Society of Japan*, **23**, 205–223, 1980.

[5] Nash, J. Non-cooperative Game. *Annals of Mathematics*, **54**, 286–295, 1951.

[6] Presman, E. L. and I. M. Sonin. Equilibrium Points in a Game Related to the Best Choice Problem. *Theory of Probability and its Applications*, **20**, 770–781, 1975.

[7] Sakaguchi, M. When to Stop: Randomly Appearing Bivariate Target Values. *Journal of the Operational Research Society of Japan*, **21**, 45–57, 1978.

[8] Szajowski, K. and M. Yasuda. Voting Procedure on Stopping Games of Markov Chain. In: *Stochastic Modeling in Innovative Manufacturing* (A. H. Christer, S. Osaki, and L. C. Thomas, eds.). Lecture Note in Economics and Mathematical System 445, Springer-Verlag, New York, 68–80, 1997.

[9] Vorobév, N. N. *Game Theory*. Springer-Verlag, New York, 1977.

[10] Yasuda, M., J. Nakagami and M. Kurano. Multivariate Stopping Problem with a Monotone Rule, *Journal of the Operational Research Society of Japan*, **25**, 334–349, 1982.

# PART IV
# General Game Theoretic Developments

# Refinement of the Nash Solution for Games with Perfect Information

Leon A. Petrosjan
Faculty of Applied Mathematics
St. Petersburg State University
St. Petersburg, Russia

**Abstract**

The $n$-person finite games with perfect information are considered. It is well known that the subgame perfectness assumption does not lead to the unique Nash equilibria (NE). Thus the problem arises of how to choose a unique subgame perfect NE. It is proved that by introducing the so-called "preference vector" for each player a unique subgame (in the sense of payoffs) NE can be constructed. The application of the approach to the "lifeline" simple pursuit game with one pursuer and two evaders enables us to find a unique subgame (in the sense of payoffs) NE in this differential game also.

## 1 Introduction

An extensive form game is said to have perfect information if the following two conditions are satisfied: there are no simultaneous moves, and at each decision point it is known which choices have previously been made. In the games with perfect information the subgame perfect equilibria can be found by dynamic programming, i.e., by inductively working backward in the game tree. But even in this case one can find a wide class of Nash equilibria [1] with different payoffs for the players. We propose the refinement of the Nash equilibrium concept based upon the preferences of the players giving the possibility of determining the unique Nash equilibrium in the sense of the player's payoffs.

The approach is used for the calculation of the unique Nash equilibrium in the differential "lifeline game" of pursuit [2]–[4] with one pursuer ($P$) and two evaders ($E_1$, $E_2$).

## 2 Game with Perfect Information in Extensive Form $\Gamma$

For simplicity, suppose that there are no random moves in $\Gamma$. Let $N$ be the set of players, let $X_1, \ldots, X_n$ be the player partition in $\Gamma$, and let $X_{n+1}$ be the set of final positions.

For each $i \in N$ define player $i$'s preference vector $F_i = \{f_i(j)\}$, $j \in N$, so that the $n$-dimensional vector, composed from absolute values of $f_i(j)$,

$$\{|f_i(j)|\}, \qquad j \in N,$$

is a permutation of the integer set $1, \dots, n$.

The interpretation of the preference vector $f_i(j)$, $j \in N$, $i \in N$, is the following: if $f_i(j) = k > 0$, then player $j$ is favourite of player $i$ of the order $k$, if $f_i(j) = k < 0$, then player $j$ is hostile to player $i$ of the order $k$, and $f_i(i) = 1$.

Every preference vector $F_i$ defines the relaiton of the player $i$ to other players in the game, or "type" of the player (here we use the terminology first introduced in [8] because we find some analogy in the problem setting).

Suppose a subgame perfect equilibrium in $\Gamma$ [7] is found by dynamic programming. First one considers all minimal subgames of $\Gamma$ (one-choice subgames) and then truncates $\Gamma$ by assuming that in any such subgame an equilibrium will be played. This procedure will be repeated until there are no subgames left. In this way one meets all subgames and a classical proof of the Nash theorem assures us that in every subgame an equilibrium results. But it may happen that in one of the subgames $\Gamma_x$ of $\Gamma$ starting from the position $x \in X_i$, player $i$, making his choice, finds that the result of this choice, for him from the point of view of his own payoff in subgames starting after $x$, is irrelevant (he gets the same payoffs in all subgames directly following $\Gamma_x$). In this case, using his preference vector, he looks at player $j$ for which $|f_i(j)| = 2$, and if $f_i(j) > 0$ he chooses his alternative at $x$ to maximize player $j$'s payoff, and if $f_i(j) < 0$ he chooses his alternative at $x$ to minimimze the payoff of player $j$.

If the payoff of player $j$ in all subgames directly following $\Gamma_x$ is the same, then he looks at player $j$ for which $|f_i(j)| = 3$ and behaves in the same manner, and so on. The strategy which includes such behavior we call type strategy. The subgame perfect Nash equilibria found by dynamic programming using type strategies we shall call Nash equilibria in type strategies. Denote by $T = (T_1, \dots, T_n)$ and by $K_i(T)$, $i \in N$, the corresponding payoffs.

Now we give the precise definition of Nash equilibrium in type strategies by induction on length $l$ of the game tree. Under the length of the game we understand the number of arcs in the maximal path of the game tree. Denote by $x(k; y)$ the next position (node) in $\Gamma$ following $y$ if in $y$ the alternative $k$ (choice) is selected.

If $l = 1$, then $\Gamma$ has only one move at the initial position $x_0$ ending with terminal position, thus $x(k; x_0) \in X_{n+1}$, for all alternatives $k$ at $x_0$. Suppose $x_0 \in X_{i_1}$. The payoffs of the players $h_i\{x(k; x_0)\}$, $i = 1, \dots, n$, are defined in the terminal nodes $x(k; x_0) \in X_{n+1}$.

Let $F(i_1)$ be the preference vector of player $i_1$, introduce the sequence

$$i_1, i_2, \dots, i_n,$$

such that

$$|f_{i_1}(i_k)| = 1 + |f_{i_1}(i_{k-1})|, \qquad k = 2, \dots, n,$$

$(f_{i_1}(i_1) = 1)$.

Denote by

$$\varphi(j) = \text{sign } f_{i_1}(i_j), \qquad j = 1, \dots, n,$$

then

$$f_{i_1}(i_k) = k[\text{sign } f_{i_1}(i_k)].$$

Define the family of sets $A_{i_1}[\varphi(j)]$

$$A_{i_1}[\varphi(1)] = \arg\left[\max_k h_{i_1}\{x(k; x_0)\}\right],$$

$$A_{i_1}[\varphi(j)] = \begin{cases} \arg\left[\max_{k \in A_{i_1}[\varphi(j-1)]} h_{i_j}\{x(k; x_0)\}\right] & \text{if} \quad \varphi(j) > 0, \\ \arg\left[\max_{k \in A_{i_1}[\varphi(j-1)]} h_{i_j}\{x(k; x_0)\}\right] & \text{if} \quad \varphi(j) < 0, \end{cases}$$

$j = 2, \dots, n$. We have $A_{i_1}[\varphi(j)] \subset A_{i_1}[\varphi(j-1)]$, $i_1 \in N$.

Denote

$$A_{i_1}^* = \bigcap_{j=1}^{n} A_{i_1}[\varphi(j)].$$

In the one-move game $\Gamma$, with $x_0 \in X_{i_1}$, a type strategy $T_{i_1}$ chooses any alternative $k \in A_{i_1}^*$, thus

$$T_{i_1}(x_0) = k, \qquad k \in A_{i_1}^*. \tag{1}$$

**Lemma 2.1.** *For any* $k_1, k_2 \in A^*$ *we have*

$$h_j\{x(k_1; x_0)\} = h_j\{x(k_2; x_0)\}, \qquad j = 1, \dots, n. \tag{2}$$

**Proof.** Suppose (2) does not hold for some $j \in N$, then ther exists $\bar{j}$ such that

$$\begin{aligned} &h_{i_j}\{x(k_1; x_0)\} = h_{i_j}\{x(k_2; x_0)\}, \qquad i_j < i_{\bar{j}}, \\ &h_{i_{\bar{j}}}\{x(k_1; x_0)\} \neq h_{i_{\bar{j}}}\{x(k_2; x_0)\}, \end{aligned} \tag{3}$$

Since $k_1, k_2 \in A^*$

$$k_1, k_2 \in A[\varphi(\bar{j})] = \begin{cases} \arg\left[\max_{k \in A_{i_1}[\varphi(\bar{j}-1)]} h_{i_{\bar{j}}}\{x(k; x_0)\}\right] & \text{if } \varphi(\bar{j}) > 0, \\ \arg\left[\min_{k \in A_{i_1}[\varphi(\bar{j}-1)]} h_{i_{\bar{j}}}\{x(k; x_0)\}\right] & \text{if } \varphi(\bar{j}) < 0. \end{cases} \tag{4}$$

In (4) max(min) of a function $h_{i_{\bar{j}}}$ is defined on the same set $A_{i_1}\left[\varphi(\bar{j}-1)\right]$. Suppose $\varphi(\bar{j}) > 0$, then by the definition

$$\max_{k \in A_{i_1}[\varphi(\bar{j}-1)]} h_{i_{\bar{j}}}\{x(k; x_0)\} = h_{i_{\bar{j}}}\{x(k_1; x_0)\} = h_{i_{\bar{j}}}\{x(k_2; x_0)\},$$

which contradicts (3). In the same manner the case $\varphi(\bar{j}) < 0$ can be considered. Lemma 2.1 is proved. □

**Corollary 2.1.** *From Lemma* 2.1 *we get that in the case* $l = 1$ *for any two pairs* $T'_{i_1}$, $T''_{i_2}$ *of type strategies the corresponding payoff vectors coincide.*

Suppose the type strategies are defined for the games of length $l' < l$. Define them in the game $\Gamma$ of length $l$.

Suppose also that in $\Gamma$, $x_0 \in X_{i_1}$. All subgames $\Gamma_x$ starting from the nodes (positions) next following $x_0$ have length $< l$. The type strategies in $\Gamma_x$ are already defined by induction hypothesis. For each $x$, fix an $n$-tuple $T^x = (T_1^x, \dots, T_n^x)$ of type strategies in $\Gamma_x$. Let $v(x) = (v_1(x), \dots, v_n(x))$ be the corresponding payoff vector in $\Gamma_x$ when $T^x$ is played. If the node $x$ corresponds to alternative $k$ at $x_0$ we shall write

$$v(x) = v\{x(k; x_0)\}$$

and

$$v_i(x) = v_i\{x(k; x_0)\}, \qquad i = 1, \dots, n.$$

Define the family of sets $A_{i_1}[\varphi(i)]$

$$A_{i_1}[\varphi(1)] = \arg\left[\max_k v_{i_1}\{x(k; x_0)\}\right],$$

$$A_{i_1}[\varphi(j)] = \begin{cases} \arg\left[\max_{k \in A_{i_1}[\varphi(j-1)]} v_{i_j}\{x(k; x_0)\}\right] & \text{if} \quad \varphi(j) > 0, \\ \arg\left[\max_{k \in A_{i_1}[\varphi(j-1)]} v_{i_j}\{x(k; x_0)\}\right] & \text{if} \quad \varphi(j) < 0, \end{cases}$$

$j = 2, \dots, n$.

Denote

$$A_{i_1}^* = \bigcap_{j=1}^{n} A_{i_1}[\varphi(j)].$$

Define the type strategy in the game $\Gamma$

$$T_{i_1}(x_0) = k, \qquad k \in A_{i_1}^*.$$
$$T_{i_1}(y) = T_i^x(y), \qquad \text{for} \quad y \in X_{i_1}^x,$$

where $X^x = \{X_1^x, \dots, X_{i_1}^x, \dots, X_n^x\}$ is a player partition in the subgame $\Gamma_x$ starting from the node $x$ next following $x_0$:

$$T_i(y) = T_i^x(y), \qquad \text{for} \quad y \in X_i^x, \quad i \neq i_1.$$

The proof of following lemma repeats the proof of Lemma 2.1 if we replace $h_i\{x(k; x_0)\}$ by $v_i\{x(k; x_0)\}$.

**Lemma 2.2.** *For any $k_1, k_2 \in A^*$ we have*

$$v_j\{x(k_1; x_0)\} = v_j\{x(k_2; x_0)\}, \qquad j = 1, \dots, n. \tag{5}$$

**Theorem 2.1.** *If the players know their own type and the types of all other players, then every game $\Gamma$ possesses a subgame perfect Nash equilibria in type strategies; and for any two different Nash equilibria in type strategies, $T = (T_1, \dots, T_n)$ and $T' = (T_1', \dots, T_n')$, the payoffs of the players coincide, thus*

$$K_i(T_1, \dots, T_n) = K_i(T_1', \dots, T_n'), \qquad i \in N. \tag{6}$$

**Proof.** Consider two different $n$-tuples in type strategies $T = (T_1, \dots, T_n)$, $T' = (T'_1, \dots, T'_n)$. From the construction of $T(T')$ it follows that any $n$-tuple of type strategies forms a subgame perfect Nash equilibrium in $\Gamma$.

Suppose now that (6) does not hold for some $i_1 \in N$,

$$K_{i_1}(T_1, \dots, T_n) \neq K_{i_1}(T'_1, \dots, T'_n). \tag{7}$$

Consider two different cases:

1.$T_{i_1}(x_0) = T'_{i_1}(x_0) = k$; and
2.$T_{i_1}(x_0) \neq T'_{i_1}(x_0)$.

In case 1 after the first move, if the $n$-tuples $T, T'$ are played, the game proceeds in the same subgame $\Gamma_x$, where $x = x(k; x_0)$ after the choice $k$ at $x_0$ is made.

By induction hypothesis the payoffs in subgame $\Gamma_x$,

$$K_i^x(T_1^x, \dots, T_n^x) = K_i^x(T_1'^x, \dots, T_n'^x) = v_i(k; x_0), \tag{8}$$

$i = 1, \dots, n$, where $T_i^x$ $(T_i'^x)$ is a trace of $T_i$ $(T'_i)$ in the subgame $\Gamma_x$. But in case 1

$$\begin{aligned} K_i(T_1, \dots, T_n) &= K_i^x(T_1'^x, \dots, T_n'^x), \\ K_i(T'_1, \dots, T'_n) &= K_i^x(T_1'^x, \dots, T_n'^x), \end{aligned} \tag{9}$$

$i = 1, \dots, n$ and (8) and (9) prove the theorem. □

Suppose in case 2

$$T_{i_1}(x_0) = k_1, \qquad T'_{i_1}(x_0) = k_2, \qquad (k_1 \neq k_2).$$

Let $x_1 = x(k_1; x_0)$, $x_2 = x(k_2; x_0)$. Then by Lemma 2.2

$$\begin{aligned} K_i^{x_1}\left(T_1^{x_1}, \dots, T_n^{x_1}\right) &= v_i(k_1; x_0) \\ &= v_i(k_2; x_0) = K_i^{x_2}\left(T_1'^{x_2}, \dots, T_n'^{x_2}\right), \quad i = 1, \dots, n. \end{aligned} \tag{10}$$

But

$$\begin{aligned} K_i\left(T_1, \dots, T_n\right) &= K_i^{x_1}\left(T_1^{x_1}, \dots, T_n^{x_1}\right), \\ K_i\left(T'_1, \dots, T'_n\right) &= K_i^{x_2}\left(T_1'^{x_2}, \dots, T_n'^{x_2}\right), \end{aligned} \tag{11}$$

$i = 1, \dots, n$, and the theorem follows from (10) and (11).

The situation is more complex if the players do not know the types of the opponents or if they know only some probability distribution over the set of all possibles types, and this is common knowledge for all of them (compare with [5]).

**Example 2.1.** There are two players ($N = \{1, 2\}$) (see Figure 1). The elements of the set $X_1$ are denoted by circles and of the set $X_2$ by rectangles. The vertexes of the game tree are denoted by double indexes.

In positions 1.2 and 1.3 both choices of player 1 are irrelevant for him. At the same time if his preference $f_1(2) = 2$ (player 2 is favorite to 1), then he will choose alternative 2 in position 1.2 and alternative 1 in position 1.3. If $f_1(2) = -2$ (player 2 is hostile to 1) he will choose 1 in position 1.2 and 2 in position 1.3.

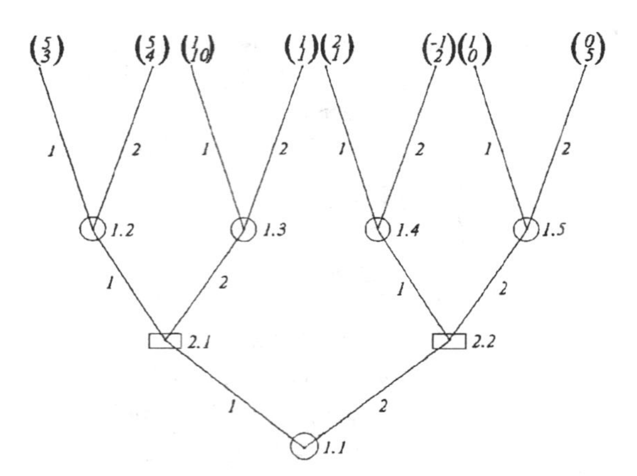


**Figure 1.**

In the case $F_1 = (1, 2)$ $(f_1(2) = 2)$ the unique subgame perfect Nash equilibrium in type strategies is $\{2, 2, 1, 1, 1), (2, 1)\}$ with payoffs $(2, 1)$, assuming that both players know the preference vector $F_1$. If this is not the case things are more complicated. Suppose player 2 thinks that player 1's preference vector is equal $F_1 = (1, -2)$ $(f_1(2) = -2)$ (being in reality $+2$) and player 1 knows the knowledge of player 2 about himself. Then using dynamic programming we shall get the strategy pair $\{(1, 2, 1, 1, 1), (1, 1)\}$ with payoffs $(5, 4)$, which is not a Nash equilibrium.

If player 1 did not know what player 2 thinks about his preference vector the solution will be different.

It is also interesting to mention that in the case of common knowledge about the preference, when $f_1(2) = -2$, the unique subgame perfect Nash equilibrium in type strategies will be $\{1, 1, 2, 1, 1), (1, 1)\}$ with payoffs $(5, 3)$. Which is very unusual and shows that sometimes being hostile is better than behaving favorably to both players.

## 3 The Three-Person "Lifeline Game" with One Pursuer $P$ and Two Evaders $E_1$, $E_2$

Consider the simple pursuit "lifeline game" $\Gamma(P, E_1, E_2)$ with one pursuer and two evaders in a given closed convex set $S$ on the plane. The velocities of the $P, E_1, E_2$ are $\alpha, \beta_1, \beta_2$, respectively, $\alpha > \max[\beta_1, \beta_2]$. The payoff of $E_i$, $i = 1, 2$, is equal to $+1$, if he reaches the "lifeline" (border $L$ of the set $S$) before being captured by the pursuer $P$ and 0 if he is captured before reaching $L$. The payoff of $P$ is equal to the number of evaders captured in the interior of $S$.

The "lifeline game" $\Gamma(P, E)$ with one pursuer and one evader was first considered by Isaaks [2] in the half-plane where the necessary conditions for optimal trajectories of the players were found. The solution of $\Gamma(P, E)$ for any convex closed $S$ was found in [3] where it was proved that the parallel pursuit strategy (II-strategy) is optimal for $P$ and in the case of pointwise capture the set of capture points coincides with the Appolonius circle.

We assume that in $\Gamma(P, E_1, E_2)$, $P$ uses parallel pursuit strategy first applied to $E_i$, and the to $E_{3-i}(i = 1, 2)$, having no possibility of changing the pursuit order in the intermediate instants of the game. Thus his strategy consists in the choice of pursuit order $E_i, E_{3-i}, i = 1, 2$, at the beginning of the game $\Gamma(P, E_1, E_2)$.

This game contains a rich class of Nash equilibria in the case when both players could not be captured in the interior of $S$, since any behavior of $E_i$ (the pursuit order being $E_i, E_{3-i}$ will be included in Nash equilibria (this equilibria may have different payoffs if from some locations of the capture point of $E_i$ the player $E_{3-i}$ may be captured in the interior of $S$ and from some locations not).

The use of the preference vector leads to unique Nash equilibrium in the sense of payoffs.

Let $f_P(E_1) = 2$ $f_P(E_2) = 3$, $f_{E_1}(P) = 3$, $f_{E_1}(E_2) = 2$, $f_{E_2}(P) = 3$, $F_{E_2}(E_1) = 2$.

Suppose that for given initial conditions by using II-strategy, $P$ can capture in the interior of $S$ both $E_1$ and $E_2$ separately; in both pursuit orders $E_1, E_2$ or $E_2$, $E_1$ there are possibilities of escaping from capture for $E_{3-i}$, $(i = 1, 2)$ depending upon the motion of player $E_i, i = 1, 2$.

From the preference vector it follows that $P$ uses the pursuit order $E_2, E_1$; $E_2$ moves with maximal velocity to his capture point such that $P$ starting the pursuit of $E_1$ from this point will not be able to capture $E_1$ inside $S$. $E_1$ moves with maximal velocity straight away from the point of capture of $E_2$ to the border of $S$. In [6] it is proved that the set of capture points of $E_1$, when $P$ uses II-strategy against him starting from the capture point of $E_2$, is a Decartus oval ($E_1$ moving from the start of the game $\Gamma(P : E_1, E_2)$), and thus $E_1$ has to go straight to the intersection of this oval with "lifeline" $L$. In [6] it is also proved that among the motions of $E_2$ leading to the escape of $E_2$ there always exists one along the straight line with maximal velocity.

If now $f_{E_2}(E_1) = 3$ or is the same $f_{E_2}(E_1) = -2$, $E_1$ will always be captured if $E_2$ hurries to his capture points such that the set of capture points of $E_1$ when $P$ uses II-strategy against him starting from the capture point $E_2$ will lie in the interior of $S$. It can be shown that for other values of preference vectors we also get unique in the sense of payoffs Nash equilibria.

**Theorem.** *Suppose in the game $\Gamma(P; E_1, E_2)$ the preference vectors are common knowledge for all players, then for any fixed values of this vector we get the family of equivalent Nash equilibria with coinciding payoffs.*

*In case of $l$-capture ($l > 0$) a slight modification of II-strategy is needed (see [4]) and the results remains true for that case also.*

*The theorem can also be generalized for the case of one pursuer and $m$ evaders.*

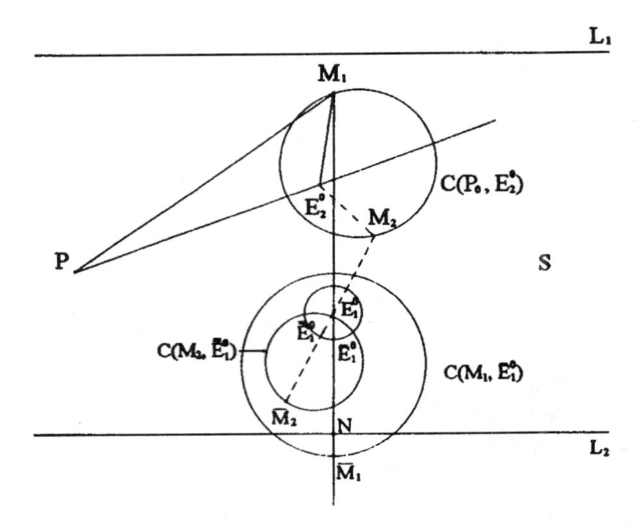


Figure 2.

**Example 3.1.** Consider the "lifeline game" on Figure 2. $S$ is here a channel bounded by two lifelines $L_1$, $L_2$. Suppose $f_P(E_1) = 2$, $f_p(E_2) = 3$.

If $f_{E_2}(E_1) = 2$, then moving to the capture point $M_1$, $E_2$ guarantees that $E_1$ will not be captured since the corresponding Appolonius circle $C(M_1, \bar{E}_1^0)$ (the set of capture points of $E_2$ when $P$ uses II-strategy) constructed for initial positions of $P$ in $M_1$ and $E_1$ in $\bar{E}_1^0$ (the point $E_1$ reaches before the capture $E_2$ occurs) intersects the "lifeline" $L_2$.

If $f_{E_1}(E_1) = 3$ (or $-2$), then moving to the capture point $M_2$, $E_2$ guarantees that capture of $E_1$ occurs in the interior of $S$ since the Appolonius circle $C(M_2 \bar{\bar{E}}_1^0)$ (the set of capture points of $E_2$ when $P$ uses II-strategy) constructed for initial positions of $P$ in $M_2$ and $E_1$ in any point $\bar{\bar{E}}_1^0$ (the point $E_1$ can reach before the capture of $E_2$ occurs in $M_2$) for all $\bar{\bar{E}}_1^0$ lies in the interior of $S$.

## 4 Conclusion

In [9] one can find a large variety of NE refinements for general $n$-person games. They essentially decrease the set of all possible NE in the game but at the same time not one of these refinements gives a unique (at least, in the sense of payoffs) NE. The approach presented in this paper is based on the idea of the introduction of a "preference vector," characterizing the type of player (behavior of player to his opponents) gives the possibility of constructing a unique, in the sense of payoffs, subgame perfect NE. But this approach has a limited application since it can be used only for games with perfect information. The example of a simple pursuit "lifeline" three-person game in the last section shows that the approach also be useful in differential games with perfect information.

## REFERENCES

[1] Nash, J.F. Equilibrium Points in $n$-Points Games. *Proceedings of the National Academy of Sciences U.S.A.*, **36**, 48–49, 1950.

[2] Isaaks, R. *Differential Games*. Wiley, New York, 1965.

[3] Petrosjan, L. A. On a Family of Differential Games of Survival in $R^n$. *Soviet Mathematics Doklady*. **161**, 1, 52–54, 1965.

[4] Dutkevich, Y. G. and L. A. Petrosjan. Games with "Life-Line." The Case of $l$-Capture. *SIAM Journal of Control*, **10**, 1, 40–47, 1972.

[5] Kreele, W. Learning in a Class of Non-Zero-Sum Two-Person Games. Part I, Discussion Paper NB-182. Bonn University, Bonn, 1991.

[6] Ambargan, V. S. The Definition of Capture Points, When $P$ Start the Pursuit with Time Delay. *Doklady of the Armenian Academy of Sciences*, 6, **92**, 4, 147–153, 1991.

[7] Selten, R. Reexamination of the Perfectness Concept for Equilibrium Points in Extensive Form Games. *International Journal of Game Theory*, **4**, 22–55, 1975.

[8] Harsanyi, J. C. Games with Incomplete Information Played by "Baysesian" Players, Parts I, II and III. Management Science, **14**, 159–182, 320-332, 468–502, 1968.

[9] Van Damme, E. E. C. *Stability and Perfection of Nash Equilibria*. Springer-Verlag, Berlin, 1991.

# A Perturbation on Two-Person Zero-Sum Games

Yutaka Kimura, Yoichi Sawasaki, and Kensuke Tanaka
Department of Mathematics and Information Science
Niigata University
Niigata, Japan

**Abstract**

In this paper, we investigate a two-person zero-sum game perturbed by constrained functions. We need, under the various conditions, to show that there exists a saddle point for the game. But, in many cases, it seems to be difficult to search directly for a saddle point. If there exists a *max-inf* (or *mini-sup*) of the perturbed game, the game has a *saddle value* and then, under some conditions, Player I (or Player II) has a weak optimal solution for the modified perturbed game with conjugate loss functions transformed by the upper envelope of the continuous affine functions from below.

Moreover, if loss functions of the original game are convex and lower semicontinuous, using its biconjugate loss functions, we show that the original perturbed game has a saddle point.

## 1 Introduction

Recently, many of the concepts and terms about game theory have been introduced and have been investigated by many authors. Both individual stability and collective stability have been studied in practical game problems. In view of individual stability in two-person zero-sum games, a saddle point was introduced as the concept of an equilibrium point in the game. Then, under the various conditions, the existence of a saddle point has been actively investigated. However, in many cases, it seems to be difficult to search directly for a saddle point. Thus, we need to consider the perturbed game. Aubin perturbs a two-person zero-sum game in [1, Chap. 6], in which a loss function is perturbed by linear functions, and discusses the various properties with respect to the perturbed game.

In this paper, we perturb a two-person zero-sum game by constrained functions. If there exists a *max-inf* (or *mini-sup*) of the perturbed game, we show that the game has a *saddle value*. Further, we investigate the modified perturbed game with conjugate loss functions transformed by the upper envelope of the continuous affine functions from below. Then, under some conditions, we show that Player I (or Player II) has a weak optimal solution with the same saddle value as the original perturbed game for the modified game. To do this, we make use of many tools in the convex analysis, for example, the separation theorem in $w^*$-topology, Fenchel's inequality, and so on. In the basis of this development, it plays an important role that each conjugate function of any loss function is convex and lower semicontinuous. Moreover, if loss functions of the original game are convex and lower semicontinuous, using its biconjugate loss functions, we show that the original perturbed game has a saddle point.

## 2 Existence of a Saddle Point in a Perturbed Game

In this paper, we describe two-person zero-sum games $(X, Y, f)$ in normal or strategic form in terms of the following items:

1. the strategy set $X$ of Player I is a Banach space;
2. the strategy set $Y$ of Player II is a Banach space;
3. the loss function $f : X \times Y \to \mathbb{R} \cup \{+\infty\}$ of Player I; and
4. the loss function $-f : X \times Y \to \mathbb{R} \cup \{-\infty\}$ of Player II.

From now, we shall study a perturbed game $(X, Y, Z, W, A, B, f, F, G)$ by perturbing the loss function $f$ of Player I by constrained functions $\{x, y\} \to F(Ax) + G(By)$, where:

1. bounded linear operator $A : X \to Z$, $Z$ is a Banach space, and $F : Z \to \mathbb{R} \cup \{+\infty\}$; and
2. bounded linear operator $B : Y \to W$, $W$ is a Banach space, and $G : W \to \mathbb{R} \cup \{-\infty\}$.

We set

$$v^\sharp := \inf_{x\in X} \sup_{y\in Y}[f(x, y) + F(Ax) + G(By)], \tag{1}$$

and

$$v^\flat := \sup_{y\in Y} \inf_{x\in X}[f(x, y) + F(Ax) + G(By)]. \tag{2}$$

If $y_0 \in Y$ satisfies that

$$\begin{aligned} v^\sharp &= \inf_{x\in X} \sup_{y\in Y}[f(x, y) + F(Ax) + G(By)] \\ &= \inf_{x\in X}[f(x, y_0) + F(Ax)] + G(By_0), \end{aligned} \tag{3}$$

and that $By_0 \in Dom(G)$, where $Dom(G) = \{w \in W \mid G(w) > -\infty\}$, then $y_0$ is said to be a *max-inf* of the perturbed game with the loss function $\{x, y\} \to f(x, y) + F(Ax) + G(By)$.

If $x_0 \in X$ satisfies

$$\begin{aligned} v^\flat &= \sup_{y\in Y} \inf_{x\in X}[f(x, y) + F(Ax) + G(By)] \\ &= \sup_{y\in Y}[f(x_0, y) + G(By)] + F(Ax_0), \end{aligned} \tag{4}$$

and that $Ax_0 \in Dom(F)$, where $Dom(F) = \{z \in Z \mid F(z) < +\infty\}$, then $x_0$ is said to be a *mini-sup* of the perturbed game.

In general, from the definitions of (1) and (2), it always holds that

$$v^\sharp \geq v^\flat. \tag{5}$$

If there exists a real number $v$ satisfying $v^\sharp = v = v^\flat$, the value $v$ is said to be a *saddle value* (in short, a *value*). Moreover, we assume that $+\infty > v^\sharp \geq v^\flat > -\infty$.

**Theorem 2.1.** *Suppose that $y_0$ is a max-inf of the perturbed game. Then,*

$$v^\sharp - G(By_0) = \inf_{x \in X}[f(x, y_0) + F(Ax)] \tag{6}$$

*and there exists a value $v$, that is,*

$$v^\flat = v = v^\sharp.$$

*Conversely, if $y_0 \in Y$ satisfies* (6)*, then $y_0$ is a max-inf of the game.*

**Proof.** Since $y_0$ is a *max-inf* of the game, from (1) and (3), we get

$$v^\sharp - G(By_0) = \inf_{x \in X}[f(x, y_0) + F(Ax)]. \tag{7}$$

Hence (6) was shown.

On the other hand, from (2) we obtain

$$v^\flat \geq \inf_{x \in X}[f(x, y_0) + F(Ax)] + G(By_0).$$

Thus we have

$$v^\flat - G(By_0) \geq \inf_{x \in X}[f(x, y_0) + F(Ax)]. \tag{8}$$

From (7) and (8),

$$\begin{aligned} v^\flat - G(By_0) &\geq v^\sharp - G(By_0) \\ v^\flat &\geq v^\sharp. \end{aligned}$$

Since, in general, $v^\sharp \geq v^\flat$ holds, the game has a value $v$, that is, $v^\sharp = v = v^\flat$.

Conversely, we assume that (6) holds. It follows that

$$\begin{aligned} v^\sharp &= \inf_{x \in X}[f(x, y_0) + F(Ax)] + G(By_0) \\ &= \inf_{x \in X} \sup_{y \in Y}[f(x, y) + F(Ax) + G(By)]. \end{aligned}$$

Thus, it is shown that $y_0$ is a *max-inf* of the game. □

By a similar argument, we can prove the following theorem:

**Theorem 2.2.** *Suppose that $x_0$ is a mini-sup of the perturbed game. Then,*

$$F(Ax_0) - v^\flat = \inf_{y \in Y}[-f(x_0, y) - G(By)] \tag{9}$$

*and there exists a value $v$, that is,*

$$v^\flat = v = v^\sharp.$$

*Conversely, if $x_0 \in X$ satisfies* (9), *then $x_0$ is a mini-sup of the game.*

For the operator $A$, we give an adjoint operator $A^* : Z^* \to X^*$ defined by

$$\langle x, A^*z^* \rangle = \langle Ax, z^* \rangle \qquad \text{for all} \quad x \in X \text{ and } z^* \in Z^*,$$

where $X^*$ and $Z^*$ denote dual spaces of $X$ and $Z$, respectively. Then, $A^*$ is a bounded linear operator. Moreover, analogously, we give an adjoint operator $B^* : W^* \to Y^*$ for $B$, where $Y^*$ and $W^*$ denote dual spaces of $Y$ and $W$, respectively.

Using the Fenchel conjugations in Rockafellar [8], [9], we have $f^*(\cdot)(y)$ on $X^*$ for $f(\cdot, y)$ defined by

$$f^*(x^*)(y) = \sup_{x \in X}[\langle x^*, x\rangle - f(x, y)]$$

and $F^*(\cdot)$ on $Z^*$ for $F$ defined by

$$F^*(z^*) = \sup_{z' \in Z}[\langle z^*, z'\rangle - F(z')].$$

Using two notations: for each $y \in Y$

$$v(y) = \inf_{x \in X}[f(x, y) + F(Ax)]$$

and

$$v_*(y) = \sup_{z^* \in Z^*}[-f^*(-A^*z^*)(y) - F^*(z^*)].$$

Roughly, we shall show the relation between spaces $X$, $Y$, $W$, and $Z$ the operator $A$ and $B$, and the relation between dual spaces of above one and the operator $A^*$ and $B^*$ by the following diagram :

$$\begin{array}{ccccccc}
Z^* & \overset{A^*}{\longmapsto} & X^* & & Y^* & \overset{B^*}{\longleftarrow\!\!\!\shortmid} & W^* \\
\uparrow \text{dual} & & \uparrow \text{dual} & & \uparrow \text{dual} & & \uparrow \text{dual} \\
Z & \overset{A}{\longleftarrow\!\!\!\shortmid} & X & & Y & \overset{B}{\longmapsto} & W
\end{array}$$

We can prove the following lemma:

**Lemma 2.1.**

$$v(y) \geq v_*(y) \qquad \textit{for each} \quad y \in Y.$$

**Proof.** By virtue of Fenchel's inequality, for any $x \in X$ and $z^* \in Z^*$, we have that

$$\begin{aligned}
f(x, y) + F(Ax) - \{-f^*(-A^*z^*)(y) - F^*(z^*)\}& \\
= f(x, y) + f^*(-A^*z^*)(y) + F(Ax) + F^*(z^*)&
\end{aligned}$$

$$\begin{aligned}&\geq \langle x, -A^*z^* \rangle + \langle Ax, z^* \rangle \\ &= -\langle Ax, z^* \rangle + \langle Ax, z^* \rangle \\ &= 0.\end{aligned}$$

This implies that

$$\inf_{x \in X}[f(x, y) + F(Ax)] - \sup_{z^* \in Z^*}[-f^*(-A^*z^*)(y) - F^*(z^*)] \geq 0.$$

Thus we get

$$v(y) \geq v_*(y). \qquad \square$$

Using $Dom(f^*(\cdot)(y)) = \{x^* \in X^* \mid f^*(x^*)(y) < +\infty\}$ for each $y \in Y$ and $Dom(F^*) = \{z^* \in Z^* \mid F^*(z^*) < +\infty\}$, we introduce a mapping $\psi : Dom(f^*(\cdot)(y)) \times Dom(F^*) \to \mathbb{R} \times X^*$ defined by

$$\psi(x^*, z^*) = (-f^*(x^*)(y) - F^*(z^*), x^* + A^*z^*),$$

together with the set $S(y) = \psi(Dom(f^*(\cdot)(y)) \times Dom(F^*)) - \mathbb{R}_+ \times \{\theta^*\}$ constructed by difference of vectors, where $\mathbb{R}_+ = [0, +\infty)$ and $\theta^*$ is the zero vector in $X^*$.

**Lemma 2.2.** *Suppose that for each $y \in Y$:*

1. $\theta \in \text{int}(A\, Dom(f(\cdot, y)) - Dom(F))$; *and*
2. $\theta^* \in A^*\, Dom(F^*) + Dom(f^*(\cdot)(y))$;

*where $\theta$ is the zero vector in $Z$. Then, for each $y$ the set $S(y)$ is convex and $w^*$-closed set in $\mathbb{R} \times X^*$, where $\mathbb{R} \times X^*$ is a dual space of $\mathbb{R} \times X$.*

**Proof.** Since the conjugate functions $f^*(\cdot)(y)$ and $F^*(\cdot)$ are always convex, by these convexities and linearity of the operator $A^*$, we can show easily that $S(y)$ is convex set.

In order to show that $S(y)$ is $w^*$-closed, we consider a sequence of vectors $(v_n^*, r_n^*)$ in $S(y)$ converging ($w^*$-topology) to $(v^*, r^*)$ in $\mathbb{R} \times X^*$. Thus, from the definition of $S(y)$, for each $n$ there exist $x_n^* \in X^*$ and $z_n^* \in Z^*$ such that, for some $c \in [0, \infty)$,

$$v_n^* = -f^*(x_n^*)(y) - F^*(z_n^*) - c, \qquad r_n^* = x_n^* + A^*z_n^*. \tag{10}$$

From (10) we have

$$v_n^* \leq -f^*(x_n^*)(y) - F^*(z_n^*), \qquad r_n^* = x_n^* + A^*z_n^*. \tag{11}$$

Since we assume that $\theta \in \text{int}(A\, Dom(f(\cdot, y)) - Dom(F))$, there exists an open ball $B_\epsilon = \{z \in Z \mid \|z\| < \epsilon\}$ such that

$$B_\epsilon \subset A\, Dom(f(\cdot, y)) - Dom(F) \subset Z.$$

Thus, for all $z \in Z$ and $0 < \epsilon' \leq \epsilon$, there exist $x \in Dom(f(\cdot, y))$ and $z' \in Dom(F)$ such that

$$\left(\frac{\epsilon'}{\|z\|}\right) z = Ax - z'.$$

Therefore, using Fenchel's inequality, we obtain

$$\begin{aligned}
\frac{\epsilon'}{||z||}\langle z, z_n^* \rangle &= \langle Ax - z', z_n^* \rangle \\
&= \langle Ax, z_n^* \rangle - \langle z', z_n^* \rangle \\
&= \langle x, A^* z_n^* \rangle - \langle z', z_n^* \rangle \\
&= \langle x, r_n^* - x_n^* \rangle - \langle z', z_n^* \rangle \\
&= \langle x, r_n^* \rangle - \langle x, x_n^* \rangle - \langle z', z_n^* \rangle \\
&\geq \langle x, r_n^* \rangle - (f(x, y) + f^*(x_n^*)(y)) - (F(z') + F^*(z_n^*)) \\
&\quad \text{(by Fenchel's inequality)} \\
&\geq \langle x, r_n^* \rangle + v_n^* - (f(x, y) + F(z')) \\
&\quad \text{(by the definition of } v_n^*).
\end{aligned}$$

Since the sequences $\{v_n^*\}$ and $\{\langle x, r_n^* \rangle\}$ converge to $v^*$ and $\langle x, r^* \rangle$, respectively, we arrive at $\inf_{n \geq 1} \langle z, z_n^* \rangle > -\infty$ for all $z \in Z$. According to the uniform boundedness theorem, it follows that $\{z_n^*\}_{n \geq 1}$ is bounded. Thus, from Alaoglu's theorem, it is $w^*$-compact. See Luenberger [6, Chap. 5] for details. Hence, there exists a subsequence $\{z_{n(j)}^*\}_{j \geq 1}$ of $\{z_n^*\}_{n \geq 1}$ which converges to $z^* \in Z^*$ in $w^*$-topology of $Z$, and consequently, a subsequence $x_{n(j)}^* = r_{n(j)}^* - A^* z_{n(j)}^*$ converges to $x^* = r^* - A^* z^*$ in $w^*$-topology of $X$.

From the construction of the conjugate function, $-f^*(\cdot)(y)$ and $-F^*(\cdot)$ are u.s.c. in the $w^*$-topology. Therefore,

$$\begin{aligned}
-f^*(x^*)(y) - F^*(z^*) &\geq \limsup_{j \to \infty}\{-f^*(x_{n(j)}^*)(y)\} + \limsup_{j \to \infty}\{-F^*(z_{n(j)}^*)\} \\
&\geq \limsup_{j \to \infty}\{-f^*(x_{n(j)}^*)(y) - F^*(z_{n(j)}^*)\} \\
&\geq \lim_{n \to \infty} v_n^* = v^*.
\end{aligned}$$

Thus we obtain

$$-f^*(x^*)(y) - F^*(z^*) \geq v^* \qquad \text{and} \qquad x^* = r^* - A^* z^*.$$

This implies that $(v^*, r^*)$ belongs to $S(y)$, which shows that $S(y)$ is $w^*$-closed. □

**Theorem 2.3.** *Suppose that for each $y \in Y$*

$$(v(y), \theta^*) \in S(y), \tag{12}$$

*where $S(y) = \psi(Dom(f^*(\cdot)(y)) \times Dom(F^*)) - [0, +\infty) \times \{\theta^*\}$.*

*Then, there holds that:*

1. $v(y) = v_*(y)$; *and*
2. *there exists* $z_0^* \in Z^*$ *such that*

$$v_*(y) = -f^*(-A^*z_0^*)(y) - F^*(z_0^*).$$

**Proof.** Since we assume that (12) holds, it follows that there exist $z_0^* \in Z^*$ such that from (11),

$$\begin{aligned} v(y) &\leq -f^*(-A^*z_0^*)(y) - F^*(z_0^*) \\ &\leq v_*(y) \\ &\quad \text{(by the definition of } v_*) \\ &\leq v(y) \\ &\quad \text{(by Lemma 2.1).} \end{aligned}$$

Consequently,

$$v(y) = v_*(y),$$

which completes the proof of the theorem. □

In order to develop main theorems, (12) in Theorem 2.3 plays a very important role. Thus, in the next theorem, we shall discuss a sufficient condition, under which (12) holds. We shall say that $f(\cdot, y)$ and $F$ are strict if $Domf(\cdot, y) = \{x \in X \mid f(x, y) < +\infty\}$ and $Dom(F) = \{z \in Z \mid F(z) < +\infty\}$ are nonempty.

**Theorem 2.4.** *Suppose that for each* $y \in Y$:

1. $f(\cdot, y) : X \to \mathbb{R} \cup \{+\infty\}$ *is strict, l.s.c., and convex on* $X$;
2. $F : Z \to \mathbb{R} \cup \{+\infty\}$ *is strict, l.s.c., and convex on* $Z$;
3. $\theta \in \text{int}(A\, Dom(f(\cdot, y)) - Dom(F))$; *and*
4. $\theta^* \in A^*\, Dom(F^*) + Dom(f^*(\cdot)(y))$.

*Then,* (12) *holds for each* $y$, *that is,* $(v(y), \theta^*) \in S(y)$.

**Proof.** We proceed by contradiction, assuming that the conclusion is false. Hence, there exists $y \in Y$ such that $(v(y), \theta^*)$ does not belong to $S(y)$ (see Luenberger [6, Chap. 5]).

Since, from Lemma 2.2, which requires hypotheses 1 and 2, $S(y)$ is convex and $w^*$-closed, $(v(y), \theta^*)$ may be strictly separated from $S(y)$. Thus, there exist $(\alpha, x_0') \in \mathbb{R} \times X$ and $\varepsilon > 0$ such that

$$\begin{aligned} \alpha v(y) &\geq \sup_{\substack{(x^*, z^*) \in X^* \times Z^* \\ c \in \mathbb{R}_+}} \left[ \langle (\alpha, x_0')(-f^*(x^*)(y) - F^*(z^*) - c, x^* + A^*z^*) \rangle \right] + \epsilon \\ &= \sup_{(x^*, z^*) \in X^* \times Z^*} \left[ \alpha\{-f^*(x^*)(y) - F^*(z^*)\} + \langle x_0', x^* + A^*z^* \rangle \right] \\ &\quad - \inf_{c \in \mathbb{R}_+} \alpha c + \epsilon. \end{aligned} \tag{13}$$

Since, $-\inf_{c \in \mathbb{R}_+} \alpha c$ is bounded from above in (13), it follows that $\inf_{c \in \mathbb{R}_+} \alpha c = 0$. If $\alpha$ is negative, $\inf_{c \in \mathbb{R}_+} \alpha c = -\infty$. Thus, $\alpha$ is positive or 0. However, $\alpha$ cannot be zero, because if $\alpha = 0$, we have

$$\sup_{(x^*,z^*)\in X^*\times Z^*} \langle x_0', x^* + A^*z^*\rangle + \varepsilon \leq 0.$$

Since there exist $x^* \in X^*$ and $z^* \in Z^*$ satisfying $x^* + A^*z^* = \theta^*$, we arrive at $0 \geq \varepsilon$, which is impossible. Thus, $\alpha$ is positive. On the other hand, since $f(\cdot, y)$ and $F(\cdot)$ are strict, l.s.c., and convex, it holds that

$$f^{**}(\cdot)(y) = f(\cdot, y) \qquad \text{and} \qquad F^{**}(\cdot) = F(\cdot), \tag{14}$$

where $f^{**}(\cdot)(y)$ and $F^{**}(\cdot)$ are the biconjugate of $f(\cdot, y)$ and $F(\cdot)$, respectively.

We divide both sides of (13) by $\alpha > 0$, and then setting $x_0 = x_0'/\alpha$ and $\eta = \varepsilon/\alpha$, we have that

$$\begin{aligned} v(y) &\geq \sup_{(x^*,z^*)\in X^*\times Z^*} \{-f^*(x^*)(y) + \langle x_0, x^*\rangle - F^*(z^*) + \langle x_0, A^*z^*\rangle\} + \eta \\ &= \sup_{x^*\in X^*} \{\langle x_0, x^*\rangle - f^*(x^*)(y)\} + \sup_{z^*\in Z^*} \{\langle x_0, A^*z^*\rangle - F^*(z^*)\} + \eta \\ &= f^{**}(x_0)(y) + F^{**}(Ax_0) + \eta \\ &= f(x_0, y) + F(Ax_0) + \eta \qquad (15) \\ &\quad \text{(by (14))}. \end{aligned}$$

Since $\eta$ is positive, from the definition of $v(y)$, it is impossible that (15) holds. Thus, it follows that (12) in Theorem 2.3 holds. □

**Theorem 2.5.** *Suppose that $y_0 \in Y$ is a max-inf and:*

1. $f(\cdot, y_0) : X \to \mathbb{R} \cup \{+\infty\}$ *is strict, l.s.c.,and convex on* $X$*;*
2. $F : Z \to \mathbb{R} \cup \{+\infty\}$ *is strict, l.s.c., and convex on* $Z$*;*
3. $\theta \in \text{int}(A\, Dom(f(\cdot, y_0)) - Dom(F))$*; and*
4. $\theta^* \in A^*\, Dom(F^*) + Dom(f^*(\cdot)(y_0))$.

*Then, the perturbed game has a value $v$ satisfying $v^\flat = v = v^\sharp$ and there exists a weak optimal solution $z_0^* \in Z^*$ for Player I such that*

$$v = -f^*(-A^*z_0^*)(y_0) - F^*(z_0^*) + G(By_0).$$

*Moreover, suppose that $\theta^* \in \text{int}(A^*\, Dom(F^*)+Dom(f^*(\cdot)(y_0)))$, then there exists an optimal solution $x_0 \in X$ for Player I such that*

$$v = f(x_0, y_0) + F(Ax_0) + G(By_0).$$

*The perturbed game has a saddle point $(x_0, y_0) \in X \times Y$.*

**Proof.** Since $y_0 \in Y$ is a *max-inf*, from Theorem 2.1, it follows that there exists a value $v$ satisfying $v^\flat = v = v^\sharp$. Then, (6) holds, that is,

$$\begin{aligned} v - G(By_0) &= \inf_{x\in X}[f(x, y_0) + F(Ax)] \\ &= v(y_0). \end{aligned}$$

On the other hand, under the conditions of the theorem, using the results of Theorem 2.3, there exists $z_0^* \in Z^*$ such that

$$v(y_0) = v_*(y_0) = -f^*(-A^* z_0^*)(y_0) - F^*(z_0^*). \tag{16}$$

Thus, from (6) and (16), we get

$$v - G(By_0) = -f^*(-A^* z_0^*)(y_0) - F^*(z_0^*).$$

This shows that the first part of the theorem holds.

Next, in order to show the second part of the theorem, for the *max-inf* $y_0 \in Y$, we put

$$v_{**}(y_0) = \inf_{x \in X} [f^{**}(x)(y_0) + F^{**}(Ax)].$$

Using similar arguments for Theorem 2.3 and Theorem 2.4, we show that there exists $x_0 \in X$ such that

$$v_*(y_0) = v_{**}(y_0) = f^{**}(x_0)(y_0) + F^{**}(Ax_0).$$

Further, under the conditions 1 and 2 of the theorem, we have

$$f^{**}(x_0)(y_0) = f(x_0, y_0), \qquad F^{**}(Ax_0) = F(Ax_0).$$

This shows that the second part of the theorem holds. □

By the analogous argument, we can prove the following theorem:

**Theorem 2.6.** *Suppose that $x_0$ is a mini-sup and:*

1. $-f(x_0, \cdot) : Y \to \mathbb{R} \cup \{-\infty\}$ *is strict, u.s.c., and concave on* $Y$*;*
2. $G : W \to \mathbb{R} \cup \{-\infty\}$ *is strict, u.s.c., and concave on* $W$*;*
3. $\theta \in \mathrm{int}(B\, Dom(g(\cdot)(x_0)) - Dom(Q))$*; and*
4. $\theta^* \in B^*\, Dom(Q^*) + Dom(g^*(\cdot)(x_0))$*;*

*where* $g(\cdot)(x_0) := -f(x_0, \cdot)$, $Q(By) := -G(By)$, $\theta$ *and* $\theta^*$ *are zero vector in* $W$ *and* $Y^*$*, respectively.*

*Then the perturbed game has a value* $v$ *satisfying* $v^\flat = v = v^\sharp$ *and there exists a weak optimal solution* $w_0^* \in W^*$ *for Player II such that*

$$v = g^*(-B^* w_0^*)(x_0) + Q^*(w_0^*) + F(Ax_0).$$

*Moreover, suppose that* $\theta^* \in \mathrm{int}(B^*\, Dom(Q^*) + Dom(g^*(\cdot)(x_0)))$*, then there exists an optimal solution* $y_0 \in Y$ *for Player II such that*

$$v = f(x_0, y_0) + F(Ax_0) + G(By_0).$$

*The perturbed game has a point* $(x_0, y_0) \in X \times Y$.

We will sketch an outline of its proof. Since $x_0$ is a *mini-sup*, by Theorem 2.2, we have a saddle value $v$ satisfying $v^\flat = v = v^\sharp$. Therefore, we have

$$F(Ax_0) - v = \inf_{y \in Y} [-f(x_0, y) - G(By)].$$

Thus, setting $g(\cdot)(x_0) = -f(x_0, \cdot)$ and $Q(B(\cdot)) = -G(B(\cdot))$, we have that $g(\cdot)(x_0)$ and $Q(B(\cdot))$ are l.s.c., and convex. Thus, using Theorems 2.3, 2.4, and 2.5, we can show that the results of the theorem hold.

## REFERENCES

[1] Aubin, J. P. *Mathematical Methods of Game and Economic Theory*, rev. ed. North-Holland, Amsterdam, 1982.

[2] Aubin, J. P. *Optima and Equilibria*. Springer-Verlag, New York, 1993.

[3] Aubin J. P. and I. Ekeland. *Applied Nonlinear Analysis*. Wiley-Interscience, New York, 1984.

[4] Aubin J. P. and H. Frankowska. *Set-Valued Analysis*. Birkhäuser, Boston, 1990.

[5] Kimura, Y. and K. Tanaka. On Duality of Dynamic Programming Model with Discrete Time. *Bulletin of Informatics and Cybernetics*, To appear.

[6] Luenberger, D. G. *Optimization by Vector Space Methods*. Wiley, New York, 1969.

[7] Von Neumann, J. and O. Moregenstern. *Theory of Games and Economic Behavior*. Princeton University Press, Princeton, NJ, 1944.

[8] Rockafellar, R. T. Extension of Fenchel's duality Theorem for Convex Functions. *Duke Mathematical Journal*, **33**, 81–89, 1966.

[9] Rockafellar, R. T. *Convex Analysis*. Princeton University Press, Princeton, NJ, 1970.

[10] Tanaka, K. On Discounted Dynamic Programming with Constraints. *Journal of Mathematical Analysis and Applications*, **155**, 264–277, 1991.

[11] Tanaka, K. *On a Perturbation of Dynamic Programming*. Lecture Notes in Economics and Mathematical Systems 419. Springer-Verlag, Berlin, pp. 275–287, 1995.

# The Linear Complementarity Problem in Static and Dynamic Games

G. S. R. Murthy
Indian Statistical Institute
Habsiguda, Hyderabad, India

T. Parthasarathy and B. Sriparna
Indian Statistical Institute
New Delhi, India

**Abstract**

It is known that the linear complementarity problem (LCP) and matrix game theory problems have interesting connections. Raghavan derived a number of LCP results using Kaplansky's theorem. We supplement this by demonstrating more connections between LCP and matrix games. A new result is obtained in which it is shown that if a matrix game is completely mixed with value 0 and if the payoff matrix is a $\boldsymbol{U}$-matrix, then all the proper principal minors of the payoff matrix are positive. A cursory mention is made regarding the applications of LCP in solving stochastic games.

## 1 Introduction

The linear complementarity problem (LCP) and matrix games have interesting connections. Raghavan [26] used von Neumann's [33] minimax theorem and Kaplansky's game theory results to derive several interesting properties of $M$-matrices that arise in LCP and gave an alternative proof of the Parron–Frobenius theorem (see [1], [27]). Nowak and Raghavan [22], Schultz [29], and Mohan, Neogy, and Parthasarathy [16], [17] used LCP formulations to solve certain classes of stochastic games. In this article we derive a number of interesting LCP results using game theory results. We shall also cite applications of LCP in solving stochastic games. In the rest of this section, we shall briefly introduce LCP and two-person zero-sum matrix games and state some relevent results including Kaplansky's theorem.

Given $A \in \boldsymbol{R}^{n\times n}$ and $q \in \boldsymbol{R}^n$ the LCP is to find vectors $w$ and $z$ such that

$$w = Az + q \geq 0, \qquad z \geq 0, \tag{1}$$

and

$$w^t z = 0. \tag{2}$$

This problem is denoted by $(q, A)$. The LCP has numerous applications both in theory and practice. Linear and quadratic programming problems are special cases of LCP. The subject has gained momentum after Lemke and Howson [14] came up with an algorithm in 1964 to solve bimatrix games by formulating the same as LCP. A rich account of LCP can be found in the excellent monograph by Cottle, Pang, and Stone [3].

Let $S(q, A) = \{z \in \boldsymbol{R}^n{}_+ : Az + q \geq 0, z^t(Az + q) = 0\}$ and let $K(A)$ be the set of all $q$ for which $S(q, A) \neq \phi$. A matrix $A \in \boldsymbol{R}^{n \times n}$ is said to be a $\boldsymbol{Q}_0$-matrix if for every $q$ for which (1) admits a solution, $S(q, A) \neq \phi$; and $A$ is said to be a $\boldsymbol{Q}$-matrix if $K(A) = \boldsymbol{R}^n$. The class of $\boldsymbol{Q}_0$-matrices ($\boldsymbol{Q}$-matrices) is denoted by $\boldsymbol{Q}_0$ ($\boldsymbol{Q}$). These are the fundamental classes of LCP. One of the most challenging open problems in LCP is to characterize the class of $\boldsymbol{Q}$-matrices. The study of LCP has led to numerous other matrix classes.

Next we shall briefly describe the two-person zero-sum games. A *two-person-zero-sum game* consists of two players, designated as Player I and Player II, each having a finite set of strategies. Let $S = \{s_1, s_2, \ldots, s_m\}$ and let $T = \{t_1, t_2, \ldots, t_n\}$ be the sets of strategies for Players I and II, respectively. The game is played as follows. Each player chooses a strategy from his set of strategies. If $(s_i, t_j)$ is the pair of strategies chosen, then Player II pays Player I \$ $a_{ij}$ ($a_{ij} < 0$ means Player II receives \$ $a_{ij}$ ). The $m \times n$ matrix $A = (a_{ij})$ is called the (Player I's) payoff matrix. The elements of $S$ and $T$ are called *pure strategies*. If there exist indices $i_0$ and $j_0$ such that $a_{i_0 j} \leq a_{i_0 j_0} \leq a_{i j_0}$ for all $i$ and $j$, then the game is said to have a solution in pure strategies with $s_{i_0}$ and $t_{j_0}$ as optimal strategies for Player I and Player II, respectively. In this case, $a_{i_0 j_0}$ is called the *value of the game* and is denoted by $v(A)$. Often the games do not have solutions in pure strategies. This forces the players to choose their strategies with some probabilities.

Any probability vectors $x$ on $S$ and $y$ on $T$ are called *mixed strategies* for the respective players. The real number $y^t Ax$ is called the expected payoff with respect to $(x, y)$. If there exist mixed strategies $\bar{x}$ and $\bar{y}$ for Players I and II, respectively, such that

$$x^t A\bar{y} \leq \bar{x}^t A\bar{y} \leq \bar{x}^t Ay$$

for all mixed strategies $x$ and $y$ for Players I and II, respectively, then the game is said to have a solution in mixed strategies. In this case, $\bar{x}$ and $\bar{y}$ are called the optimal mixed strategies for Players I and II, respectively, and $\bar{y}^t A\bar{x}$ is called the *value of the game* and is denoted by $v(A)$.

The games described above are also called matrix games (often we refer to the game by the matrix itself). The fundamental minimax theorem due to von Neumann [33] asserts that every matrix game has a solution in mixed strategies. We shall now present two important theorems on matrix games. For proofs and other details, see Kaplansky [11] and Owen [23]. The following proposition is easy to prove [18], [19]:

**Proposition 1.1.** *For any nonsigular matrix A, $v(A) > 0$ ($v(A) < 0$) if, and only if, $v(A^{-1}) > 0$ ($v(A^{-1} < 0)$).*

A mixed strategy $x$ is said to be completely mixed if $x > 0$. A matrix game is said to be completely mixed if every optimal mixed strategy (of either player) is completely mixed. Below we state a beautiful result due to Kaplansky [11].

**Theorem 1.1** (Kaplansky, [11]). *Let A denote the payoff matrix of order $m \times n$ of a two-person game. We then have the following:*

(i) *If Player I has a completely mixed optimal strategy, then any optimal strategy $q^0 = (q_1^0, q_2^0, \ldots, q_n^0)$ for Player II satisfies $\sum_j a_{ij} q_j^0 = v(A)$, $\forall i = 1, 2, \ldots, m$.*
(ii) *If $m = n$ and if the game is not completely mixed, then both players have optimal strategies that are not completely mixed.*
(iii) *A game with $v(A) = 0$ is completely mixed if and only if: (a) $m = n$ and the rank of $A = n - 1$; and (b) all the cofactors $A_{ij}$ of A are different from zero and have the same sign.*
(iv) *Suppose A is a completely mixed game. Then $v(A) = |A|/\sum_i \sum_j A_{ij}$, where $|A|$ stands for the determinant of A and $A_{ij}$'s are the cofactors.*
(v) *Let $V = (V_{ij})$ denote the matrix of order $m \times n$ where $V_{ij}$ is the value of a game whose payoff matrix is obtained from A by omitting its $i$th row and $j$th column. Then the game with payoff matrix A is completely mixed if and only if the game with payoff matrix V has no solution in pure strategies.*

**Remark 1.1.** The proof of Theorem 1.1 may be found in [11]. If the game $A$ is completely mixed, then the game with payoff matrix $A^t$ (transpose of $A$) is also completely mixed and $v(A) = v(A^t)$. If $A$ is a nonsingular matrix with $A^{-1} < 0$ (every entry in $A^{-1}$ is negative), then the game with payoff matrix $A$ is completely mixed.

## 2 Game-Theoretic Results in LCP

In this section, we shall illustrate: (i) the usefulness of the minimax theorem and Kaplansky's results in deriving LCP results; and (ii) the connections between LCP and matrix games.

One of the largest matrix classes encountered in LCP is the class of semi-monotone matrices (introduced by Eaves [7]), $\boldsymbol{E}_0$. A matrix $A$ is in this class if $(q, A)$ has unique solution for every $q > 0$. In a game-theoretic context, it turns out that $A \in \boldsymbol{E}_0$ if, and only if, all the matrix games with payoff matrices as $A_{\alpha\alpha}$ have nonnegative value where $A_{\alpha\alpha}$ is the principal submatrix obtained by deleting rows and columns of $A$ corresponding to the indices not in $\alpha$.

**Theorem 2.1.** *Let $A \in \boldsymbol{R}^{n\times n}$. Then the following two conditions are equivalent.*

(i) $A \in \boldsymbol{E}_0$; *and*
(ii) $A^t \in \boldsymbol{E}_0$.

**Proof.** Assume (i). If $A \in \boldsymbol{E}_0$, then $v(A_{\alpha\alpha}) \geq 0, \forall\alpha$. It is enough to show that $v(A^t_{\alpha\alpha}) \geq 0, \forall\alpha$. We prove this by induction on $n$. If $n = 1$, the result is immediate. So, we will assume the result to be true for all $\alpha$ with cardinality of $\alpha \leq n - 1$. That is, we have $v(A_{\alpha\alpha}) \geq 0, \forall\alpha$, and $v(A^t_{\alpha\alpha}) \geq 0, \forall\alpha$, with $|\alpha| \leq n - 1$. We need to show that $v(A^t) \geq 0$. Suppose $v(A^t) < 0$. It means the game $A^t$ is completely mixed and, consequently, $v(A^t) = v(A) < 0$ leading to a contradiction. Hence, $A^t \in \boldsymbol{E}_0$. Proof of (ii) $\Rightarrow$ (i) follows from the fact that $(A^t)^t = A$. □

**Proposition 2.1.** *The following statements hold:*

(i) *Every $\boldsymbol{Q}_0$-matrix $A$ is in $\boldsymbol{Q}$ if, and only if, $v(A) > 0$.*
(ii) *For any real square matrix $A$, all its principal submatrices are $\boldsymbol{Q}$-matrices if, and only if, the value of $A_{\alpha\alpha}$ is positive for all $\alpha$.*

**Proof.** (i) Let $A$ be a $\boldsymbol{Q}_0$-matrix. Assume that $A \in \boldsymbol{Q}$. Let $q$ be such that all its coordinates are equal to $-1$. Since $A \in \boldsymbol{Q}$, there exists a nonnegative vector $z$ such that $Az + q \geq 0$. Since $q < 0$, $Az > 0$. Without loss of generality we may assume that $z$ is a probability vector. Note that for any mixed strategy $x$ for Player I, $x^t Az > 0$. This implies that the $v(A) > 0$. Conversely, assume $v(A) > 0$. Then there exists a probability vector $x$ such that $Ax > 0$. Let $q$ be any arbitrary vector. We can choose a large positive number $\lambda$ such that $Az + q \geq 0$, where $z = \lambda x$. From the definition of $\boldsymbol{Q}_0$ it follows that $A \in \boldsymbol{Q}$.

(ii) This result is due to Cottle [2]. We omit the proof and request the interested readers to refer to Sections 3.9 and 3.10 of [3]. See also [12]. □

A diagonal matrix is called a signature matrix if its diagonal entries are equal to 1 or $-1$. For $A \in \boldsymbol{R}^{n\times n}$ and any signature matrix $S$ of order $n$, $SAS$ is called the signature transformation of $A$ with respect to $S$. Often one is interested in finding out whether a particular property is invariant under signature transformations. For example, if $A$ is such that $S(q, A)$ is convex for all $q$ (matrices which satisfy this property are known as *column sufficient matrices*), can we say that this property is also true for all signature transformations of $A$? In fact, this is true (see [2]). On the other hand, suppose we ask the question: if $v(A) > 0$, then is it true that $v(SAS) > 0$ for any signature matrix $S$? The answer to this question is "no." Take, for example, $A = \begin{bmatrix} 1 & 1 \\ 1 & 1 \end{bmatrix}$ and $S = \begin{bmatrix} -1 & 0 \\ 0 & 1 \end{bmatrix}$. It is easy to check that $v(A) = 1$ but $v(SAS) = 0$.

A square matrix $A$ is called an $\boldsymbol{N}$-matrix ($\boldsymbol{P}$-matrix) if every principal minor of $A$ is negative (positive). It is interesting to note that these matrix classes are invariant under signature transformations. In general, signature transformations of $\boldsymbol{Q}$-matrices need not necessarily be $\boldsymbol{Q}$-matrices. But when $A$ is a $\boldsymbol{P}$-matrix, then all its signature transformations are $\boldsymbol{Q}$-matrices; and when $A$ is a negative $\boldsymbol{N}$-matrix, then all but two signature transformations are $\boldsymbol{Q}$-matrices. Our next result gives a characterization of $\boldsymbol{N}$-matrices through signature transformations and LCP.

**Theorem 2.2.** *Let $A < 0$. Then the following statements are equivalent:*

(i) *$A$ is an $N$-matrix.*
(ii) *For every signature matrix $S \neq I$ or $-I$, there exists a nonnegative vector $x$, such that $SASx > 0$, that is, $v(SAS) > 0$.*
(iii) *$A$ does not reverse the sign of any non-unisigned vector, that is, $(Ax)_i x_i \leq 0, \forall i$ implies that $x \geq 0$ or $x \leq 0$.*
(iv) *For all $q > 0$, $(q, A)$ has exactly two solutions.*
(v) *$SAS \in \boldsymbol{Q}$ for every $S \neq I$ or $-I$.*

**Proof.** Equivalence of the first four statements ((i) to (iv)) is already established in [24]. Hence it suffices to show that (ii) and (v) are equivalent. Assume (ii) holds. Let $S$ be any signature matrix other than $I$ and $-I$. From Proposition 1.1 and the fact that $(SAS)^{-1} = SA^{-1}S$, it follows that $v(SA^{-1}S) > 0$. Write $M = SA^{-1}S$. It is easy to check that every proper principal submatrix of $M$ is a $\boldsymbol{P}$-matrix and hence has value positive (see Corollary 3.3.5 of [3]). Hence $M$ and all its principal submatrices have value positive. From Proposition 2.1 (ii), $M$ is a $\boldsymbol{Q}$-matrix. It is a well-known result in LCP that if a nonsingular matrix is a $\boldsymbol{Q}$-matrix ($\boldsymbol{Q}_0$-matrix), then so is its inverse. From this it follows that $SAS$ is a $\boldsymbol{Q}$-matrix. Conversely, assume (v). From Proposition 2.1, the value of every $\boldsymbol{Q}$-matrix is positive. Note that (ii) is a direct consequence of this. □

A similar result can be given for $\boldsymbol{P}$-matrices. Interested readers may consult [20].

An equivalent characterization of $\boldsymbol{Q}_0$-matrices is that $K(A)$ is convex (see [8]). Below we give an alternative proof of a result (Theorem 2.3) due to Eagambaram and Mohan [6] using Kaplansky's theorem.

**Lemma 2.1.** *Let $A \in \boldsymbol{R}^{n\times n} \cap \boldsymbol{Q}_0$ with rank$(A) = n - 1$. Assume that there exists a vector $\delta > 0$ such that $A\delta = 0$. Then there exists an optimal stategy $\pi$ for Player I (in the matrix game A) such that $K(A)$ is equal to the half-space $H = \{q : \pi^t q \geq 0\}$.*

**Proof.** It is easy to check that the hypothesis of the lemma implies that $v(A) = 0$. Hence $\delta$ is a completely mixed strategy for Player II. Let $\pi$ be any optimal strategy for Player I. From Theorem 1.1(i), $\pi^t A = 0$. We will show that $K(A) = H$. Let $q \in K(A)$. Then there exists an $x \geq 0$ such that $Ax + q \geq 0$. This implies, as $\pi \geq 0$ and $\pi^t A = 0$, $\pi^t q \geq 0$. Hence $q \in H$. Conversely, let $q \in H$. Suppose the system $Ax + q \geq 0,\ x \geq 0$ has no feasible solution. Then, from a theorem of alternatives, there exists a $y$ such that $y^t A \leq 0,\ y \geq 0$, and $q^t y < 0$. Without loss of generality, we may assume that $y$ is a probability vector. Clearly, $y$ is an optimal stategy different from $\pi$ as $q^t \pi \geq 0$. But from Theorem 1.1(i), $y^t A = 0$ and as rank$(A) = n - 1$, it follows that $y = \pi$, a contradiction. It follows that

$Ax + q \geq 0,\ x \geq 0$ has a feasible solution and as $A \in \boldsymbol{Q}_0$, $q \in K(A)$. Thus $K(A) = H$. □

**Theorem 2.3.** *Let $A \in \boldsymbol{R}^{n\times n}$. Suppose $A^t\pi = 0 = A\delta$, where $\pi > 0, \delta > 0$, and rank of $A = n - 1$. Then $K(A) = \{q : \pi^t q \geq 0\}$. Furthermore, if $\pi^t q = 0$ (i.e., $q$ is on the boundary of $K(A)$), then $(q, A)$ has infinitely many solutions.*

**Proof.** First, we show that $A \in \boldsymbol{Q}_0$. Let $q$ be feasible, that is, there exists $x \geq 0$ such that $Ax + q \geq 0$. This implies $\pi^t q \geq 0$. We will prove $q \in K(A)$. If $\pi^t q = 0$, then, as $\pi > 0$, $Ax + q \geq 0$ implies $Ax + q = 0$, Hence, $q \in K(A)$. So assume $\pi^t q > 0$. From Kaplansky's results (Theorem 1.1), the matrix game $A$ is completely mixed with value zero and the cofactors $A_{ij}$ are all different from zero and are of the same sign. For each positive integer $k$, define

$$B_k = \begin{bmatrix} a_{11} + 1/k & a_{12} & \dots & a_{1n} \\ a_{21} & a_{22} & \dots & a_{2n} \\ \vdots & \vdots & \vdots & \vdots \\ a_{n1} & a_{n2} & \dots & a_{nn} \end{bmatrix}.$$

It can be checked that $|B_k| = (1/k)A_{22}$ (see determinantal formula in page 60 of [3]). For all large $k$, the cofactors of $B_k$ have the same sign as those of $A$. Consequently, $B_k^{-1} > 0$ for all large $k$. Then, from Proposition 2.1(i), $B_k^{-1}$ and hence $B_k$ are $\boldsymbol{Q}$-matrices for all large $k$ (also see [20]). Let $w^k,\ x^k$ be a solution of $(q, B_k)$. Then

$$\begin{bmatrix} a_{11} + 1/k & a_{12} & \dots & a_{1n} \\ a_{21} & a_{22} & \dots & a_{2n} \\ \vdots & \vdots & \vdots & \vdots \\ a_{n1} & a_{n2} & \dots & a_{nn} \end{bmatrix} \begin{bmatrix} x_1^k \\ x_2^k \\ \vdots \\ x_n^k \end{bmatrix} + \begin{bmatrix} q_1 \\ q_2 \\ \vdots \\ q_n \end{bmatrix} = w^{(k)}. \tag{3}$$

If $\{x^k\}$ is bounded, then clearly $(q, A)$ has a solution. So assume $x^k$ is unbounded. Without loss of generality, we may assume $x^k/\sum_i x_i^k \to \bar{x}$. Clearly $0 \neq \bar{x} \geq 0$. Hence $\bar{x}$ is an optimal strategy for Player II. Since $A$ is completely mixed, $\bar{x} > 0$. Hence, $x^k > 0$ for all large $k$. Consequently,

$$Ax^k + \begin{bmatrix} q_1 + x_1^k/k \\ q_2 \\ \vdots \\ q_n \end{bmatrix} = 0.$$

This is impossible for

$$0 = \pi^t A x^k + \pi^t \begin{bmatrix} q_1 + x_1^k/k \\ q_2 \\ \vdots \\ q_n \end{bmatrix} = \pi^t A x^k + \pi^t q + \pi_1 \frac{x_1^k}{k} > 0,$$

because $\pi^t q > 0$. It follows that $x^k$ cannot be unbounded. Thus $(q, A)$ has a solution. We conclude that $A \in \boldsymbol{Q}_0$. From Lemma 2.1, we have $K(A) = \{q : \pi^t q \geq 0\}$.

If $q$ is on the boundary of $K(A)$, then $\pi^t q = 0$ and hence $Ax + q = 0$ for any $x \in S(q, A)$. Note that $x + \lambda\delta \in S(q, A)$ for all $\lambda > 0$. Thus $(q, A)$ has infinitely many solutions in this case. □

A matrix $A$ is called a $\boldsymbol{U}$-matrix if $(q, A)$ has a unique solution for every $q$ in the interior of $K(A)$; and $A$ is called a $\boldsymbol{P}_1$-matrix if exactly one of its principal minors is zero and all the other principal minors are positive; $A$ is called a $\boldsymbol{P}_0$-matrix if all its principal minors are nonnegative. Stone [31] showed that every $\boldsymbol{Q}_0 \cap \boldsymbol{U}$-matrix is a $\boldsymbol{P}_0$-matrix.

It is known that $\boldsymbol{P}_1$-matrices are in $\boldsymbol{Q}_0$. Cottle and Stone [4] showed that if $A \in \boldsymbol{P}_1 \backslash \boldsymbol{Q}$, then $A \in U$ and that $K(A)$ is a half-space. They further showed that if in addition $|A| = 0$, then the support of the half-space can be chosen to be a positive vector. It easy to check that if $A$ is a $\boldsymbol{P}_1$-matrix with $|A| = 0$, then the matrix game $A$ is completely mixed with $v(A) = 0$. Hence we can elaborate Cottle and Stone's result as follows: If $A \in \boldsymbol{P}_1$ with $|A| = 0$, then $A$ is a completely mixed game and that $A \in U$. A relevent question in this context is: Is the converse true, that is, if $A$ is a completely mixed game with $v(A) = 0$ and if $A \in \boldsymbol{U}$, is it true that $A \in \boldsymbol{P}_1$? The answer is "yes." This is the content of our next theorem.

**Theorem 2.4.** *Let $A \in \boldsymbol{R}^{n \times n}$ be a completely mixed game with $v(A) = 0$. Then the following two statements are equivalent:*

(i) *$A \in \boldsymbol{P}_1$ with $|A| = 0$; and*
(ii) *$A \in U$.*

**Proof.** The first implication, (i) ⇒ (ii), is already proved by Cottle and Stone (see Theorem 4.1.13 of [3] for a proof). We shall show that (ii) implies (i). Since $A$ is completely mixed with $v(A) = 0$, it follows from Theorem 1.1, rank$(A) = n - 1$ and that there exist positive vectors $\pi$ and $\delta$ satisfying the hypothesis of Theorem 2.3. Hence $A$ is a $\boldsymbol{Q}_0$-matrix. Since $A \in \boldsymbol{U}$, $A \in \boldsymbol{P}_0$ (see [4]). To complete the proof, we need to show that every proper principal minor of $A$ is positive.

Suppose $a_{11} = 0$. Then,

$$\begin{bmatrix} 0 & a_{12} & \dots & a_{1n} \\ a_{21} & a_{22} & \dots & a_{2n} \\ \vdots & \vdots & \vdots & \vdots \\ a_{n1} & a_{n2} & \dots & a_{nn} \end{bmatrix} \begin{bmatrix} x \\ 0 \\ \vdots \\ 0 \end{bmatrix} + \begin{bmatrix} 0 \\ q_2 \\ \vdots \\ q_n \end{bmatrix} = w,$$

$$\begin{bmatrix} 0 & a_{12} & \dots & a_{1n} \\ a_{21} & a_{22} & \dots & a_{2n} \\ \vdots & \vdots & \vdots & \vdots \\ a_{n1} & a_{n2} & \dots & a_{nn} \end{bmatrix} \begin{bmatrix} 2x \\ 0 \\ \vdots \\ 0 \end{bmatrix} + \begin{bmatrix} 0 \\ q_2 \\ \vdots \\ q_n \end{bmatrix} = w'.$$

In other words, we can choose $q_2, \dots, q_n$ large enough with $\pi^t q > 0$ where $q = (0, q_2, \dots, q_n)^t$. Clearly, such a $q$ is in the interior of $K(A) = \{q : \pi^t q \geq 0\}$ (follows from Theorem 2.3) and $(q, A)$ will have two solutions contradicting our hypothesis, namely $A \in U$. It follows that $a_{ii} > 0$ for every $i$. We will now show that $2 \times 2$ principal minors are positive. Suppose $\det \begin{bmatrix} a_{11} & a_{12} \\ a_{21} & a_{22} \end{bmatrix} = 0$. Also, there exist reals $t_1$ and $t_2$, not both zero, such that $\begin{bmatrix} a_{11} & a_{12} \\ a_{21} & a_{22} \end{bmatrix} \begin{bmatrix} t_1 \\ t_2 \end{bmatrix} = 0$. Choose $q_1$ and $q_2$ so that $\begin{bmatrix} a_{11} & a_{12} \\ a_{21} & a_{22} \end{bmatrix} \begin{bmatrix} 1 \\ 1 \end{bmatrix} + \begin{bmatrix} q_1 \\ q_2 \end{bmatrix} = 0$. We can choose $q_3, \dots, q_n$ so large satisfying: (a) $\pi^t q > 0$ and (b) $(q, A)$ has at least two solutions, namely, $(1, 1, 0 \dots 0)$ and $(1+t_1, 1+t_2, 0 \dots 0)$. This leads to a contradiction to the hypothesis, namely $A \in U$. It follows that the determinant of every $2 \times 2$ principal submatrix of $A$ is positive. Similarly, we can show that all proper principal minors of $A$ are positive. Thus, $A \in \boldsymbol{P}_1$. □

Say that a matrix $A \in \boldsymbol{E}'$ if $|S(q, A)| = 1$ for all nonzero nonnegative $q$ and if $|S(0, A)| \geq 2$. Danao [5] attributes the following conjecture to Cottle: Let $A \in \boldsymbol{R}^{n \times n} \cap \boldsymbol{P}_0$. Then $A \in \boldsymbol{E}'$ if, and only if, $A \in \boldsymbol{P}_1 \backslash \boldsymbol{Q}$. One can verify that if $A \in \boldsymbol{P}_1 \backslash \boldsymbol{Q}$, then $A \in \boldsymbol{E}'$ (see Danao [5]). However, the converse is not true. In other words, $A \in \boldsymbol{E}' \cap \boldsymbol{P}_1$ need not imply $A \in \boldsymbol{P}_1$. Below we give a counterexample.

**Example 2.1.** We now give an example of a matrix $A$ with the following property. $A \in \boldsymbol{P}_0 \cap \boldsymbol{E}'$ but $A \notin U$ and, consequently, by Theorem 2.4, $A \notin \boldsymbol{P}_1$. Let

$$A = \begin{bmatrix} 1 & 1 & -1 & -1 \\ 1 & 1 & -2 & 0 \\ 1 & 2 & 1 & -4 \\ -3 & -4 & 2 & 5 \end{bmatrix}.$$

Here, $K(A) = \{q \in \boldsymbol{R}^4 : q_1 + q_2 + q_3 + q_4 \geq 0\}$.

Let $q_0 = (-4, -4, 20, 100)$. Then $q_0$ is in the interior of $K(A)$. Note that $((q, A))$ has (at least) two solutions namely $x^1 = (4, 0, 0, 0)^t$ and $x^2 = (0, 4, 0, 0)^t$ with $Ax^i + q^0 \geq 0$ and $x^i(Ax^i + q_0) = 0$ for $i = 1, 2$. Thus, $A \notin U$. Also, observe that column sums and row sums are zero and that rank$(A) = 3$. Invoking Theorem 2.4, we infer $A \notin \boldsymbol{P}_1$. It is not hard to check that $A \in \boldsymbol{E}' \cap \boldsymbol{P}_0$.

In the next section we shall make a cursory mention of some applications of LCP in solving certain classes of game theory problems and briefly introduce the notion of polystochastic games.

## 3 LCP in Stochastic Games

When Nash introduced the concept of equilibrium points for noncooperative games [21], one of the major open problems in the bimatrix games was to compute the equilibrium points. The original proof by Nash was existential in nature, and it used either Brouwer's fixed point theorem or Kakutani's fixed point theorem. It was Lemke and Howson [14] who came up with a remarkable (finite) algorithm to get an equilibrium point for bimatrix games. This algorithm is basically to solve an LCP, and the solutions to bimatrix games are obtained by formulating the bimatrix games as LCPs. Though the algorithm works fairly well with general LCPs in practice, it is known to work conclusively only for certain known classes of LCPs. These known classes are pretty large and include problems that arise from bimatrix game formulations. Consequently, there have been attempts to formulate game theory problems as LCPs.

Schultz [29] fomulates discounted switching controller stochastic zero-sum games as LCPs and shows that these games have order field property. However, it is not proved that the LCPs encountered in his formulations can be processed by Lemke's algorithm.

In a similar attempt, Nowak and Raghavan [22] formulate nonzero-sum two-person stochastic games as LCPs and show that these LCPs are indeed processable by Lemke's algorithm. However, the approach followed here involves computation of limiting average expected costs at each possible combination of pure strategies before arriving at the LCP formulation.

In contrast, the works of Mohan, Neogy, and Parthasarathy [16], [17] consider the polystochastic games with the criterion as either total discounted expected costs or the limiting average expected costs. These authors formulate the problems as LCPs and the formulations depend on the data directly arising from games and no additional computations are involved. We shall briefly outline the recent works of Mohan, Neogy, and Parthasarathy [16], [17], relating stochastic games to LCP.

A polystochastic game is defined by the objects, $(N, S, N_i(s), q(t/s, a_n)$, for $s \in S, i \neq j, i, j \in N)$.

Here, $N = \{1, 2, \ldots, n\}$ denotes the set of players, $S = \{1, 2, \ldots, m\}$ denotes the states of a system, $N_i(s)$ stands for the actions available to Player $i$ in state $s$, the matrix $A_{ij}(s)$ denotes the matrix of partial costs incurred by Player $i$ depending on the actions of players $i$ and $j$, $i \neq j$, and $q(t/s, a_n)$ is the probability that the game moves to state $t$ given the game is played in state $s$ and Player $n$ chooses action $a_n \in N_n(s)$.

Suppose $x_i(s)$ denotes the vector of probabilities over $N_i(s)$ used by Player $i$. Then the total expected cost incurred by Player $i$ on any given day is given by $x_i(s)^t \sum_{i \neq j} A_{ij}(s) x_j(s)$. Suppose the game is played over the infinite horizon. Let $f_i = \{x_i(s) : s \in S\}$ be a sequence of mixed strategies played by Player $i$. Such strategies $f_i$ are called stationary strategies.

Usually the payoffs are considered either with discount or without discount. Therefore, the criteria for computation of costs are of two types: (1) $\beta$-discounted payoff ($\beta$ is the discounting factor); and (2) undiscounted payoff.

(1) $\beta$-discounted payoff: For the $n$ players and $\beta \in [0, 1)$, the $\beta$-discounted payoff vector is given by

$$I^i_\beta(f_1, f_2, \dots, f_n) = \sum_{r=0}^{\infty} \beta^r Q^r(f_1, f_2, \dots, f_n)c,$$

where $c$ stands for the cost vector, $(f_1, f_2, \dots, f_n)$ for the strategies used by the $n$ players, $Q^0 = I$ (Identity matrix), $Q^r = Q \times Q \times \cdots \times Q$ (product of $Q$ taken $r$ times), and $Q$ is the Markov matrix whose elements are given by $q(t/s, f_1(s), \dots, f_n(s))$, the probability that the system moves to state $t$ from $s$ given that $f_1(s), \dots, f_n(s)$ are the actions taken by the players.

(2) Undiscounted payoff: For the $i$th player's payoff is given by

$$(f_1, f_2, \dots, f_n) = Q^* c,$$

where $Q^* = \lim_{i \to \infty}((1/k) \sum_{i=1}^{k} Q^i)$ and $c$ is the cost vector.

The objective of the game is that each player tries to minimize his cost. Call $(f_1^*, f_2^*, \dots, f_n^*)$ an equilibrium point if

$$I^1_\beta(f_1^*, f_2^*, \dots, f_n^*)(s) \leq I^1_\beta(f_1, f_2^*, \dots, f_n^*)(s), \forall f_1 \text{ and } s,$$
$$I^2_\beta(f_1^*, f_2^*, \dots, f_n^*)(s) \leq I^2_\beta(f_1^*, f_2, \dots, f_n^*)(s), \forall f_2 \text{ and } s,$$
$$\vdots \quad \vdots \quad \vdots$$
$$I^n_\beta(f_1^*, f_2^*, \dots f_n^*)(s) \leq I^n_\beta(f_1^*, \dots, f_{n-1}^*, f_n)(s), \forall f_n \text{ and } s.$$

A similar definition can be given for the undiscounted case.

The existence of equilibrium points is known. See [10] and [32]. Is it possible to give an efficient algorithm to get an equilibrium point to polystochastic games? It was shown recently that indeed one can compute an equilibrium point through a Lemke-type algorithm. The details can be found in [16] and [17]. A word of caution: In the case of undiscounted payoff, it was possible to give a Lemke-type algorithm under the additional assumption that the transition probability matrix is irreducible. The general problem is open.

We urge the interested reader to see the references to get a feel of stochastic games. In particular, see [16], [17], [22], [25], and [28]. Readers interested in stochastic games may also look at the interesting papers by Filar and Vrieze [9] and Maitra and Sudderth [15].

## Acknowledgments

We are grateful to Professors T. E. S. Raghavan, S. R. Mohan, S. K. Neogy, and R. Sridhar for several helpful comments. We record our sincere thanks to the anonymous referees whose valuable suggestions and comments have helped us in the complete restructuring of the original version of this paper.

## REFERENCES

[1] Bapat, R. B. and T. E. S. Raghavan. *Nonnegative Matrices and Applications*. Cambridge University Press, Cambridge, UK, 1996.

[2] Cottle, R. W. A Note on Completely $Q$-Matrices. *Mathematical Programming*, **19**, 347–351, 1980.

[3] Cottle, R. W., J. S. Pang, and R. E. Stone. *The Linear Complementarity Problem*. Academic Press, New York, 1992.

[4] Cottle, R. W. and R. E. Stone. On the Uniqueness of Solutions to Linear Complementarity Problems. *Mathematical Programming*, **27**, 191–213, 1983.

[5] Danao, R. A. A Note on $E'$-Matrices. *Linear Algebra and its Applications*, **259**, 299–305, 1997.

[6] Eagambaram, N. and S. R. Mohan. On Some Classes of Linear Complementarity Problems with Matrices of Order $n$ and Rank $(n-1)$. *Mathematical Operations Research*, **15**, 243–257, 1990.

[7] Eaves, B. C. The Linear Complementarity Problem. *Management Science*, **17**, 612–634, 1971.

[8] Eaves, B. C. On Quadratic Programming. *Management Science*, **17**, 698–711, 1971.

[9] Filar, J. and K. Vrieze. *Competetive Markov Decision Processes*. Springer-Verlag, New York, 1996.

[10] Fink, A. M. Equilibrium in a Stochastic $n$-Person Game. *Journal of Science, Hiroshima University Ser., A-1*, **28**, 89–93, 1964.

[11] Kaplansky, I. A Contribution to von Neumann's Theory of Games. *Annals of Mathematics*, **46**, 474–479, 1945.

[12] Karamardian, S. The Complementarity Problem. *Mathematical Programming*, **2**, 107–129, 1962.

[13] Lemke, C. E. Bimatrix Equilibrium Points and Mathematical Programming. *Management Science*, **11**, 681–689, 1965.

[14] Lemke, C. E. and J. T. Howson. Equilibrium Points of Bimatrix Games. *SIAM Journal of Applied Mathematics*, **12**, 413–423, 1961.

[15] Maitra, A. P. and W. D. Sudderth. *Discrete Gambling and Stochastic Games*. Springer-Verlag, New York, 1996.

[16] Mohan, S. R., S. K. Neogy, and T. Parthasarathy. Linear Complementarity and Discounted Polystochastic Game when One Player Controls Transitions. *Proceedings of the International Conference on Complementarity Problem*. Editors M.C. Ferris and J.S. Pang. SIAM, Philadelphia, PA (to appear).

[17] Mohan, S. R., S. K. Neogy, and T. Parthasarathy. Linear Complementarity and the Irreducible Polystochastic Game with the Average Cost Criterion when One Player Controls Transitions. *Proceedings of the International Conference on Game Theory and Its Applications*, Kluwer Academic, Amsterdam, 153–170, 1997.

[18] Mohan, S. R., T. Parthasarathy, and R. Sridhar. The Linear Complementarity Problem with Exact Order Matrices. *Mathematics of Operations Research*, **19**, 618–644, 1994.

[19] Murthy, G. S. R., T. Parthasarathy, and G. Ravindran. On Copositive, Semimonotone ***Q***-Matrices. *Mathematical Programming*, **68**, 187–203, 1995.

[20] Murthy, K. G. *Linear Complementarity, Linear and Nonlinear Programming*. Heldermann Verlag, Berlin, 1998.

[21] Nash, J. Non-Cooperative Games. *Annals of Mathematics*, **54**, 286–295, 1951.

[22] Nowak, A. S. and T. E. S. Raghavan. A Finite Step Algorithm via a Bimatrix to a Single Controller Nonzero Stochastic Game. *Mathematical Programming*, **59**, 249–259, 1993.

[23] Owen, G. *Game Theory*. Academic Press, New York, 1982.

[24] Parthasarathy, T. and G. Ravindran. $N$-Matrices. *Linear Algebra and its Applications*, **139**, 89–102, 1990.

[25] Parthasarathy, T. and T. E. S. Raghavan. An Orderfield Property for Stochastic Games when One Player Controls Transition Probabilities. *Journal of Optimization Theory and its Applications*, **33**, 375–392, 1981.

[26] Raghavan, T. E. S. Completely Mixed Games and ***M***-Matrices. *Linear Algebra and its Applications*, **21**, 35–45, 1978.

[27] Raghavan, T. E. S. Some Remarks on Matrix Games and Nonnegative Matrices. *SIAM Journal of Applied Mathematics*, **36**, 83–85, 1979.

[28] Raghavan, T. E. S. and J. A. Filar. Algorithms for Stochastic Games, A Survey. *ZOR—Methods and Models of Operations Research*, **35**, 437–472, 1995.

[29] Schultz, T. A. Linear Complementarity and Discounted Switching Controller Stochastic Games. *Journal of Optimization. Theory and its Applications*, **73**, No. 1, 89–99, 1992.

[30] Shapley, L. S. Stochastic Games. *Proceedings of the National Academy of Sciences, USA*, **28**, 95–99, 1964.

[31] Stone, R. E. Geometric Aspects of Linear Complementarity Problem. Ph.D. thesis, Department of Operations Research, Stanford University, Stanford, CA, 1981.

[32] Takahashi, M. Equilibrium Points of Stochastic Noncooperative $n$-Person Game. *Journal of Science, Hiroshima University A-I*, **28**, 95–99, 1964.

[33] Von Neumann, J. and O. Morgenstern. *Theory of Games and Economic Behaviour*. Princeton University Press, Princeton, NJ, 1944.

# Weighted Discounted Stochastic Games with Perfect Information

Eitan Altman
INRIA B.P.93
Sophia-Antipolis Cedex, France

Eugene A. Feinberg
Harriman School for Management and Policy
SUNY at Stony Brook
Stony Brook, New York, USA

Adam Shwartz
Technion-Israel Institute of Technology
Department of Electrical Engineering
Haifa, Israel

**Abstract**

We consider a two-person zero-sum stochastic game with an infinite-time horizon. The payoff is a linear combination of expected total discounted rewards with different discount factors. For a model with a countable state space and compact action sets, we characterize the set of persistently optimal (subgame perfect) policies. For a model with finite state and action sets and with perfect information, we prove the existence of an optimal pure Markov policy, which is stationary from some epoch onward, and we describe an algorithm to compute such a policy. We provide an example that shows that an optimal policy, which is stationary after some step, may not exist for weighted discounted sequential games with finite state and action sets and without the perfect information assumption. We also provide examples of similar phenomena of nonstationary behavior for the following two classes of problems with weighted discounted criteria: (i) models with one controller and with finite state and compact action sets, and (ii) nonzero-sum games with perfect information and with finite state and action sets.

## 1 Introduction

Several problems in finance, project management, budget allocation, production, and management of computer and telecommunication systems lead to sequential decision problems where the objective functions are linear combinations of total expected discounted rewards, each with its own discount factor. Some of these problems are described in [4], [5], [6], [15], [24].

Many applications of these criteria arise from the conflict of short- and long-term interests, since discount factors may be viewed as related to time-scales. For example, in the development or selection of technologies, ecological consequences and objectives have significantly larger time-scales than direct economical goals. This is also the case in applications to pavement management in roads [12] and building maintenance [28]. Other applications are related to control of systems in which different components have different reliability. Often, different system or performance measures such as costs of items change over time in different ways (see [6]). Production costs often change due to learning or experience (see [5]), at a rate which is unrelated to the change in cost of different parts, etc.

Markov Decision Processes with weighted criteria have been studied in [3], [4], [5], [6], [8], [15]. Even in this case, when there is just one player, the results for problems with weighted discounted rewards differ significantly from the results for standard discounted models. For example, stationary optimal policies may not exist for weighted discounted problems with finite state and action sets [4]. However, in the case of one player, there exist optimal pure Markov policies that are stationary from some epoch onward [4].

Stochastic two-person zero-sum weighted discounted games have been studied in [9] where the existence of $\epsilon$-optimal policies which are stationary from some epoch onward is proved. The existence of a value was established in [4], [9] by reducing weighted discounted games to standard discounted games with expanded state spaces. The existence of the value can also be proved by using the results of [21], which studies nonstationary stochastic games.

This paper deals with a two-person zero-sum stochastic game with weighted discounted payoffs. The main goal is to study finite state and action models with perfect information.

Perfect information means that at any state either the action set of Player 1 or the action set of Player 2 is a singleton (see [11]). We show that for each player there exists an optimal pure Markov policy which is stationary from some epoch onward. We also provide an algorithm that computes such a policy in a finite number of steps. An optimal policy that is stationary after some step may not exist for standard weighted discounted games with finite state and action sets.

The paper is organized as follows. The model definition and the notation are given in Section 2. Section 3 introduces and studies lexicographical games. In particular, in Section 3 we describe the structure of lexicographically optimal policies for discounted stochastic games with countable state and compact action sets. In Section 4 we describe general results on countable state weighted discounted games. Section 5 deals with finite state and action stochastic games with perfect information. We prove the existence of an optimal pure Markov policy which is stationary from some step onward, describe the sets of all optimal policies, and formulate an algorithm for their computation. Section 6 deals with counterexamples.

As counterexamples in [4] and in Section 6 show, the main results of the paper, presented in Section 5, hold just for games with perfect information, with finite state and action sets, and a zero-sum assumption on the costs. The results on lexicographical games and on the existence of optimal Markov policies, presented in Sections 3 and 4, hold for countable state games with compact action sets introduced in Section 2.

## 2 Definitions and Notation

Consider a two-person zero-sum stochastic game with a finite or countable state space $\mathbf{X}$, and two metric spaces of actions $\mathbf{A}$ and $\mathbf{B}$ for Players 1 and 2, respectively. The transition probability $p(y \mid x, a, b)$ is the probability to go to state $y$ given state $x$, given that actions $a$ and $b$ are chosen by the players. Let $\mathbf{A}(x)$ and $\mathbf{B}(x)$ denote the set of actions available for Players 1 and 2 at state $x$. We assume that $\sum_{y \in \mathbf{X}} p(y \mid x, a, b) = 1$ for all $x \in \mathbf{X}$, $a \in \mathbf{A}(x)$, and $b \in \mathbf{B}(x)$. In contrast with standard discounted stochastic games (with a single payoff function that Player 2 pays Player 1 and one discount factor), here we have $K$ payoff functions $r_k : \mathbf{X} \times \mathbf{A} \times \mathbf{B} \to \mathbb{R}$, $k = 1, \dots, K$, and $K$ discount factors $\beta_k \in [0, 1[$, $k = 1, \dots, K$, where $K$ is a given positive integer. Throughout the paper the following is in force:

**Technical Assumptions.** For each $x \in \mathbf{X}$ the sets $\mathbf{A}(x)$ and $\mathbf{B}(x)$ are compact. For each $x, y \in \mathbf{X}$, the transition probabilities $p(y \mid x, a, b)$ are continuous in $a$ and $b$, and each function $r_k(x, a, b)$, $k = 1, \dots, K$, is bounded, upper semicontinuous in $a \in \mathbf{A}(x)$, and lower semicontinuous in $b \in \mathbf{B}(x)$.

Without loss of generality, assume $\beta_1 > \beta_2 > \cdots > \beta_K$. (If the discount factors are not ordered, we reorder them. If $\beta_i = \beta_j$ for some $i > j$, we rename $r_j := r_j + r_i$, eliminate the $i$th component, and obtain a model with $K - 1$ payoff functions and $K - 1$ discount factors.) Define histories $h_t = x_0, a_0, b_0, x_1, a_1, b_1, \dots, x_t$, where $t = 0, 1, \dots$. Let $\mathcal{U}$ and $\mathcal{V}$ be the set of (behavioral) policies available to Players 1 and 2, respectively. Policies $u \in \mathcal{U}$ and $v \in \mathcal{V}$ are sequences $u = u_0, u_1, \dots$ and $v = v_0, v_1, \dots$, where $u_t$ and $v_t$ are probability distributions, respectively, on $\mathbf{A}(x_t)$ and $\mathbf{B}(x_t)$ conditioned on $h_t$. The randomizations used by the two players are assumed to be independent.

A policy $u$ for Player 1 is (randomized) Markov if at any time $t$, $u_t$ depends only on the current state $x_t$. A Markov policy is called stationary if it is time homogeneous, i.e., $u_0 = u_1 = \cdots$. A policy $u$ for Player 1 is called pure if the distribution $u_t(\cdot \mid h_t)$ is concentrated at one point $u_t(h_t)$ for each history $h_t$, $t = 0, 1, \dots$. Pure stationary policies are called deterministic. For $N = 0, 1, \dots$, a pure Markov policy $u$ for Player 1 is called $(N, \infty)$-stationary if $u_t = u_N$ for all $t \geq N$. The notions of $(0, \infty)$-stationary and deterministic policies coincide. Special classes of policies for Player 2 are defined analogously.

Given an initial state $x$, each pair of policies $(u, v)$ defines a probability measure $P_x^{u,v}$ on the set of trajectories $x_0, a_0, b_0, x_1, a_1, b_1, \dots$. We denote by $E_x^{u,v}$ the expectation with respect to this measure.

The discounted payoff associated with the one-step payoff $r_k$ and discount factor $\beta_k$ for an initial state $x$, where the players use policies $u$ and $v$, is defined to be

$$V_k(x, u, v) = E_x^{u,v} \sum_{t=0}^{\infty} (\beta_k)^t r_k(x_t, a_t, b_t). \tag{1}$$

The weighted discounted payoff corresponding to the initial state $x$, and policies $u$ and $v$ is then given by

$$V(x, u, v) = \sum_{k=1}^{K} V_k(x, u, v). \tag{2}$$

Player 1 wishes to maximize $V(x, u, v)$, and Player 2 wishes to minimize it.

Remark 2.8 in [4] reduces this game to a game with one discount factor and with a countable state space. Countable state discounted games with compact action sets have values; see, e.g., [7], [19], or Theorem 3.1. Therefore, countable state games with compact action sets and with weighted discounted payoffs have values as well. A reduction to a game with one discount factor but with a continuous state space was described in [9]. Define $V(x)$ to be the value of the weighted discounted game and, for $k = 1, \ldots, K$, let $V_k(x)$ denote the value of the game with criterion $V_k(x, \cdot, \cdot)$.

A policy $u^*$ is said to be optimal for Player 1 in game (2) if and only if for any $u \in \mathcal{U}$, $\inf_v V(x, u, v) \leq \inf_v V(x, u^*, v)$ (where the latter is equal to $V(x)$) for all $x \in \mathbf{X}$. Optimality of a policy for the second player is defined similarly.

For a policy $u \in \mathcal{U}$ and a history $\tilde{h}_n = \tilde{x}_0, \tilde{a}_0, \tilde{b}_0, \ldots, \tilde{x}_n, \tilde{a}_n, \tilde{b}_n \in (\mathbf{X} \times \mathbf{A} \times \mathbf{B})^{n+1}$, $n = 0, 1, \ldots$, we define the shifted policy $\tilde{h}_n u$ as the policy which uses, in response to a history $h_m = x_0, a_0, b_0, \ldots, x_m$, the action that $u$ would use at epoch $(n+m)$ if the history $\tilde{h}_n h_m = \tilde{x}_0, \tilde{a}_0, \tilde{b}_0, \ldots, \tilde{x}_n, \tilde{a}_n, \tilde{b}_n, x_0, a_0, \ldots, x_m$ is observed. A similar definition holds for $v \in \mathcal{V}$. We will also use the formal notation $\tilde{h}_{-1} u = u$ and $\tilde{h}_{-1} v = v$. For a Markov policy $u$ of any player $\tilde{h}_n u = (u_{n+1}, u_{n+2}, \ldots)$ does not depend on $\tilde{h}_n$. If $u$ is stationary, $\tilde{h}_n u = u$.

The total expected weighted discounted reward incurred from epoch $(n+1)$, $n = 0, 1, \ldots$, onward if the players use policies $u$, $v$, a history $\tilde{h}_n$ took place and $x_{n+1} = x$ is

$$V(x, n+1, u, v) = (\beta_1)^{-(n+1)} \sum_{k=1}^{K} (\beta_k)^{n+1} V_k(x, \tilde{u}, \tilde{v}), \tag{3}$$

where $\tilde{u} = \tilde{h}_n u$, $\tilde{v} = \tilde{h}_n v$. We also set $V(x, 0, u, v) = V(x, u, v)$. We introduce the normalization constant $(\beta_1)^{-n}$ in (3) just in order to have $V_1(x) = V(x, n) + o(1)$.

So any history $\tilde{h}_n$ defines a new stochastic game that starts at epoch $(n+1)$. Let $V(x, n)$ be the value of the zero-sum game that starts at epoch $n = 0, 1, \ldots$ with the payoffs (3). The existence of this value follows from Remark 2.8 in [4]. Since both players know the history and therefore have the same information about the past, this value does not depend on the history before epoch $n$. A policy $u^*$

($v^*$) for Player 1 (2) is called persistently optimal [4] if it is optimal and for any $\tilde{h}_n \in (\mathbf{X} \times \mathbf{A})^{n+1}$, $n = 0, 1, 2, \ldots$, the policy $\tilde{h}_n u$ ($\tilde{h}_n v$) is optimal (with respect to the cost (3)) as well. We will apply the definition of persistent optimal policies to both criteria $V_k$, $k = 1, \ldots, K$, and $V$.

The main objective of this paper is to study games with perfect information which are a special case of stochastic games (see, e.g., [11], [16]). We say that a game is with perfect information if there exist two sets of states $\mathbf{Y}$ and $\mathbf{Z}$ such that: (i) $\mathbf{Y} \cup \mathbf{Z} = \mathbf{X}$; (ii) $\mathbf{Y} \cap \mathbf{Z} = \emptyset$; and (iii) the sets $\mathbf{A}(z)$ and $\mathbf{B}(y)$ are singletons for all $z \in \mathbf{Z}$ and for all $y \in \mathbf{Y}$. In particular, if $p(\mathbf{Y} \mid y, a, b) = p(\mathbf{Z} \mid z, a, b) = 0$ for all $y \in \mathbf{Y}$, $z \in \mathbf{Z}$, $a \in \mathbf{A}$, and $b \in \mathbf{B}$ in a game with perfect information, then the players make their moves sequentially.

A particular, important example is a stochastic game where the players make their moves simultaneously, but Player 2 knows the decision of Player 1 at each epoch [16]. In other words, $v_t$ may depend on $(h_t, a_t)$, not just on $h_t$. In this game, all definitions of special policies should be modified by replacing $x_t$ with $(x_t, a_t)$ in all conditional distributions $v_t$. Let us define an equivalent game with perfect information. All objects in this new model are marked with "$\tilde{\ }$." Let $\tilde{\mathbf{Y}} = \mathbf{X}$, $\tilde{\mathbf{Z}} = \mathbf{X} \times \mathbf{A}$, $\tilde{\mathbf{X}} = \tilde{\mathbf{Y}} \cup \tilde{\mathbf{Z}}$, and $\tilde{\mathbf{A}}(x) = \mathbf{A}(x)$, $\tilde{\mathbf{B}}(x, a) = \mathbf{B}(x)$, $\tilde{\mathbf{A}}(x, a) = \{a\}$, and $\tilde{\mathbf{B}}(x)$ be any singleton, where $x \in \mathbf{X}$ and $a \in \mathbf{A}$. We define transition probabilities $\tilde{p}$ and payoff functions $\tilde{r}_k$ which do not depend on the action of Player 1 (2) on $\mathbf{Z}$ ($\mathbf{Y}$). For $a \in \mathbf{A}$ and $b \in \mathbf{B}$ we set

$$\tilde{p}(\tilde{y} \mid \tilde{x}, a, b) = \begin{cases} 1 & \text{if } \tilde{x} = x \in \mathbf{X},\ \tilde{y} = (x, a), \\ p(y \mid x, a, b) & \text{if } \tilde{x} = (x, a) \in \mathbf{X} \times \mathbf{A},\ \tilde{y} = y \in \mathbf{X}, \\ 0 & \text{otherwise,} \end{cases}$$

and

$$\tilde{r}_k(\tilde{x}, a, b) = \begin{cases} r_k(x, a, b) & \text{if } \tilde{x} = (x, a) \in \mathbf{X} \times \mathbf{A}, \\ 0 & \text{otherwise.} \end{cases}$$

Each step in the original model corresponds to two sequential steps in the new one. In order to get the same total payoffs for initial states from $\mathbf{X}$, we set $\tilde{\beta}_k = \sqrt{\beta}_k$. It is easy to see that, for all initial points from $\mathbf{X}$, there is a one-to-one correspondence between policies in these two models and $\tilde{V}_k(x, u, v) = V_k(x, u, v)$ and therefore $\tilde{V}(x, u, v) = V(x, u, v)$ for all policies $u$, $v$ and for all states $x \in \mathbf{X}$.

## 3 Lexicographical Stochastic Discounted Games

Denote by $\mathrm{B}(E)$ the Borel $\sigma$-field on a metric space $E$, and by $\mathcal{P}(E)$ the set of probability distributions on $(E, \mathrm{B}(E))$, endowed with the weak topology. If $E$ is compact, then $\mathcal{P}(E)$ is compact in this topology [22].

Fix $x$ and let $f(a, b)$ be a bounded function on $\mathbf{A} \times \mathbf{B}$ which is upper semicontinuous in $a$ on $\mathbf{A}(x)$ and lower semicontinuous in $b$ on $\mathbf{B}(x)$. Let $b(a)$ and $f$ satisfy $f(a) = \min_b f(a, b) = f(a, b(a))$. Since $a_n \to a$ implies $f(a) = f(a, b(a)) \geq$

$\limsup_{n\to\infty} f(a_n, b(a)) \geq \limsup_{n\to\infty} f(a_n, b(a_n)) = \limsup_{n\to\infty} f(a_n)$, we conclude that $f(a)$ is upper semicontinuous. Similarly $f(a) = \max_a f(a, b)$ is lower semicontinuous. The function $f(p, q) = \int_{\mathbf{A}(x)} \int_{\mathbf{B}(x)} f(a, b)p(da)q(db)$ is upper semicontinuous on $\mathcal{P}(\mathbf{A}(x))$ and it is lower semicontinuous on $\mathcal{P}(\mathbf{B}(x))$ [2, p. 17], convex in $p$, and concave in $q$ (actually, it is linear in each of these coordinates).

For each $x \in \mathbf{X}$ let $\mathcal{A}(x)$ in $\mathcal{P}(\mathbf{A}(x))$ and $\mathcal{B}(x)$ in $\mathcal{P}(\mathbf{B}(x))$ be nonempty convex compact subsets. In particular, one may consider $\mathcal{A}(x) = \mathcal{P}(\mathbf{A}(x))$ and $\mathcal{B}(x) = \mathcal{P}(\mathbf{B}(x))$. We denote by $\mathcal{U}_{\mathcal{A}}$ ($\mathcal{V}_{\mathcal{B}}$) the set of policies for Player 1 (2) such that $u_t(\cdot \mid h_t) \in \mathcal{A}(x_t)$ ($v_t(\cdot \mid h_t) \in \mathcal{B}(x_t)$) for all $h_t = x_0, a_0, b_0, \ldots, x_t, t = 0, 1, \ldots$. We notice that $\mathcal{U} = \mathcal{U}_{\mathcal{A}}$ if and only if $\mathcal{A}(x) = \mathcal{P}(\mathbf{A}(x))$ for all $x \in \mathbf{X}$. Similarly, $\mathcal{V} = \mathcal{V}_{\mathcal{B}}$ if and only if $\mathcal{B}(x) = \mathcal{P}(\mathbf{B}(x))$ for all $x \in \mathbf{X}$.

By Theorem 3.4 in Sion [27]

$$\max_{p\in\mathcal{A}(x)} \min_{q\in\mathcal{B}(x)} f(x, p, q) = \min_{q\in\mathcal{B}(x)} \max_{p\in\mathcal{A}(x)} f(x, p, q) \tag{4}$$

and the appropriate minima and maxima exist in (4). We denote

$$\mathbf{val}\ f(x, \mathcal{A}, \mathcal{B}) = \min_{q\in\mathcal{B}(x)} \max_{p\in\mathcal{A}(x)} f(x, p, q). \tag{5}$$

Let

$$\mathbf{P}(x, f, \mathcal{A}, \mathcal{B}) = \{p \in \mathcal{A}(x) : \min_{q\in\mathcal{B}(x)} f(x, p, q) = \mathbf{val}\ f(x)\}, \tag{6}$$

$$\mathbf{Q}(x, f, \mathcal{A}, \mathcal{B}) = \{q \in \mathcal{B}(x) : \max_{p\in\mathcal{A}(x)} f(x, p, q) = \mathbf{val}\ f(x)\}. \tag{7}$$

When $\mathcal{A}(y) = \mathcal{P}(\mathbf{A}(y))$ and $\mathcal{B}(y) = \mathcal{P}(\mathbf{B}(y))$ for all $y \in \mathbf{X}$ we use the notation $\mathbf{val}\ f(x) = \mathbf{val}\ f(x, \mathcal{A}, \mathcal{B})$, $\mathbf{P}(x, f) = \mathbf{P}(x, f, \mathcal{A}, \mathcal{B})$, and $\mathbf{Q}(x, f) = \mathbf{Q}(x, f, \mathcal{A}, \mathcal{B})$.

Since $\min_q f(x, p, q)$ is upper semicontinuous in $p$, $\mathbf{P}(x, f, \mathcal{A}, \mathcal{B})$ are nonempty and compact. In addition, they are convex; see Lemma 2.1.1 in Karlin [14] for the proof of a similar statement. Similarly, $\mathbf{Q}(x, f, \mathcal{A}, \mathcal{B})$ are nonempty convex compact sets. We have that $p \in \mathcal{A}(x)$ ($q \in \mathcal{B}(x)$) is an optimal policy for Player 1 (2) in a zero-sum game with the payoff function $f(x, a, b)$ and sets of decisions limited to randomized decisions from $\mathcal{A}(x)$ and $\mathcal{B}(x)$ if and only if $p \in \mathbf{P}(x, f, \mathcal{A}, \mathcal{B})$ ($q \in \mathbf{Q}(x, f, \mathcal{A}, \mathcal{B})$). If $\mathbf{A}(x)$ and $\mathbf{B}(x)$ are finite, then $\mathbf{P}(x, f)$ and $\mathbf{Q}(x, f)$ are nonempty polytopes. This fact is known as the Shapley–Snow theorem [23], [26].

For stochastic games with one discount factor ($K = 1$), we omit the subscripts $k = 1$ from the notation. In that case, the game reduces to a standard stationary (i.e., with a time-homogeneous structure) stochastic game, for which a very rich theory exists see, e.g., [10], [13], [18], [19], [20], [23], [25]. We shall make use of the following known theorem:

**Theorem 3.1.** *Let $K = 1$. Consider a zero-sum discounted stochastic game such that the set of policies for Player 1 (2) is $\mathcal{U}_{\mathcal{A}}(\mathcal{V}_{\mathcal{B}})$, where $\mathcal{A}(x)$ ($\mathcal{B}(x)$) are nonempty convex compact subsets of $\mathcal{P}(\mathbf{A}(x))$ ($\mathcal{P}(\mathbf{B}(x))$) defined for all $x \in \mathbf{X}$.*

(i) *The game has a value* $V(x) = V(x, \mathcal{A}, \mathcal{B})$ *which is the unique bounded solution of*

$$V(x, \mathcal{A}, \mathcal{B}) = \mathbf{val}\, F(x, \mathcal{A}, \mathcal{B}), \qquad x \in \mathbf{X}, \tag{8}$$

*with*

$$F(x, a, b) = r(x, a, b) + \beta \sum_{z \in \mathbf{X}} p(z \mid x, a, b) V(z). \tag{9}$$

(ii) *A stationary policy* $u^*(v^*)$ *for Player* 1 (2) *is optimal if and only if* $u^*(\cdot \mid x) \in \mathbf{P}(x, F, \mathcal{A}, \mathcal{B})$ $(v^*(\cdot \mid x) \in \mathbf{Q}(x, F, \mathcal{A}, \mathcal{B}))$ *for all* $x \in \mathbf{X}$.
(iii) *A policy* $u^*(v^*)$ *for Player* 1 (2) *is persistently optimal if and only if* $u_n^*(\cdot \mid h_n) \in \mathbf{P}(x_n, F, \mathcal{A}, \mathcal{B})$ $(v_n^*(\cdot \mid h_n) \in \mathbf{Q}(x_n, F, \mathcal{A}, \mathcal{B}))$ *for all* $h_n = x_0, a_0, b_0, \ldots, x_n$, $n = 0, 1, \ldots$.

**Proof.** (i), (ii) For the case of a standard game when $\mathcal{A}(x) = \mathcal{P}(\mathbf{A}(x))$ and $\mathcal{B}(x) = \mathcal{P}(\mathbf{B}(x))$ for all $x \in \mathbf{X}$, these statements are well known. Indeed, (i) and the first part of (ii) are a particular case of a corresponding result for games with Borel state spaces; see, e.g., [17], [19]. Note that the immediate costs in [17] are assumed to be continuous; however, the proof extends in a straightforward way to our case, since it is based on minimax results that hold also in our case. Note also that it is stated in Theorem 1 in [17] that $V$ is the unique solution of (8); this is in general not true. However, the proof of Theorem 1 in [17] shows that $V$ is the unique *bounded* solution of (8). The "only if" part of (ii) follows from Theorem 2.3(i) in [1] (again, continuous immediate costs are considered, but the proof holds also under our assumptions).

If $\mathcal{A}(x) \neq \mathcal{P}(\mathbf{A}(x))$ or $\mathcal{B}(x) \neq \mathcal{P}(\mathbf{B}(x))$ for some $x \in \mathbf{X}$, we consider the game with the action sets $\mathcal{A}(x)$ and $\mathcal{B}(x)$, $x \in \mathbf{X}$. Since the reward function $r(x, p, q)$ is concave in $p$ and convex in $q$, this game has a solution in pure policies. Therefore, statements (i) and (ii) for the case $\mathcal{A}(x) = \mathcal{P}(\mathbf{A}(x))$ and $\mathcal{B}(x) = \mathcal{P}(\mathbf{B}(x))$ for all $x \in \mathbf{X}$ imply the corresponding statements for nonempty convex compact sets $\mathcal{A}(x)$ and $\mathcal{B}(x)$, $x \in \mathbf{X}$.

(iii) Let $u^*$ and $v^*$ be as stated. Consider a game with a finite horizon $T$ and terminal cost $V$, which is the unique bounded solution of (8). Consider the value of the game from time $n$ till time $T$, which we call the $(n, T)$ game, given that a history $\tilde{h}_{n-1}$ took place and $x_n = x$. It easily follows from a backward induction argument that the value of this game is $V$, and that $u^*$ and $v^*$ are optimal. Since this holds for any $T$ and since the immediate cost is bounded, a standard limiting argument shows that the value of the $(n, \infty)$ game is also $V$, and that $u^*$ and $v^*$ are optimal for the $(n, \infty)$ game as well. This establishes the "if" part.

To show the "only if" part, consider some history $\tilde{h}_{n-1}$ and some state $x_n = x$, and let $u$ be a policy for Player 1 which does not satisfy the condition on $u^*$ for that history and $x$. Then

$$V(x, n, u, v^*) \leq E^{u,v^*} \left[ r(x, A_n, B_n) + \beta \sum_{z \in \mathbf{X}} p(z \mid x, A_n, B_n) V(z) \right]$$

$$< \mathbf{val} F(x, \mathcal{A}, \mathcal{B}) = V.$$

This establishes the "only if" part for Player 1. A symmetric argument leads to the result for Player 2. □

For one-step as well as for stochastic zero-sum games, let us describe the notions of lexicographical games and lexicographical values: the formal definitions are given below. Consider a game with sets $\mathcal{U}_1$ and $\mathcal{V}_1$ of policies for Players 1 and 2 and with a vector of payoffs $(\tilde{V}_1(x, u, v), \ldots, \tilde{V}_K(x, u, v))$. We say that a vector $\tilde{V}(x) = (\tilde{V}_1(x), \ldots, \tilde{V}_K(x))$ is a lexicographical value of this game if:

(i) $\tilde{V}_1(x)$ is the value for the game $\Gamma_1$ with sets of policies $\mathcal{U}_1$ and $\mathcal{V}_1$ for Players 1 and 2 and with payoff $\tilde{V}_1(x, u, v)$; and
(ii) for $k = 1, \ldots, K-1$, $\tilde{V}_{k+1}(x)$ is the value of the game $\Gamma_{k+1}$ whose set of policies consists of those policies which are optimal for the game $\Gamma_k$, and whose payoff function is $\tilde{V}_{k+1}(x, u, v)$.

A policy is called lexicographically optimal if it is optimal for the game $\Gamma_K$.

First, we give definitions for one-step games and construct the sets of lexicographically optimal policies for Players 1 and 2. Then we shall define lexicographically persistently optimal policies for stochastic games.

Consider $K$ payoff functions $f_1, \ldots, f_K$, where $f_k = f_k(x, a, b)$ with $a \in \mathbf{A}$, $b \in \mathbf{B}$. All these functions are assumed to be bounded, upper semicontinuous in $a$, and lower semicontinuous in $b$. Given $x \in \mathbf{X}$, we define lexicographically optimal policies for games with these payoffs. The sets of policies for Players 1 and 2 in game $\Gamma_1$ are, respectively, $\mathcal{P}(\mathbf{A}(x))$ and $\mathcal{P}(\mathbf{B}(x))$ which are nonempty convex compact sets in the weak topology. Therefore, the sets of optimal policies for this game with payoff $f_1$ are nonempty convex compact subsets of $\mathcal{P}(\mathbf{A}(x))$ and $\mathcal{P}(\mathbf{B}(x))$. We consider the game $\Gamma_2$ with these sets of policies and payoff function $f_2$. The sets of optimal policies for this game are also nonempty, convex, and compact. By repeating this procedure, we define the vector value of the lexicographical game and the set of optimal policies. Combining lexicographical optimal policies for one-step games with Theorem 3.1, we shall construct the value and the sets of persistently optimal policies for a standard stochastic discounted game.

Now we give formal definitions. We start with a one-step game. Consider an arbitrary $x \in \mathbf{X}$. We denote $\mathbf{val}_1\ f_1(x) = \mathbf{val}\ f(x)$, $\mathbf{P}_1(x, f_1) = \mathbf{P}(x, f_1)$, and $\mathbf{Q}_1(x, f_1) = \mathbf{Q}(x, f_1)$. For fixed $k = 1, \ldots, K-1$, suppose that the value $\mathbf{val}_k\ f_k(x)$ and two nonempty convex compact sets $\mathbf{P}_k(x, f_k) \subseteq \mathcal{P}(\mathbf{A}(x))$ and $\mathbf{Q}_k(x, f_k) \subseteq \mathcal{P}(\mathbf{B}(x))$ are given.

Let $\mathcal{A}_k(x) = \mathbf{P}_k(x, f_k)$ and $\mathcal{B}_k(x) = \mathbf{Q}_k(x, f_k)$. We define the value

$$\mathbf{val}_{k+1}\ f_{k+1}(x) = \mathbf{val}\ f_{k+1}(x, \mathcal{A}_k, \mathcal{B}_k)$$

and the sets of optimal policies

$$\mathbf{P}_{k+1}(x, f_{k+1}) = \mathbf{P}(x, f_{k+1}, \mathcal{A}_k, \mathcal{B}_k),$$
$$\mathbf{Q}_{k+1}(x, f_{k+1}) = \mathbf{Q}(x, f_{k+1}, \mathcal{A}_k, \mathcal{B}_k),$$

which are nonempty, convex, and compact. This construction implies that $(\mathbf{val}_1\ f_k(x), \ldots, \mathbf{val}_K\ f_K(x))$ is a lexicographical value of the one-step game and the nonempty convex compact sets $\mathbf{P}_K(x, f_K)$ and $\mathbf{Q}_K(x, f_K)$ are the sets of lexicographically optimal policies for Players 1 and 2 respectively.

Now we consider a stochastic game with $K$ discount factors $\beta_k$, with $K$ one-step payoff functions $r_k$, $k = 1, \ldots, K$, and with the payoff criterion $(V_1, \ldots, V_K)$ defined by (1). Here we do not need the assumption $\beta_1 > \cdots > \beta_K$. First we consider this stochastic game with the reward function $r_1$ and discount factor $\beta_1$. In view of Theorem 3.1 (i), this game has a unique value function which we denote by $V_1(x)$, $x \in \mathbf{X}$. Let $F_1$ be defined by (9) with $r = r_1, \beta = \beta_1$, and $V = V_1$. Theorem 3.1 (iii) implies that $\mathcal{U}_{\mathcal{A}_1}$ and $\mathcal{V}_{\mathcal{B}_1}$ are the sets of lexicographically optimal policies for Players 1 and 2 respectively where $\mathcal{A}_1(x) = \mathbf{P}(x, F_1)$ and $\mathcal{B}_1(x) = \mathbf{Q}(x, F_1)$ for all $x \in \mathbf{X}$. In addition, $\mathcal{A}_1(x)$ and $\mathcal{B}_1(x)$ are nonempty convex compact sets for all $x \in \mathbf{X}$.

For fixed $k = 1, \ldots, K-1$, suppose that the value $V_k(x)$ and nonempty convex compact sets $\mathcal{A}_k(x)$ and $\mathcal{B}_k(x)$ are defined for all $x \in \mathbf{X}$. We consider a stochastic game with a set of policies $\mathcal{U}_{\mathcal{A}_k}$ for Player 1, a set of policies $\mathcal{V}_{\mathcal{B}_k}$ for Player 2, and with the payoffs $V_{k+1}(x, u, v)$. Theorem 3.1(i) implies that this game has a unique value function $V_{k+1}$ which is a unique solution of $V_{k+1}(x) = \mathbf{val}\ F_{k+1}(x, \mathcal{A}_k, \mathcal{B}_k)$ for all $x \in \mathbf{X}$, where $F_{k+1}$ is defined by (9) with $r = r_{k+1}, \beta = \beta_{k+1}$, and $V = V_{k+1}$. We denote $\mathcal{A}_{k+1}(x) = \mathbf{P}(x, F_{k+1}, \mathcal{A}_k, \mathcal{B}_k)$ and $\mathcal{B}_{k+1}(x) = \mathbf{Q}(x, F_{k+1}, \mathcal{A}_k, \mathcal{B}_k)$. By Theorem 3.1(iii), $\mathcal{U}_{\mathcal{A}_{k+1}}$ and $\mathcal{V}_{\mathcal{B}_{k+1}}$ are the sets of persistently optimal policies for this game.

We say that a policy $u$ $(v)$ for Player 1 (2) is lexicographically persistent optimal if $u \in \mathcal{U}_{\mathcal{A}_K}$ $(v \in \mathcal{V}_{\mathcal{B}_K})$. The above construction and Theorem 3.1 lead to the following theorem in which we do not assume that $\beta_1 > \cdots > \beta_K$:

**Theorem 3.2.** *Consider a stochastic zero-sum game with $K$ reward functions $r_1, \ldots, r_K$, and with $K$ discount factors $\beta_1, \ldots, \beta_K$.*

(i) *This game has a lexicographical value $V_1, \ldots, V_K$, where $V_k$, $k = 1, \ldots, K$, is a unique solution of $V_k(x) = \mathbf{val}_k\ F_k(x)$ with $F_k(x)$ defined for all $x \in \mathbf{X}$ by (9) with $r = r_k$, $V = V_k$, and with $\beta = \beta_k$, $k = 1, \ldots, K$.*
(ii) *A stationary policy $u^*$ $(v^*)$ for Player 1 (2) is lexicographically optimal if and only if $u^*(\cdot \mid x) \in \mathbf{P}_K(x, F_K)$ $(v^*(\cdot \mid x) \in \mathbf{Q}_K(x, F_K))$ for all $x \in \mathbf{X}$.*
(iii) *A policy $u^*$ $(v^*)$ for Player 1 (2) is lexicographically persistently optimal if and only if $u^*_n(\cdot \mid h_n) \in \mathbf{P}_K(x_n, F_K)$ $(v^*_n(\cdot \mid h_n) \in \mathbf{Q}_K(x_n, F_K))$ for all $h_n = x_0, a_0, b_0, \ldots, x_n$, $n = 0, 1, \ldots$.*

Now we consider a game with perfect information. Without loss of generality, we can consider the situation that the singletons $\mathbf{B}(x)$ $(\mathbf{A}(x))$ coincide for all $x \in \mathbf{Y}$ $(x \in \mathbf{Z})$. If we write a triplet $(x, a, b)$, this means that $\{a\} = \mathbf{A}(x)$ for $x \in \mathbf{Z}$ and $\{b\} = \mathbf{B}(x)$ for $x \in \mathbf{Y}$. For a stochastic discounted zero-sum game with perfect information (8) can be rewritten in the following form:

$$V(x) = \mathbf{val}\ F(x) = \begin{cases} \max_{a\in\mathbf{A}(x)} F(x,a,b) & \text{if } x \in \mathbf{Y}, \\ \min_{b\in\mathbf{B}(x)} F(x,a,b) & \text{if } x \in \mathbf{Z}, \end{cases}$$

where $F$ is defined in (9).

Let

$$\mathbf{A}(x, F) = \{a \in \mathbf{A}(x) : F(x,a,b) = \mathbf{val}\ F(x)\}, \qquad x \in \mathbf{Y}, \tag{10}$$

$$\mathbf{B}(x, F) = \{b \in \mathbf{B}(x) : F(x,a,b) = \mathbf{val}\ F(x)\}, \qquad x \in \mathbf{Z}. \tag{11}$$

We observe that if $x \in \mathbf{Y}$, then $\mathbf{Q}(x, F)$ is a measure concentrated at the singleton $\mathbf{B}(x)$ and $\mathbf{P}(x, F) = \mathcal{P}(\mathbf{A}(x, F))$. If $x \in \mathbf{Z}$, then $\mathbf{P}(x, F)$ is a measure concentrated at the singleton $\mathbf{A}(x)$ and $\mathbf{Q}(x, F) = \mathcal{P}(\mathbf{B}(x, F))$.

Since a stochastic discounted game with perfect information is a particular case of a general stochastic discounted game, Theorem 3.2 is applicable to games with perfect information. Since the nonempty convex compact sets of optimal policies at each step for games with perfect information are the sets of all randomized policies on subsets of action sets, optimal pure policies exist for games with perfect information. We get the following theorem from Theorem 3.2:

**Theorem 3.3.** *Consider a zero-sum discounted stochastic game with perfect information.*

(i) *A deterministic policy $u^*$ ($v^*$) for Player* 1 (2) *is optimal if and only if $u^*(x) \in \mathbf{A}(x, F)$ for all $x \in \mathbf{Y}$ ($v^*(x) \in \mathbf{B}(x, F)$ for all $x \in \mathbf{Z}$).*
(ii) *A pure policy $u^*$ ($v^*$) for Player* 1 (2) *is persistently optimal if and only if $u_t^*(h_t) \in \mathbf{A}(x_t, F)$ whenever $x_t \in \mathbf{Y}$ ($v_t^*(h_t) \in \mathbf{B}(x_t, F)$ whenever $x_t \in \mathbf{Z}$) for all $h_t = x_0, a_0, b_0, \ldots, x_t$, $t = 0, 1, \ldots$.*

Now we consider a lexicographical zero-sum stochastic discounted game with perfect information. We define the sets $\mathbf{A}_1(x, F_1) = \mathbf{A}(x, F)$ and $\mathbf{B}_1(x, F_1) = \mathbf{B}(x, F)$ with $r = r_1$ and $\beta = \beta_1$. We also set for $k = 1, \ldots, K-1$

$$\mathbf{A}_{k+1}(x, F_{k+1}) = \left\{a' \in \mathbf{A}_k(x) : F_{k+1}(x, a', b) = \max_{a\in\mathbf{A}_k(x)} F_{k+1}(x,a,b)\right\}, \quad x \in \mathbf{Y}, \tag{12}$$

$$\mathbf{B}_{k+1}(x, F_{k+1}) = \left\{b' \in \mathbf{B}_k(x) : F_{k+1}(x, a, b') = \min_{b\in\mathbf{B}_k(x)} F_{k+1}(x,a,b)\right\}, \quad x \in \mathbf{Z}. \tag{13}$$

Then if $x \in \mathbf{Y}$ we have that $\mathbf{Q}_k(x, F_k)$ is a measure concentrated at the singleton $\mathbf{B}(x)$ and $\mathbf{P}_k(x, F_k) = \mathcal{P}(\mathbf{A}_k(x, F_k))$, $k = 1, \ldots, K$. If $x \in \mathbf{Z}$, then $\mathbf{P}_k(x, F_k)$ is a measure concentrated at the singleton $\mathbf{A}(x)$ and $\mathbf{Q}_k(x, F_k) = \mathcal{P}(\mathbf{B}_k(x, F_k))$, $k = 1, \ldots, K$.

For games with perfect information, lexicographically optimal policies can be selected among pure policies. The following corollary follows from Theorem 3.2 and it does not assume that $\beta_1 > \cdots > \beta_K$.

**Corollary 3.1.** *Consider a stochastic zero-sum game with perfect information, with $K$ reward functions $r_1, \ldots, r_K$, and with $K$ discount factors $\beta_1, \ldots, \beta_K$.*

(i) *A deterministic policy $u^*$ ($v^*$) for Player 1 (2) is lexicographically optimal if and only if $u^*(x) \in \mathbf{A}_K(x, F_K)$ for all $x \in \mathbf{Y}$ ($v^*(x) \in \mathbf{B}_K(x, F_K)$ for all $x \in \mathbf{Z}$).*

(ii) *A pure policy $u^*$ ($v^*$) for Player 1 (2) is lexicographically persistently optimal if and only if $u_t^*(h_t) \in \mathbf{A}_K(x_t, F_K)$ for all $x_t \in \mathbf{Y}$ ($v_t^*(h_t) \in \mathbf{B}_K(x_t, F_K)$ for all $x_t \in \mathbf{Z}$) for all $h_t = x_0, a_0, b_0, \ldots, x_t$, $t = 0, 1, \ldots$.*

## 4 Countable State Weighted Discounted Games

Weighted discounted stochastic games are special cases of nonstationary stochastic games, as defined in [21]. One can thus use the results in [21] to establish the existence of a value. More insight into the structure of the game can be gained, however, by reducing the game into a standard equivalent discounted stochastic game, as was established in Feinberg and Shwartz [4, Remark 2.8], and in Filar and Vrieze [9]. They showed that a weighted discounted stochastic game can be reduced to a standard discounted stochastic game with discount factor $\beta_1$, state space $\overline{\mathbf{X}} = \mathbf{X} \times \{0, 1, \ldots\}$, action sets $\overline{\mathbf{A}}(x, n) = A(x)$, and $\overline{\mathbf{B}}(x, n) = B(x)$, one-step payoffs

$$\bar{r}(x, n, a, b) = \sum_{k=1}^{K} \left(\frac{\beta_k}{\beta_1}\right)^n r_k(x, a, b), \qquad n = 0, 1, \ldots, \tag{14}$$

and transition probabilities

$$\bar{p}((x, n), a, b, (y, k)) = \begin{cases} p(x, a, b, y) & \text{if } k = n + 1, \\ 0 & \text{otherwise.} \end{cases}$$

Actually, the new game is equivalent to the set of original games that starts at all possible epochs $n = 0, 1, \ldots$, not just at $n = 0$. For a one-step game with a payoff function $f$ that depends on a parameter other than $x$, we also consider a notion of a value. In particular, we consider $f = f(x, n)$. We also can consider the sets of optimal policies $\mathbf{P}(x, n, f)$ and $\mathbf{Q}(x, n, f)$ defined by (6) and (7) when $f$ depends on $n$. The following theorem follows from Theorem 3.1 above and from Remark 2.8 in [4]:

**Theorem 4.1.** *Consider a weighted discounted zero-sum Markov game.*

(i) *Each of the games that start at epochs $n = 0, 1, \ldots$ has a value $V(x, n)$ which is the unique solution of*

$$V(x, n) = \mathbf{val}\, F(x, n), \qquad x \in \mathbf{X}, \quad n = 0, 1, \ldots, \tag{15}$$

*with*

$$F(x, n, a, b) = \bar{r}(x, n, a, b) + \beta_1 \sum_{z \in \mathbf{X}} p(z \mid x, a, b) V(z, n+1). \tag{16}$$

(ii) *A policy $u^*$ ($v^*$) for Player 1 (2) is persistently optimal if and only if $u_t^*(\cdot \mid h_t) \in \mathbf{P}(x_t, t, F)$ ($v_t^*(\cdot \mid h_t) \in \mathbf{Q}(x_t, t, F)$) for all $h_t = x_0, a_0, b_0, \ldots, x_t$, $t = 0, 1, \ldots$.*

Since policies $u$ and $v$ may be selected to be Markov in Theorem 4.1(ii), this theorem implies the following result:

**Corollary 4.1.** *In a weighted discounted zero-sum stochastic game each player has an optimal Markov policy.*

Now we consider a weighted discounted zero-sum stochastic game with perfect information. We observe that the game with state space $\overline{\mathbf{X}}$ is also a game with perfect information, with $\overline{\mathbf{Y}} = \mathbf{Y} \times \{0, 1, \ldots\}$ and $\overline{\mathbf{Z}} = \mathbf{Z} \times \{0, 1, \ldots\}$. Since $F$ depends on $n$, the sets of optimal actions defined in (10) and (11) also depend on $n$. We write $\mathbf{A}(x, n, F)$ and $\mathbf{B}(x, n, F)$. We notice that if $x \in \mathbf{Y}$ then $\mathbf{Q}(x, n, F)$ is a measure concentrated at the singleton $\mathbf{B}(x)$ and $\mathbf{P}(x, n, F) = \mathcal{P}(\mathbf{A}(x, n, F))$. If $x \in \mathbf{Z}$ then $\mathbf{P}(x, n, F)$ is a measure concentrated at the singleton $\mathbf{A}(x)$ and $\mathbf{Q}(x, n, F) = \mathcal{P}(\mathbf{B}(x, n, F))$. Therefore, Theorem 4.1 implies the following result:

**Theorem 4.2.** *Consider a zero-sum weighted discounted stochastic game with perfect information. A pure policy $u^*$ ($v^*$) for Player 1 (2) is persistently optimal if and only if $u_t^*(h_t) \in \mathbf{A}(x_t, t, F)$ whenever $x_t \in \mathbf{Y}$ ($v_t^*(h_t) \in \mathbf{B}(x_t, t, F)$ whenever $x_t \in \mathbf{Z}$) for all $h_t = x_0, a_0, b_0, \ldots, x_t$, $t = 0, 1, \ldots$.*

Theorem 4.2 implies the following result which is similar to Corollary 4.1:

**Corollary 4.2.** *In a weighted discounted zero-sum stochastic game with perfect information, each player has an optimal pure Markov policy.*

## 5 The Finite Case with Perfect Information: Main Results

This section describes the structure of persistently optimal policies in weighted discounted stochastic games with perfect information and with finite state and action sets. By Theorem 4.2 there are sets $\mathbf{A}(x, n, F)$ and $\mathbf{B}(x, n, F)$ such that policies for Players 1 and 2 are persistently optimal if and only if at each step they select actions from these sets. The following theorem claims that there exists a finite integer $N$ such that at each state the optimal sets of actions for each player coincide at all steps $n \geq N$. Furthermore, these sets of actions for $n \geq N$ are the sets of lexicographically optimal actions described in Corollary 3.1.

Feinberg and Shwartz [5, Definition 5.4], define a funnel as the set of all policies with the following properties. (Note that there was a typo in that definition: in condition (ii), $\mathbf{A}_n(z)$ should be replaced with $\mathbf{A}_N(z)$.) Suppose we are given action sets that depend on the current state and also on time, but for some $N$, the action sets do not depend on time $n$, whenever $n \geq N$. The funnel (associated with these action sets) is then the set of all policies that select actions from these action sets.

The following theorem shows that for each player the set of optimal policies is a funnel. Algorithm 5.1 provides a method for computing optimal policies; by Remark 5.1, the algorithm can be used to compute the (time-dependent) optimal action sets, and thus obtain the funnel of optimal policies. Recall the assumption $\beta_l > \beta_{l+1}, l = 1, \ldots, K-1$.

**Theorem 5.1.** *Consider a weighted discounted stochastic game with perfect information and with finite state and action sets. There exists a finite integer $N$ such that $A(x, n, F) = A_K(x, F_K)$ and $B(x, n, F) = B_K(x, F_K)$ for all $x \in \mathbf{X}$ and for all $n \geq N$.*

**Proof.** Let for $k = 1, \ldots, K$

$$C_k = \sup\{|r_k(x, a, b)| : x \in \mathbf{X}, a \in \mathbf{A}(x), b \in \mathbf{B}(x)\}. \tag{17}$$

For $n = 0, 1, \ldots$ and for $l = 1, \ldots, K-1$, we define

$$\delta_{n,l} = (\beta_l)^{-n} \sum_{k=l+1}^{K} \frac{(\beta_k)^n C_k}{1 - \beta_k} \tag{18}$$

and

$$\gamma_{n,l} = (\beta_l)^{-n} \sum_{k=l+1}^{K} (\beta_k)^n C_k. \tag{19}$$

Observe that $\gamma_{n,l} + \beta_l \delta_{n+1,l} = \delta_{n,l}$ and that $\delta_{n,l} \to 0$ as $n \to \infty$ for all $l = 1, \ldots, K-1$.

First we show that there exists $N_{1,1}$ such that $\mathbf{A}(x, t, F) \subseteq \mathbf{A}_1(x, F_1)$ for all $t \geq N_{1,1}$ and for all $x \in \mathbf{X}$. Since all sets $\mathbf{A}(x)$ are singletons for all $x \in \mathbf{Z}$, we have to prove this just for $x \in \mathbf{Y}$.

We observe that for any $x \in \mathbf{X}$, any $n = 0, 1, \ldots$, any couple of policies $(u, v)$ for Players 1 and 2, respectively, and for any history $\tilde{h}_n = \tilde{x}_0, \tilde{a}_0, \tilde{b}_0, \ldots, \tilde{x}_{n-1}, \tilde{a}_{n-1}, \tilde{b}_{n-1}$,

$$|V(x, n, \tilde{h}_{n-1}u, \tilde{h}_{n-1}v) - V_1(x, \tilde{h}_{n-1}u, \tilde{h}_{n-1}v)| \leq \delta_{n,1}. \tag{20}$$

Therefore, $|V(x, n) - V_1(x)| \leq \delta_{n,1}$. From (16), (9), and (14) we have that

$$|F(x, n, a, b) - F_1(x, a, b)| \leq \gamma_{n,1} + \beta_1 \delta_{n+1,1} = \delta_{n,1}. \tag{21}$$

We recall that $\mathbf{B}(x) = \{b\}$ for $x \in \mathbf{Y}$ and $\mathbf{A}(x) = \{a\}$ for $x \in \mathbf{Z}$. Let $x \in \mathbf{Y}$. If $\mathbf{A}(x) = \mathbf{A}_1(x, F_1)$, we set $N_{1,1}(x) = 0$: in particular, $N_{1,1}(x) = 0$ for $x \in \mathbf{Z}$. If

$\mathbf{A}(x) \neq \mathbf{A}_1(x, F_1)$, we set

$$N_{1,1}(x) = \min\left\{n = 0, 1, \dots : \min_{a \in \mathbf{A}(x) \setminus \mathbf{A}_1(x, F_1)} \{V_1(x) - F_1(x, a, b)\} > 2\delta_{n,1}\right\}. \tag{22}$$

Then for any $a \in \mathbf{A}(x) \setminus \mathbf{A}_1(x, F_1)$ and for any $n \geq N_{1,1}(x)$,

$$\begin{aligned} V(x,n) - F(x,n,a,b) &> V(x,n) - F(x,n,a,b) - V_1(x) + F_1(x,a,b) + 2\delta_{n,1} \\ &\geq (V(x,n) - V_1(x)) + (F_1(x,a,b) - F(x,n,a,b)) + 2\delta_{n,1} \\ &\geq 0. \end{aligned} \tag{23}$$

By (23), if $n \geq N_{1,1}(x)$ and $a \notin \mathbf{A}_1(x, F_1)$, then $a \notin \mathbf{A}(x, n, F)$. In other words, $\mathbf{A}_1(x, F_1) \supseteq \mathbf{A}(x, n, F)$ for $n \geq N_{1,1}(x)$.

We repeat the above construction for the second player. For $x \in \mathbf{Z}$ such that $\mathbf{B}_1(x, F_1) \neq \mathbf{B}(x)$, we define

$$N_{2,1}(x) = \min\left\{n = 0, 1, \dots : \min_{b \in \mathbf{B}(x) \setminus \mathbf{B}_1(x, F_1)} \{F_1(x, a, b) - V_1(x)\} > 2\delta_{n,1}\right\}. \tag{24}$$

We set $N_{2,1}(x) = 0$ for all other $x$. Then $\mathbf{B}_1(x, F_1) \supseteq \mathbf{B}(x, n, F)$ for all $n \geq N_{2,1}(x)$, $x \in \mathbf{X}$. We define $N_{i,1} = \max_{x \in \mathbf{X}} N_{i,1}(x)$, $i = 1, 2$.

We set $N_1 = \max\{N_{1,1}, N_{1,2}\}$. Then $\mathbf{A}_1(x, F_1) \supseteq \mathbf{A}(x, n, F)$ and $\mathbf{B}_1(x, F_1) \supseteq \mathbf{B}(x, n, F)$ for all $n \geq N_1$ and for all $x \in \mathbf{X}$. We observe that for $n \geq N_1$ and for any history $\tilde{h}_{n-1}$

$$V_1(x, n, \tilde{h}_{n-1}u, \tilde{h}_{n-1}v) = V_1(x) \tag{25}$$

for any policies $u$ and $v$ that, whenever $t \geq N_1$, select actions from the sets $\mathbf{A}_1(x_t, F_1)$ and $\mathbf{B}_1(x_t, F_1)$.

We consider our game for $n \geq N_1$ and with action sets $\mathbf{A}(\cdot)$ reduced to $\mathbf{A}_1(\cdot, F_1)$ and action sets $\mathbf{B}(\cdot)$ reduced to $\mathbf{B}_1(\cdot, F_1)$. Since in the new model, the component $V_1(x, n, u, v)$ of the payoff function $V(x, n, u, v)$ is constant with respect to policies $u$ and $v$ that start at epoch $n$, we can remove the first criterion (i.e., we can set $r_1 \equiv 0$).

We thereby obtain a model with $(K - 1)$ criteria. We repeat this procedure at most $(K-2)$ times and eventually obtain a model with a single payoff function $r_K$. At each step $l = 2, \dots, K$, for each $x \in \mathbf{Y}$ such that $\mathbf{A}_{l-1}(x, F_{l-1}) \neq \mathbf{A}_l(x, F_l)$, we define

$$N_{1,l}(x) = \min\left\{n \geq N_{l-1} : \min_{a \in \mathbf{A}_{l-1}(x, F_{l-1}) \setminus \mathbf{A}_l(x, F_l)} \{F_l(x, a, b) - V_l(x)\} > 2\delta_{n,l}\right\} \tag{26}$$

and $N_{1,l}(x) = N_{l-1}$ for all other $x$. For each $x \in \mathbf{Z}$ such that $\mathbf{B}_{l-1}(x, F_{l-1}) \neq \mathbf{B}_l(x, F_l)$, we also define

$$N_{2,l}(x) = \min\left\{n \geq N_{l-1} : \min_{b \in \mathbf{B}_{l-1}(x, F_{l-1}) \setminus \mathbf{B}_l(x, F_l)} \{V_l(x) - F_l(x, a, b)\} > 2\delta_{n,l}\right\} \tag{27}$$

and $N_{2,l}(x) = N_{l-1}$ for all other $x$. We also set $N_{i,l} = \max_{x\in\mathbf{X}} N_{i,l}(x)$, where $i = 1, 2$ and $l = 2, \ldots, K$, and $N_l = \max\{N_{1,l}, N_{2,l}\}$.

After iteration $K$ we have $\mathbf{A}_K(x, F_K) \supseteq \mathbf{A}(x, n, F)$ and $\mathbf{B}_K(x, F_K) \supseteq \mathbf{B}(x, n, F)$ for all $n \geq N_K$ and for all $x \in \mathbf{X}$. In addition, for any history $\tilde{h}_{n-1}$

$$V(x, n, \tilde{h}_{n-1}u, \tilde{h}_{n-1}v) = (\beta_1)^{-n} \sum_{k=1}^{K} (\beta_k)^n V_k(x) \tag{28}$$

for $n \geq N_K$ and for any policies $u$ and $v$ that use actions from the sets $\mathbf{A}_K(x_t, F_K)$ and $\mathbf{B}_K(x_t, F_K)$ for all $t \geq N_K$. Therefore $\mathbf{A}(x, n, F) = \mathbf{A}_K(x, F_K)$ and $\mathbf{B}(x, n, F) = \mathbf{B}_K(x, F_K)$ for all $x \in \mathbf{X}$ and for all $n \geq N = N_K$. □

**Corollary 5.1.** *In a weighted discounted zero-sum stochastic game with finite state and action sets, for some $N < \infty$ each player has a persistently optimal $(N, \infty)$-stationary policy.*

**Algorithm 5.1.**

0. Set $k = 1$.

1. Compute $V_k(x)$, $\mathbf{A}_k(x)$, and $\mathbf{B}_k(x)$ for all $x \in \mathbf{X}$. Compute $N_k$.

2. If $\mathbf{A}_k(x)$ and $\mathbf{B}_k(x)$ are singletons for all $x \in \mathbf{X}$ or $k = K$, set $\tilde{\mathbf{A}}(x) = \mathbf{A}_k(x)$ and $\tilde{\mathbf{B}}(x) = \mathbf{B}_k(x)$ for all $x \in \mathbf{X}$ and continue to the next step. Otherwise increase $k$ by one and repeat from Step 1.

3. Fix stationary policies $\tilde{u}$ and $\tilde{v}$ for Players 1 and 2, respectively, where $\tilde{u}(x) \in \tilde{\mathbf{A}}(x)$ and $\tilde{v}(x) \in \tilde{\mathbf{B}}(x)$ for all $x \in \mathbf{X}$.

4. Compute $F_N(x) = \sum_{k=1}^{K} (\beta_k)^N V_k(x, \tilde{u}, \tilde{v})$ for all $x \in \mathbf{X}$, where $N = N_K$.

5. Compute $N$-stage optimal pure Markov policies $(u, v)$ by solving the $N$-stage stochastic zero-sum game with perfect information with state space $\mathbf{X}$, action sets $\mathbf{A}(x)$ and $\mathbf{B}(x)$ for Players 1 and 2, respectively, transition probabilities $p$, and rewards $r_t = \sum_{k=1}^{K} (\beta_k)^t r_k$. Since $\mathbf{A}(x)$ ($\mathbf{B}(x)$) are singletons for $x \in \mathbf{Z}$ ($x \in \mathbf{Y}$), $u_t(x)$ ($v_t(x)$) are defined in a unique way for $x \in \mathbf{Z}$ ($x \in \mathbf{Y}$), $t = 0, \ldots, N-1$. For $t = 0, \ldots, N-1$ the policies $u$, $v$ can be defined by

$$\begin{aligned} F_t(x) &= \max_{a\in\mathbf{A}(x)} \left\{ r_t(x, a, b) + \sum_{z\in\mathbf{X}} p(z \mid x, a, b) F_{t+1}(z) \right\} \\ &= r_t(x, u_t(x), b) + \sum_{z\in\mathbf{X}} p(z \mid x, u_t(x), b) F_{t+1}(z) \end{aligned} \tag{29}$$

for $x \in \mathbf{Y}$ and by

$$\begin{aligned} F_t(x) &= \min_{b\in\mathbf{B}(x)} \left\{ r_t(x, a, b) + \sum_{z\in\mathbf{X}} p(z \mid x, a, b) F_{t+1}(z) \right\} \\ &= r_t(x, a, v_t(x)) + \sum_{z\in\mathbf{X}} p(z \mid x, a, v_t(x)) F_{t+1}(z) \end{aligned} \tag{30}$$

for $x \in \mathbf{Z}$.

6. $(N, \infty)$-stationary policies $u$ and $v$ for Players 1 and 2 are optimal, where Step 5 defines $u_t(\cdot)$ and $v_t(\cdot)$ for $t = 0, \ldots, N-1$ and $u_t(\cdot) = \tilde{u}(\cdot)$, $v_t(\cdot) = \tilde{v}(\cdot)$ for $t \geq N$.

**Remark 5.1.** A minor modification of the algorithm leads to the computation of the sets of persistently optimal policies described in Theorems 4.2 and 5.1. The algorithm computes $N$, $\mathbf{A}(x, N, F_K) = \tilde{\mathbf{A}}(x)$, $\mathbf{B}(x, N, F_K) = \tilde{\mathbf{B}}(x)$. A minor modification of Step 5 leads to the computation of $\mathbf{A}(x, t, F_K)$ and $\mathbf{B}(x, t, F_K)$ as sets of actions at which maximums in (29) and minimums in (30) are attained, $t = 0, \ldots, N-1$.

**Remark 5.2.** The number $N$ which the algorithm computes is an upper bound for the actual threshold after which optimal policies must take actions from the sets $A_K(x, F_K)$ and $B_K(x, F_K)$. In fact, it compares the loss over one step due to an action which is nonoptimal for criterion $l = 1, \ldots, K-1$, to the maximum gain from the next step onward due to criteria $l+1, \ldots, K$. Since this gain decreases faster than the losses, after some step the one-step loss cannot be compensated by payoffs with smaller discount factors. The numbers $\delta_{n,l}$ provide an upper estimate for this compensation. It is possible to sharpen this estimate by using the difference between values of MDPs when both players maximize and minimize their payoffs. This approach was used in Algorithm 3.7 in Feinberg and Shwartz [4]. It provides a better upper estimate for $N$, but requires solutions of up to $K(K-1)$ additional MDPs.

## 6 Counterexamples

The first example describes a stochastic game with weighted discounted payoffs and with finite state and action sets in which there is no optimal policy that is stationary from some epoch onward. This shows that the perfect information structure is essential. The existence of $\epsilon$-optimal policies with this property was proved in Filar and Vrieze [9].

**Example 6.1.** Consider a single state, which will be omitted from the notation below (we are thus in the framework of repeated games). Let $\mathbf{A} = \{1, 2\}$, $\mathbf{B} = \{1, 2\}$. Let $r_1(a, b) = 1\{a = b\}$, and $r_2(a, b) = 1\{b = 2\}$. Assume $1 > \beta_1 > \beta_2 > 0$, and define the total payoff:

$$V(u, v) = E^{(u,v)} \left[ \sum_{t=0}^{\infty} \beta_1^t r_1(a_t, b_t) + \sum_{t=0}^{\infty} \beta_2^t r_2(a_t, b_t) \right].$$

Then the optimal policy for Player 1 (that controls the actions $\mathbf{A}$) for all $t$ large enough is to use action 1 with probability

$$(1 + [\beta_2/\beta_1]^t)/2. \tag{31}$$

This converges to a limit $1/2$, but not in finite time.

To obtain (31), we note that for any $2 \times 2$ matrix game $R$, for which a dominating policy does not exist for either player, the optimal policy is the one that results in the indifference to the other players policy. (This is true whether Player 1 minimizes or maximizes.) Hence, the optimal policy of Player 1 in the matrix game, $u(1)$ and $u(2)$, satisfies

$$R_{11}u(1) + R_{21}u(2) = R_{12}u(1) + R_{22}u(2)$$

and hence

$$u(1) = \frac{R_{22} - R_{21}}{R_{11} - R_{12} - R_{21} + R_{22}}.$$

In our repeated game, $R$ is the matrix

$$\beta_1^t \left\{ \begin{vmatrix} 1 & 0 \\ 0 & 1 \end{vmatrix} + \frac{\beta_2^t}{\beta_1^t} \begin{vmatrix} 0 & 1 \\ 0 & 1 \end{vmatrix} \right\}.$$

In the next example, we consider a Markov Decision Process with one state, a compact set of actions, and continuous payoffs. Example 3.16 in Feinberg and Shwartz [4] shows that optimal $(N, \infty)$-stationary policies may not exist under these conditions. In the next example, there is a unique optimal Markov policy $u$, but the sequence $u_t$ does not have a limit.

**Example 6.2.** Here again we assume a single state, and the actions are $\mathbf{A} = [-1, 1]$. Assume $1 > \beta_1 > \beta_2 > 0$. Let $r_1(a) = |a|$.

The single controller maximizes the expected total reward:

$$V(u) = E^u \left[ \sum_{t=0}^{\infty} \beta_1^t r_1(a_t) + \sum_{t=0}^{\infty} \beta_2^t r_2(a_t) \right].$$

Let $r_2(a) = \sqrt{1 - |a|}$. If the component $r_2$ did not exist, the optimal actions would be $|a| = 1$. Now in the presence of $r_2$, since the derivative of $r_2$ at $a = -1$ (and $a = 1$) is $\infty$ ($-\infty$, resp.), the optimal policies $u_t$ satisfy $|u_t| < 1$ for all $t$. Here again the convergence to the limit optimal policy $|a| = 1$ does not take finite time. Note that for any $t$, there are two optimizing actions.

Next we modify $r_2$ slightly so as to destroy the convergence. For positive $a$, we replace $r_2(a)$ with a linear interpolation of $\sqrt{1 - |a|}$ between points $a = 1 - (2n)^{-1}, n = 1, 2, \ldots$. For negative $a$, we replace $r_2(a)$ with a linear interpolation of $\sqrt{1 - |a|}$ between points $a = 1 - (2n + 1)^{-1}, n = 1, 2, \ldots$. As a result, for each $t$ there will be just one optimizing action, $u_t$. We have that $\lim_{t \to \infty} |u_t| = 1$, but $u_t$ is going to be infinitely often close to $-1$, and infinitely often close to 1. Hence, it does not converge. (Note however that in the sense of set convergence $\limsup_{t \to \infty} \{u_t\} = \{-1, 1\}$.)

It is well known that non-zero-sum games with perfect information need not have deterministic equilibria policies. This was illustrated by Federgruen in [7, Sect. 6.6]. A natural question is whether for weighted discounted stochastic games with perfect information, there exist equilibrium policies which are stationary from some epoch $n$ onward.

**Example 6.3.** The following counterexample shows that the answer is negative. Consider a game with two players. Let the payoff function of Player $i$ be

$$W_i(x, u, v) = V_{i,1}(x, u, v) + V_{i,2}(x, u, v), \tag{32}$$

where

$$V_{ik} = E_x^{u,v} \sum_{t=0}^{\infty} (\beta_k)^t r_{ik}(x_t, a_t, b_t)$$

with $i, k = 1, 2$ and $\beta_1 > \beta_2$.

Let $X = \{1, 2\}$, $A(1) = B(2) = \{1, 2\}$, and $A(2) = B(1) = \{1\}$. We also have $p(1 \mid 1, 1, 1) = p(1 \mid 2, 1, 1) = \frac{2}{3}$ and $p(1 \mid 1, 2, 1) = p(1 \mid 2, 1, 2) = \frac{1}{3}$. The one-step rewards are $r_{21}(1, 1, 1) = r_{11}(2, 1, 1) = 1$, $r_{22}(1, 2, 1) = r_{22}(2, 1, 2) = -1$, $r_{12}(2, 1, 1) = r_{12}(2, 1, 2) = r_{22}(1, 1, 1) = r_{22}(1, 2, 1) = 1$ and all other rewards are 0. We remark that if we remove the second summand from (32), we get an example from Federgruen [7, Sect. 6.6], in which there is no equilibrium deterministic policies for a standard discounted non-zero-sum game with perfect information.

Any stationary policy $u$ $(v)$ of Player 1 (2) is defined by a probability $p = u(1 \mid 1)$ $(q = v(1 \mid 2))$, $p, q \in [0, 1]$. Let $P(p, q)$ be a matrix of transition probabilities of a Markov chain defined on $X$ by a couple of policies $(p, q)$,

$$P(p, q) = \begin{pmatrix} \dfrac{1+p}{3} & \dfrac{2+p}{3} \\ \dfrac{1+q}{3} & \dfrac{2-q}{3} \end{pmatrix}.$$

A straightforward computation leads to

$$(I - \beta P(p, q))^{-1} = [(1-\beta)(3-\beta(p-q))]^{-1} \begin{pmatrix} 3 - \beta(2-q) & \beta(2-p) \\ \beta(1+q) & 3 - \beta(1+p) \end{pmatrix}.$$

Let also $r_{ik}(p, q)$ be a one-step expected payoff vector if a couple of policies $(p, q)$ is used

$$r_{11}(p, q) = \begin{pmatrix} 0 \\ 2q - 1 \end{pmatrix}, \qquad r_{21}(p, q) = \begin{pmatrix} 2p - 1 \\ 0 \end{pmatrix},$$

$$r_{12}(p, q) = \begin{pmatrix} 0 \\ 1 \end{pmatrix}, \qquad r_{12}(p, q) = \begin{pmatrix} 1 \\ 0 \end{pmatrix}.$$

Let there exist an equilibrium policy which is stationary from some epoch $N$ onward. Since all transition probabilities are positive, this means that there exists a couple $(p^*, q^*)$ which is equilibrium for any objective vector

$(W_1(x, n, p, q), W_2(x, n, p, q))$, $n = N, N+1, \ldots, x \in X$,

$$W_i(x, n, p, q) = V_{i1}(x, p, q) + \left(\frac{\beta_2}{\beta_1}\right)^n V_{i2}(x, p, q),$$

where $i = 1, 2$.

We have that $W_i(x, n, p, q) \to V_{i1}(x, p, q)$ and therefore $(p^*, q^*)$ is an equilibrium pair of policies for the standard discounted game with the payoff vector $(V_{11}(x, p, q), V_{21}(x, p, q))$. This game is described in section 6.6 of Federgruen [7] and it has a unique stationary equilibrium solution $p^* = q^* = 0.5$.

We also have that

$$W_1(1, n, p, q) = (2q-1) f(\beta_1, p, q) + \left(\frac{\beta_2}{\beta_1}\right)^n f(\beta_2, p, q),$$

$$W_2(1, n, p, q) = (2p-1) g(\beta_1, p, q) + \left(\frac{\beta_2}{\beta_1}\right)^n g(\beta_2, p, q),$$

where

$$f(\beta, p, q) = \frac{\beta(2-p)}{(1-\beta)(3-\beta(p-q))},$$

$$g(\beta, p, q) = \frac{(3-\beta(2-q))}{(1-\beta)(3-\beta(p-q))}.$$

We have that

$$\left.\frac{\partial W_1(1, n, p, q)}{\partial p}\right|_{p=q=.5} < 0 \quad \text{and} \quad \left.\frac{\partial W_2(1, n, p, q)}{\partial q}\right|_{p=q=.5} > 0.$$

Therefore, there is no equilibrium policy which is stationary from some epoch $n$ onward.

## Acknowledgments

Research of the second author was partially supported by NSF Grant DMI-9500746. Research of the third author was supported in part by the Israel Science Foundation, administered by the Israel Academy of Sciences and Humanities, and in part by the fund for promotion of research at the Technion.

## REFERENCES

[1] Altman, E., A. Hordijk, and F. M. Spieksma. Contraction Conditions for Average and $\alpha$-Discounted Optimality in Countable State, Markov Games with Unbounded Payoffs. *Mathematics of Operations Research*, **22**, No. 3, 588–618, 1997.

[2] Billingsley, P. *Convergence of Probability Measures.* Wiley, New York, 1968.

[3] Feinberg, E. A. Controlled Markov Processes with Arbitrary Numerical Criteria. *Theory Probability and its Applications*, **27**, 486–503, 1982.

[4] Feinberg, E. A. and A. Shwartz. Markov Decision Models with Weighted Discounted Criteria. *Mathematics of Operations Research*, **19**, 152–168, 1994.

[5] Feinberg, E. A. and A. Shwartz. Constrained Markov Decision Models with Weighted Discounted Criteria. *Mathematics of Operations Research*, **20**, 302–320, 1994.

[6] Feinberg, E. A. and A. Shwartz. Constrained Dynamic Programming with Two Discount Factors: Applications and an Algorithm. *IEEE Transactions on Automatic Control*. **44**, 628–630, 1990.

[7] Federgruen, A. On $N$-Person Stochastic Games with Denumerable State Space. *Advances in Applied Probability*, **10**, 452–471, 1978.

[8] Fernández-Gaucherand, E., M. K. Ghosh, and S. I. Marcus. Controlled Markov Processes on the Infinite Planning Horizon: Weighted and Overtaking Cost Criteria. *ZOR—Methods and Models of Operations Research*, **39**, 131–155, 1994.

[9] Filar, J. A. and O. J. Vrieze. Weighted Reward Criteria in Competitive Markov Decision Processes. *ZOR—Methods and Models of Operations Research*, **36**, 343–358, 1992.

[10] Filar, J. and O.J. Vrieze. *Competitive Markov Decision Processes*. Springer-Verlag, New York, 1996.

[11] Gillette, D. Stochastic Games with Zero Stop Probabilities. *Contribution to the Theory of Games*, vol. III (M. Dresher, A. W. Tucker, and P. Wolfe, eds.). Princeton University Press, Princeton, NJ, pp. 179–187, 1957.

[12] Golabi, K., R. B. Kulkarni, and G. B. Way. A Statewide Pavement Management System. *Interfaces*, **12**, 5–21, 1982.

[13] Himmelberg, C. J., T. Parthasarathy, and T. E. S. Raghavan. Existence of $p$-Equilibrium and Optimal Stationary Strategies in Stochastic Games. *Proceedings of the American Mathematical Society*, **60**, 245–251, 1976.

[14] Karlin, S. *Mathematical Methods and Theory in Games, Programming, and Economics*. Volume II: *The Theory of Infinite Games*. Addison-Wesley, New York, 1959

[15] Krass, D., J. A. Filar, and S. S. Sinha. A Weighted Markov Decision Process. *Operations Research*, **40**, 1180–1187, 1992.

[16] Küenle, H.-U. *Stochastiche Spiele und Entscheidungesmodelle*. Tebuner-Texte, Leipzig, Band 89, 1986.

[17] Kumar, P. R. and T. H. Shiau. Existence of Value and Randomized Strategies in Zero-Sum Discrete-Time Stochastic Dynamic Games. *SIAM Journal on Control and Optimization*, **19**, 617–634, 1981.

[18] Maitra, A., and W. D. Sudderth. *Discrete Gambling and Stochastic Games*. Springer-Verlag, New York, 1996.

[19] Nowak, A. S. On Zero-Sum Stochastic Games with General State Space I. *Probability and Mathematical Statistics*, **4**, 13–32, 1984.

[20] Nowak, A. S. Universally Measurable Strategies in Stochastic Games. *Annals of Probability*, **13**, 269–287, 1985.

[21] Nowak, A. S. Semicontinuous Nonstationary Stochastic Games. *Journal of Mathematical Analysis and Applications*, **117**, 84–99, 1986.

[22] Parthasarathy, K. R. *Probability Measures on Metric Spaces*. Academic Press, New York, 1967.

[23] Parthasarathy, T. and E. S. Raghavan. *Some Topics in Two-Person Games*. Elsevier, New York, 1971.

[24] Reiman, M. I. and A. Shwartz. Call Admission: A New Approach to Quality of Service. CC Pub. 216, Technion, and Bell Labs Manuscript, 1997.

[25] Shapley L. S. Stochastic Games. *Proceedings of the National Academy of Sciences U.S.A.*, **39**, 1095–1100, 1953.

[26] Shapley L. S. and R. N. Snow. Basic Solutions of Discrete Games. *Contribution to the Theory of Games*, I (H. W. Kuhn and A. W. Tucker, eds.). Princeton University Press, Princeton, NJ, pp. 27–35, 1957.

[27] Sion, M. On General Minimax Theorems. *Pacific Journal of Mathematics*, **8**, 171–176, 1958.

[28] Winden, C. V. and R. Dekker. Markov Decision Models for Building Maintenance: A Feasibility Study. Report 9473/A, ERASMUS University Rotterdam, The Netherlands, 1994.

# Stochastic Games with Complete Information and Average Cost Criteria

Heinz-Uwe Küenle
Institut für Mathematik Brandenburgische
Technische Universität Cottbus
03013 Cottbus, Germany

**Abstract**

Two-person nonzero-sum stochastic games with complete information and average cost criteria are treated. It is shown that there is a quasi-stationary deterministic $2\epsilon$-equilibrium pair if in two related zero-sum games stationary deterministic $\epsilon$-equilibrium pairs exist and a certain ergodicity property is fulfilled. This result allows us to present semicontinuity and compactness conditions sufficient for the existence of such $\epsilon$-equilibrium pairs.

## 1 Introduction

The two-person stochastic game with complete information can be described in the following way: The state $x_n$ of a dynamic system is periodically observed at times $n = 1, 2, \ldots$. After an observation at time $n$ the first player chooses an action $a_n$ from the action set $\boldsymbol{A}(x_n)$ dependent on the complete history of the system at this time, and afterward the second player chooses an action $b_n$ from the action set $\boldsymbol{B}(x_n, a_n)$ dependent on the complete history of the system including the action $a_n$. The first player must pay cost $k^1(x_n, a_n, b_n)$, the second player must pay $k^2(x_n, a_n, b_n)$, and the system moves to a new state $x_{n+1}$ from the state space $\mathbf{X}$ according to the transition probability $q(\cdot \mid x_n, a_n, b_n)$. We consider in this paper stochastic games with standard Borel state space, standard Borel action spaces, and an average cost criterion.

In Section 2 we give the definition of a Markov game with complete information, and formulate the criterion. The nonzero-sum case is treated in Section 3. Several results are presented which follow essentially from [9]. In Section 4 it is shown that in the nonzero-sum case there is a quasi-stationary $2\epsilon$-equilibrium pair if in the two corresponding zero-sum games with the stage cost functions of the $i$th player ($i = 1, 2$) stationary deterministic $\epsilon$-equilibrium pairs exist and a certain ergodicity property is fulfilled. By using the results of Section 3 it is then possible to present semicontinuity and compactness conditions under which for every $\epsilon > 0$ there are quasi-stationary $\epsilon$-equilibrium pairs.

The strategy pairs considered in Section 4 are called $\epsilon$-intimidation strategy pairs or threat strategy pairs. Such strategy pairs are also used, e.g., in [13] or [10] and [11] to get similar results for the average cost criterion with finite state and action spaces or the criterion of total discounted cost.

## 2 Stochastic Games with Complete Information

Stochastic games considered in this paper are defined by eight objects:

**Definition 2.1.** $\mathcal{M} = ((\mathbf{X}, \sigma_{\mathbf{X}}), (\mathbf{A}, \sigma_{\mathbf{A}}), \boldsymbol{A}, (\mathbf{B}, \sigma_{\mathbf{B}}), \boldsymbol{B}, q, k^1, k^2)$ is called a *stochastic game with complete information* if the elements of this tuple have the following meaning:

— $(\mathbf{X}, \sigma_{\mathbf{X}})$ is a standard Borel space, called the *state space.*
— $(\mathbf{A}, \sigma_{\mathbf{A}})$ is a standard Borel space and $\boldsymbol{A} : \mathbf{X} \to \sigma_{\mathbf{A}}$ is a set-valued map which has a $\sigma_{\mathbf{X}}$–$\sigma_{\mathbf{A}}$-measurable selector. $\mathbf{A}$ is called the *action space of the first player* and $\boldsymbol{A}(x)$ is called the *admissible action set of the first player at state $x \in \mathbf{X}$.*
— $(\mathbf{B}, \sigma_{\mathbf{B}})$ is a standard Borel space and $\boldsymbol{B} : \mathbf{X} \times \mathbf{A} \to \sigma_{\mathbf{B}}$ is a set-valued map which has a $\sigma_{\mathbf{X}\times\mathbf{A}}$–$\sigma_{\mathbf{B}}$-measurable selector. $\mathbf{B}$ is called the *action space of the second player* and $\boldsymbol{B}(x, a)$ is called the *admissible action set of the second player at state $x \in \mathbf{X}$ and action $a \in \boldsymbol{A}(x)$ of the first player.*
— $q$ is a transition probability from $\sigma_{\mathbf{X}\times\mathbf{A}\times\mathbf{B}}$ to $\sigma_{\mathbf{X}}$, the *transition law.*
— $k^i$, $i = 1, 2$, are $\sigma_{\mathbf{X}\times\mathbf{A}\times\mathbf{B}}$-measurable functions, called *stage cost functions.* We assume that there is a constant $C$ with $|k^i(x, a, b)| \leq C/2$ for all $x \in \mathbf{X}$, $a \in \mathbf{A}$, $b \in \mathbf{B}$.

Assume that $(\mathbf{Y}, \sigma_{\mathbf{Y}})$ is a standard Borel space. Then we denote by $\overline{\sigma}_{\mathbf{Y}}$ the $\sigma$-algebra of the $\sigma_{\mathbf{Y}}$-universally measurable sets. Let $\mathbf{G}_n = (\mathbf{X} \times \mathbf{A} \times \mathbf{B})^n$ and $\mathbf{H}_n = \mathbf{G}_n \times \mathbf{X}$ for $n \geq 1$, $\mathbf{H}_0 = \mathbf{X}$. $(g, x) \in \mathbf{H}_0$ means $x \in \mathbf{X}$. $h \in \mathbf{H}_n$ is called the *history at time $n$.*

A transition probability $\pi_n$ from $\overline{\sigma}_{\mathbf{H}_n}$ to $\overline{\sigma}_{\mathbf{A}}$ with

$$\pi_n(\boldsymbol{A}(x_n) \mid x_0, a_0, b_0, \ldots, x_n) = 1$$

for all $(x_0, a_0, b_0, \ldots, x_n) \in \mathbf{H}_n$ is called a *decision rule of the first player at time $n$.*

A transition probability $\rho_n$ from $\overline{\sigma}_{\mathbf{H}_n\times\mathbf{A}}$ to $\overline{\sigma}_{\mathbf{B}}$ with $\rho_n(\boldsymbol{B}(x_n, a_n) \mid x_0, a_0, b_0, \ldots, x_n, a_n) = 1$ for all $(x_0, a_0, b_0, \ldots, x_n, a_n) \in \mathbf{H}_n \times \mathbf{A}$ is called a *decision rule of the second player at time $n$.*

A decision rule of the first [second] player is called *deterministic* if a function $e_n : \mathbf{H}_n \to \mathbf{A}$ [$f_n : \mathbf{H}_n \times \mathbf{A} \to \mathbf{B}$] exists with $\pi_n(e_n(h_n) \mid h_n) = 1$ for all $h_n \in \mathbf{H}_n$ [$\rho_n(f_n(h_n, a_n) \mid h_n, a_n) = 1$ for all $(h_n, a_n) \in \mathbf{H}_n \times \mathbf{A}$ ]. (Notation: We write also $\delta_{e_n}$ or simply $e_n$ for $\pi_n$ [$\delta_{f_n}$ or $f_n$ for $\rho_n$].)

A decision rule of the first [second] player is called *Markov* iff a transition probability $\tilde{\pi}_n$ from $\overline{\sigma}_{\mathbf{H}_n}$ to $\overline{\sigma}_{\mathbf{A}}$ [$\tilde{\rho}_n$ from $\overline{\sigma}_{\mathbf{H}_n\times\mathbf{A}}$ to $\overline{\sigma}_{\mathbf{B}}$] exists with $\pi_n(\cdot \mid x_0, a_0, b_0, \ldots, x_n) = \tilde{\pi}_n(\cdot \mid x_n)$ [$\rho_n(\cdot \mid x_0, a_0, b_0, \ldots, x_n, a_n) = \tilde{\rho}_n(\cdot \mid x_n, a_n)$] for all $(x_0, a_0, b_0, \ldots, x_n, a_n) \in \mathbf{H}_n \times \mathbf{A}$. (Notation: We identify $\pi_n$ with $\tilde{\pi}_n$ and $\rho_n$ with $\tilde{\rho}_n$.)

A sequence $\Pi = (\pi_n)$ or $P = (\rho_n)$ of decision rules of the first or second player is called a *strategy* of that player. Strategies are called *deterministic*, or *Markov* iff all their decision rules have the corresponding property.

A Markov strategy $\Pi = (\pi_n)$ or $P = (\rho_n)$ is called *stationary* iff $\pi_0 = \pi_1 = \pi_2 = \cdots$ or $\rho_0 = \rho_1 = \rho_2 = \cdots$. (Notation: $\Pi = \pi^\infty$ or $P = \rho^\infty$.)

Let $\Omega = \mathbf{X} \times \mathbf{A} \times \mathbf{B} \times \mathbf{X} \times \mathbf{A} \times \mathbf{B} \times \ldots$ and $\mathcal{A} = \sigma_\mathbf{X} \otimes \sigma_\mathbf{A} \otimes \sigma_\mathbf{B} \otimes \sigma_\mathbf{X} \otimes \sigma_\mathbf{A} \otimes \sigma_\mathbf{B} \otimes \ldots$. Then for every $x \in \mathbf{X}$ and for every strategy pair $(\Pi, P)$, $\Pi = (\pi_n)$, $P = (\rho_n)$, there is a unique probability measure $\mathsf{P}_{x,\Pi,P}$ on $\mathcal{A}$ defined by the transition probabilities $\pi_n$, $\rho_n$ and $q$ (see [5]). Let $K^{i,n,N}(\omega) = \sum_{j=n}^N k^i(x_j, a_j, b_j)$ for $\omega = (x_0, a_0, b_0, x_1, \ldots)$. We set

$$V_{\Pi P}^{i,0,N}(x) = \int_\Omega K^{i,0,N}(\omega)\mathsf{P}_{x,\Pi,P}(d\omega) \tag{1}$$

and

$$\Phi_{\Pi P}^i(x) = \liminf_{N\to\infty} \frac{1}{N+1} V_{\Pi P}^{i,0,N}(x). \tag{2}$$

**Definition 2.2.** Let $\epsilon \geq 0$. A strategy pair $(\Pi^*, P^*)$ is called an $\epsilon$*-equilibrium pair* iff

$$\Phi_{\Pi^* P^*}^1 \leq \Phi_{\Pi P^*}^1 + \epsilon,$$
$$\Phi_{\Pi^* P^*}^2 \leq \Phi_{\Pi^* P}^2 + \epsilon,$$

for all strategy pairs $(\Pi, P)$. A 0-equilibrium pair is called a *Nash equilibrium pair*.

We are interested in finding Nash equilibrium pairs or $\epsilon$-equilibrium pairs of simple structure. It is helpful to introduce several operators which correspond to the well-known reward operators in stochastic dynamic programming. Let $u$ be a real function on $\mathbf{H}_n$, $\gamma \in \mathbb{R}$ and let $\tilde{r}$ be a real function on $\mathbf{X}$. Then we denote by $u + \gamma\tilde{r}$ the function with $[u+\gamma\tilde{r}](h_n) = u(h_n) + \gamma\tilde{r}(x_n)$ for all $h_n = (x_0, a_0, b_0, \ldots, x_n) \in \mathbf{H}_n$. Let $(\mathbf{Y}, \sigma_\mathbf{Y})$ be a standard Borel space. $\mathcal{U}(\mathbf{Y})$ denotes the set of all bounded $\overline{\sigma}_\mathbf{Y}$-measurable functions $u$. We introduce on $\mathcal{U}(\mathbf{Y})$ the supremum norm. Then we can consider $\mathcal{U}(\mathbf{Y})$ as a complete metric space. Let $\pi_n$, $\rho_n$ be arbitrary decision rules at time $n$ of the first and second player. We set first

$$k_{\pi_n\rho_n}^i(g_n, x) = \int_\mathbf{A} \pi_n(da \mid g_n, x) \int_\mathbf{B} \rho_n(db \mid g_n, x, a) k^i(x, a, b),$$

$$k_{\cdot\rho_n}^i(g_n, x, a) = \int_\mathbf{B} \rho_n(db \mid g_n, x, a) k^i(x, a, b),$$

$$k_{\pi_n\cdot}^i(g_n, x, b) = \int_\mathbf{A} \pi_n(da \mid g_n, x) k^i(x, a, b),$$

for all $(g_n, x) \in \mathbf{H}_n$, $a \in \mathbf{A}$, $b \in \mathbf{B}$, $i = 1, 2$.

Then we define the operators $q_{\pi_n\rho_n}$, $q_{\cdot\rho_n}$, $q_{\pi_n\cdot}$, $T^i_{\pi_n\rho_n}$, $T^i_{\cdot\rho_n}$, and $T^i_{\pi_n\cdot}$ $(i = 1, 2)$ in the following way:

$q_{\pi_n\rho_n} : \mathcal{U}(\mathbf{H}_{n+1}) \to \mathcal{U}(\mathbf{H}_n)$ with

$$q_{\pi_n\rho_n}u(g_n, x) = \int_{\mathbf{A}} \pi_n(da \mid g_n, x) \int_{\mathbf{B}} \rho_n(db \mid g_n, x, a) \times \int_{\mathbf{X}} q(dy \mid x, a, b) u(g_n, x, a, b, y),$$

$q_{\cdot\rho_n} : \mathcal{U}(\mathbf{H}_{n+1}) \to \mathcal{U}(\mathbf{H}_n \times \mathbf{A})$ with

$$q_{\cdot\rho_n}u(g_n, x, a) = \int_{\mathbf{B}} \rho_n(db \mid g_n, x, a) \int_{\mathbf{X}} q(dy \mid x, a, b) u(g_n, x, a, b, y),$$

$q_{\pi_n\cdot} : \mathcal{U}(\mathbf{H}_{n+1}) \to \mathcal{U}(\mathbf{H}_n \times \mathbf{B})$ with

$$q_{\pi_n\cdot}u(g_n, x, b) = \int_{\mathbf{A}} \pi_n(da \mid g_n, x) \int_{\mathbf{X}} q(dy \mid x, a, b) u(g_n, x, a, b, y),$$

$$T^i_{\pi_n\rho_n} : \mathcal{U}(\mathbf{H}_{n+1}) \to \mathcal{U}(\mathbf{H}_n) \quad \text{with } T^i_{\pi_n\rho_n}u = k^i_{\pi_n\rho_n} + q_{\pi_n\rho_n}u,$$
$$T^i_{\cdot\rho_n} : \mathcal{U}(\mathbf{H}_{n+1}) \to \mathcal{U}(\mathbf{H}_n \times \mathbf{A}) \quad \text{with } T^i_{\cdot\rho_n}u = k^i_{\cdot\rho_n} + q_{\cdot\rho_n}u,$$
$$T^i_{\pi_n\cdot} : \mathcal{U}(\mathbf{H}_{n+1}) \to \mathcal{U}(\mathbf{H}_n \times \mathbf{B}) \quad \text{with } T^i_{\pi_n\cdot}u = k^i_{\pi_n\cdot} + q_{\pi_n\cdot}u,$$

for all $u \in \mathcal{U}(\mathbf{H}_{n+1})$, $(g_n, x) \in \mathbf{H}_n$, $a \in \mathbf{A}$, $b \in \mathbf{B}$, $i = 1, 2$.

The following properties of these operators are well known (see, e.g., [9]):

**Lemma 2.1.** *Let* $(\Pi, P)$ *with* $\Pi = (\pi_n)$, $P = (\rho_n)$ *be an arbitrary strategy pair. Then for every* $n \in \mathbb{N}_0$, $u \in \mathcal{U}(\mathbf{H}_{n+1})$ *and* $v \in \mathcal{U}(\mathbf{H}_{n+1})$:

(a) *The operators* $q_{\pi_n\rho_n}$, $q_{\cdot\rho_n}$, $q_{\pi_n\cdot}$, $T^i_{\pi_n\rho_n}$, $T^i_{\cdot\rho_n}$, *and* $T^i_{\pi_n\cdot}$ $(i = 1, 2)$ *are order preserving.*

(b)

$$\begin{aligned} T^i_{\pi_n\rho_n}(u + v) &= T^i_{\pi_n\rho_n}u + q_{\pi_n\rho_n}v, \\ T^i_{\cdot\rho_n}(u + v) &= T^i_{\cdot\rho_n}u + q_{\cdot\rho_n}v, \\ T^i_{\pi_n\cdot}(u + v) &= T^i_{\pi_n\cdot}u + q_{\pi_n\cdot}v. \end{aligned} \tag{3}$$

(c)

$$T^i_{\pi_0\rho_0}T^i_{\pi_1\rho_1}\cdots T^i_{\pi_n\rho_n}u = V^{i,0,n}_{\Pi P} + q_{\pi_0\rho_0}q_{\pi_1\rho_1}\cdots q_{\pi_n\rho_n}u. \tag{4}$$

## 3 The Zero-Sum Case

In this section the case $k^1 = -k^2$ is considered. It holds then $V^{1,0,N}_{\Pi P} = -V^{2,0,N}_{\Pi P}$ but it need not be $\Phi^1_{\Pi P} = -\Phi^2_{\Pi P}$. We use the term “zero-sum” however.

**Lemma 3.1.** *Let* $(\Pi, P)$ *with* $\Pi = (\pi_n)$, $P = (\rho_n)$ *be an arbitrary strategy pair. If there is a constant* $g$ *and a function* $u \in \mathcal{U}(\mathbf{X})$ *with*

$$g + u \le T^1_{\pi_n\rho_n}u \qquad (g + u \ge T^1_{\pi_n\rho_n}u),$$

*then it holds*

$$g \le \Phi^1_{\Pi P} \quad and \quad g \le -\Phi^2_{\Pi P}, \qquad (g \ge \Phi^1_{\Pi P} \quad and \quad g \ge -\Phi^2_{\Pi P}).$$

**Proof.** From Lemma 2.1 it follows

$$\begin{aligned} g &\le \frac{1}{N+1}(T^1_{\pi_0\rho_0}\cdots T^1_{\pi_N\rho_N}u - u) \\ &= \frac{1}{N+1}(V^{1,0,N}_{\Pi P} + q_{\pi_0\rho_0}\cdots q_{\pi_N\rho_N}u - u) \\ &= \frac{1}{N+1}(-V^{2,0,N}_{\Pi P} + q_{\pi_0\rho_0}\cdots q_{\pi_N\rho_N}u - u). \end{aligned}$$

Since $u$ is bounded we get the statement for $N \to \infty$. □

**Theorem 3.1.** *If there is a constant $g$, a function $v \in \mathcal{U}(\mathbf{X})$, and a pair $(e^\infty_\epsilon, f^\infty_\epsilon)$ of stationary deterministic strategies such that for all $x \in \mathbf{X}$, $a \in \boldsymbol{A}(x)$,*

$$g + v(x) = \inf_{a\in\boldsymbol{A}(x)} \sup_{b\in\boldsymbol{B}(x,a)} \left\{k(x,a,b) + \int_{\mathbf{X}} q(d\xi \mid x,a,b)v(\xi)\right\}, \tag{5}$$

$$\begin{aligned} & k(x,a,f_\epsilon(x,a)) + \int_{\mathbf{X}} q(d\xi \mid x,a,f_\epsilon(x,a))v(\xi) + \epsilon \\ \ge & \sup_{b\in\boldsymbol{B}(x,a)} \left\{k(x,a,b) + \int_{\mathbf{X}} q(d\xi \mid x,a,b)v(\xi)\right\}, \end{aligned} \tag{6}$$

$$\begin{aligned} & \sup_{b\in\boldsymbol{B}(x,e_\epsilon(x))} \left\{k(x,e_\epsilon(x),b) + \int_{\mathbf{X}} q(d\xi \mid x,e_\epsilon(x),b)v(\xi)\right\} - \epsilon \\ \le & \, g + v(x), \qquad x \in \mathbf{X}, \end{aligned} \tag{7}$$

*then $(e^\infty_\epsilon, f^\infty_\epsilon)$ is a $2\epsilon$-equilibrium pair.*

**Proof.** From (6) and (7) we get for all $\Pi = (\pi_n)$, $P = (\rho_n)$

$$\begin{aligned} T^1_{\pi_n f_\epsilon} + \epsilon &\ge g + v, \\ T^1_{e_\epsilon \rho_n} - \epsilon &\le g + v. \end{aligned}$$

From Lemma 3.1 it follows

$$\begin{aligned} \Phi^1_{e^\infty_\epsilon f^\infty_\epsilon} - \epsilon \le g \le \Phi^1_{\Pi f^\infty_\epsilon} + \epsilon, \\ \Phi^2_{e^\infty_\epsilon f^\infty_\epsilon} - \epsilon \le -g \le \Phi^2_{e^\infty_\epsilon P} + \epsilon. \end{aligned}$$ □

The following strong ergodicity assumption can be found in several papers on Markov decision processes and Markov games, see [3], [4], and [9] for examples. A somewhat stricter condition is considered in [14].

**Assumption 3.1.** There is a measure $\psi$ on $\overline{\sigma}_{\mathbf{X}}$ with $\psi(\mathbf{X}) > 0$ and $\psi \le q(\cdot \mid x, a, b)$ for all $x \in \mathbf{X}$, $a \in \boldsymbol{A}(x)$, $b \in \boldsymbol{B}(x, a)$.

We set $\tilde{q} = q - \psi$. Obviously, $\tilde{q}$ is a substochastic kernel.

In the next section we use the following result:

**Theorem 3.2.** *Let $e$ be a Markov deterministic decision rule of the first player and let $f$ be a Markov deterministic decision rule of the second player. Under Assumption 3.1 the functional equation*

$$g + u = T^i_{ef} u \tag{8}$$

*has a solution $(g, u)$ with $g = const$ and $u \in \mathcal{U}(\mathbf{X})$ and it holds*

$$g = \Phi^i_{e^\infty f^\infty}$$

*for $i \in \{1, 2\}$.*

**Proof.** Let $\tilde{T}^i_{ef} u = T^i_{ef} u - \int_{\mathbf{X}} \psi(d\xi)u(\xi) = k^i_{ef} + \tilde{q}^i_{ef} u$. From Lemma 2.1 it follows that $\tilde{T}^i_{ef}$ is contractive on $\mathcal{U}(\mathbf{X})$. By means of Banach's fixed point theorem we find that the functional equation $u = \tilde{T}^i_{ef} u$ there has a unique solution $u$. $(g, u)$ with $g = \int_{\mathbf{X}} \psi(d\xi)u(\xi)$ fulfils then (8). The rest of the proof follows from Lemma 3.1. □

The proof of the following theorem is also easy and and can be omitted here:

**Theorem 3.3.** *Let Assumption 3.1 be satisfied. If $v \in \mathcal{U}(\mathbf{X})$ is a solution of the functional equation*

$$v(x) = \inf_{a \in \boldsymbol{A}(x)} \sup_{b \in \boldsymbol{B}(x,a)} \{k(x, a, b) + \int_{\mathbf{X}} \tilde{q}(d\xi \mid x, a, b)v(\xi)\} \qquad \text{for all } x \in \mathbf{X}, \tag{9}$$

*then $(g, v)$ with $g = \int_{\mathbf{X}} \psi(d\xi)v(\xi)$ is a solution of (5). If, furthermore, $f_\epsilon$ is a Markov deterministic decision rule of the second player with*

$$k(x, a, f_\epsilon(x, a)) + \int_{\mathbf{X}} \tilde{q}(d\xi \mid x, a, f_\epsilon(x, a))v(\xi) + \epsilon$$
$$\geq \sup_{b \in \boldsymbol{B}(x,a)} \left\{k(x, a, b) + \int_{\mathbf{X}} \tilde{q}(d\xi \mid x, a, b)v(\xi)\right\}, \qquad x \in \mathbf{X}, \quad a \in \boldsymbol{A}(x),$$

*then $f_\epsilon$ fulfils also (6).*

*If $e_\epsilon$ is a Markov deterministic decision rule of the first player with*

$$sup_{b \in \boldsymbol{B}(x, e_\epsilon(x))} \left\{k(x, e_\epsilon(x), b) + \int_{\mathbf{X}} \tilde{q}(d\xi \mid x, e_\epsilon(x), b)v(\xi)\right\} - \epsilon \leq v(x) \qquad \text{for all} \quad x \in \mathbf{X},$$

*then $e_\epsilon$ also fulfills (7).*

We remark that we can consider (9) as the optimality equation in a Markov game with total cost criterion and contraction property. In [9] it is shown that (9) has a unique solution $v \in \mathcal{U}(\mathbf{X})$ if one of the spaces $\mathbf{X}$, $\mathbf{A}$, or $\mathbf{B}$ is countable. In [9] also further conditions are given under which the assumptions of Theorem 3.3 are fulfilled. The following Assumption 3.2 will be helpful to formulate such conditions.

In the following set of assumptions semicontinuity of set-valued maps means semicontinuity in the sense of Kuratowski (see, e.g., [1]):

**Assumption 3.2.**

(A1) There is a countable set $G$ of Borel measurable selectors $g$ of $\boldsymbol{A}$ such that $\{g(x) : g \in G\}$ is dense in $\boldsymbol{A}(x)$ for each $x \in \mathbf{X}$.
(A2) $\boldsymbol{A}$ is compact-valued.
(A3) $\boldsymbol{A}$ is lower semicontinuous.
(A4) $\boldsymbol{A}$ is upper semicontinuous.
(B1) There is a countable set $G$ of Borel measurable selectors $g$ of $\boldsymbol{B}$ such that $\{g(x, a) : g \in G\}$ is dense in $\boldsymbol{B}(x, a)$ for all $x \in \mathbf{X}$, $a \in \mathbf{A}(x)$.
(B2) $\boldsymbol{B}$ is lower semicontinuous.
(B3) $\boldsymbol{B}$ is upper semicontinuous.
(B4) $\boldsymbol{B}(x, \cdot)$ is lower semicontinuous for every $x \in \mathbf{X}$.
(B5) $\boldsymbol{B}(x, \cdot)$ is upper semicontinuous for every $x \in \mathbf{X}$.
(C1) $q$ is weak continuous (i.e., $\int_{\mathbf{X}} u(\xi) q(d\xi \mid \cdot)$ is continuous for each continuous $u \in \mathcal{U}(\mathbf{X})$).
(C2) $q$ is strong continuous (i.e., $\int_{\mathbf{X}} u(\xi) q(d\xi \mid \cdot)$ is continuous for each $u \in \mathcal{U}(\mathbf{X})$).
(D1.$i$) $k^i$ is lower semicontinuous ($i = 1, 2$).
(D2.$i$) $k^i$ is upper semicontinuous ($i = 1, 2$).
(D3.$i$) $k^i(x, \cdot)$ is lower semicontinuous for every $x \in \mathbf{X}$ ($i = 1, 2$).
(D4.$i$) $k^i(x, \cdot)$ is upper semicontinuous for every $x \in \mathbf{X}$ ($i = 1, 2$).

Using Theorem 3.3 the next theorem follows by means of the results of [9] concerning stochastic games with the criterion of expected total costs. (See [9, Sect. 5.5]. A view of these results is on p. 69.) Some weaker theorems can also be found in [7] and [8]. The proofs are based on dynamic programming methods and use the properties of semianalytic and semicontinuous functions. Related results for general zero-sum stochastic games (with independent action choice of both players) can be found in [9] and in a recent paper of Rieder [12], for example. In this case, it follows only the existence of stationary but not of stationary deterministic $\epsilon$-optimal strategies.

**Theorem 3.4.** *Let Assumption 3.1 be satisfied. Then there exists a stationary deterministic $\epsilon$-equilibrium pair for every $\epsilon > 0$ if one of the following sets of assumptions is fulfilled:*

(a) *One of the spaces* $\mathbf{X}$, $\mathbf{A}$, *or* $\mathbf{B}$ *is countable.*
(b) (A2), (B1), (C2), *and* (D3.1).
(c) (A1), (B5), (C2), *and* (D4.1).
(d) (A2), (A4), (B2), (C1), *and* (D1.1).
(e) (A3), (B3), (C1), *and* (D2.1).

## 4 The Nonzero-Sum Case

In this section we introduce first some further special classes of strategy pairs.

**Definition 4.1.** A strategy pair $(\Pi^1, P^1)$ is called an *$\epsilon$-I-strategy pair* iff it is an $\epsilon$-equilibrium pair in the stochastic game

$$\mathcal{M} = ((\mathbf{X}, \sigma_{\mathbf{X}}), (\mathbf{A}, \sigma_{\mathbf{A}}), \boldsymbol{A}, (\mathbf{B}, \sigma_{\mathbf{B}}), \boldsymbol{B}, q, k^1, -k^1).$$

A strategy pair $(\Pi^2, P^2)$ is called an *$\epsilon$-II-strategy pair* iff it is an $\epsilon$-equilibrium pair in the stochastic game $\mathcal{M} = ((\mathbf{X}, \sigma_{\mathbf{X}}), (\mathbf{A}, \sigma_{\mathbf{A}}), \boldsymbol{A}, (\mathbf{B}, \sigma_{\mathbf{B}}), \boldsymbol{B}, q, -k^2, k^2)$.

Obviously, the strategy pair $(\Pi^1, P^1)$ is an $\epsilon$-equilibrium pair in the zero-sum game with the cost of the first player and $(\Pi^2, P^2)$ is an $\epsilon$-equilibrium pair in the zero-sum game with the cost of the second player. In this section we assume that $(\Pi^1, P^1) = (e^{1\infty}, f^{1\infty})$ is a stationary $\epsilon$-I-strategy pair and that $(\Pi^2, P^2) = (e^{2\infty}, f^{2\infty})$ is a stationary $\epsilon$-II-strategy pair. We denote for $n \geq 1$:

$$\begin{aligned} \mathbf{G}_n^* &= \{(x_0, a_0, b_0, \ldots, x_{n-1}, a_{n-1}, b_{n-1}) \in \mathbf{G}_n : \\ &\qquad a_i = e^1(x_i),\ b_i = f^2(x_i, a_i),\ i = 0, \ldots, n-1\}, &(10)\\ \overline{\mathbf{G}}_n^* &= \mathbf{G}_n \backslash \mathbf{G}_n^*, &(11)\\ \mathbf{H}_n^* &= \mathbf{G}_n^* \times \mathbf{X}, \qquad \mathbf{H}_0^* = \mathbf{X}, &(12)\\ \overline{\mathbf{H}}_n^* &= \mathbf{H}_n \backslash \mathbf{H}_n^* = \overline{\mathbf{G}}_n^* \times \mathbf{X}. &(13) \end{aligned}$$

**Definition 4.2.** A pair of deterministic strategies $(\Pi, P)$ with $\Pi = (e_n)$, $P = (f_n)$ is called an *$\epsilon$-intimidation strategy pair connected with the deterministic strategies $\hat{\Pi} = (\hat{e}_n)$, $\hat{P} = (\hat{f}_n)$ (and $(\Pi^1, P^1)$, $(\Pi^2, P^2)$)* iff for all $n \in \mathbb{N}_0$:

$$e_n(g, x) = \begin{cases} \hat{e}_n(g, x) & \text{for } (g, x) \in \mathbf{H}_n^*, \\ e^2(x) & \text{otherwise,} \end{cases}$$

$$f_n(g, x, a) = \begin{cases} \hat{f}_n(g, x, a) & \text{for } (g, x) \in \mathbf{H}_n^*, \quad a = \hat{e}_n(g, x), \\ f^1(x, a) & \text{otherwise.} \end{cases}$$

Such $\epsilon$-intimidation strategy pairs are treated in [10] (for $\epsilon = 0$) and [11]. It is known that there are stochastic games with complete information and finite state and action spaces which have no stationary deterministic equilibrium pairs (see [2]). For this reason, we consider a larger class of strategy pairs, called quasi-stationary strategy pairs. Quasi-stationary strategy pairs were introduced in [11].

**Definition 4.3.** A strategy pair $(\Pi^*, P^*)$ is called *quasi-stationary* iff a stationary strategy pair $(\pi^\infty, \rho^\infty)$ exists with

$$\mathsf{P}_{x,\Pi^*,P^*} = \mathsf{P}_{x,\pi^\infty,\rho^\infty} \qquad \text{for all} \quad x \in \mathbf{X}.$$

We denote by $\delta_l$ the transition probability from $\overline{\sigma}_{\mathbf{H}_l}$ to $\overline{\sigma}_{\mathbf{X}}$ given by $\delta_l(\tilde{X} \mid g, x) = I_{\tilde{X}}(x)$ for all $(g, x) \in \mathbf{H}_l$, $\tilde{X} \in \overline{\sigma}_{\mathbf{X}}$. Let $(\Pi, P)$ with $\Pi = (\pi_n)$, $P = (\rho_n)$ be an arbitrary strategy pair. For $l \geq m$ we define a transition probability from $\overline{\sigma}_{\mathbf{H}_l}$ to $\overline{\sigma}_{\mathbf{H}_{m-l+1}}$ by

$$\mathsf{P}_{\Pi P}^{l,m} = \delta_l \otimes \pi_l \otimes \rho_l \otimes q \otimes \cdots \otimes \pi_m \otimes \rho_m \otimes q.$$

Then for $M \in \overline{\sigma}_{\mathbf{H}_{m-l+1}}$

$$\begin{aligned}
&\mathsf{P}_{\Pi P}^{l,m}(M \mid g, x)\\
&\quad = \int_{\mathbf{A}} \pi_l(da_l \mid g, x) \int_{\mathbf{B}} \rho_l(db_l \mid g, x, a_l) \int_{\mathbf{X}} q(dx_{l+1} \mid x, a_l, b_l) \ldots\\
&\qquad \times \int_{\mathbf{X}} q(dx_m \mid x_{m-1}, a_{m-1}, b_{m-1}) \int_{\mathbf{A}} \pi_m(da_m \mid g, x, a_l, b_l, \ldots, x_m)\\
&\qquad \times \int_{\mathbf{B}} \rho_m(db_m \mid g, x, a_l, b_l, \ldots, x_m, a_m) \int_{\mathbf{X}} q(dx_{m+1} \mid x_m, a_m, b_m)\\
&\qquad \times I_M(x, a_l, b_l, \ldots, x_m, a_m, b_m, x_{m+1})\\
&\quad = q_{\pi_l \rho_l} q_{\pi_{l+1} \rho_{l+1}} \cdots q_{\pi_m \rho_m} I_M(g, x).
\end{aligned}$$

For $h = (x_l, a_l, b_l, \ldots, x_m, a_m, b_m, x_{m+1}) \in \mathbf{H}_{m-l+1}$ let

$$K^{i,l,m}(h) = \sum_{j=l}^{m} k^i(x_j, a_j, b_j).$$

We set for $h_l \in \mathbf{H}_l$

$$V_{\Pi P}^{i,l,m}(h_l) = \int_{\mathbf{H}_{m-l+1}} \mathsf{P}_{\Pi P}^{l,m}(dh \mid h_l) K^{i,l,m}(h). \tag{14}$$

We remark that this definition of $V_{\Pi P}^{i,l,m}$ is consistent with (1).

It is well known from stochastic dynamic programming that

$$V_{\Pi P}^{i,l,m} = T_{\pi_l \rho_l}^{i} V_{\Pi P}^{i,l+1,m} \tag{15}$$

for $l < m$ and furthermore

$$\Phi_{\Pi P}^{i}(x) = \liminf_{N \to \infty} \frac{1}{N+1} \int_{\mathbf{H}_n} \mathsf{P}_{\Pi P}^{0,n-1}(dh \mid x) V_{\Pi P}^{i,n,N}(h) \tag{16}$$

for each $n \geq 1$.

**Lemma 4.1.** *Let $(\Pi^*, P^*)$, $\Pi^* = (e_n^*)$, $P^* = (f_n^*)$, be the $\epsilon$-intimidation strategy pair connected with $(e^{1\infty}, f^{2\infty})$ (and $(e^{1\infty}, f^{1\infty})$, $(e^{2\infty}, f^{2\infty})$). Let, furthermore, $\Pi = (\pi_n)$ be an arbitrary strategy of the first player. Then for $m = 1, \ldots, N-1$, $h \in \mathbf{H}_m^*$,*

$$V_{\Pi^* P^*}^{1,m,N}(h) \leq V_{\Pi P^*}^{1,m,N}(h) + (N - m + 1)C \cdot \mathsf{P}_{\Pi P^*}^{m,N}(\overline{\mathbf{H}}_{N-m+1}^* \mid h). \tag{17}$$

**Proof.** We remark that (17) is equivalent to

$$\begin{aligned}
V_{\Pi^* P^*}^{1,m,N}(h) \leq{}& V_{\Pi P^*}^{1,m,N}(h)\\
&+ (N - m + 1)C \cdot q_{\pi_m f_m^*} q_{\pi_{m+1} f_{m+1}^*} \cdots q_{\pi_N f_N^*} I_{\overline{\mathbf{H}}_{N-m+1}^*}(h).
\end{aligned} \tag{18}$$

For the further proof we use mathematical induction. Let $\boldsymbol{A}'(x) = \boldsymbol{A}(x)\backslash\{e^1(x)\}$. We show first that (18) is true for $m = N \geq 1$. Let $g \in \mathbf{G}_N^*$, $x \in \mathbf{X}$.

$$\begin{aligned}
V_{\Pi P^*}^{1,N,N}(g,x) &= \pi_N(\{e^1(x)\} \mid g,x)k^1_{\cdot f^2}(x, e^1(x)) + \int_{\boldsymbol{A}'(x)} \pi_N(da \mid g,x)k^1_{\cdot f^1}(x,a) \\
&= k^1_{\cdot f^2}(x, e^1(x)) + \int_{\boldsymbol{A}'(x)} \pi_N(da \mid g,x)(k^1_{\cdot f^1}(x,a) - k^1_{\cdot f^2}(x, e^1(x))) \\
&\geq V_{\Pi^* P^*}^{1,N,N}(g,x) - C \cdot \pi_N(\boldsymbol{A}'(x) \mid g,x) \\
&= V_{\Pi^* P^*}^{1,N,N}(g,x) - C \cdot q_{\pi_N f_N^*} I_{\overline{\mathbf{H}}_1^*}(g,x).
\end{aligned}$$

We assume now that (18) holds for $m = l+1$ ($l = 0, 1, \ldots, N-1$). Then it follows for $(g,x) \in \mathbf{H}_l^*$ by means of (3) and (15)

$$\begin{aligned}
V_{\Pi P^*}^{1,l,N}(g,x) &= \int_{\boldsymbol{A}(x)} \pi_l(da \mid g,x) T^1_{\cdot f_l^*} V_{\Pi P^*}^{1,l+1,N}(g,x,a) \\
&= \pi_l(\{e^1(x)\} \mid g,x) T^1_{e^1 f^2} V_{\Pi P^*}^{1,l+1,N}(g,x)) \\
&\quad + \int_{\boldsymbol{A}'(x)} \pi_l(da \mid g,x) T^1_{\cdot f_l^*} V_{\Pi P^*}^{1,l+1,N}(g,x,a) \\
&\geq \pi_l(\{e^1(x)\} \mid g,x)(T^1_{e^1 f^2} V_{\Pi^* P^*}^{1,l+1,N}(g,x) \\
&\quad - (N-l)C \cdot q_{e^1 f^2} q_{\pi_{l+1} f_{l+1}^*} \cdots q_{\pi_N f_N^*} I_{\overline{\mathbf{H}}_{N-l}^*}(g,x)) \\
&\quad + \int_{\boldsymbol{A}'(x)} \pi_l(da \mid g,x)(T^1_{\cdot f_l^*} V_{\Pi P^*}^{1,l+1,N}(g,x,a) - T^1_{e^1 f^2} V_{\Pi^* P^*}^{1,l+1,N}(g,x)) \\
&\quad + \int_{\boldsymbol{A}'(x)} \pi_l(da \mid g,x) T^1_{e^1 f^2} V_{\Pi^* P^*}^{1,l+1,N}(g,x) \\
&\geq T^1_{e^1 f^2} V_{\Pi^* P^*}^{1,l+1,N}(g,x) \\
&\quad - (N-l)C \cdot q_{\pi_l f_l^*}(I_{\mathbf{G}_1^*} q_{\pi_{l+1} f_{l+1}^*} \cdots q_{\pi_N f_N^*} I_{\overline{\mathbf{H}}_{N-l}^*})(g,x) \\
&\quad - (N-l+1)C\pi_l(\boldsymbol{A}'(x) \mid g,x) \\
&\geq V_{\Pi^* P^*}^{1,l,N}(g,x) - (N-l+1)C \\
&\quad \times q_{\pi_l f_l^*} q_{\pi_{l+1} f_{l+1}^*} \cdots q_{\pi_N f_N^*}(I_{\mathbf{G}_1^* \times \overline{\mathbf{H}}_{N-l}^*} + I_{\overline{\mathbf{G}}_1^* \times \mathbf{H}_{N-l}})(g,x) \\
&= V_{\Pi^* P^*}^{1,l,N}(g,x) - (N-l+1)C \cdot q_{\pi_l f_l^*} q_{\pi_{l+1} f_{l+1}^*} \cdots q_{\pi_N f_N^*} I_{\overline{\mathbf{H}}_{N-l+1}^*}(g,x).
\end{aligned}$$

□

**Lemma 4.2.** *Let $(\Pi^*, P^*)$, $\Pi^* = (e_n^*)$, $P^* = (f_n^*)$, be the $\epsilon$-intimidation strategy pair connected with $(e^{1\infty}, f^{2\infty})$ (and $(e^{1\infty}, f^{1\infty})$, $(e^{2\infty}, f^{2\infty})$). Let, furthermore, $P = (\rho_n)$ be an arbitrary strategy of the second player. Then for $m = 1, \ldots, N-1$, $h \in \mathbf{H}_m^*$,*

$$V_{\Pi^* P^*}^{2,m,N}(h) \leq V_{\Pi^* P}^{2,m,N}(h) + (N-m+1)C \cdot \mathsf{P}_{\Pi^* P}^{m,N}(\overline{\mathbf{H}}_{N-m+1}^* \mid h). \qquad (19)$$

**Proof.** (19) is equivalent to

$$V_{\Pi^* P^*}^{2,m,N}(h) \leq V_{\Pi^* P}^{2,m,N}(h)$$

$$+ (N - m + 1)C \cdot q_{e^*_m \rho_m} q_{e^*_{m+1}\rho_{m+1}} \cdots q_{e^*_N \rho_N} I_{\overline{\mathbf{H}}^*_{N-m+1}}(h). \tag{20}$$

Analogously to the proof of Lemma 4.1 we use mathematical induction. We put $\boldsymbol{B}'(x) = \boldsymbol{B}(x, e^1(x)) \setminus \{f^2(x, e^1(x))\}$, $f'^2(x) = f^2(x, e^1(x))$, $\rho'_n(\cdot \mid g, x) = \rho_n(\cdot \mid g, x, e^1(x))$ for all $x \in \mathbf{X}$, $(g, x) \in \mathbf{H}_n$, $n \in \mathbb{N}_0$. We show first that (20) is true for $m = N \geq 1$. Let $g \in \mathbf{G}^*_N$, $x \in \mathbf{X}$,

$$\begin{aligned}
V^{2,N,N}_{\Pi^* P}(g, x) &= \rho'_N(\{f'^2(x)\} \mid g, x) k^2_{e^1 f^2}(x) + \int_{\boldsymbol{B}'(x)} \rho'_N(db \mid g, x) k^2_{e^1 \cdot}(x, b) \\
&= k^2_{e^1 f^2}(x) + \int_{\boldsymbol{B}'(x)} \rho'_N(db \mid g, x)(k^2_{e^1 \cdot}(x, b) - k^2_{e^1 f^2}(x)) \\
&\geq V^{2,N,N}_{\Pi^* P^*}(g, x) - C \cdot \rho'_N(\boldsymbol{B}'(x) \mid g, x) \\
&= V^{2,N,N}_{\Pi^* P^*}(g, x) - C \cdot q_{e^*_N \rho_N} I_{\overline{\mathbf{H}}^*_1}(g, x).
\end{aligned}$$

We assume now that (20) holds for $m = l + 1$ $(l = 0, 1, \ldots, N - 1)$. Then it follows for $(g, x) \in \mathbf{H}^*_l$ by means of (3) and (15)

$$\begin{aligned}
V^{2,l,N}_{\Pi^* P}(g, x) &= \rho'_l(\{f'^2(x)\} \mid g, x) T^2_{e^1 f^2} V^{2,l+1,N}_{\Pi^* P}(g, x)) \\
&\quad + \int_{\boldsymbol{B}'(x)} \rho'_l(db \mid g, x) T^2_{e^1_l \cdot} V^{2,l+1,N}_{\Pi^* P}(g, x, b) \\
&\geq \rho'_l(\{f'^2(x)\} \mid g, x)(T^2_{e^1 f^2} V^{2,l+1,N}_{\Pi^* P^*}(g, x) \\
&\quad - (N - l)C \cdot q_{e^1 f^2} q_{e^*_{l+1}\rho_{l+1}} \cdots q_{e^*_N \rho_N} I_{\overline{\mathbf{H}}^*_{N-l}}(g, x)) \\
&\quad + \int_{\boldsymbol{B}'(x)} \rho'_l(db \mid g, x)(T^2_{e^1 \cdot} V^{2,l+1,N}_{\Pi^* P}(g, x, b) - T^2_{e^1 f^2} V^{2,l+1,N}_{\Pi^* P^*}(g, x)) \\
&\quad + \int_{\boldsymbol{B}'(x)} \rho'_l(db \mid g, x) T^2_{e^1 f^2} V^{2,l+1,N}_{\Pi^* P^*}(g, x) \\
&\geq T^2_{e^1 f^2} V^{2,l+1,N}_{\Pi^* P^*}(g, x) \\
&\quad - (N - l)C \cdot q_{e^*_l \rho_l}(I_{\mathbf{G}^*_1} q_{e^*_{l+1}\rho_{l+1}} \cdots q_{e^*_N \rho_N} I_{\overline{\mathbf{H}}^*_{N-l}})(g, x) \\
&\quad - (N - l + 1)C \cdot \rho'_l(\boldsymbol{B}'(x) \mid g, x) \\
&\geq V^{2,l,N}_{\Pi^* P^*}(g, x) - (N - l + 1)C \\
&\quad \times q_{e^*_l \rho_l} q_{e^*_{l+1}\rho_{l+1}} \cdots q_{e^*_N \rho_N}(I_{\mathbf{G}^*_1 \times \overline{\mathbf{H}}^*_{N-l}} + I_{\overline{\mathbf{G}}^*_1 \times \mathbf{H}_{N-l}})(g, x) \\
&= V^{2,l,N}_{\Pi^* P^*}(g, x) - (N - l + 1)C \cdot q_{e^*_l \rho_l} q_{e^*_{l+1}\rho_{l+1}} \cdots q_{e^*_N \rho_N} I_{\overline{\mathbf{H}}^*_{N-l+1}}(g, x).
\end{aligned}$$

□

**Theorem 4.1.** *Let Assumption 3.1 be satisfied. Let $\epsilon \geq 0$. Assume that there is an $\epsilon$-I-strategy pair $(\Pi^1, P^1) = (e^{1\infty}, f^{1\infty})$ and an $\epsilon$-II-strategy pair $(\Pi^2, P^2) = (e^{2\infty}, f^{2\infty})$. Then there is an $\epsilon$-intimidation strategy pair $(\Pi^*, P^*)$ connected with $(\Pi^1, P^2)$ which is a quasi-stationary $2\epsilon$-equilibrium pair.*

**Proof.** Let

$$\tilde{\mathbf{G}}^*_n = \mathbf{G}^*_n \times \Omega \quad \text{for} \quad n \in \mathbb{N}_0,$$

$$\mathbf{H}^* = \{\omega = (x_0, a_0, b_0, x_1, a_1, b_1, \ldots) \in \Omega : \\ a_i = e^1(x_i),\ b_i = f^2(x_i, a_i),\ i \in \mathbb{N}_0\}.$$

Then it holds $\mathbf{H}^* = \bigcap_{n\in\mathbb{N}_0} \tilde{\mathbf{G}}_n^*$, hence $\lim_{n\to\infty} \mathsf{P}_{x,\Pi,P^*}(\tilde{\mathbf{G}}_n^* \backslash \mathbf{H}^*) = 0$. We consider an arbitrary $\eta > 0$ and assume $\mathsf{P}_{x,\Pi,P^*}(\mathbf{G}_n^* \backslash \mathbf{H}^*) < \eta$. It follows

$$\mathsf{P}_{\Pi P^*}^{0,N}(\mathbf{G}_n^* \times \overline{\mathbf{H}}_{N-n+1}^* \,|\, x) = \mathsf{P}_{x,\Pi,P^*}(\mathbf{G}_n^* \times \overline{\mathbf{G}}_{N-n+1}^* \times \Omega) < \eta, \tag{21}$$

since $\mathbf{G}_n^* \times \overline{\mathbf{G}}_{N-n+1}^* \times \Omega \subseteq \tilde{\mathbf{G}}_n^* \backslash \mathbf{H}^*$. From Theorem 3.2 it follows that $\Phi^1_{\Pi^1 P^2}$ is constant. We put $\Phi^1_{\Pi^1 P^2} = \Phi^{1*}$. For every strategy $\tilde{\Pi}$ of the first player we have

$$\Phi^1_{\tilde{\Pi} P^1} \geq \Phi^1_{\Pi^1 P^1} - \epsilon \geq \Phi^1_{\Pi^1 P^2} - 2\epsilon = \Phi^{1*} - 2\epsilon. \tag{22}$$

We denote by $x_i(h)$ the $x_i$-coordinate of $h = (x_0, a_0, b_0, x_1, a_1, b_1, \ldots, x_n) \in \mathbf{H}_n$ and by $\Pi_h = (\tilde{\pi}_m)$ that strategy of the first player with $\tilde{\pi}_m(\cdot) = \pi_{m+n}(h, \cdot)$ for every fixed $h \in \mathbf{H}_n$. It follows then

$$\begin{aligned}
\Phi^1_{\Pi P^*}(x) &= \liminf_{N\to\infty} \frac{1}{N+1} \int_{\mathbf{H}_n} \mathsf{P}_{\Pi P^*}^{0,n-1}(dh \,|\, x) V_{\Pi P^*}^{1,n,N}(h) \quad \text{(see (16))} \\
&= \liminf_{N\to\infty} \frac{1}{N+1} \int_{\overline{\mathbf{H}}_n^*} \mathsf{P}_{\Pi P^*}^{0,n-1}(dh \,|\, x) V_{\Pi P^*}^{1,n,N}(h) \\
&\quad + \liminf_{N\to\infty} \frac{1}{N+1} \int_{\mathbf{H}_n^*} \mathsf{P}_{\Pi P^*}^{0,n-1}(dh \,|\, x) V_{\Pi P^*}^{1,n,N}(h) \quad \text{(see (12), (13))} \\
&\geq \liminf_{N\to\infty} \frac{1}{N+1} \int_{\overline{\mathbf{H}}_n^*} \mathsf{P}_{\Pi P^*}^{0,n-1}(dh \,|\, x) V_{\Pi P^*}^{1,n,N}(h) \\
&\quad + \liminf_{N\to\infty} \frac{1}{N+1} \int_{\mathbf{H}_n^*} \mathsf{P}_{\Pi P^*}^{0,n-1}(dh \,|\, x) V_{\Pi^* P^*}^{1,n,N}(h) \\
&\quad - \liminf_{N\to\infty} \frac{1}{N+1} (N-m+1) C \\
&\quad \int_{\mathbf{H}_n^*} \mathsf{P}_{\Pi P^*}^{0,n-1}(dh \,|\, x) \mathsf{P}_{\Pi P^*}^{n,N}(\overline{\mathbf{H}}_{N-n+1}^* \,|\, h) \quad \text{(see Lemma 4.1)} \\
&\geq \int_{\overline{\mathbf{H}}_n^*} \mathsf{P}_{\Pi P^*}^{0,n-1}(dh \,|\, x) \liminf_{N\to\infty} \frac{1}{N+1} V_{\Pi P^1}^{1,n,N}(h) \\
&\quad + \int_{\mathbf{H}_n^*} \mathsf{P}_{\Pi P^*}^{0,n-1}(dh \,|\, x) \liminf_{N\to\infty} \frac{1}{N+1} V_{\Pi^* P^*}^{1,n,N}(h) - C\eta \\
&\qquad \text{(see Fatou's Lemma and (21))} \\
&= \int_{\overline{\mathbf{H}}_n^*} \mathsf{P}_{\Pi P^*}^{0,n-1}(dh \,|\, x) \Phi^1_{\Pi_h P^1}(x_n(h)) \\
&\quad + \int_{\mathbf{H}_n^*} \mathsf{P}_{\Pi P^*}^{0,n-1}(dh \,|\, x) \Phi^1_{\Pi^1 P^2}(x_n(h)) - C\eta \text{(see (16))} \\
&\geq \Phi^{1*} - 2\epsilon - C\eta \quad \text{(see (22))} \\
&= \Phi^1_{\Pi^1 P^2}(x) - C\eta - 2\epsilon = \Phi^1_{\Pi^* P^*}(x) - 2\epsilon - C\eta
\end{aligned}$$

Since $\eta$ is arbitrary, we get for all strategies $\Pi$ of the first player

$$\Phi^1_{\Pi^* P^*} \leq \Phi^1_{\Pi P^*} + 2\epsilon.$$

Analogously, it follows for all strategies $P$ of the second player

$$\Phi^2_{\Pi^* P^*} \leq \Phi^2_{\Pi^* P} + 2\epsilon. \qquad \square$$

Theorems 3.4 and 4.1 imply the following result:

**Theorem 4.2.** *Let Assumption 3.1 be satisfied. Then there exists a quasi-stationary deterministic $\epsilon$-equilibrium pair for every $\epsilon > 0$ if one of the following sets of conditions given in Assumption 3.2 holds:*

(a) *One of the spaces* **X**, **A**, *or* **B** *is countable.*
(b) (A2), (B1), (C2), (D3.1), *and* (D4.2).
(c) (A1), (B5), (C2), (D4.1), *and* (D3.2).
(d) (A2), (A4), (B2), (C1), (D1.1), *and* (D2.2).
(e) (A3), (B3), (C1), (D2.1), *and* (D1.2).

**Acknowledgments**

The author thanks the anonymous referees for many important comments.

## REFERENCES

[1] Berge, C. *Topological Spaces*. Oliver and Boyd, Edinburgh, 1963.

[2] Federgruen, A. On $N$-Person Stochastic Games with Denumerable State Space. *Advances in Applied Probability*, **10**, 452–471, 1978.

[3] Gubenko, L. G. Multistage Stochastic Games. *Theory Probability and Mathematical Statistics*, **8**, 31–44, 1975.

[4] Gubenko, L. G. and E. S. Statland. Controlled Markov processes with Discrete Time Parameter (Russian). *Teoriya Veroyatnosteĭ i Matematichiskaya Statistic*, **7**, 51–64, 1972.

[5] Hinderer, K. *Foundations of Non-Stationary Dynamic Programming with Discrete Time Parameter.* Lecture Notes in Operations Research and Mathematical Systems 33. Springer-Verlag, Berlin, 1970.

[6] Küenle, H.-U. Über die Optimalität von Strategien in stochastischen dynamischen Minimax-Entscheidungsmodellen I. *Mathematische Operationsforschung and Statistik, Series Optimization*, **12**, 421–435, 1981.

[7] Küenle, H.-U. Über die Optimalität von Strategien in stochastischen dynamischen Minimax-Entscheidungsmodellen II. *Mathematische Operationsforschung und Statistik, Series Optimization*, **12**, 301–313, 1983.

[8] Küenle, H.-U. *On $\epsilon$-Optimal Strategies in Discounted Markov Games*. Mathematical Control Theory. Banach Center Publications, Vol. 14, pp. 263–276, Warsaw, 1985.

[9] Küenle, H.-U. *Stochastische Spiele und Entscheidungsmodelle.* Teubner-Texte zur Mathematik 89. Teubner-Verlag, Leipzig, 1986.

[10] Küenle, H.-U. On Nash Equilibrium Solutions in Non-Zero Sum Stochastic Games with Complete Information. *International Journal of Game Theory*, **23**, 303–324, 1994.

[11] Küenle, H.-U. On Equilibrium Strategies in Stochastic Games with Complete Information. Preprint M3/95. Technische Universität Cottbus, Institut für Mathematik, Cottbus, 1995.

[12] Rieder, U. Average Optimality in Markov Games with General State Space. *Proceedings of the 3rd International Conference on Approximation and Optimization*, Puebla, 1995.

[13] Thuijsman, F., and T. E. S. Raghavan. Perfect Information Stochastic Games and Related Classes. *International Journal of Game Theory*, 1997, **26**, 403–408, 1997.

[14] Yushkevich, A. A. Blackwell Optimal Policies in a Markov Decision Procress with a Borel State Space. *ZOR—Mathematical Methods of Operations Research*, **40**, 253–288, 1994.

# PART V
# Application

# Crime and Law Enforcement: A Multistage Game

Herbert Dawid
Department of Economics
University of Southern California
Los Angeles, California, USA

Gustav Feichtinger
Department of Operations Research and Systems Theory
Vienna University of Technology
Vienna, Austria

Steffen Jørgensen
Department of Management
Odense University
Odense, Denmark

**Abstract**

In this paper a conflict between a potential criminal offender and a law enforcement agency is studied. The model is a two-stage extensive form game with imperfect information. We identify all relevant sequential equilibria of the game and show how the equilibrium strategies and game values of both players depend on the parameter values. Further, it is shown that in equilibrium the offense rate and the law enforcement rate in the first period are always less than or equal to that in the second period. It is also established for multistage games that these two rates are monotonously nonincreasing; however, this property disappears if recidivistic behavior is present.

## 1 Introduction

Becker [1] revitalized the utilitarian approach to crime and punishment and gave an economic analysis of the problem of how many resources society should allocate and how much punishment should be inflicted upon offenders to enforce various kinds of legislation concerning violation of personal and property rights. This work started a neoclassical oriented theory of crime, punishment, and optimal law enforcement. For the offender, Becker introduced the notion of a crime supply function such that the crime rate (the supply) is affected by the probabilities of apprehension and conviction, the severity of punishment, and the gain to an offender if crime goes unpunished. Predominantly, this literature took a static view of the problem and studied various aspects of crime supply and law enforcement from a one-sided point of view. Later on, dynamic approaches to crime and law enforcement have modified these static analyses. Dynamic optimization problems are studied in Sethi [15], Davis [2], Leung [9], [10], [11] from the point of view of an offender, and in Gerchak and Parlar [6] from the point of view of the enforcement authorities. Introducing the offender and the authorities as rational adversaries requires a game-theoretic setup. Dynamic games between an offender and a law enforcement agency are analyzed in continuous time by Feichtinger [4], Jørgensen [8], or Dawid and Feichtinger [3].

This paper takes a dynamic game-theoretic view at the conflict between an offender and the authorities but models the problem as a two-stage game with imperfect information. With respect to the structural dynamics of the model, a two-stage game is considerably simpler than, for example, the differential games of Feichtinger [4] and Jørgensen [8]. A two-stage approach, on the other hand, allows for richer strategic dynamics. This means that we can identify strategies that are more satisfying from a game-theoretic point of view. In differential games of some complexity reasons of tractability imply the confinement to open-loop strategies. It is well known that from a strategic point of view such strategies are less satisfactory than the closed-loop strategies considered in this paper. Moreover, [11] pointed out that the majority of the optimal control and differential game models essentially have a two-stage structure. This derives from the fact that they use a hazard rate type of modeling of the process that leads to apprehension and conviction of an offender. What matters in these models is whether we look at the game at an instant of time *before* (stage one) or *after* (stage two) the apprehension and conviction of the offender.

The paper proceeds as follows. Section 2 presents the model and its assumptions. The set of interesting sequential equilibria is identified in Section 3. Section 4 provides a characterization of equilibrium behavior of the two players. This section also states some sensitivity results. Section 5 briefly deals with extensions to multistage games with or without recidivism. Section 6 concludes.

## 2 The Model

The game is played over two periods (two stages) between an offender (Player 1) and a law enforcement authority (Player 2). The fundamentals of the game are the same in both stages. There are two pure strategies available in each period to the offender: to act legally ($H$) or to be criminal ($C$).[1] Similarly, the authorities have two pure strategies: to enforce the law ($E$) or to do nothing ($N$).[2]

[1]The restriction to only two pure strategies may not be that limiting as we are going to introduce behavioral strategies. In linear models the probability associated with the pure strategy $C$ may be interpreted as an offense rate.

[2]The probability associated with the pure strategy $E$ may be interpreted as an enforcement rate.

If the offender in any of the two stages chooses strategy $H$, nothing is earned from criminal activity and the offender only earns his lawful income of $o > 0$ (dollars). If the offender chooses strategy $C$ in stage one, an illegal income of the amount of $\pi > 0$ (dollars) may be earned in that stage in addition to his legal income. If the offender is apprehended at stage one the following is assumed. Apprehension leads inevitably to conviction and the offender is punished by having to pay a fine of the amount of $f > 0$ (dollars) and is imprisoned until the end of the game. When in prison, the offender earns no income at all. Thus, if convicted he foregoes the chance of having any income in stage two. The corresponding opportunity costs imply that being apprehended is more severe for the offender at stage 1 than at stage 2. To motivate this assumption we do not necessarily have to think of life-long prison terms but may also consider examples like bankrupcy offences which jeopardize future business opportunities and thus are the more costly the earlier they appear. Such a temporal effect is also present in the analyses of [11].

In any period, if the authorities choose strategy $E$, the probability of convicting the offender equals $v \in (0, 1)$ and the authorities incur a per-period cost of $e > 0$ (dollars).[3] If (and only if) the offender commits a crime that goes unpunished, a loss of $c > 0$ (dollars) per period is inflicted upon society. On the other hand, if strategy $N$ is chosen, the offender is caught with probability zero but then enforcement costs are also zero.

Given that the offender chooses $C$, we assume that strategy $E$ yields a positive expected payoff to the authorities: $vc > e$. If this condition is not imposed, enforcement would never be optimal and then the problem would be trivial.

The setup is an extensive-form game with imperfect information. In period one, both players choose simultaneously their action from the sets $\{H, C\}$ and $\{E, N\}$, respectively. In three of the four possible outcomes the offender is not caught and punished and the period-one game is played again in period two. Only if the strategy pair $(C, N)$ is chosen, the offender is caught and punished with a positive probability. To model this we insert a chance move to represent the two possibilities that the offender is convicted or not. Hence with probability $v$ the offender is convicted and cannot carry out any activity in period two. With probability $1 - v$ the offender is not convicted in period one and has the option of earning legal or illegal income in period two. In this case, the authorities have to bear both enforcement and social costs, that is, $e + c$.

Figure 1 shows a single stage of the two-stage game. To simplify the notation we have introduced payoff-pairs $(p_i, q_i)$ as follows:

$$(p_1, q_1) = (-f, -e),$$
$$(p_2, q_2) = (o + \pi, -e - c),$$
$$(p_3, q_3) = (o + \pi, -c),$$

[3] It is sensible to exclude the two extreme probabilities $v = 0$ and $v = 1$.

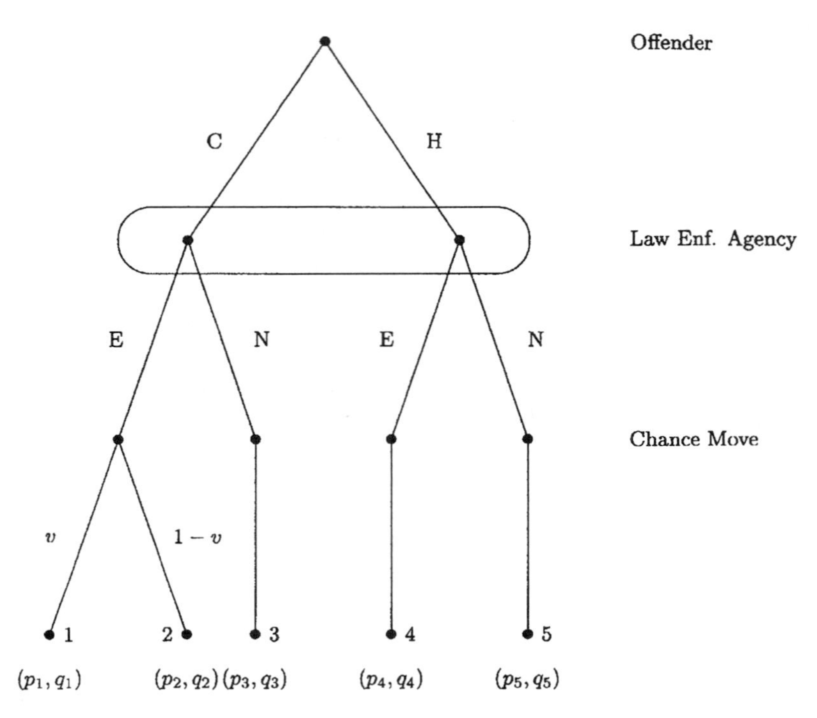


**Figure 1**: Extensive form of a single stage of the two stage game.

$$(p_4, q_4) = (o, -e),$$
$$(p_5, q_5) = (o, 0).$$

To illustrate, consider the first pair. The strategy combination $(C, E)$ is chosen and the payoff-pair $(p_1, q_1)$ is the outcome given that the offender is caught and punished. The latter happens with probability $v$. The offender incurs a direct cost of $f$ and does not receive any income in this (and also in the subsequent) period. The authorities incur the enforcement cost $e$ but escape the social costs $c$.

The two decision nodes of Player 2 define an information set, indicating that action has to be taken by Player 2 without knowing which of the two nodes is actually reached. To get the extensive form for the whole two period game we have to paste an identical tree as the one in Figure 1 below any of the five terminal nodes but the one to the left. Recall that this node corresponds to the situation where the criminal is arrested in the first period. Thus, this node becomes a terminal node of the two-stage game. Note further that the law enforcement agency does not know in the second stage which strategy the offender used in the first stage unless he was arrested. Accordingly there are only three different information sets for the law enforcement agency in the second stage: one where the agency enforced in the first period and apprehended an offender, one where it enforced without success, and one where it did not enforce in the first period. Figure 2 shows the game tree

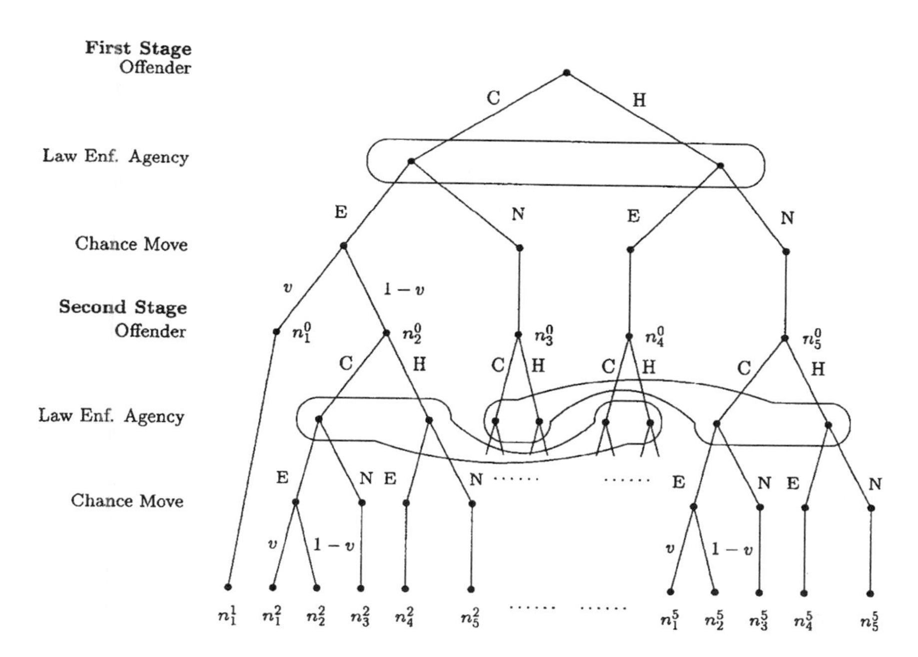


**Figure 2**: Game tree of the two-stage game.

with seven layers of nodes and 21 terminal nodes. We denote the terminal node $i$ of the second stage game, starting from node $j$ of the first stage game, by $n_i^j$; node $i$ of the first stage game is denoted by $n_i^0$. The payoff that the offender receives in the first stage if node $n_i^0$ is reached is denoted by $p_i^0$, the payoff of the second player is denoted by $q_i^0$. The overall payoffs of the two players when terminal node $n_i^j$ is reached are $p_i^j$ and $q_i^j$, respectively. Second period payoffs are not discounted.

## 3 Solution of the Game

In this section we identify Nash equilibria in behavioral strategies of the imperfect information game.[4] A behavioral strategy assigns to each information set a probability distribution over the actions available at this set. We write a behavioral strategy of the offender as a vector with five components, such that the first component gives the probability that the first pure strategy is played in the first stage and the remaining four components give the probability for playing the first pure strategy in nodes $n_2^0$, $n_3^0$, $n_4^0$, and $n_5^0$, respectively. The behavioral strategy of the law enforcement agency contains only three components, one for the first stage and two for the two relevant information sets in the second. Let us first assume that for both players the move probabilities in the second stage do not depend on

[4]See [5] or [12] for an introduction to extensive form games.

the information sets they are in. It is easy to see that if the offender is not arrested in the first stage the opponents play in the second stage a normal form game with payoff matrix[5]

$$(A^2, B^2) = \begin{bmatrix} ((1-v)(\pi+o) - vf), -e-(1-v)c) & (\pi+o, -c) \\ (o, -e) & (o, 0) \end{bmatrix}, \quad (1)$$

where $a_{ij}^2$ and $b_{ij}^2$, $i \in \{C, H\}$, $j \in \{E, N\}$, are the payoffs of the two players in the second stage if the offender uses strategy $i$ and the law enforcement agency uses $j$. Taking into account our assumption $e < vc$, three cases have to be distinguished in this game.

First, if

$$\pi > \pi^1 := \frac{v}{1-v}(f+o) \quad (2)$$

there is a unique equilibrium in pure strategies, where the offender plays $C$ and the authorities $E$. We have the game values $\lambda^2 = (1-v)(\pi+o) - vf$ for the first player and $\mu^2 = -e-(1-v)c$ for the second player.

Second, if the inequality in (2) is reversed there is a unique Nash equilibrium in mixed strategies $((x^2, 1-x^2), (y^2, 1-y^2))$ with probabilities

$$x^2 = \frac{e}{vc}, \qquad y^2 = \frac{\pi}{v(\pi+f+o)}. \quad (3)$$

We get these values by solving the equations $(y^2, 1-y^2)A^T = \lambda^2(1,1)$ and $(x^2, 1-x^2)B = \mu^2(1,1)$ with respect to $x^2$, $y^2$, $\lambda^2$, and $\mu^2$ where $\lambda^2$ and $\mu^2$ are the game values for the two players of this normal form game. The game value for the first player is $\lambda^2 = o$ and the second has $\mu^2 = -e/v$.

Third, in the case that

$$\pi = \pi^1 \quad (4)$$

we have a continuum of equilibria. If the second player plays $E$ the first player is indifferent between being noncriminal and criminal. Thus, any strategy profile where the first player plays a mixed strategy such that $E$ is the best response and the second player plays $E$ is an equilibrium. It is easy to see that such profiles are given by $\{((z, 1-z), (1,0)) \mid z \geq x^2\}$. The game value of the first player is $\lambda^2 = o$, whereas the game value of the second player, $\mu^2 = -e - z(1-v)c$, may be any value in the interval $[-e-(1-v)c, -e/v]$.

Let us now deal with the case where the players use different move probabilities in different information sets. Neglecting the border case $\pi = \pi_1$ there is always a unique Nash equilibrium of game 1. For $\pi > \pi^1$ both opponents have to use their pure equilibrium strategies. For $\pi < \pi^1$ the law enforcement agency has to use the equilibrium strategy in both information sets in stage two. This implies that both strategies of the offender have the same payoff. Accordingly, we

[5]The superscript "2" indicates that we are in the second stage.

can have sequential equilibrium profiles where the offender uses different moving probabilities in nodes $n_2^0$ to $n_5^0$ as long as the law enforcement agency's consistent expectation about the offenders' strategy in the second period equals the equilibrium value $(x_2, 1-x_2)$. Considering the information set below $\{n_2^0, n_4^0\}$ consistent beliefs imply that the law enforcement agency expects an offense with probability $[x^1(1-v)/(1-x^1v)]s_2^2 + [(1-x^1)/(1-x^1v)]s_4^2$, where $x^1$ is the equilibrium probability of an offense in the first stage and $s_2^2$ and $s_4^2$ are the offense probabilities in nodes $n_2^0$ and $n_4^0$. Accordingly, any profile $((x_1^1, s_2^2, s_2^3, s_2^4, s_2^5), (y^1, y^2, y^2))$, where the probability above equals $x^2$ given in (3) and

$$x^1 s_3^2 + (1-x^1)s_5^2 = x^2$$

yields a sequential equilibrium if $x^1$ and $y^1$ are equilibrium probabilities in the first stage. However, since the second stage game value in all those equilibria equals the game value of (1) and the expected offense probability in the second stage is always $x^2$ we do not deal separately with those equilibria, and restrict in what follows our attention to the equilibria where the second stage behavior does not depend on the information set the player is in.

It follows from the analysis above that given equilibrium behavior in the second stage the payoffs at nodes $n_i^0$, $i \geq 2$, are given by $(p_i^0 + \lambda^2, q_i^0 + \mu^2)$ and that $(p_1^0, q_1^0)$ is the payoff at $n_1^0$. Taking the expected value over the outcomes of the chance move determines the Nash strategies in the first stage by solving a $2 \times 2$ normal form game. Direct calculation shows that the payoff matrix of this game is given by $(\tilde{A}^1, \tilde{B}^1)$, with elements

$$\begin{aligned}
\tilde{a}^1_{CE} &= (1-v)(\pi + o + \lambda^2) - vf,\\
\tilde{b}^1_{CE} &= -e - (1-v)(c - \mu^2),\\
\tilde{a}^1_{CN} &= \lambda^2 + \pi + o,\\
\tilde{b}^1_{CN} &= \mu^2 - c,\\
\tilde{a}^1_{HE} &= \lambda^2 + o,\\
\tilde{b}^1_{HE} &= \mu^2 - e,\\
\tilde{a}^1_{HN} &= \lambda^2 + o,\\
\tilde{b}^1_{HN} &= \mu^2.
\end{aligned}$$

It is easier to solve an equivalent game with payoff matrix[6]

$$(A^1, B^1) = \begin{bmatrix} ((1-v)(\pi + o) - v(f + \lambda^2), -e - (1-v)c - v\mu^2) & (\pi + o, -c) \\ (o, -e) & (o, 0) \end{bmatrix}. \quad (5)$$

[6]We subtract $\lambda^2$ from all payoffs of the first player and $\mu^2$ from all payoffs of the second player.

and again we distinguish three cases ($\pi > \pi^1, \pi = \pi^1, \pi < \pi^1$) which yield different values for $\lambda^2$ and $\mu^2$. We get five different parameter constellations in which qualitatively different Nash equilibria occur.

If (2) holds, $\lambda^2 = (1-v)(\pi + o) - vf$ and $\mu^2 = -e - (1-v)c$. Inserting these values into (5) we obtain a similar situation as in the game analyzed above. If

$$(1-v)(\pi+o) - v(f+\lambda^2) - o = (1-v)^2\pi - v(1-v)(f+o) - vo \geq 0 \quad (6)$$

the strategy profile $CE$ is a Nash equilibrium in pure strategies. As (6) implies (2) we have shown that if

$$\pi > \pi^2 := \frac{v}{1-v}(f+o) + \frac{v}{(1-v)^2}o \quad (7)$$

the strategy profile $(s_1, u_1)$ with

$$s_1 = (1,1,1,1,1), \qquad u_1 = (1,1,1), \quad (8)$$

yields a sequential[7] equilibrium of the two-stage game. In this equilibrium the offender always plays strategy $C$, whereas the authorities always enforce. The game values $(\lambda^1, \mu^1)$ of the two players are given by the values of $\tilde{a}^1_{CE}$ and $\tilde{b}^1_{CE}$, respectively. Thus

$$\lambda^1_1 = (2-v)((1-v)\pi - vf + (1-v)o), \qquad \mu^1_1 = (2-v)(-e-(1-v)c). \quad (9)$$

The game value $\mu^1_1$ for the law enforcement agency is negative and the game value $\lambda^1_1$ for the offender is strictly positive.

On the other hand, if (2) holds with the inequality in (7) reversed, the root game (5) has no equilibrium in pure strategies. But it has a unique equilibrium in mixed strategies $((x^1_2, 1-x^1_2), (y^1_2, 1-y^1_2))$ such that

$$x^1_2 = \frac{e}{v(e+(2-v)c)}, \qquad y^1_2 = \frac{\pi}{v((2-v)\pi + (1-v)(f+o)+o)}. \quad (10)$$

Hence, if

$$\pi^1 < \pi < \pi^2 \quad (11)$$

the strategy profile $(s_2, u_2)$ with

$$s_2 = (x^1_2, 1, 1, 1, 1), \qquad u_2 = (y^1_2, 1, 1), \quad (12)$$

yields a Nash equilibrium of the two-stage game. The game values are

$$\lambda^1_2 = (1-v)\pi - vf + (2-v)o, \qquad \mu^1_2 = -e-(1-v)c - \frac{ec}{v(e+(2-v)c)}. \quad (13)$$

In the case where

$$\pi = \pi^2 \quad (14)$$

[7] We only write down the moving probabilities since the corresponding consistent beliefs are trivially equal to the actual moving probabilities in this setup.

there is a continuum of equilibria for the first stage game and correspondingly a continuum of equilibria of the whole game, given by all profiles of the form

$$s_3 = (z, 1, 1, 1, 1), \qquad u_3 = (1, 1, 1), \tag{15}$$

where $z \in [x_2^1, 1]$. Direct calculations show that the game value of the first player is given by $\lambda_3^1 = [(2-v)/(1-v)]o$.[8] The value of the game for the second player is $\mu_3^1 = -e - z((1-v)c - v\mu^2)$.

Having dealt with all cases in which only pure strategies are played in the second stage we now analyze the root game if (4) holds. We know that the value of $\mu^2$ is not uniquely determined in this case but there exists a continuum of equilibria of the second stage game and the corresponding values of $\mu^2$ are in the interval $[-e-(1-v)c, -e/v]$. Due to the non-uniqueness of the equilibrium of game (1) we may also consider Nash equilibria of the two-stage game where different equilibrium strategies are played in nodes $n_i^0$, $i \geq 2$. This would imply that the value of $\mu^2$ depends on which node is reached in the first stage play and the transformation of the first stage game into the form (5) is no longer admissible. However, recalling that the parameter constellation considered in (4) is a very special one, we do not work out these solutions in detail but consider only equilibria where both players' strategies in the second stage are independent of which information set the players are in. Making this assumption we realize that $b_{CE}^1 > b_{CN}^1$ for any possible value of $\mu^2$ and $a_{CE}^1 < a_{HE}^1$, which implies that there is a unique equilibrium in mixed strategies for the root game (5). In equilibrium, the probabilities of playing the first pure strategy are given by

$$x_4^1 = \frac{e}{v(c-\mu^2)}, \qquad y_4^1 = \frac{f+o}{f+(2-v)o}. \tag{16}$$

Accordingly, the equilibrium strategies and the game values (which depend on $\mu^2$) are

$$s_4 = (x_4^1, z, z, z, z), \qquad u_4 = (y_4^1, 1, 1), \tag{17}$$

where $z = -[(\mu^2 + e)/(1-v)c]$ and

$$\lambda_4^1 = 2o, \qquad \mu_4^1 = -\frac{ec}{v(c-\mu^2)}. \tag{18}$$

To complete the analysis we have to deal with the case

$$\pi < \pi^1. \tag{19}$$

Inserting the corresponding values $\lambda^2 = o$ and $\mu^2 = -e/v$ into the payoff matrix (5) shows that also in this case the root game has a unique Nash equilibrium in mixed strategies with

$$x_5^1 = \frac{e}{vc+e}, \qquad y_5^1 = \frac{\pi}{v(\pi+f+2o)}. \tag{20}$$

---

[8]Notice that this value coincides with $\lambda_1^1$ and $\lambda_2^1$.

For the two-stage game we get the equilibrium strategies

$$s_5^1 = (x_5^1, x^2, x^2, x^2, x^2), \qquad s_5^2 = (y_5^1, y^2, y^2), \tag{21}$$

and the game values

$$\lambda_5^1 = 2o, \qquad \mu_5^1 = -\frac{e(2vc + e)}{v(vc + e)}, \tag{22}$$

where $x^2$ and $y^2$ are given by (3).

## 4 Analysis of the Equilibria

In Section 3 we calculated the relevant sequential Nash equilibria of the two-stage game. In cases where the equilibrium strategies imply the occurrence of combination $CE$ with probability one in the first stage there also exist Nash equilibria which contain irrational actions in the information sets below nodes $n_i^0$, $i \geq 3$. However, it is obvious that those Nash equilibria are not sequential equilibria and we do not deal with them here. For all parameter constellations where a mixed strategy is played in the first stage there are no other Nash equilibria than the ones given here.

In this section we look at the qualitative properties of the sequential equilibria. We start by showing that in equilibrium both players play the first strategy in the first stage with a probability which is less than or equal to the probability that they play the first strategy in the second stage.

**Proposition 4.1.** *Let $(s, u)$ with*

$$s = (s^1, s^2, s^2, s^2, s^2), \qquad u = (u^1, u^2, u^2),$$

*be a Nash equilibrium of the two-stage game. Then $s^1 \leq s^2$ and $u^1 \leq u^2$.*

**Proof.** If $\pi > \pi^1$ the result of the proposition follows trivially as $s^2 = u^2 = 1$ in all equilibria. For $\pi = \pi^1$ we have $s^1 = x_4^1$ given by (16) and $s^2 = -[(\mu^2 + e)/(1 - v)c]$. Thus we have to show that

$$\frac{e}{v(c - \mu^2)} \leq -\frac{\mu^2 + e}{(1 - v)c}$$

for all $\mu^2 \in [-e - (1 - v)c, -e/v]$. It is easy to see that the left-hand side of the inequality attains its maximum and the right-hand side its minimum for $\mu^2 = -e/v$. This leaves us showing that

$$\frac{e}{v(c + e/v)} \leq \frac{(e/v - e)}{(1 - v)c} \quad \Leftrightarrow \quad \frac{1}{c + e/v} \leq \frac{1}{c}.$$

As $e > 0$ and $v > 0$ the right-hand inequality is always fulfilled. In the case $\pi < \pi^1$ we have $s^1 = x_5^1$, $s^2 = x^2$ and $u^1 = y_5^1$, $u^2 = y^2$. Recalling the expressions for these variables the proposition follows directly from the assumed positivity of $e$, $o$, and $v$. □

Proposition 4.1 has a simple interpretation. For the offender it is more risky to commit a crime in the first stage than in the second stage for if he is caught in the first stage he is sure to lose his income in the second stage. Thus, it is more likely that he will play strategy $C$ in the second stage than in the first stage. The law enforcement authorities react to this behavior by investing more effort in enforcement in the second stage than in the first stage.

We have seen that the characteristics of the equilibrium depend crucially on the constellation of parameters. Hence, some sensitivity analyses are in order. We show how equilibrium behavior changes if we increase the profit of crime $\pi$ and keep all other parameters constant. Denote by $(s(\pi) = (s^1(\pi), s^2(\pi), s^2(\pi), s^2(\pi), s^2(\pi))$ and $u(\pi) = (u^1(\pi), u^2(\pi), u^2(\pi))$ the equilibrium strategies of the two players as functions of $\pi$. Select the equilibria $(s(\pi), u(\pi)) = ((x_2^1, 1, 1, 1, 1), (y_2^1, 1, 1))$ for $\pi = \pi^1$ and $(s(\pi), u(\pi)) = ((1, 1, 1, 1, 1), (1, 1, 1))$ for $\pi = \pi^2$ out of the continuum of equilibria existing for these parameter values. Thus $s(\pi)$ and $u(\pi)$ are single-valued functions. Denote the corresponding game values by $\lambda^1(\pi)$ and $\mu^1(\pi)$. Proposition 4.2 characterizes the strategy and the game value of the offender as depending on $\pi$.

**Proposition 4.2.** *The game value of the offender is continuous and nondecreasing in* $\pi$.[9] *The strategy* $s^2(\pi)$ *has an upward jump at* $\pi^1$ *whereas* $s^1(\pi)$ *has a downward jump at* $\pi^1$ *and an upward jump at* $\pi^2$.

**Proof.** To show that $\lambda^1(\pi)$ is continuous it suffices to show that it is continuous at $\pi^1$ and $\pi^2$. This can easily be ascertained by inserting $\pi^2$ into the expressions for $\lambda_1^1$ and $\lambda_2^1$ given by (9) and (13), and showing that inserting $\pi^1$ into $\lambda_2^1$ gives $2o$. Obviously, $\lambda^1(\pi)$ is constant and equal to $2o$ on $[0, \pi^1]$ and strictly increasing afterward. The function $s^2(\pi)$ has the value $x^2 = e/(vc)$ on the interval $[0, \pi^1]$ and is equal to 1 for all $\pi > \pi^1$. Due to the assumption $e < vc$, we have $x^2 < 1$ and $s^2(\pi)$ has an upward jump from $x^2$ to 1 at $\pi = \pi^1$. The function $s^1(\pi)$ is given by

$$s^1(\pi) = \begin{cases} \dfrac{e}{vc+e}, & \pi < \pi^1, \\ \dfrac{e}{v(e+(2-v)c)}, & \pi^1 \le \pi < \pi^2, \\ 1 & \pi \ge \pi^2, \end{cases}$$

and all we have to show is

$$\begin{aligned} & \frac{e}{vc+e} > \frac{e}{v(e+(2-v)c)}, \\ \Leftrightarrow \quad & e < ve + v(1-v)c, \\ \Leftrightarrow \quad & e < vc. \end{aligned}$$

[9]The game value of the first player is also continuously differentiable everywhere with the exception of the points $\pi = \pi^1$ and $\pi = \pi^2$.

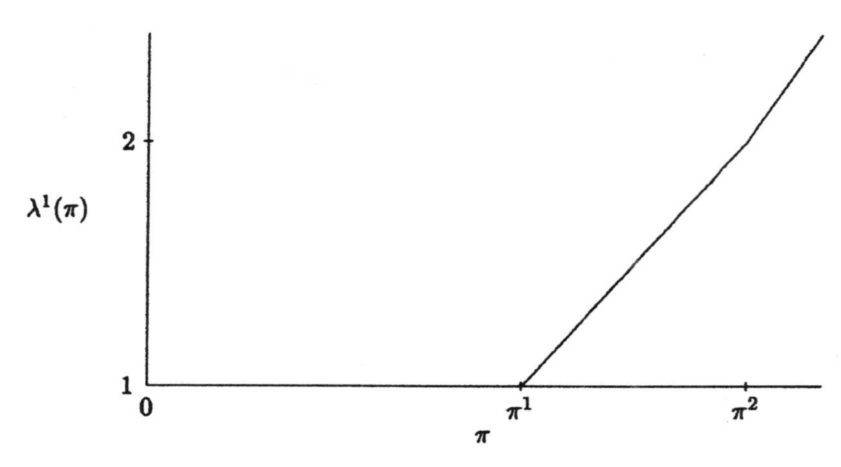


**Figure 3**: The game value of the offender for $e = 1, c = 2, f = 2, o = 0.5, v = \frac{2}{3}$.

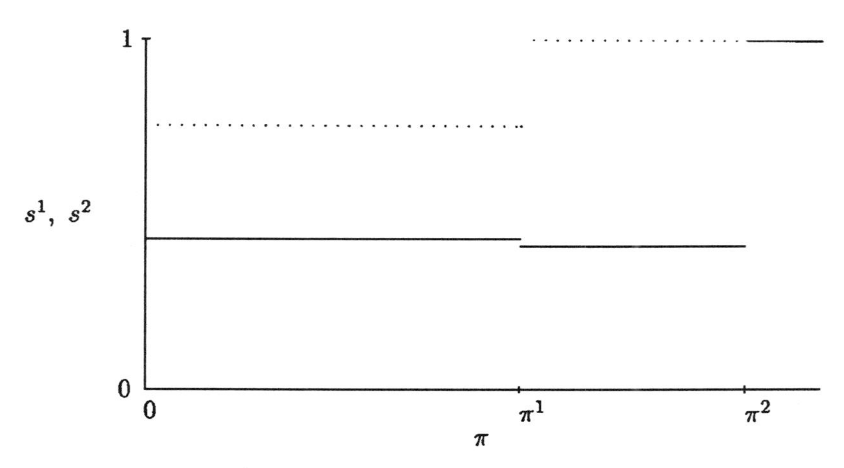


**Figure 4**: The functions $s^1(\pi)$ (solid line) and $s^2(\pi)$ (dashed line) for the same parameter values as in Figure 3.

As the last inequality holds by assumption, we have proved the proposition. □

The results of this proposition confirm intuition. The payoff of the offender does not decrease if the profit from his criminal activities increases. We have illustrated the game value function $\lambda^1(\pi)$ in Figure 3 and the corresponding strategy functions $s^1(\pi)$ and $s^2(\pi)$ in Figure 4. These figures show how equilibrium behavior differs between the three areas in the parameter space defined by the hyperplanes $\pi = \pi^1$ and $\pi = \pi^2$. If $\pi < \pi^1$ it is possible for the authorities to use enough enforcement effort to make the offender indifferent between legal and illegal activities in both periods. We have an equilibrium if the offense rate in both stages is such that the authorities are indifferent between enforcing and not enforcing. An increase in $\pi$ leads to such an increase of the enforcement effort that the expected payoff of a crime stays zero. Accordingly, the game value for the offender equals his legal income on $[0, \pi^1]$. However, if $\pi$ exceeds the threshold $\pi^1$, the authorities are no longer able to make the offender indifferent between legal and illegal activities in a single stage. The offender changes to the pure strategy $C$ in the second period. Note, however, that crime still does not pay in the first period as the costs of being arrested are higher than in the second period. When being a criminal is strictly better than being lawful, the game value increases continuously with the profit from a successful criminal action.

To understand the downward jump of $s^1(\pi)$ at $\pi = \pi^1$, recall that the authorities randomize between enforcing and nonenforcing in the first period. In equilibrium both actions must have the same expected payoff. As the game values for the second player at the second stage jump downward at $\pi = \pi^1$, the offense rate at the first stage must decrease to prevent that enforcing gets more attractive than nonenforcing.

The changes in the equilibrium outcome of the game for the law enforcement authorities are described in Proposition 4.3.

**Proposition 4.3.** *The strategy functions $u^1(\pi)$ and $u^2(\pi)$ are continuous nondecreasing functions. The game value $\mu^1(\pi)$ has a downward jump at $\pi^1$ and another jump at $\pi^2$. The second jump is upward if*

$$(1-v)ve^2 - (1 - v(1-v)(3-2v))ec + v(1-v)^2(2-v)c^2 > 0, \qquad (23)$$

*and downward if the inequality in* (23) *is reversed.*

**Proof.** To prove that $u^2(\pi)$ is continuous and nondecreasing we claim that $y^2$ given by (3) equals one if $\pi^1$ is inserted for $\pi$. Direct calculations show that this holds true. It is easy to see that when $\pi^1$ is inserted into $y_2^1$ and $y_5^1$, this gives the same value of $(f + o)/[f + (2 - v)o]$, and $\pi^2$ inserted into $y_2^1$ gives 1. Hence both strategy functions are continuous and as all segments are nondecreasing, the functions themselves are nondecreasing. Since $\pi$ does not occur in the expressions for $\mu_1^1$, $\mu_2^1$ or $\mu_5^1$, the game value $\mu^1(\pi)$ is constant on $[0, \pi^1)$, $[\pi^1, \pi^2)$, and $[\pi^2, \infty)$. To show that the first jump is downward, we have to show that $\mu_2^1 < \mu_5^1$ for $\pi = \pi^1$. Direct calculations show that this inequality holds if and only if

$$\Psi(e) = v(e+(1-v)c)(e+(2-v)c)(vc+e)+ec(vc+e)-e(2vc+e)(e+(2-v)c) > 0$$

for all $e < vc$. Straightforward, but rather lengthy calculations show that $\Psi(0) = v^2(1-v)(2-v)c^3 > 0$, $\Psi(vc) = 0$ and $\Psi''(e) < 0, \forall e$. Thus $\Psi(e)$ must be positive on $[0, vc)$ and the jump at $\pi = \pi^1$ is downward. On the other hand, $\mu_1^1 > \mu_2^1$ holds if and only if (23) is fulfilled. There exist parameter constellations such that this inequality holds (but also some where it does not hold). □

The proposition shows that, in equilibrium, the law enforcement authorities increase their effort if the illegal payoff of the offender increases. They have to increase the probability of being arrested in order to equalize the expected payoff

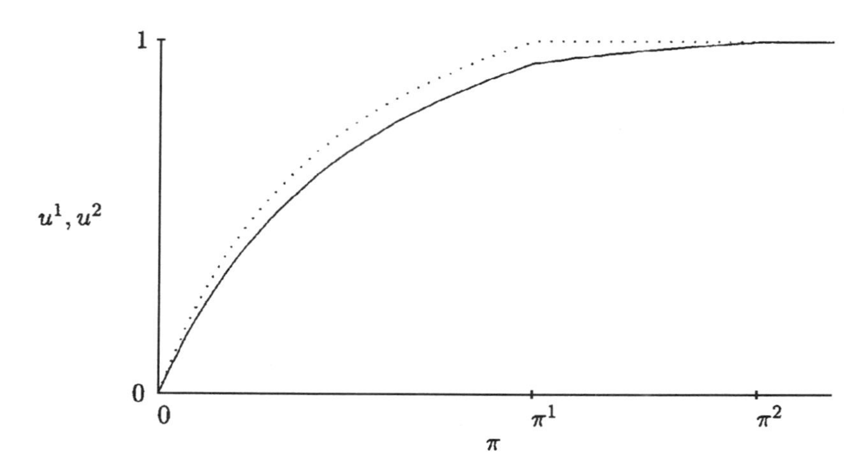


**Figure 5**: The equilibrium probability of enforcement in the first (solid line) and second stage (dashed line) of the game. Same parameter values as in Figures 3 and 4.

of crimes with the legal income (cf. Figure 5). If $\pi$ is sufficiently small and the offender plays mixed strategies in both periods, this can be done at no cost in an equilibrium in which the authorities are indifferent between enforcement and no enforcement. However, if $\pi$ exceeds $\pi^1$, the offense rate jumps to 1 in the second stage, causing an increase in the expected social costs. If $\pi$ also exceeds $\pi^2$, the offender plays the pure strategy $C$ in both periods. This has two opposite effects. On the one hand, the expected social costs are increased even more but, on the other hand, the higher criminal activities in the first stage increase the probability that the offender is arrested in the first period. In this case, enforcement is not needed in the second stage and enforcement costs will be lower. Which of these two effects prevails depends on the sign of the expression on the left-hand side of (23). In Figure 6 the second effect is the stronger one which means that an increase in the offense rate in the first stage leads to a reduction of the expected costs to the authorities. Figure 7 shows a contrary case where the first effect dominates and both jumps of $\mu^1(\pi)$ are downward. This implies that, at least under certain circumstances, it will be beneficial for the authorities if the profit of criminal activities increases.

Similar results to the above apply if other parameters in the model are varied. In particular, the jumps in the strategies of the first player and in the value function of the second player always occur if a parameter changes from one of the three areas in the parameter space into another.

## 5 Extensions of the Model

Having completed the analysis of the two-stage game we briefly discuss the case of an $N$-stage game. We distinguish two cases. First, we consider a straight-

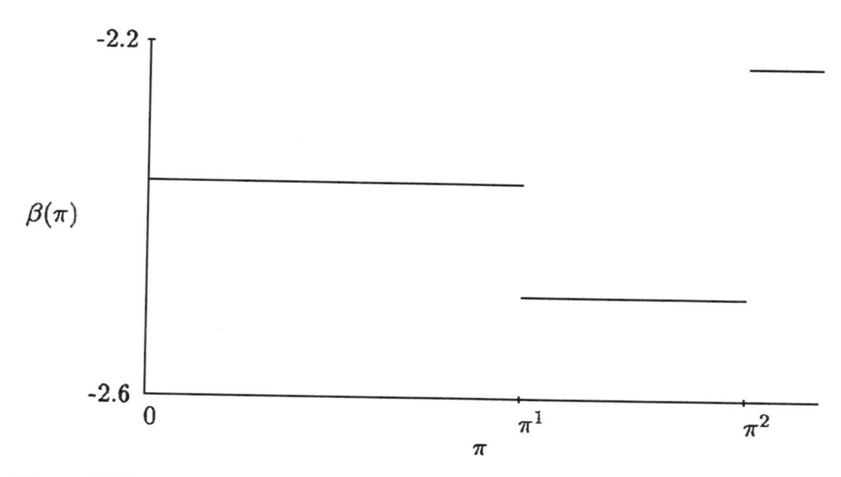


**Figure 6**: The game value for the second player. Parameter values as in Figures 3–5.

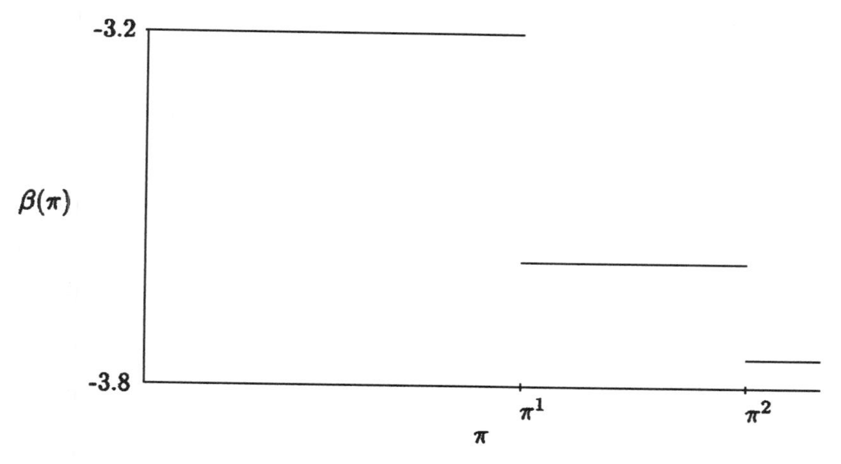


**Figure 7**: The game value for the second player. Parameter values $e = 1$, $c = 3$, $f = 2$, $o = 0.5$, $v = 0.5$.

forward extension of the two-stage game where an offender who is convicted stays in prison for all subsequent periods. Second, we sketch the implications of a model with recidivism; a convicted offender stays in prison for only one period and may again return to criminal activity afterwards.

Assume that the game described in Section 2 is played for $N$ periods. In each period the offender and the authorities play a mixed strategy and whenever convicted the offender is unable to gain any income in all subsequent periods. On the other hand, the costs to the authorities are zero for all periods after the conviction of the offender. Obviously, the game tree gets quite large for larger values of $N$ and since it is a straightforward extension of the two-stage tree given in Figure 2 we refrain from presenting it here. In this multistage game the basic structure is very similar to the one of the two-stage game. Denoting by $\lambda^n$, $\mu^n$, $n = 1, \ldots, N$, the game values of the $(N+1-n)$-stage game we get the following entries in the payoff matrix for the game in period $n$:

$$\begin{aligned}
\tilde{a}^n_{CE} &= (1-v)(\lambda^{n+1} + \pi + o) - vf,\\
\tilde{b}^n_{CE} &= -e - (1-v)(c - \mu^{n+1}),\\
\tilde{a}^n_{CN} &= \lambda^{n+1} + \pi + o,\\
\tilde{b}^n_{CN} &= \mu^{n+1} - c,\\
\tilde{a}^n_{HE} &= \lambda^{n+1} + o,\\
\tilde{b}^n_{HE} &= \mu^{n+1} - e,\\
\tilde{a}^n_{HN} &= \lambda^{n+1} + o,\\
\tilde{b}^n_{HN} &= \mu^{n+1}.
\end{aligned}$$

Equilibrium behavior in the $n$th stage is given by the Nash equilibrium of the $2 \times 2$ normal form game with payoff matrix

$$(A^n, B^n) = \begin{bmatrix} ((1-v)(\pi+o) - v(f+\lambda^{n+1}), -e-(1-v)c - v\mu^{n+1}) & (\pi+o, -c) \\ (o, -e) & (o, 0) \end{bmatrix}. \tag{24}$$

Given this, it is easy to establish a monotonicity result similar to Proposition 4.1.

**Proposition 5.1.** *Let $s^n$ and $u^n$ be the probabilities that in equilibrium the first and second player, respectively, play the first pure strategy in period n. Then both sequences $\{s^n\}_{n=1}^N$ and $\{u^n\}_{n=1}^N$ are nondecreasing.*

**Proof.** First note that $\{\lambda^n\}$ is nonincreasing and that $\{\mu^n\}$ is nondecreasing. This is obvious, since one-period payoffs for the second player are nonpositive and the first player can enforce a one-period payoff of $o > 0$ by playing his second pure strategy. If $a^n_{CE} \geq o$ the equilibrium of the game is given by $s^n = u^n = 1$.[10] This condition is met if and only if

$$\pi \geq \pi^n := \frac{v}{1-v}(f + o + \lambda^{n+1}). \tag{25}$$

It follows from the monotonicity of $\lambda^n$ that $\pi \geq \pi^n$ implies $\pi \geq \pi^{n+1}$. Accordingly, $s^{n+1} = s^n = 1$ and $u^{n+1} = u^n = 1$ for $\pi \geq \pi^n$. For $\pi < \pi^n$ we have

$$s^n = \frac{e}{v(c - \mu^{n+1})},$$

$$\text{and} \quad u^n = \frac{\pi}{v(\pi + f + o + \lambda^{n+1})}.$$

[10]For simplicity, we consider only the equilibrium in pure strategies for $a^n_{CE} = o$.

On the other hand, $\pi \geq \pi^{n+1}$ implies $s^{n+1} = u^{n+1} = 1$ and for $\pi < \pi^{n+1}$ we get

$$s^{n+1} = \frac{e}{v(c - \mu^{n+2})},$$
$$\text{and} \quad u^{n+1} = \frac{\pi}{v(\pi + f + o + \lambda^{n+2})}.$$

Then monotonicity of $\lambda^n$ and $\mu^n$ establishes the claim of the proposition. □

As noted, the structural properties of the multistage setup are very similar to the ones of the two-stage game and we do not provide a detailed analysis of the multistage game. However, we point out that the monotonicity of the offense and enforcement rate may be violated if recidivism is introduced into the model. Most of the models in the field have ignored recidivistic behavior, and the importance of this phenomenon has just recently been stressed by Leung [11]. To introduce recidivism in our model we consider the multistage setup and assume that an offender who is convicted in period $n$ stays in prison in period $n+1$ but reenters the game at stage $n+2$. We do not carry out a complete analysis of this model but only show that strategies $\{s^n\}$ and $\{u^n\}$ now might be nonmonotonic.

Consider a three-stage game with recidivism and look at the situation in the first period. If the offender is not convicted in this period, both players get their immediate payoffs plus their game values for the subsequent two-stage game $\lambda^2$, $\mu^2$. Note that these values equal the values $\lambda^1$, $\mu^1$ calculated in Section 3. If the offender is convicted in period one he receives no income in period two but reenters the game in period three and gains an additional expected payoff of $\lambda^3$ (which is equal to $\lambda^2$ in Section 3). The second player gets zero payoff in the second period and $\mu^3$ (which is equal to $\mu^2$ in Section 3) in the third. Accordingly, we get the following entries in the payoff matrix:

$$\begin{aligned}
\tilde{a}^1_{CE} &= (1-v)(\lambda^2 + \pi + o) + v(\lambda^3 - f), \\
\tilde{b}^1_{CE} &= -e - (1-v)(c - \mu^2) + v\mu^3, \\
\tilde{a}^1_{CN} &= \lambda^2 + \pi + o, \\
\tilde{b}^1_{CN} &= \mu^2 - c, \\
\tilde{a}^1_{HE} &= \lambda^2 + o, \\
\tilde{b}^1_{HE} &= \mu^2 - e, \\
\tilde{a}^1_{HN} &= \lambda^2 + o, \\
\tilde{b}^1_{HN} &= \mu^2.
\end{aligned}$$

and the corresponding transformed normal form game becomes

$$(A^1, B^1) = \begin{bmatrix} ((1-v)(\pi + o) - v(f + \lambda^2 - \lambda^3), -e - (1-v)c - v(\mu^2 - \mu^3) & (\pi + o, -c) \\ (o, -e) & (o, 0) \end{bmatrix}. \tag{26}$$

Suppose, for instance, that $\pi^1 < \pi < \pi^2$ holds. Using the values for $\lambda^2$ and $\lambda^3$ (calculated in Section 3) we realize that their difference is $o$ and that this game has an equilibrium in pure strategies if

$$\pi \geq \tilde{\pi}^3 := \frac{v}{1-v}(f+o) + \frac{v}{1-v}o.$$

Since $\pi^1 < \tilde{\pi}^3 < \pi^2$ we get for $\tilde{\pi}_3 < \pi < \pi^2$ the equilibrium values $s^1 = s^3 = 1$, $u^1 = u^3 = 1$, and $s^2 < 1, u^2 < 1$. This shows, at least in a specific case, that offense and enforcement no longer change monotonically from stage to stage if recidivistic behavior is present in the model.

## 6 Conclusions

This paper has analyzed a multistage game of a conflict between an offender and the law enforcement authorities. Although the game has a simple structure, it captures the essential elements of such a conflict: the probability of conviction, the offender's reward gained by criminal activity, a direct punishment in the form of a fine, and the enforcement and social costs incurred by the authorities. We obtained a complete characterization of sequential equilibria in the two-stage game and provided results about equilibrium strategies that might have an interest from a policy-making point of view. The main result of this paper was that equilibrium offense and enforcement rates are nondecreasing over time, but we also showed that this result is not necessarily true if recidivism is taken into account. An interesting result of the sensitivity analysis was that an increase in the criminal income could lead to a decrease in enforcement.

Many extensions of the simple setup are possible. One which we find particularly interesting is to consider the multistage game with recidivism and allow punishment to depend on the offender's criminal record (instead of having fixed punishment irrespective of the number of previous offences). This would mean that the offender "stays out of the game" not for exactly one period (as we assumed above), but that the time in prison as well as the size of the fine, would depend on the number of former offenses. Some work in this area has been made, see, for example, Polinsky and Rubinfeld [13] for a two-period setup, and Jørgensen et al. [8] for a differential game.

## Acknowledgments

H. Dawid's research was supported by the Austrian Science Foundation under contract No. J01281-SOZ. Helpful comments of Gernot Tragler and two anonymous referees are gratefully acknowledged.

## REFERENCES

[1] Becker, G. Crime and punishment: An Economic Approach. *Journal of Political Economy*, **76**, 169–217, 1968.

[2] Davis, M. Time and Punishment: An Intertemporal Model of Crime. *Journal of Political Economy*, **96**, 383–390, 1988.

[3] Dawid, H. and G. Feichtinger. Optimal Allocation of Drug Control Efforts: A Differential Game Analysis. *Journal of Optimization Theory and Applications*, **91**, 279–297, 1996.

[4] Feichtinger, G. A Differential Game Solution to a Model of Competition Between a Thief and the Police. *Management Science*, **29**, 686–699, 1983.

[5] Fudenberg, D. and J. Tirole. *Game Theory*. MIT Press, Cambridge, MA, 1991.

[6] Gerchak, Y. and M. Parlar. Optimal Control Analysis of a Simple Criminal Prosecution Model. *Optimal Control Applications and Methods*, **6**, 305–312, 1985.

[7] Harris, J. On the Economics of Law and Order. *Journal of Political Economy*, **78**, 165–174, 1970.

[8] Jørgensen, S., G. Feichtinger, and H. Dawid. Economic Crimes and Law Enforcement: A Differential Game, Working Paper, Odense University, 1996.

[9] Leung, S. How to Make the Fine Fit the Corporate Crime, *Journal of Public Economics*, **45**, 243–256, 1991.

[10] Leung, S. An Economic Analysis of the Age-Crime Profile. *Journal of Economic Dynamics and Control*, **18**, 481–497, 1994.

[11] Leung, S. Dynamic Deterrence Theory, *Economica*, **62**, 65–87, 1995.

[12] Myerson, R. B. *Game Theory: Analysis of Conflict*. Harvard University Press, Cambridge, MA, 1991.

[13] Polinsky, A. M. and D. L. Rubinfeld. A Model of Optimal Fines for Repeated Offenders. *Journal of Public Economics*, **46**, 291–306, 1991.

[14] Polinsky, A. M. and S. Shavell. Enforcement Costs and the Optimal Magnitude and Probability of Fines. *Journal of Law and Economics*, **XXXV**, 133–148, 1992.

[15] Sethi, S. Optimal Pilfering Policies for Dynamic Continuous Thieves. *Management Science*, **25**, 535–542, 1979.

[16] Stigler, G. The Optimum Enforcement of Laws. *Journal of Political Economy*, **78**, 526–536, 1970.

# Global Analysis of a Dynamic Duopoly Game with Bounded Rationality

Gian Italo Bischi
Istituto di Scienze Economiche
Università di Urbino
Urbino, Italy

Ahmad Naimzada
Università "L. Bocconi"
Milano, Italy

**Abstract**

A dynamic Cournot duopoly game, characterized by firms with bounded rationality, is represented by a discrete-time dynamical system of the plane. Conditions ensuring the local stability of a Nash equilibrium, under a local (or myopic) adjustment process, are given, and the influence of marginal costs and speeds of adjustment of the two firms on stability is studied. The stability loss of the Nash equilibrium, as some parameter of the model is varied, gives rise to more complex (periodic or chaotic) attractors. The main result of this paper is given by the exact determination of the basin of attraction of the locally stable Nash equilibrium (or other more complex bounded attractors around it), and the study of the global bifurcations that change the structure of the basin from a simple to a very complex one, with consequent loss of predictability, as some parameters of the model are allowed to vary. These bifurcations are studied by the use of critical curves, a relatively new and powerful method for the study of noninvertible two-dimensional maps.

## 1 Introduction

The static Cournot oligopoly model, in which each firm, given the optimal production decisions of the other firms selling the same homogeneous good, sets its optimal production, is a fully rational game based on the following assumptions:

(i) each firm, in taking its optimal production decision, must know beforehand all its rivals' production decisions taken at the same time;
(ii) each firm has a complete knowledge of the market demand function.

Under these conditions of full information the system moves straight (in one shot) to a Nash equilibrium, if it exists, independently of the initial status of the market, so that no dynamic adjustment process is needed. Dynamic Cournot oligopoly models arise from more plausible assumptions of partial information, so that the players' behavior is not fully rational. For example, a dynamic model is obtained if assumption (i) is replaced by some kind of expectation on the rivals' outputs. The simplest kind of expectation, based on Cournot's assumption that each firm, in taking its optimal decision, guesses that the output of the other firms remains at the same level as in the previous period, has given rise to a flourishing literature on dynamic oligopoly models, starting from the seminal paper of Teocharis [18] (see, e.g., McManus and Quandt [11], Fisher [6], Hahn [10], and Okuguchi [15]).

Models where also assumption (ii) is relaxed have recently been proposed by many authors, e.g., Bonanno and Zeeman [3], Bonanno [2], and Sacco [16]. In these models firms are supposed to have a knowledge of demand function limited to a small neighborhood of the status of the market in which they operate, and use such local knowledge to update their production strategy by a local profit maximization. Since they play the game repeatedly, they can gradually adjust their production over time. In these dynamic models a Nash equilibrium, if it exists, is a stationary state, i.e., if all the firms have outputs at the Nash equilibrium, their production decisions will remain the same forever. Instead, if the oligopoly system is outside a Nash equilibrium, then the repeated local adjustment process may converge to the Nash equilibrium, where there is no further possibility for improvement, or may move around it by a periodic or aperiodic time evolution, or may irreversibly depart from it. Thus the main question addressed in the literature on dynamic oligopoly models is that of the stability of the Nash equilibria, and how such stability is influenced by the model structure and the values of the parameters which characterize the model. Results of global asymptotic stability, which means that the adjustment process converges to the Nash equilibrium independently of the initial condition, have been given both for linear models, by the analysis of the eigenvalues of the model, and for nonlinear models by the second Lyapunov method, as in Rosen [17]. On the contrary, the question of stability extent in models, in which global stability does not hold, has been rather neglected in the literature. In fact, for nonlinear models, the analysis is often limited to the study of the linear approximation, but these results are in general quite unsatisfactory for practical purposes, since they determine the attractivity of a Nash equilibrium only for games starting in some region around the equilibrium, and such a region may be so small that every practical meaning of the mathematical concept of stability is lost. In these cases the question of the stability extent, that is, the delimitation of the basin of attraction of a locally stable equilibrium, becomes crucial for any practical stability result. In fact, only an exact determination of the boundaries of the basin of attraction can give a clear idea of the robustness of an attractor with respect to exogenous perturbations, always present in real systems, since it permits one to understand if a given shock of finite amplitude can be recovered by the endogenous dynamics of the system, or if it will cause an irreversible departure from the Nash equilibrium.

The present paper moves toward this less explored direction. We propose a nonlinear, discrete-time, duopoly model, where a Nash equilibrium exists that is, under given conditions on the model's parameters, locally asymptotically stable, but not globally stable.

The adjustment mechanism considered in this paper is based on the pseudo-gradient of the profit functions, i.e., each player changes its own production so as to obtain the maximum rate of change of its own profit with respect to a change in its own strategy. Such an adjustment has been proposed, in a continuous time model, by Rosen [17], and similar mechanisms are also considered by Furth [8], Sacco [16], Varian [19], and Flam [7]. In the paper by Rosen it is shown that, under the assumption of strict diagonal concavity of the payoffs, the unique equilibrium is globally asymptotically stable. In this paper we show that if a discrete-time model is considered the situation is more complex, since even if the conditions for stability required by Rosen are satisfied, local stability does not imply global stability.

The plan of the paper is as follows. A detailed description of the model is given in Section 2. In Section 3 the existence and local stability of the equilibrium points of the model are studied. The main results of the paper are given in Section 4, where the exact delimitation of the basin of attraction of the Nash equilibrium is obtained. The occurrence, as some parameter is allowed to vary, of some global bifurcations causing qualitative changes in the structure of the basins is studied by the use of critical curves, a powerful tool for the analysis of noninvertible maps of the plane. In this section we also show that even when the Nash equilibrium becomes unstable, the process may be characterized by periodic or chaotic trajectories which are confined in a bounded region around the Nash equilibrium, so that the duopoly system continues to have an asymptotic behavior which is not far from optimality. In any case, the global analysis of the dynamical system reveals that bifurcations can occur, as some parameter is left to vary, that cause qualitative changes in the structure of the basins of attraction.

## 2 The Duopoly Model: Assumptions and Notations

We consider an industry consisting of two quantity-setting firms, labeled by $i = 1, 2$, producing the same good for sale on the market. Production decisions of both firms occur at discrete-time periods $t = 0, 1, 2, \ldots$. Let $q_i(t)$ represent the output of the $i$th firm during period $t$, at a production cost $C_i(q_i)$. The price prevailing in period $t$ is determined by the total supply $Q(t) = q_1(t)+q_2(t)$ through a demand function

$$p = f(Q) \tag{1}$$

from which the single-period profit of $i$th firm is given by

$$\Pi_i(q_1, q_2) = q_i f(Q) - C_i(q_i). \tag{2}$$

As stressed in Section 1, we assume that each duopolist does not have a complete knowledge of the demand function, and tries to infer how the market will respond to its production changes by an empirical estimate of the marginal profit. This estimate may be obtained by market research or by brief experiments of small (or local) production variations performed at the beginning of period $t$ (see Varian [19]) and we assume that even if the firms are quite ignorant about the market demand, they are able to obtain a correct empirical estimate of the marginal profits, i.e.,

$$\Phi_i(t) = \left(\frac{\partial \Pi_i}{\partial q_i}\right)^{(e)} = \frac{\partial \Pi_i}{\partial q_i}(q_1, q_2), \qquad i = 1, 2. \tag{3}$$

Of course, this local estimate of expected marginal profits is much easier to obtain than a global knowledge of the demand function (involving values of $Q$ that may be very different from the current ones). With this kind of information the producers behave as local profit maximizers, the local adjustment process being one where a firm increases its output if it perceives a positive marginal profit $\Phi_i(t)$, and decreases its production if the perceived $\Phi_i(t)$ is negative. This adjustment mechanism has been called by some authors *myopic* (see Dixit [5] and Flam [7]). Let $G_i(\cdot)$, $i = 1, 2$, be an increasing function, such that

$$\operatorname{sgn} G_i(\cdot) = \operatorname{sgn}(\cdot), \qquad i = 1, 2. \tag{4}$$

Then the dynamic adjustment mechanism can be modeled as

$$q_i(t+1) = q_i(t) + \alpha_i(q_i)\, G_i(\Phi_i), \qquad i = 1, 2, \tag{5}$$

where $\alpha_i(q_i)$ is a positive function which gives the extent of production variation of the $i$th firm following a given profit signal $\Phi_i$. It is important to note that a Nash equilibrium, if it exists, is also a fixed point of the dynamical system (5). In fact, a Nash equilibrium is located at the intersection of the reaction curves, defined by $(\partial \Pi_i / \partial q_i)(q_1, q_2) = 0$, $i = 1, 2$ (as noticed by Dixit [5], the term "reaction curve" is not appropriate in models like (5), since they describe a simultaneous-move game, but we follow the tradition of using the same term also in this context). Since (4) implies $G_i(0) = 0$, $i = 1, 2$, the dynamic process (5) is stationary if the strategy point $(q_1, q_2)$ is at a Nash equilibrium. The converse is not necessarily true, that is, stationary points of (5) that are not Nash equilibria can exist, as we shall see in the particular model studied in the following.

An adjustment mechanism similar to (5) has been proposed by some authors, mainly with continuous time and constant $\alpha_i$ (see, e.g., Rosen [17], Furth [8], Sacco [16], Varian [19], and Flam [7]). However, we believe that a discrete-time decision process is more realistic since in real economic systems production decisions cannot be revised at every time instant. We also assume that $\alpha_i$ are increasing functions of $q_i$ (hence of the "size" of the firm). This assumption captures the fact that in the presence of a positive profit signal $\Phi_i > 0$, a bigger firm has greater capacity to make investments in order to increase its production, whereas in the presence of a negative profit signal a bigger producer must reduce more drastically its production to avoid bankruptcy risks.

In the following we shall assume, for sake of simplicity, a linear relation

$$\alpha_i(q_i) = v_i q_i, \qquad i = 1, 2, \tag{6}$$

where $v_i$ is a positive constant that will be called *speed of adjustment*. We also assume a linear demand function

$$f(Q) = a - bQ \tag{7}$$

with $a,\ b$ positive constants, and linear cost functions

$$C_i(q_i) = c_i q_i, \qquad i = 1, 2, \tag{8}$$

where the positive constants $c_i$ are the marginal costs. With these assumptions

$$\Pi_i(q_1, q_2) = q_i \left[a - b(q_1 + q_2) - c_i\right], \qquad i = 1, 2, \tag{9}$$

and the marginal profit for firm $i$ at the point $(q_1, q_2)$ of the strategy space is

$$\Phi_i = \frac{\partial \Pi_i}{\partial q_i} = a - c_i - 2bq_i - bq_j, \qquad i, j = 1, 2, \quad j \neq i. \tag{10}$$

If, as in the quoted papers of Varian, Furth, Flam, and Sacco we consider a linear adjustment function

$$G(\Phi) = \Phi \tag{11}$$

the model (5), with the above assumptions, gives rise to the following two-dimensional nonlinear map $T\ (q_1, q_2) \to \left(q_1', q_2'\right)$ defined as

$$T: \begin{cases} q_1' = (1 + v_1(a - c_1))q_1 - 2bv_1 q_1^2 - bv_1 q_1 q_2, \\ \\ q_2' = (1 + v_2(a - c_2))q_2 - 2bv_2 q_2^2 - bv_2 q_1 q_2, \end{cases} \tag{12}$$

where $'$ denotes the unit-time advancement operator, that is, if the right-hand side variables are productions of period $t$, then the left-hand ones represent productions of period $(t + 1)$.

The map (12) is a noninvertible map of the plane, that is, starting from some nonnegative initial production strategy

$$\left(q_{1_0}, q_{2_0}\right) \tag{13}$$

the iteration of (12) uniquely defines the trajectory $(q_1(t), q_2(t)) = T^t\left(q_{1_0}, q_{2_0}\right)$, $t = 1, 2, \ldots$, whereas the backward iteration of (12) is not uniquely defined. In fact, a point $\left(q_1', q_2'\right)$ of the plane may have several preimages, obtained by solving the fourth-degree algebraic system (12) with respect to $q_1$ and $q_2$ (see Mira et al. [12] for a complete treatment of the properties of noninvertible maps of the plane). The study of the dynamical properties of (12) allows us to have information on the long-run behavior of a bounded rationality adjustment process starting from a given initial condition (13), and how this is influenced by the parameters of the model.

## 3 Equilibrium Points and Local Stability

We define *equilibrium point* (or *stationary point*) of the dynamic duopoly game as a nonnegative fixed point of the map (12), i.e., a solution of the algebraic system

$$\begin{cases} q_1(a - c_1 - 2bq_1 - bq_2) = 0, \\ q_2(a - c_2 - bq_1 - 2bq_2) = 0, \end{cases} \tag{14}$$

obtained by setting $q_i^{'} = q_i$ , $i = 1, 2$, in (12). We can have at most four fixed points: $E_0 = (0, 0)$, $E_1 = [(a - c_1)/2b, 0]$ if $c_1 < a$, $E_2 = [0, (a - c_2)/2b]$ if $c_2 < a$, which will be called *boundary equilibria*, and the fixed point $E_* = (q_1^*, q_2^*)$, with

$$q_1^* = \frac{a + c_2 - 2c_1}{3b}, \qquad q_2^* = \frac{a + c_1 - 2c_2}{3b}, \tag{15}$$

provided that

$$\begin{cases} 2c_1 - c_2 < a, \\ 2c_2 - c_1 < a. \end{cases} \tag{16}$$

It is easy to verify that the equilibrium point $E_*$, when it exists, is the unique Nash equilibrium, located at the intersection of the two reaction curves given by the two straight lines which represent the locus of points of vanishing marginal profits (10). In the following we shall assume that (16) are satisfied, so that the Nash equilibrium $E_*$ exists.

An important feature of the map (12) is that it can generate unbounded (i.e., divergent) trajectories (this can also be expressed by saying that (12) has an attracting set at infinite distance). In fact, unbounded (and negative) trajectories are obtained if the initial condition (13) is taken sufficiently far from the origin, i.e., in a suitable neighborhood of infinity, since if $q_{i0} > (1 + a - c_i)/bv_i$, $i = 1, 2$, then the first iterate of (12) gives negative values $q_i^{'} < 0$, $i = 1, 2$, so that the successive iterates give negative and decreasing values because $q_i^{'} = q_i + v_i q_i \left(a - c_i - 2bq_i - bq_j\right) < q_i$ being $(a - c_i) > 0$ if (16) hold. This implies that any attractor at finite distance cannot be globally attracting in $\mathbb{R}_+^2$, since its basin of attraction cannot extend out of the rectangle $[0, (1 + a - c_1)/bv_1] \times [0, (1 + a - c_2)/bv_2]$.

The study of the local stability of the fixed points is based on the localization, on the complex plane, of the eigenvalues of the Jacobian matrix of (12)

$$J(q_1, q_2) = \begin{bmatrix} 1 + v_1(a - c_1 - 4bq_1 - bq_2) & -v_1 bq_1 \\ -v_2 bq_2 & 1 + v_2(a - c_2 - bq_1 - 4bq_2) \end{bmatrix}. \tag{17}$$

It is easy to prove that whenever the equilibrium $E_*$ exists (i.e., (16) are satisfied), the boundary fixed points $E_i, i = 0, 1, 2$, are unstable. In fact, at $E_0$ the Jacobian matrix becomes a diagonal matrix

$$J(0,0) = \begin{bmatrix} 1 + v_1(a - c_1) & 0 \\ 0 & 1 + v_2(a - c_2) \end{bmatrix}, \tag{18}$$

whose eigenvalues, given by the diagonal entries, are greater than 1 if $c_1 < a$ and $c_2 < a$. Thus $E_0$ is a repelling node with eigendirections along the coordinate axes. At $E_1$ the Jacobian matrix becomes a triangular matrix

$$J\left(\frac{a - c_1}{2b}, 0\right) = \begin{bmatrix} 1 - v_1(a - c_1) & -(v_1/2)(a - c_1) \\ 0 & 1 + (v_2/2)(a - 2c_2 + c_1) \end{bmatrix} \tag{19}$$

whose eigenvalues, given by the diagonal entries, are $\lambda_1 = 1 - v_1(a - c_1)$, with eigenvector $\mathbf{r}_1^{(1)} = (1, 0)$ along the $q_1$-axis, and $\lambda_2 = 1 + (v_2/2)(a - 2c_2 + c_1)$, with eigenvector $\mathbf{r}_1^{(2)} = (1, 2[1 - v_1(a - c_1)]/v_1(a - c_1))$. When (16) are satisfied $E_1$ is a saddle point, with local stable manifold along the $q_1$-axis and the unstable one tangent to $\mathbf{r}_1^{(2)}$, if

$$v_1 < \frac{2}{a - c_1}, \tag{20}$$

otherwise $E_1$ is an unstable node. The bifurcation occurring at $v_1 = 2/(a - c_1)$ is a flip bifurcation at which $E_1$ from attracting becomes repelling along the $q_1$-axis, on which a saddle cycle of period 2 appears.

The same arguments hold for the other boundary fixed point $E_2$. It is a saddle, with local stable manifold along the $q_2$-axis and the unstable one tangent to $\mathbf{r}_2^{(2)} = (1, 2[1 - v_2(a - c_2)]/v_2(a - c_2))$, if

$$v_2 < \frac{2}{a - c_2}, \tag{21}$$

otherwise it is an unstable node. Also, in this case, the bifurcation that transforms the saddle into the repelling node is a flip bifurcation creating a 2-cycle saddle on the $q_2$-axis.

To study the local stability of the Nash equilibrium we consider the Jacobian matrix at $E_*$

$$J(q_1^*, q_2^*) = \begin{bmatrix} 1 - 2v_1 b q_1^* & -v_1 b q_1^* \\ -v_2 b q_2^* & 1 - 2v_2 b q_2^* \end{bmatrix}. \tag{22}$$

Its eigenvalues are real because the characteristic equation $\lambda^2 - \mathrm{Tr}\,\lambda + \mathrm{Det} = 0$, where Tr represents the trace and Det the determinant of (22), has positive discriminant

$$\mathrm{Tr}^2 - 4\,\mathrm{Det} = 4b^2\left[\left(v_1 q_1^* - v_2 q_2^*\right)^2 + v_1 v_2 q_1^* q_2^*\right] > 0.$$

It is easy to realize that $\lambda_i < 1$, $i = 1, 2$, since $1 - \mathrm{Tr} + \mathrm{Det} > 0$ when (16) hold, thus a sufficient condition for the local asymptotic stability of $E_*$ is

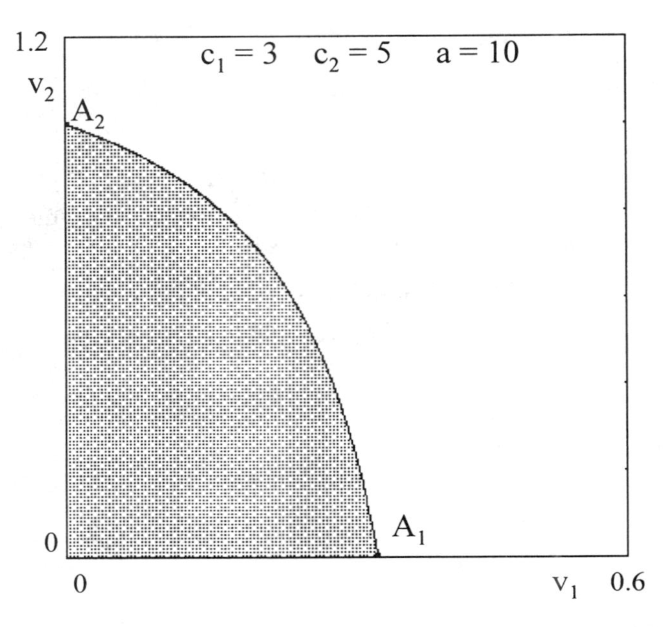


**Figure 1**: The shaded area represents, in the plane of speeds of adjustment $(v_1, v_2)$, the region of local asymptotic stability of the Nash equilibrium. The values of the other parameters are $c_1 = 3$, $c_2 = 5$, $a = 10$.

$1 + \text{Tr} + \text{Det} > 0$, which ensures $\lambda_i > -1$, $i = 1, 2$ (see, e.g., Gumowski and Mira [9, p. 159]). This condition, which becomes

$$3b^2 q_1^* q_2^* v_1 v_2 - 4bq_1^* v_1 - 4bq_2^* v_2 + 4 < 0, \tag{23}$$

defines a region of stability in the plane of the speeds of adjustment $(v_1, v_2)$ whose shape is like the shaded area of Figure 1. This stability region is bounded by the portion of hyperbola, with positive $v_1$ and $v_2$, whose equation is given by the vanishing of the left-hand side of (23). For values of $(v_1, v_2)$ inside the stability region the Nash equilibrium $E_*$ is a stable node, and the hyperbola represents a bifurcation curve at which $E_*$ looses its stability through a period doubling (or *flip*) bifurcation. This bifurcation curve intersects the axes $v_1$ and $v_2$ in the points $A_1$ and $A_2$, respectively, whose coordinates are given by

$$A_1 = \left(\frac{3}{a + c_2 - 2c_1}, 0\right) \quad \text{and} \quad A_2 = \left(0, \frac{3}{a + c_1 - 2c_2}\right). \tag{24}$$

From these results we can obtain information on the effects of the model's parameters on the local stability of $E_*$. For example, an increase of the speeds of adjustment, with the other parameters held fixed, has a destabilizing effect. In fact, an increase of $v_1$ and/or $v_2$, starting from a set of parameters which ensures the local stability of the Nash equilibrium, can bring the point $(v_1, v_2)$ out of the stability region, crossing the flip bifurcation curve. This destabilizing effect has already been evidenced by many authors (see, e.g., Fisher [6] and Flam [7]).

Similar arguments apply if the parameters $v_1, v_2, c_1, c_2$ are fixed and the parameter $a$, which represents the maximum price of the good produced, is increased. In this case, the stability region becomes smaller, as can easily be deduced from (24), and this can cause a loss of stability of $E_*$ when the moving boundary is crossed by the point $(v_1, v_2)$. An increase of the marginal cost $c_1$, with $c_2$ held fixed, causes a displacement of the point $A_1$ to the right and of $A_2$ downward. Instead, an increase of $c_2$, with $c_1$ held fixed, causes a displacement of $A_1$ to the left and of $A_2$ upward. In both cases the effect on the local stability of $E_*$ depends on the position of the point $(v_1, v_2)$. In fact, if $v_1 < v_2$, i.e., the point $(v_1, v_2)$ is above the diagonal $v_1 = v_2$, an increase of $c_1$ can destabilize $E_*$, whereas an increase of $c_2$ reinforces its stability. The situation is reversed if $v_1 > v_2$.

From these arguments the combined effects due to simultaneous changes of more parameters can be deduced. For example, if $E_*$ becomes unstable because of a price increase (due to a shift of the demand curve), its stability can be regained by a reduction of the speeds of reaction, whereas an increase of a marginal cost $c_i$ can be compensated for by a decrease of the corresponding $v_i$, i.e., in the presence of a high marginal cost stability is favored by a more prudent behavior (i.e., lower reactivity to profit signals).

Another important property of the map (12) is that each coordinate axis $q_i = 0$, $i = 1, 2$, is trapping, that is, mapped into itself, since $q_i = 0$ gives $q_i' = 0$ in (12). This means that starting from an initial condition on a coordinate axis (*monopoly case*) the dynamics is confined in the same axis for each $t$, governed by the restriction of the map $T$ to that axis. Such a restriction is given by the following one-dimensional map, obtained from (12) with $q_i = 0$

$$q_j = f_j(q_j) = (1 + v_j(a - c_j))q_j - 2bv_jq_j^2, \qquad j = 1, 2, \quad j \neq i. \tag{25}$$

This map is conjugate to the standard logistic map

$$x' = \mu x\,(1 - x) \tag{26}$$

through the linear transformation

$$q_j = \frac{1 + v_j(a - c_j)}{2bv_j} x \tag{27}$$

from which we obtain the relation

$$\mu = 1 + v_j(a - c_j). \tag{28}$$

This means that the dynamics of (25) can be obtained from the well-known dynamics of (26). A brief description of the main features of the map (25) is given in Appendix B, because the dynamic behavior of the restrictions of $T$ to the invariant axes plays an important role in the understanding of the global properties of the duopoly model.

# 4 Basin Boundaries and Their Bifurcations

In Section 3 we have shown that if the conditions (16) are satisfied then the Nash equilibrium $E_* = (q_1^*, q_2^*)$ exists, and it is locally asymptotically stable provided that (23) holds true. In this section we consider the question of the stability extent of the Nash equilibrium, or of different bounded attracting sets around it. In the following we call *attractor at finite distance*, denoted by $\mathcal{A}$, a bounded attracting set (which may be the Nash equilibrium $E_*$, a periodic cycle, or some more complex attractor around $E_*$) in order to distinguish it from the limit sets at infinite distance, i.e., the unbounded trajectories, which represent exploding (or collapsing) evolutions of the duopoly system. We denote by $\mathcal{D}(\mathcal{A})$ the basin of attraction of an attractor $\mathcal{A}$, defined as the open set of points $(q_1, q_2)$ of the phase plane whose trajectories $T^t(q_1, q_2)$ have limit sets belonging to $\mathcal{A}$ as $t \to +\infty$. We also denote by $\mathcal{D}(\infty)$ the *basin of infinity*, defined as the set of points which generate unbounded trajectories. Let $\mathcal{F}$ be the boundary (or frontier) separating $\mathcal{D}(\mathcal{A})$ from $\mathcal{D}(\infty)$. An exact determination of $\mathcal{F}$ is the main goal of this section. Indeed, this boundary may be rather complex, as evidenced by the numerical results shown in Figure 2. In Figure 2(a) the attractor at finite distance is the Nash equilibrium $E_*$, and its basin of attraction is represented by the white area, whereas the grey-shaded area represents the basin of infinity. Two typical trajectories are also shown in Figure 2(a), one converging to $E_*$ and one divergent. Notice that in Figure 2(a) the adjustment process which starts from the grey region, and consequently exhibits an irreversible departure from the Nash equilibrium, starts from an initial production strategy which is closer to the Nash equilibrium than the convergent one, a rather counterintuitive result. In the situation shown in Figure 2(a), the boundary separating $\mathcal{D}(\mathcal{A})$ from $\mathcal{D}(\infty)$ has a fractal boundary, as will be explained below. In Figure 2(b) the bounded attractor $\mathcal{A}$ is a chaotic set, with a multiply connected (or connected with holes) basin of attraction. The same property can be expressed by saying that $\mathcal{D}(\infty)$ is a nonconnected set, with nonconnected regions given by the holes inside $\mathcal{D}(\mathcal{A})$ (see Mira et al. [12] or Mira et al. [13]). In this situation there is a great uncertainty about the long-run behavior of a given adjustment process, since a small change in the initial strategy of the game may cause a crossing of $\mathcal{F}$.

## 4.1 Determination of the Basin Boundaries

The boundary $\mathcal{F} = \partial\mathcal{D}(\mathcal{A}) = \partial\mathcal{D}(\infty)$ behaves as a repelling line for the points near it, since it acts as a watershed for the trajectories of the map $T$. Points belonging to $\mathcal{F}$ are mapped into $\mathcal{F}$ both under forward and backward iteration of $T$, that is, the boundary is invariant for application of $T$ and $T^{-1}$. More exactly $T(\mathcal{F}) \subseteq \mathcal{F}$, $T^{-1}(\mathcal{F}) = \mathcal{F}$ (see Mira et al. [12] and Mira and Rauzy [14]). This implies that if a saddle point, or a saddle cycle, belongs to $\mathcal{F}$, then $\mathcal{F}$ must also contain all the preimages of such singularities, and it must also contain the whole stable

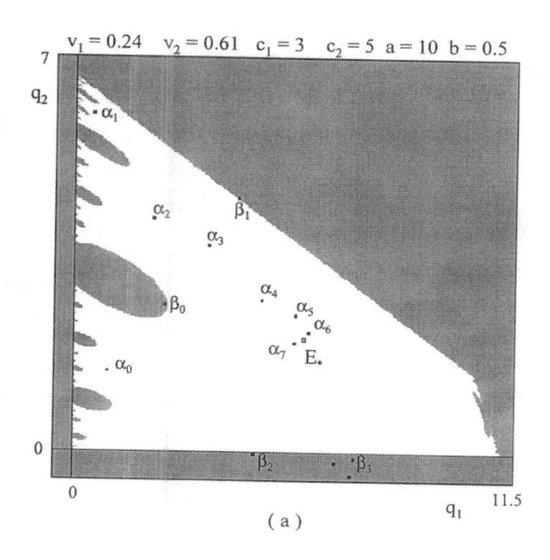


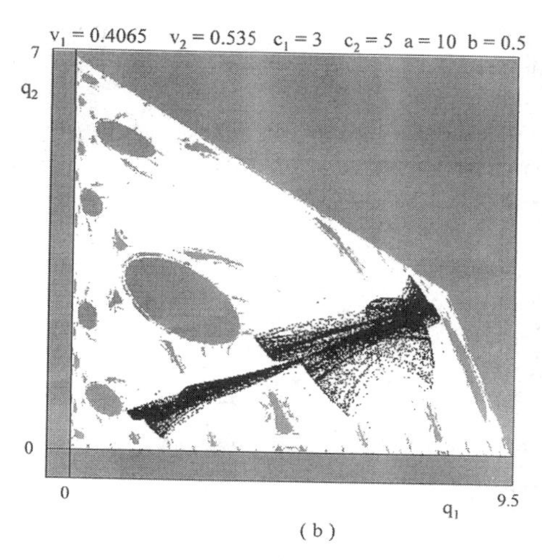


**Figure 2**: Numerical representation of the basins of attraction for the duopoly map. The two figures are obtained by taking a grid of initial conditions $(q_{10}, q_{20})$ and generating, for each of them, a numerically computed trajectory of the duopoly map. If the trajectory is diverging (i.e., if it reaches large negative values) then a grey dot is painted in the point corresponding to the initial condition, otherwise a white dot is painted. In Figure **(a)** the white region represents the basin of attraction of the Nash equilibrium, which is the only attractor at finite distance for that set of parameters. In this figure also the early points of two typical trajectories, one convergent to $E_*$, denoted by $\{\alpha_0, \alpha_1, ...\}$, and one divergent, denoted by $\{\beta_0, \beta_1, ...\}$,are represented. In Figure **(b)** the attractor at finite distance is given by a chaotic attractor surrounding the unstable Nash equilibrium.

manifold $W^s$ (see Gumowski and Mira [9] and Mira et al. [13]). For example, the saddle fixed points (or the saddle cycles, if (20) or (21) no longer hold) located on the coordinate axes belong to $\mathcal{F}$, and also the invariant coordinate axes, which form the local stable manifold (or inset) of the saddles, are part of $\mathcal{F}$.

Let us consider the two segments $\omega_j = \left[0, 0_{-1}^{(j)}\right]$, where $0_{-1}^{(j)}$, $j = 1, 2$, is the rank-1 preimage of the origin computed according to the restriction (25), i.e.,

$$0_{-1}^{(j)} = \frac{1 + v_j \left(a - c_j\right)}{2bv_j}, \qquad j = 1, 2. \tag{29}$$

Instead, negatively divergent trajectories along the invariant axis $q_j$ are obtained starting from an initial condition out of the segment $\omega_j$. The segments $\omega_1$ and $\omega_2$ on the two coordinate axes play an important role in the determination of $\mathcal{F}$. In fact:

(a) from the computation of the eigenvalues of the cycles belonging to $\omega_1$ and $\omega_2$ we have that the direction transverse to the coordinate axes is always repelling; and

(b) a point $\left(q_{1_0}, q_{2_0}\right)$ generates a divergent trajectory if $q_{1_0} < 0$ or $q_{2_0} < 0$.

From (a) and (b) it follows that $\omega_1$ and $\omega_2$ belong to $\mathcal{F}$, as well as their preimages of any rank. From these arguments the following proposition can be stated, that gives an exact determination of $\mathcal{F}$.

**Proposition 4.1.** *Let* $\omega_1 = \left[0, 0_{-1}^{(1)}\right]$ *and* $\omega_2 = \left[0, 0_{-1}^{(2)}\right]$ *be the segments of the coordinate axes* $q_1$ *and* $q_2$*, respectively, with* $0_{-1}^{(j)}$, $j = 1, 2$*, defined in* (29). *Then*

$$\mathcal{F} = \left(\bigcup_{n=0}^{\infty} T^{-n}(\omega_1)\right) \cup \left(\bigcup_{n=0}^{\infty} T^{-n}(\omega_2)\right). \tag{30}$$

*where* $T^{-n}$ *represents the set of all the preimages of rank-n.*

In order to compute the preimages in (30) let us consider a point $P = (0, p) \in \omega_2$. Its preimages are the real solutions of the algebraic system obtained from (12) with $(q_1^{'}, q_2^{'}) = (0, p)$:

$$\begin{cases} q_1 \left[1 + v_1(a - c_1) - 2bv_1q_1 - bv_1q_2\right] = 0, \\ \\ (1 + v_2(a - c_2))\, q_2 - 2bv_2q_2^2 - bv_2q_1q_2 \ = p. \end{cases} \tag{31}$$

From the first of (31) we obtain $q_1 = 0$ or

$$1 + v_1(a - c_1) - 2bv_1q_1 - bv_1q_2 = 0, \tag{32}$$

which means that if the point $P$ has preimages, then they must be located either on the same invariant axis or on the line of (32). With $q_1 = 0$ the second equation becomes a second degree algebraic equation which has two distinct, coincident or no real solutions if the discriminant

$$(1 + v_2\,(a - c_2))^2 - 8bv_2p \tag{33}$$

is positive, zero, or negative, respectively. A similar conclusion holds if (32) is used to eliminate a state variable in the first equation of (31). From this we can deduce that the point $P$ can have no preimages or two preimages on the same axis (which are the same obtained by the restriction (25) of $T$ to the axis $q_2$) or four preimages, two on the same axis and two on the line of (32). This implies that the set of the rank-1 preimages of the $q_2$-axis belongs to the same axis and to the line (32). Following the same arguments we can state that the other invariant axis, $q_1$, has preimages on itself and on the line of equation

$$1 + v_2(a - c_2) - bv_2q_1 - 2bv_2q_2 = 0. \tag{34}$$

It is straightforward to see that the origin $O = (0, 0)$ always has four preimages: $O_{-1}^{(0)} = (0, 0)$, $O_{-1}^{(1)} = (q_1^{o-1}, 0)$, $O_{-1}^{(2)} = (0, q_2^{o-1})$, where $q_j^{o-1}$, $j = 1, 2$, are given by (39), and

$$O_{-1}^{(3)} = \left(q_1^* + \frac{2v_2 - v_1}{3bv_1v_2}, q_2^* + \frac{2v_1 - v_2}{3bv_1v_2}\right),$$

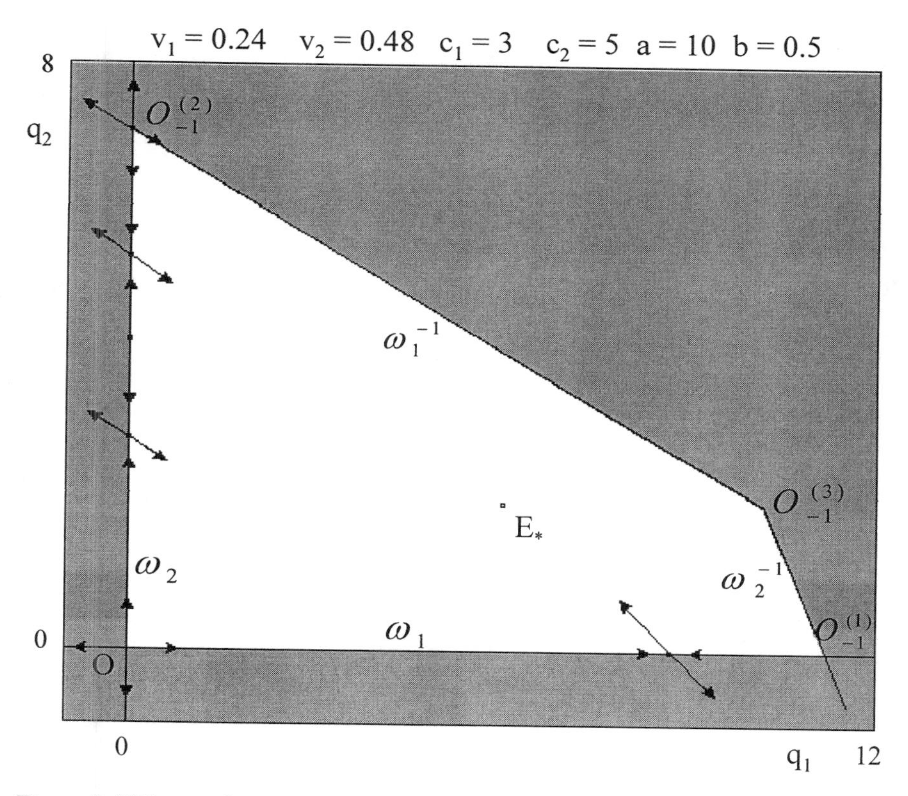


**Figure 3**: With $c_1 = 3$, $c_2 = 5$, $a = 10, b = 0.5$, $v_1 = 0.24$, $v_2 = 0.48$, the boundary of the basin of attraction of the Nash equilibrium $E_*$ is formed by the invariant axes, denoted by $\omega_1$ and $\omega_2$, and their rank-1 preimages $\omega_1^{-1}$ and $\omega_2^{-1}$. For this set of parameters the boundary fixed point $E_1$ is a saddle point with local stable manifold along the $q_1$-axis, $E_2$ is a repelling node with a saddle cycle of period two around it, since $v_2 > \frac{2}{a-c_2}$.

located at the intersection of the lines (32) and (34) (see Figure 3). In the situation shown in Figure 3 the segments $\omega_2$ and $\omega_1$ of the coordinate axes, together with their rank-1 preimages, belonging to the lines (32) and (34), and labeled by $\omega_2^{-1}$ and $\omega_1^{-1}$, respectively, delimitate the quadrilateral region $O\,O_{-1}^{(1)}\,O_{-1}^{(3)}O_{-1}^{(2)}$ of the strategy space $(q_1, q_2)$ which is exactly the basin of attraction of $E_*$.

These four sides, given by the segments $O\,O_{-1}^{(1)}$ and $O\,O_{-1}^{(2)}$ of the coordinate axes and their rank-1 preimages, constitute the whole boundary $\mathcal{F}$ because no preimages of higher rank exist, since $\omega_{-1}^{1}$ and $\omega_{-1}^{2}$ belong to the region $Z_0$ of the plane whose points $(q_1', q_2')$ have no preimages, i.e., the fourth degree algebraic system has no real solutions. This fact can be characterized through the study of the critical curves of the noninvertible map (12) (some basic definitions and properties of the critical curves are given in Appendix A; see Mira et al. [13], for a more complete treatment).

Since the map $T$ is continuously differentiable, the critical curve $LC_{-1}$ is the locus of points in which the determinant of $J(q_1, q_2)$, given in (17), vanishes, and the critical curve $LC$, locus of points having two coincident rank-1 preimages, can be obtained as the image, under $T$, of $LC_{-1}$(see Appendix A). For the map (12) $LC_{-1}$ is formed by the two branches of an hyperbola, denoted by $LC_{-1}^{(a)}$ and $LC_{-1}^{(b)}$ in Figure 4(a) (its equation is given in Appendix A). Thus also $LC = T(LC_{-1})$ consists of two branches, $LC^{(a)} = T(LC_{-1}^{(a)})$ and $LC^{(b)} = T(LC_{-1}^{(b)})$, represented by the thicker curves of Figure 4(a). These two branches of $LC$ separate the phase plane into three regions, denoted by $Z_0$, $Z_2$, and $Z_4$, whose points have 0, 2, and 4 distinct rank-1 preimages, respectively. It can be noticed that, as already stressed above, the origin always belongs to the region $Z_4$. It can also be noticed that the line $LC_{-1}$ intersects the axis $q_j$, $j = 1, 2$, in correspondence of the critical point $c_{-1}$ of the restriction (25) of $T$ to that axis, whose coordinate is given by (36), and that the line $LC$ intersects each axis in correspondence of the critical values of (25), given by (41).

### 4.2 Contact Bifurcations

In order to understand how complex basin boundaries, like those shown in Figure 2, are obtained, we start from a situation in which $\mathcal{F}$ has a much simpler shape, and then we study the sequence of bifurcations that cause the main qualitative changes in the structure of the basin boundaries as some parameter is varied. Such bifurcations, typical of noninvertible maps, can be characterized by contacts of the basin boundaries with the critical curves (see Mira et al. [13], and references therein).

The simple shape that the frontier $\mathcal{F}$ assumes for values of the parameters like those used in figure 4(a), where the basin of attraction of $E_*$ is a simply connected set, is due to the fact that the preimages of the invariant axes, denoted in Figure 4(a) by $\omega_i^{-1}$, $i = 1, 2$, are entirely included inside the region $Z_0$, so that no preimages of higher rank exist. The situation is different when the values of the parameters are such that some portions of these lines belong to the regions $Z_2$ or $Z_4$. In this case, preimages of higher order of the invariant coordinate axes are obtained, which form new arcs of the frontier $\mathcal{F}$, so that its shape becomes more complex. The switch between these two qualitatively different situations can be obtained by a continuous variation of some parameters of the model, and determines a global (or nonclassical) bifurcation (see Mira et al. [13]). The occurrence of these global bifurcations can be revealed by the study of critical curves. In order to illustrate this, in the rest of this section we fix the marginal costs and the parameters of the demand function at the same values as those used to obtain figures 2, 3, i.e., $c_1 = 3$, $c_2 = 5$, $a = 10$, $b = \frac{1}{2}$, and we vary the values of the speeds of adjustment $v_1$ and $v_2$. However, similar bifurcation sequences can be obtained with fixed values of $v_1$ and $v_2$ and changing the other parameters. For example,

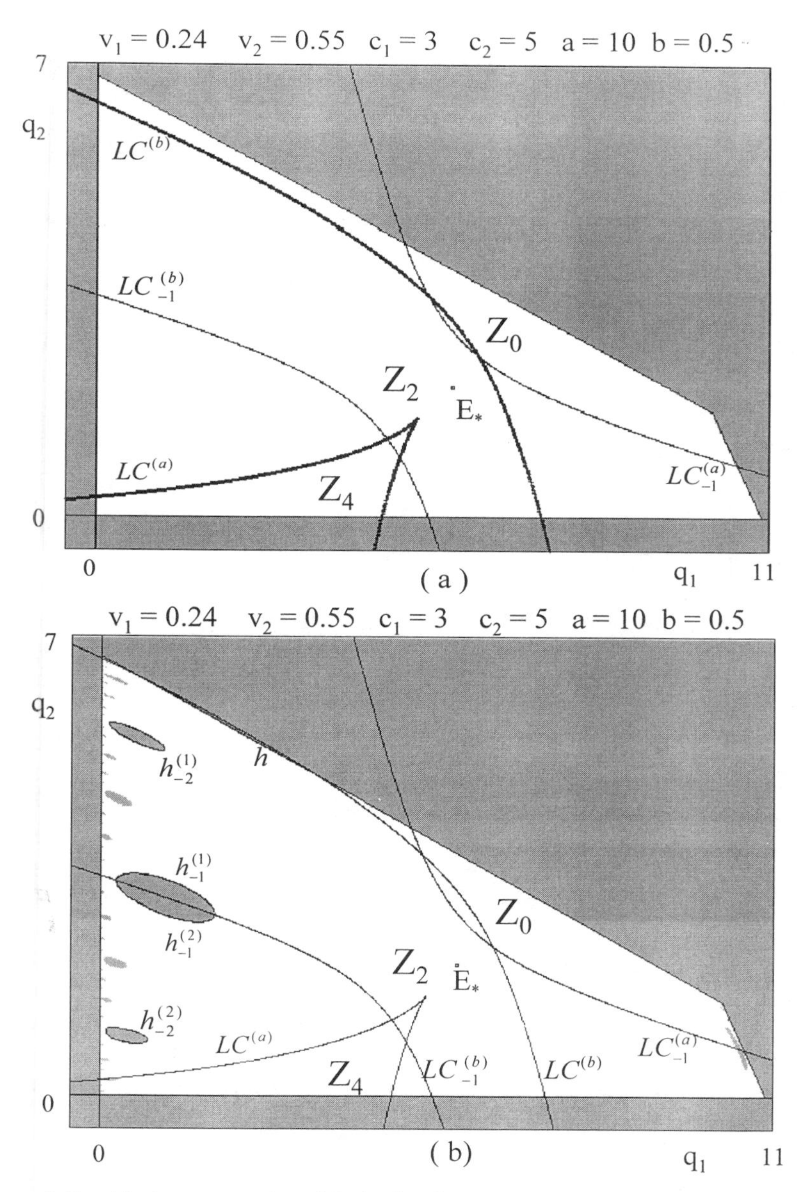


**Figure 4**: Graphical representation of the basin of attraction of the Nash equilibrium (white region) and the basin $\mathcal{D}(\infty)$ of unbounded trajectories (grey region) together with the basic critical curve $LC_{-1}$, formed by the two branches of an equilateral hyperbola and the critical curve $LC$ (represented by heavy lines). The values of parameters $c_1, c_2, a, b$ are the same as in Figure 2, whereas in Figure **(a)** $v_1 = 0.24$ and $v_2 = 0.55$ , in Figure **(b)** $v_1 = 0.24$ and $v_2 = 0.596$ (just after the contact of $LC$ with $\omega_1^{-1}$).

the same effect of increasing $v_1$ can be obtained by decreasing the corresponding marginal cost $c_1$ or by increasing the marginal cost $c_2$ of the other firm, whereas a simultaneous increase of both $v_1$ and $v_2$ is equivalent to a decrease of the parameter $a$ in the demand function.

If, starting from the parameters' values used to obtain the simple basin structure of Figure 4(a), the parameter $v_2$ is increased, the two branches of the critical curve $LC$ move upward. The first global bifurcation of the basin occurs when the branch of $LC$, which separates the regions $Z_0$ and $Z_2$, becomes tangent to $\mathcal{F}$, that is, to one of the lines (34) or (32). In Figure 4(b) it can be seen that just after the bifurcation value of $v_2$, at which $LC^{(b)}$ is tangent to the line $\omega_1^{-1}$ of (34), a portion of $\mathcal{D}(\infty)$, say $H_0$ (bounded by the segment $h$ of $\omega_1^{-1}$ and $LC$) that before the bifurcation was in region $Z_0$, enters into $Z_2$. The points belonging to $H_0$ have two distinct preimages, located at opposite sides with respect to the line $LC_{-1}$, with the exception of the points of the curve $LC^{(b)}$ inside $D(\infty)$ whose preimages, according to the definition of $LC$, merge on $LC_{-1}^{(b)}$. Since $H_0$ is part of $\mathcal{D}(\infty)$ its preimages also belong to $\mathcal{D}(\infty)$. The locus of the rank-1 preimages of $H_0$, bounded by the two preimages of $h$, is composed by two areas joining along $LC_{-1}$ and forms a *hole* (or lake) of $\mathcal{D}(\infty)$ nested inside $\mathcal{D}(E_*)$. This is the largest hole appearing in Figure 4(b), and is called the *main hole*. It lies entirely inside region $Z_2$, hence it has two preimages, which are smaller holes bounded by preimages of rank 3 of the $q_1$-axis. Even these are both inside $Z_2$. So each of them has two further preimages inside $Z_2$, and so on. Now the boundary $\mathcal{F}$ is given by the union of an external part, formed by the coordinate axes and their rank-1 preimages (34) and (32), and the boundaries of the holes, which are sets of preimages of higher rank of the $q_1$-axis. Thus the global bifurcation just described transforms a *simply connected* basin into a *multiply connected* one, with a countable infinity of holes, called an *arborescent sequence of holes*, inside it (see Mira et al. [12] for a rigorous treatment of this type of global bifurcation and Abraham et al. [1] for a simpler and charming exposition).

As $v_2$ is further increased $LC$ continues to move upward and the holes become larger. This fact causes a sort of predictability loss, since a greater uncertainty is obtained with respect to the destiny of games starting from an initial strategy falling in zone of the holes. If $v_2$ is further increased a second global bifurcation occurs when $LC$ crosses the $q_2$-axis at $O_{-1}^{(2)}$. This happens when condition (40) holds, that is, $v_2 = 3/(a - c_2)$, as in Figure 5(a). After this bifurcation all the holes reach the coordinate axis $q_2$, and the infinite contact zones are the intervals of divergence of the restriction (25), which are located around the critical point (36) and all its preimages under (25) (compare Figure 5(a) with Figure B1(b)). After this bifurcation the basin $\mathcal{D}(E_*)$ becomes simply connected again, but its boundary $\mathcal{F}$ now has a fractal structure, since its shape, formed by infinitely many peninsulas, have a self-similarity property.

The sequence of pictures shown in Figure 5 is obtained with $v_1 = 0.24$ (as in Figure 4) and increasing values of $v_2$. Along this sequence the point $(v_1, v_2)$

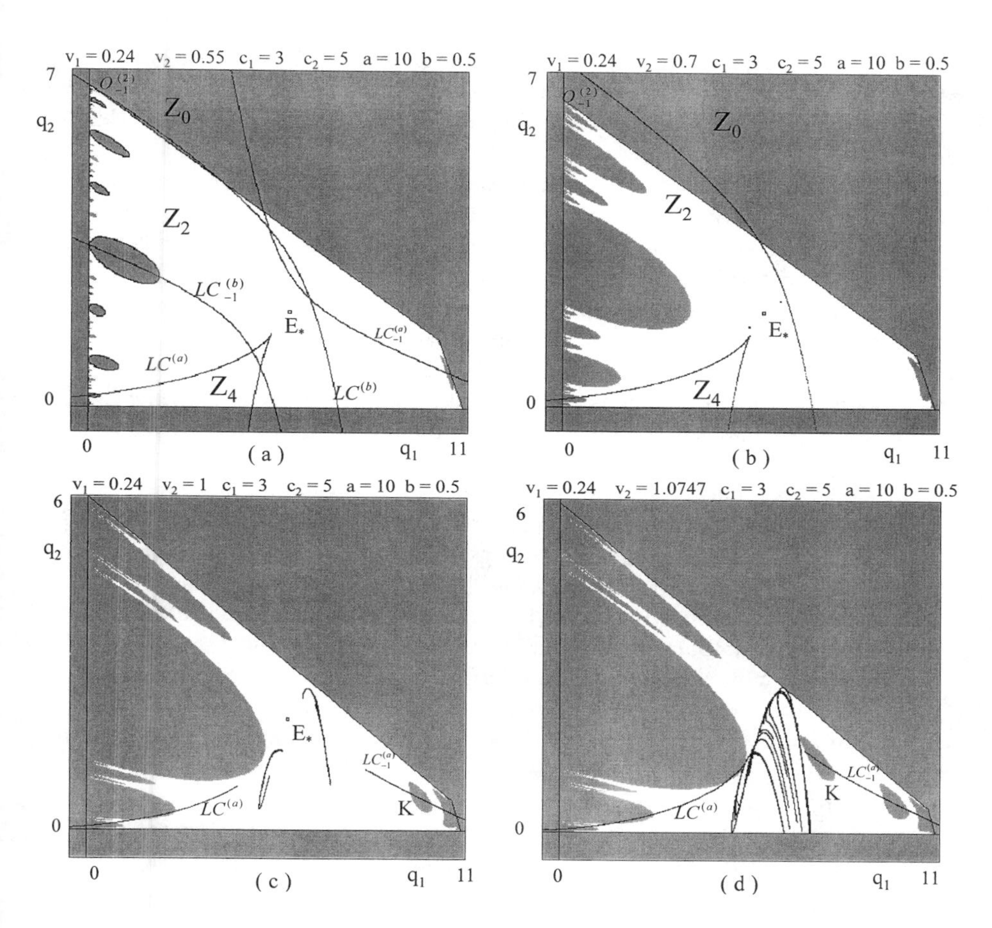


**Figure 5**: Sequence of numerical simulations of the duopoly map, obtained with fixed parameters $c_1 = 3,\ c_2 = 5,\ a = 10, b = 0.5,\ v_1 = 0.24$, and increasing values of $v_2$.

reaches, in the plane of adjustment speeds, the line of flip bifurcations. When this line is crossed the Nash equilibrium $E_*$ becomes a repelling saddle point, and an attracting cycle of period 2, say $\mathcal{C}_2$, is created near it (as in Figure 5(b)). The flip bifurcation opens a cascade of period doublings, that creates a sequence of attracting cycles of period $2^n$ followed by the creation of chaotic attractors, which may be cyclic chaotic areas, like the 2-cyclic one shown in Figure 5(c), or a unique chaotic area like that of Figure 5(d).

If $v_2$ is further increased, new holes, like that denoted by $K$ in Figure 5(c), appear. These are formed by the rank-1 preimages of portions of $\mathcal{D}(\infty)$ which cross $LC^{(a)}$ passing from $Z_2$ to $Z_4$, like those evidenced in Figures 5(c,d). Even in this case, the holes are created after contact between $LC$ and $\mathcal{F}$, but, differently from the hole $H_{-1}$, the hole $K$ does not generate an arborescent sequence of holes since it has no preimages, belonging entirely to the region $Z_0$.

In Figure 5(d) the chaotic area collides with the boundary of $\mathcal{D}(\infty)$. This contact bifurcation is known as *final bifurcation* (Mira et al. [13] and Abraham et al. [1]), and causes the destruction of the attractor at finite distance. After this contact bifurcation the generic initial strategy generates an unbounded trajectory, that is, the adjustment process is not able to approach the Nash equilibrium, independently of the initial strategy of the duopoly game.

It is worth noting that, in general, there are no relations between the bifurcations which change the qualitative properties of the basins and those which change the qualitative properties of the attractor at finite distance. In other words, we may have a simple attractor, like a fixed point or a cycle, with a very complex basin structure, or a complex attractor with a simple basin. Both these sequences of bifurcations, obtained by increasing the speeds of adjustment $v_i$, cause a loss of predictability. After the local bifurcations the myopic duopoly game no longer converges to the global optimal strategy, represented by the Nash equilibrium $E_*$, and even if the game starts from an initial strategy very close to $E_*$ the duopoly system goes toward a different attractor, which may be periodic or aperiodic. These bifurcations cause, in general, a loss of predictability about the asymptotic behavior of the duopoly system: for example, in the sequence shown in Figure 5 the situation of convergence to the unique Nash equilibrium, as in the static Cournot game, is replaced by asymptotic convergence to a periodic cycle, with predictable output levels, and then by a cyclic behavior with output levels that are not well predictable since the fall inside cyclic chaotic areas and, finally, a situation of erratic behavior, inside a large area of the strategy space, with no apparent periodicity. Instead, the global bifurcations of the basin boundaries cause an increasing uncertainty with respect to the destiny of a duopoly game starting from a given initial strategy, since a small change in the initial condition of the duopoly, or a small exogenous shock during the adjustment process, may cause a great modification to the long-run behavior of the system. Similar bifurcation sequences can also be obtained by increasing the parameter $v_1$ by a fixed value of $v_2$. In this case, a contact between $LC$ and $\omega_2^{-1}$, rank-1 preimage of the $q_2$-axis, gives the first bifurcation that transforms the basin $\mathcal{D}(\mathcal{A})$ from a simply connected into a multiply connected set, with holes near the $q_1$-axis. Situations with values of $v_1$ and $v_2$ both near the critical values $v_i = 3/(a - c_i)$, $i = 1, 2$, can give complex basin boundaries near both the coordinate axes, with two arborescent sequences of holes, generated by contacts of $LC$ with the lines (32) and (34). In any case, the computation of the preimages of the coordinate axes allows us to obtain, according to (30), the exact delimitation of the basin boundary also in these complex situations. For example, in Figure 6 the preimages of the $q_1$-axis, up to rank-6, are represented for the same set of parameters as that used in Figure 2(b). It can be noticed that some preimages of rank 5 and 6 bound holes that enter the region $Z_4$, thus giving a faster exponential growth of the number of higher-order preimages. This is the cause for the greater complexity of the basin boundary which is clearly visible in Figure 2(b).

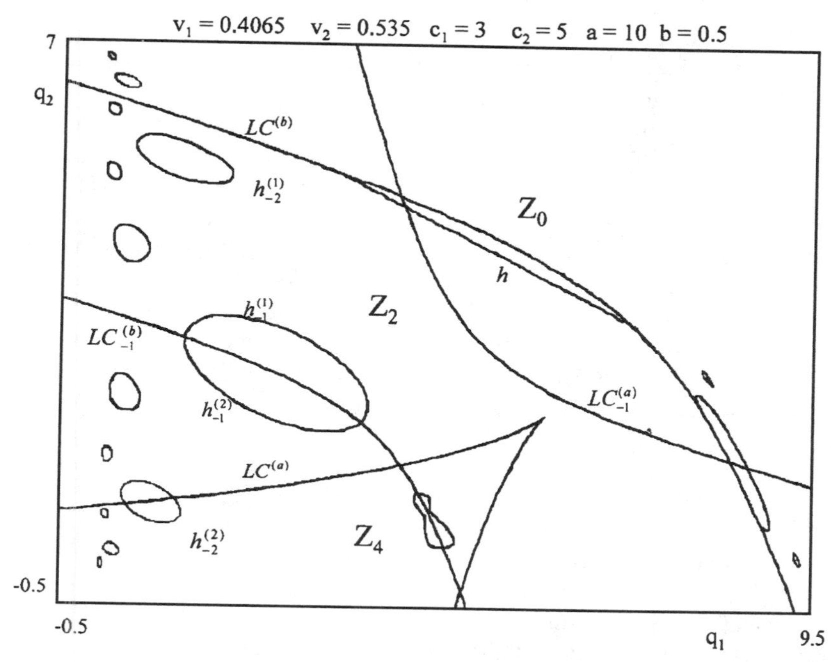


**Figure 6**: Preimages of the $q_1$-axis, up to rank 6, obtained with the same set of parameters as those used in Figure 2b.

## 5 Conclusions

In this paper we have proposed a dynamic Cournot duopoly model where the competing firms do not have a complete knowledge of the demand function. Owing to this lack of information they behave as local (or myopic) profit maximizers, following an adjustment process based on local estimates of marginal profit. If the marginal costs of both producers are not too high, a noncooperative Nash equilibrium exists, and if the behavior of the firms is characterized by relatively low speeds of adjustment such an equilibrium solution is locally asymptotically stable, that is, the local adjustment process converges to the unique Nash equilibrium provided that the duopoly game starts from an initial production strategy inside a well-defined bounded region around the Nash equilibrium.

For higher values of speeds of adjustment the Nash equilibrium becomes unstable and, through period doubling bifurcations, more complex bounded attractors are created around it, which may be periodic cycles or chaotic sets. When the dynamics of the duopoly system become so complicated, the assumption that producers are unable to gain a complete understanding of the structure of the market, and consequently behave myopically, would be even more justified. This adjustment process can also generate unbounded trajectories. Of course, the occurrence of divergent sequences of production strategies is a very unrealistic evolution of the duopoly game, that simply means that the bounded rationality adjustment mechanism, based on the profit gradients, is completely inadequate to reach optimal, or quasi-optimal, production strategies.

The main result of this paper is the exact determination of the basin of attraction of the attracting sets, wether it be the Nash equilibrium or a more complex attractor around it. This basin of attraction can have a simple shape, but it can also assume, after some global bifurcations, a very complex structure, which causes a sort of indeterminacy about the destiny of the dynamic game starting from a given initial strategy. In general, an increase in the firms' reactivity to profit signals, measured by the speeds of adjustment, can have two different effects on the dynamical properties of the duopoly model. The first one, already studied in the literature, is given by the destabilization of the Nash equilibrium, as discussed above. The second effect, that as far as we know has not yet been studied in the literature, is given by qualitative changes in the structure of the basins of attraction, which can only be revealed by an analysis of the global properties of the nonlinear dynamical system. An exact delimitation of the basin of an attractor of a nonlinear dynamical system is very important in applied models since it gives quantitative information on the possible effects of exogenous shocks of finite amplitude on the evolution of the system. Thus the determination of the global bifurcations, that cause qualitative modifications of the structure of the basins, is important to characterize the robustness of an attractor with respect to external disturbances. In this paper such bifurcations have been studied in detail using critical curves, a relatively new and powerful tool for the study of the global behavior of noninvertible two-dimensional discrete dynamical systems. In the model studied in this paper the main qualitative changes of the global structure of the basins, that for nonlinear maps are generally studied only by numerical methods, can be obtained analytically, through the exact determination of the curves bounding the basin and the knowledge of the critical curves. For this reason the model studied in this paper may also be considered as a pedagogical example for the study of a nonlinear discrete dynamical system of the plane.

Even if in this paper the analysis is limited to the duopoly case with a particular choice of demand and cost functions, we believe that the main conclusions on the attractors and on the structure of their basins of attraction are indicative of what may happen in more general models, and can be seen as a starting point for the study of similar phenomena in oligopoly models with more than two firms and with more general demand and cost functions.

## Appendices

### A. Critical Curves

In this appendix we give some basic definitions and properties, and a minimal vocabulary, about the theory of noninvertible maps of the plane and the method of critical curves. We also give the analytical expression of the basic critical curve $LC_{-1}$ of the map (12).

A two-dimensional map can be written in the general form

$$\mathbf{x}' = T(\mathbf{x}) = (g_1(\mathbf{x}), g_2(\mathbf{x})), \tag{35}$$

where $\mathbf{x} = (x_1, x_2) \in \mathbb{R}^2$ is the phase variable, $g_1$ and $g_2$ are real valued continuous functions and $\mathbf{x}' = \left(x_1', x_2'\right)$ is called a rank-1 image of $\mathbf{x}$ under $T$. The point $\mathbf{x}_t = T^t(\mathbf{x})$, $t \in \mathbb{N}$, is called the image (or forward iterate) of rank-$t$ of the point $\mathbf{x}$, and the sequence $\{\mathbf{x}_t\}$ of all these images is the trajectory generated by the initial condition $\mathbf{x}$ (or $\mathbf{x}_0$, since $T^0$ is identified with the identity map). The fact that the map $T$ is single valued does not imply the existence and uniqueness of its inverse $T^{-1}$. Indeed, for a given $\mathbf{x}'$ the rank-1 preimage (or backward iterate) $\mathbf{x} = T^{-1}(\mathbf{x}')$ may not exist or it may be multivalued. In such cases $T$ is said to be a noninvertible map. The duopoly model (12) belongs to this class, because if in (12) the point $(q_1, q_2)$ is computed in terms of a given $\left(q_1', q_2'\right)$ a fourth-degree algebraic system is obtained, that can have four, two, or no solutions. As the point $\left(q_1', q_2'\right)$ varies in the plane $\mathbb{R}^2$ the number of solutions, i.e., the number of its real rank-1 preimages, can change. Generally, pairs of real preimages appear or disappear as the point $\left(q_1', q_2'\right)$ crosses the boundary separating regions characterized by a different number of preimages. Such boundaries are characterized by the presence of two coincident (merging) preimages. This leads to the definition of the critical curves, one of the distinguishing features of noninvertible maps. The critical curve of rank-1, denoted by $LC$, is defined as the locus of points having two, or more, coincident rank-1 preimages, located on a set called $LC_{-1}$. $LC$ is the two-dimensional generalization of the notion of critical value (local minimum or maximum value) of a one-dimensional map, and $LC_{-1}$ is the generalization of the notion of the critical point (local extremum point). Arcs of $LC$ separate the regions of the plane characterized by a different number of real preimages.

When in (35) $g_1$ and $g_2$ are continuously differentiable functions, $LC_{-1}$ is generally given by the locus of points where the Jacobian determinant of $T$ vanishes (i.e., the points where $T$ is not locally invertible):

$$LC_{-1} = \left\{\mathbf{x} \in \mathbb{R}^2 \mid \det J = 0\right\},$$

and $LC$ is the rank-1 image of $LC_{-1}$ under $T$, i.e., $LC = T(LC_{-1})$.

For the map (12) studied in this paper, from the expression of $J$ given in (17), the condition $\det J = 0$ becomes

$$q_1^2 + q_2^2 + 4q_1q_2 - \alpha_1 q_1 - \alpha_2 q_2 + \beta = 0,$$

where

$$\alpha_i = \frac{4(1 + v_j(a - c_j)bv_i) + 1 + v_i(a - c_i)bv_j}{4b^2 v_1 v_2}, \qquad i = 1, 2, \quad j \neq i,$$

and

$$\beta = \frac{(1 + v_1(a - c_1)bv_1)(1 + v_2(a - c_2)bv_2)}{4b^2 v_1 v_2}.$$

This is an hyperbola in the plane $(q_1, q_2)$ with symmetry center in the point $((2\alpha_2 - \alpha_1)/3, (2\alpha_1 - \alpha_2)/3)$ and asymptotes of angular coefficients $\left(-2 \pm \sqrt{3}\right)$. Thus $LC_{-1}$ is formed by two branches, denoted by $LC_{-1}^{(a)}$ and $LC_{-1}^{(b)}$ in Section 4. This implies also that $LC$ is the union of two branches, denoted by $LC^{(a)} = T(LC_{-1}^{(a)})$ and $LC^{(b)} = T(LC_{-1}^{(b)})$. Each branch of the critical curve $LC$ separates the phase plane of $T$ into regions whose points possess the same number of distinct rank-1 preimages. In the case of the map (12) $LC^{(b)}$ separates the region $Z_0$, whose points have no preimages, from the region $Z_2$, whose points have two distinct rank-1 preimages, and $LC^{(a)}$ separates the region $Z_2$ from $Z_4$, whose points have four distinct preimages. In order to study the action of the multivalued inverse relation $T^{-1}$ it is useful to consider a region $Z_k$ of the phase plane as the superposition of $k$ sheets, each associated with a different inverse. Such a representation is known as *foliation* of the plane (see Mira et al. [13]). Different sheets are connected by folds joining two sheets, and the projections of such folds on the phase plane are arcs of $LC$. The foliation associated with the map (12) is qualitatively represented in Figure 7. It can be noticed that the cusp point of $LC$ is characterized by three merging preimages at the junction of two folds.

## B. Dynamics on the Invariant Axes

In the following we recall some of the properties of the dynamic behavior of the map (25). Such properties are well known since such a map is conjugate to the standard logistic map $x' = \mu x(1 - x)$ through the transformation (27).

The map (25) is a unimodal map with the unique critical point $C_{-1}$ (see Figure 8(a) of coordinate

$$q_j^{C_{-1}} = \frac{1 + v_j(a - c_j)}{4bv_j}, \qquad j = 1, 2, \tag{36}$$

(conjugate to the critical point $x = \frac{1}{2}$ of (26)) and two fixed points $O$ and $E_j$ of coordinates

$$q_j^o = 0, \qquad q_j^{E_j} = \frac{a - c_j}{2b}, \qquad j = 1, 2, \tag{37}$$

(conjugate to the fixed points $x = 0$ and $x = (1-1/\mu)$, respectively) corresponding to the boundary fixed points of the duopoly map $T$. The fixed point $O$ is always repelling for (25), whereas $E_j$ is attracting for $0 < v_j\left(a - c_j\right) < 2$. When

$$v_j\left(a - c_j\right) = 2 \tag{38}$$

a flip bifurcation occurs which starts the well-known Feigenbaum (or Myrberg) cascade of period doubling bifurcations leading to the chaotic behavior of (25).

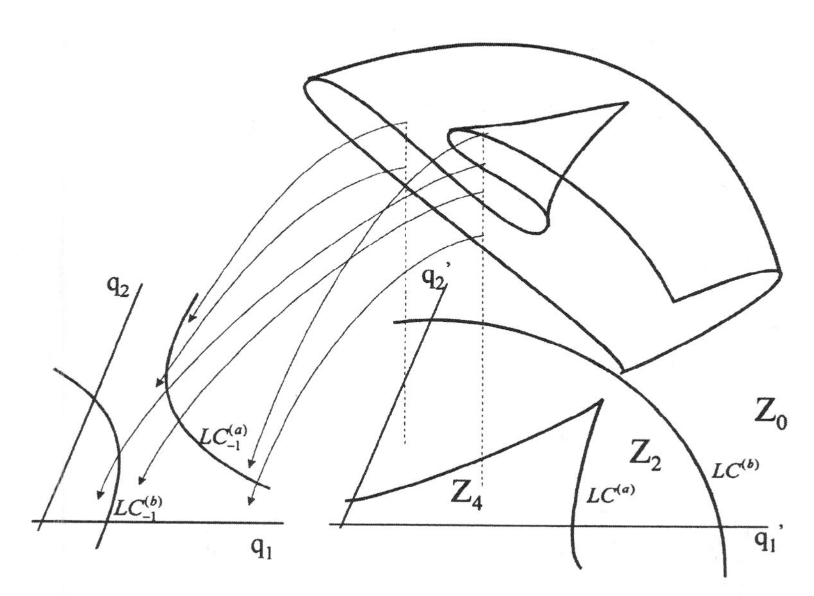


**Figure 7**: Qualitative graphical representation of the foliation associated with the fourth degree noninvertible map (12). The projections of the folds joining two superimposed sheets are the branches of the critical curve $LC$. The arrows show the relation between the foliation and the $k$ distinct rank-one preimages of a point belonging to a region $Z_k$.

Of course, the bifurcation occurring at (38) corresponds to the flip bifurcation of the map $T$ described above, which transforms the saddle point $E_j$ of $T$ into a repelling node. As $v_j$ is further increased, or $c_j$ decreased, cycles of (25) of any order are created: every attracting cycle of (25) corresponds to a saddle cycle of $T$, located on the line $q_i = 0$, with the attracting branch along the invariant axis, and every repelling cycle of (25) corresponds to a repelling node cycle for $T$. For any given value of $v_j(a - c_j) \in (2, 3)$ we can have only one attractor, that may be a cycle or a cyclic-invariant chaotic interval (as for the standard logistic (26) with $\mu \in (3, 4)$) whose basin of attraction is bounded by the unstable fixed point $O$ and its preimage $O^j_{-1}$, of coordinate

$$q_j^{o-1} = \frac{1 + v_j(a - c_j)}{2bv_j} \tag{39}$$

(conjugate to the point $x = 1$ of the standard logistic). Any trajectory of (25) starting from an initial point taken out of the interval $\left[0, q_j^{o-1}\right]$ is divergent toward $-\infty$. At

$$v_j(a - c_j) = 3 \tag{40}$$

the whole interval $\left[0, q_j^{o-1}\right]$ is an invariant chaotic interval. For $v_j(a - c_j) > 3$, as in Figure 8(b), the generic trajectory of (25) is divergent (see, e.g., Devaney, [4,

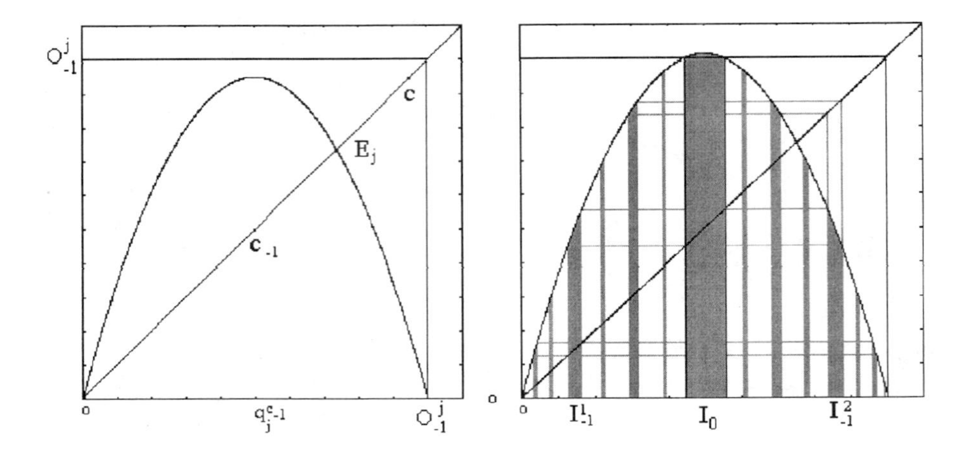


**Figure 8**: Graph of the map (25), conjugate to the logistic map (26). In **(a)** $v_j(a - c_j) < 3$. In this case $q_j^c < q_j^{o-1}$; hence all points inside $I = [0, q_j^{o-1}]$ remain inside $I$ under iteration of (25). In **(b)** $v_j(a - c_j) > 3$. In this case $q_j^c > q_j^{o-1}$ and an interval $I_0$ exists around $q_j^c$ whose points have images out of $I$, so that their trajectories are divergent. The preimages of $I_0$ are two smaller intervals, denoted in figure by $I_{-1}^1$ and $I_{-1}^2$, whose points escape interval $I$ after two iterations; these two intervals have four intervals as preimages, etc. The union of all the preimages, of any rank, of $I_0$, is an open and dense set on $I$ whose points generate unbounded and negative trajectories. Its complement in $I$ has zero measure and is homeomorphic to a Cantor set.

p. 34]). This final bifurcation occurring when (40) holds, is characterized by the collision (or merging) of the critical value $c$, whose coordinate

$$q_j^C = \frac{\left[1 + v_j(a - c_j)\right]^2}{8bv_j} \tag{41}$$

is given by the image of the critical point $c_{-1}$, with the basin boundary at $q_j^{o-1}$. In fact, $q_j^{o-1} = q_j^C = 4/2bv_j$ when (40) holds.

## Acknowledgments

This work has been performed as an activity of the national research project "Dinamiche non lineari ed applicazioni alle scienze economiche e sociali," MURST, Italy, and under the auspices of CNR, Italy.

## REFERENCES

[1] Abraham, R., L. Gardini, and C. Mira. *Chaos in Discrete Dynamical Systems (A Visual Introduction in Two Dimensions)*. Springer-Verlag, New York, 1996.

[2] Bonanno, G. Oligopoly Equilibria When Firms Have Local Knowledge of Demand. *International Economic Review*, **29**, 45–55, 1988.

[3] Bonanno, G. and C. Zeeman. Limited Knowledge of Demand and Oligopoly Equilibria. *Journal of Economic Theory*, **35**, 276–283, 1985.

[4] Devaney, R. L. *An Introduction to Chaotic Dynamical Systems*. Benjamin/Cummings, Menlo Park, CA, 1987.

[5] Dixit, A. Comparative Statics for Oligopoly. *International Economic Review*, **27**, 107–122, 1986.

[6] Fisher, F. M. The Stability of the Cournot Oligopoly Solution: The Effect of Speeds of Adjustment and Increasing Marginal Costs. *Review of Economic Studies*, **28**, 125–135, 1961.

[7] Flam, S. D. Oligopolistic Competition: From Stability to Chaos. In: F. Gori, L. Geronazzo and M. Galeotti (eds.), *Nonlinear Dynamics in Economics and Social Sciences*. Lecture Notes in Economics and Mathematical Systems, vol. 399. pp. 232–237, Springer-Verlag, Berlin, 1993.

[8] Furth, D. Stability and Instability in Oligopoly. *Journal of Economic Theory*, **40**, 197–228, 1986.

[9] Gumowski, I. and C. Mira. *Dynamique Chaotique*. Cepadues Editions, Toulose, 1980.

[10] Hahn, F. The Stability of the Cournot Oligopoly Solution. *Review of Economic Studies*, **29**, 329–331, 1962.

[11] McManus, M. and R. E. Quandt. Comments on the Stability of the Cournot Oligopoly Model. *Review of Economic Studies*, **27**, 136–139, 1961.

[12] Mira, C., L. Gardini, A. Barugola, and J. C. Cathala. *Chaotic Dynamics in Two-Dimensional Noninvertible Maps*. World Scientific, Singapore, 1996.

[13] Mira, C., D. Fournier-Prunaret, L. Gardini, H. Kawakami, and J. C. Cathala. Basin Bifurcations of Two-Dimensional Noninvertible Maps: Fractalization of Basins. *International Journal of Bifurcations and Chaos*, **4**, 343–381, 1994.

[14] Mira, C. and C. Rauzy. Fractal Aggregation of Basin Islands in Two-Dimensional Quadratic Noninvertible Maps. *International Journal of Bifurcations and Chaos*, **5**, 991–1019, 1995.

[15] Okuguchi, K. Adaptive Expectations in an Oligopoly Model. *Review of Economic Studies*, **37**, 233–237, 1970.

[16] Sacco, P. L. Adaptive Response and Cournotian Behavior. *Economic Notes*, **20**, 474–496, 1991.

[17] Rosen, J. B. Existence and Uniqueness of Equilibrium Points for Concave $n$-Person Games. *Econometrica*, **33**, 520–534, 1965.

[18] Teocharis, R. D. On the Stability of the Cournot Solution of the Oligopoly Problem. *Review of Economic Studies*, **27**, 133–134, 1960.

[19] Varian, H. R. *Microeconomic Analysis*, 3rd ed. W. W. Norton, 1992.

# A Multistage Supergame of Downstream Pollution

Leon A. Petrosjan
Department of Applied Mathematics
St. Petersburg State University
St. Petersburg, Russia

Georges Zaccour
GERAD-École des HEC
Montréal, Canada

**Abstract**

A multistage supergame model for the management of a downstream pollution dispute between two neighboring regions is constructed. An important feature of this model is that it allows for renegotiation of the contractual terms agreed on up to the current date. The set of Nash equilibria is characterized. It is shown that the Nash bargaining outcomes are self-enforcing. A time-consistent procedure to share the cooperative outcome between the two players is devised.

## 1 Introduction

Environmental problems often involve different sovereign countries and need to be addressed collectively, since that any unilateral emissions reduction has a very small impact on the total accumulation of pollutants. If the parties agree to cooperate, then one usually seeks to devise a mechanism to share the emissions reduction and cleaning costs among the players, which has some desirable properties (e.g., fairness). It is empirically observed that reaching a multilateral, or even a bilateral, agreement acceptable by every one is a lengthy negotiation process. There is one instance, downstream pollution, where such negotiations are inherently more complicated. Indeed, since the polluter country or region does not suffer from its emissions, it will be very reluctant to pay for the damage incurred by his neighbor. Downstream pollution is a commonly observed problem. For instance, Canada and Norway suffer from acid rain due to industrial activities in the United States and Great Britain, and the pollution of the Rhine River by France and Germany damages the environment in the Netherlands. In an idealistic world where the players are altruistic and fair, the polluted region computes the right cost of the environmental damages and the polluter pays the bill. In a more realistic world, it may be hopeless to expect that the polluter will reduce significantly its emissions or pay for them, at least in the short run, simply because the cost would be very high in terms of economic development.

The aim of this paper is to investigate in a game setting the problem of two neighboring regions facing a downstream pollution dispute. We assume for simplicity that the polluter (region 2) does not suffer at all from its emissions that, however, cause important damages to its neighbor (region 1). We suppose that the population of this last region cares about damages to the environment, generated also in part by its own emissions, and is willing to incur a cost to achieve a better environmental standard. Following Kaitala and Pohjola [8], we shall refer to region 2 as the nonvulnerable player and to region 1 as the vulnerable one. Schematically, the vulnerable player has, given the selfish behavior of his neighbor, two options. The first one is to manage his environmental problem alone by taking it for granted that his neighbor will not do anything to reduce his emissions while optimizing his objective function. The second one is to provide an incentive to the nonvulnerable player to reduce its emissions. Given a certain environmental target, this last option can be economically wise if the cost of emissions reduction by player 2 is lower than the damage and emission reduction costs incurred by player 1. In what follows, these options are modeled as noncooperative and cooperative games.

Tackling environmental problems is typically a long-run task and therefore any model aiming to represent them adequately has to be dynamic. Further, it is often the case in international agreements to renegotiate the contractual terms, at given intermediate dates, to take into account the evolution of the system. We retain these features in our game model.

Dynamic game models have been widely used during the last years to represent and solve environmental problems (see, for instance, Carraro and Siniscalo [2], Dockner and Wirl [5], Filar and Gaetner [6], Haurie and Zaccour [7], Kaitala and Pohjola [8], Kaitala and al. [9], Krawczyk [10], Krawczyk and Zaccour [11], Petrosjan and Zakharov [13], Tolwinski and Martin [14], Tulkens and al. [15], and van der Ploeg and de Zeeuw [16],[17]). Our approach is different. Indeed, the situation is represented by a supergame that is, to the best of our knowledge, new in this area. This structure permits us to handle the issue of renegotiation. Further, the solution derived for any stage of given duration has the property of being time consistent, which means that no one player will find it preferable to switch from the cooperative strategy to the noncooperative one. Put differently, our solution is self-enforcing. From this perspective, our approach shares the idea pursued in the developing literature (see, for instance, Barrett [1], Carraro and Siniscalco [3], Chander and Tulkens [4], and Tulkens [15]) on stable coalitions and self-enforcing agreements in the context of international negotiation on transnational environmental problems (e.g., greenhouse gas emissions, acid rains, etc.). The main differences between this literature and our approach are:

(1) this literature considers one-shot games whereas our model is fully dynamic; and

(2) this literature is interested in deriving self-enforcing cost-sharing agreements among $N > 3$ countries whereas our problem involves only two (asymmetric) players and the focus is on the design of a distribution mechanism of the fruit of cooperation which is dynamically time consistent.

The paper is organized as follows: In Section 2, we consider a simple production and pollution emissions model which captures however the main qualitative ingredients of the situation of interest. In Section 3 the case where the players cannot revise their decisions is stated. In Section 4, we define the structure of the multistage supergame and derive some results for the one-stage supergame. In Section 5, we characterize the solution of the multistage supergame. In Section 6, we give some concluding remarks.

## 2 The Production and Pollution Model

Let us refer to the vulnerable country as Player (or region) 1 and to the non-vulnerable one as Player (or region) 2. As in Kaitala and Pohjola (1995) and van der Ploeg and de Zeeuw [16], [17], we adopt the following model of production and pollution emissions. Denote by $q_i(t)$ the industrial production in region $i$ by time $t \in [0, T]$, and by $U_i(q_i)$ the net utility (i.e., utility minus production cost) derived from this production. We assume that $U_i(q_i)$ is a nonnegative concave function, increasing up to a certain finite value and decreasing afterward. This assumption can be justified, for instance, by supposing that production cost increases more rapidly than the utility derived from this production and consequently the net utility will start to decrease at a certain production level. Denote by $e_i(t)$ the inevitable byproducts flow of pollutants. The latter is taken as a proportion, which in turn is a function depending on the capacity of clean technology (abatement capital) $K_i(t)$, of the production level, i.e., $e_i(t) = \alpha_i\big(K_i(t)\big)q_i(t)$. This capacity can be increased through an investment schedule $I_i(t)$. In order to take into account marginal decreasing returns, we assume that $\alpha_i(\cdot)$ is a convex function satisfying $d\alpha_i(K_i)/dK_i \leq 0$, and bounded as follows:

$$\alpha_i(0) = \overline{\alpha}_i, \qquad \alpha_i(\infty) = \overline{\overline{\alpha}}_i, \qquad \text{and} \qquad 0 < \overline{\overline{\alpha}}_i < \alpha_i(\cdot) < \overline{\alpha}_i < 1.$$

The investment cost function depends on the physical investment $I_i(t)$ and is denoted $F_i(I_i)$. We assume that $F_i(I_i)$ is a nonnegative increasing convex function.

The evolution of the stock of abatement capital of Player $i$ is governed by the differential equation:

$$\begin{aligned} \dot{K}_i &= I_i - \mu_i K_i, \\ K_i(0) &= K_i^0, \end{aligned} \tag{1}$$

where $\mu_i \geq 0$ denotes the rate of physical depreciation of this stock. The evolution of the stock of pollutants $S(t)$ is described by the differential equation:

$$\begin{aligned} \dot{S}(t) &= \sum\nolimits_{i=1}^{2} \alpha_i(K_i)q_i - \delta S, \\ S(0) &= S^0, \end{aligned} \tag{2}$$

where $\delta$ is the natural absorbtion rate. Denote by $D_1(S)$ the nonnegative increasing convex damage cost incurred by region 1. This cost is taken identically equal to zero for the nonvulnerable Player 2. We assume that each player seeks to maximize the total stream of welfare. Formally, the optimization problem of Player $i$, $i = 1, 2$, can be written as follows:

$$\begin{aligned} &\max \quad w_i(S^0, K^0, q, I) = \int_0^T \{U_i(q_i) - F_i(I_i) - D_i(S)\}\, dt \\ &\text{s.t. (1)–(2)} \qquad q_i(t), I_i(t) \geq 0, \; t \in [0, T]. \end{aligned} \tag{3}$$

Notice that in order to avoid unnecessary complications, we have retained only a few ingredients of what should be a full model of pollution emissions. Indeed, we focused only on one type of capital good, namely the abatement capacity. Obviously, in an implementable model, we should introduce production capacity, possibly under the form of a vector of different types of technology, each one characterized by an emission to production ratio as is actually done in energy-environment mathematical programming models (see Haurie and Zaccour [7]). Further, we can add, on top of the natural absorption rate $\delta$, some cleaning activities that regions may undertake and introduce a consequent cost in the objective functions. We feel that these additions would come at a considerable mathematical complication without adding any qualitative insights to the problem at hand, which is the construction of a time consistent mechanism to share the benefit of cooperation.

## 3 The Once for All Cooperative and Noncooperative Games

It is apparent in the above inherently conflicting formulation that the two players are involved in a differential game. Let us denote by $\Gamma(S^0, K^0; T)$ the game under consideration.

Let us assume, for the moment, that the players can choose at the starting date to cooperate or not. Once this choice is made, it cannot be changed before the final date of the planning horizon $T$. If a noncooperative mode of play is selected, then Player 2 will obviously choose the dominant strategy $I_2(t) = 0, t \in [0, T]$ (since $D_2(S) = 0$). Assuming that $K_2(0) = 0$, then the flow of pollutants emitted by this player will be given by $\tilde{e}_2(t) = \overline{\alpha}_2\tilde{q}_2(t)$, where $\tilde{q}_2(t)$ is the optimal production level at period $t$, obtained by solving problem (3).

Knowing that Player 2 will stick to this selfish position in the absence of any incentive to behave differently, region 1 will maximize its welfare index subject to (1) and (2). In this case, (2) must be written as follows:

$$\dot{S}(t) = \alpha_1(K_1)q_1 + \overline{\alpha}_2\tilde{q}_2 - \delta S. \tag{4}$$

The outcome of this optimization problem is a production and an investment schedule $\{\tilde{q}_1(t), \tilde{I}_1(t); t \in [0, T]\}$ and an optimal total welfare $\tilde{w}_1$. Consequently, the stocks of clean technology and pollutants will be given by $\tilde{K}_1(t)$ and $\tilde{S}(t)$, $t \in [0, T]$. Denote by $\tilde{w}_i(S^0, K^0, \tilde{I}, \tilde{q})$ the payoff of Player $i$, $i = 1, 2$, in this noncooperative game.

If the players decide to cooperate, then they will seek such open-loop strategies $\hat{I}_i(t)$, $\hat{q}_i(t)$ which maximize the sum of payoffs

$$w_1(S^0, K^0; \hat{I}, \hat{q}) + w_2(S^0, K^0; \hat{I}, \hat{q}) = \max_{I(t), q(t)} \sum_{i=1}^{2} w_i(S^0, K^0; I, q) = \hat{v}(S^0, K^0, T).$$

We assume that the game is essential, i.e., $\hat{v}(S^0, K^0, T - t_0) > \tilde{w}_1[S^0, K^0, T - t_0] + \tilde{w}_2[S^0, K^0, T - t_0]$, where $\tilde{w}_i[S^0, K^0, T - t_0] = \tilde{w}_i(S^0, K^0; \tilde{I}, \tilde{q})$, $i = 1, 2$ and $t_0 = 0$. Let us refer by $\hat{\Gamma}$ to the cooperative game and by $\tilde{\Gamma}$ to the noncooperative one. Denote by $\hat{S}(t)$, $\hat{K}(t)$ the cooperative trajectory resulting from the implementation of $\hat{I}_i(t)$, $\hat{q}_i(t)$ in $\hat{\Gamma}$. The issue of sharing the total payoff obtained under cooperation will be addressed later on.

The first-order necessary conditions of the joint maximization problem are stated and interpreted in the Appendix.

## 4 The Multistage Supergame

The restriction imposed in the last section on the behavior of the players is clearly a very strong one. Actually, international agreements are usually reconsidered at different time moments. In order to allow for renegotiation, we will construct a multistage supergame of duration $T$, denoted $G(S^0, K^0, T)$, as follows. The planning horizon $[0, T]$, is divided into $n$ time intervals $[t_k, t_{k+1}[$, $k = 0, \ldots, n-1$, with $t_0 = 0$ and $t_n = T$. Let us denote by $\hat{X}(t_{k-1}) = \big(\hat{K}(t_{k-1}), \hat{S}(t_{k-1})\big)$, $k = 1, \ldots, n$, the state of the game at the moment $t_{k-1}$ (which is the beginning of stage $k$). At $t_{k-1}$, Player 1, moves first and selects a number $\gamma_k \in \big[0, \hat{v}\big(\hat{X}(t_{k-1}), T - t_{k-1}\big)\big]$ and offers to Player 2 the following amount on the time interval $[t_{k-1}, t_k[$ if he agrees to cooperate on this time interval (i.e., play the cooperative differential game $\hat{\Gamma}\big(\hat{X}(t_{k-1}), T - t_{k-1}\big)\big)$:

$$\frac{\gamma_k}{\hat{v}\big(\hat{X}(t_{k-1}), T - t_{k-1}\big)}\big[\hat{v}\big(\hat{X}(t_{k-1}), T - t_{k-1}\big) - \hat{v}\big(\hat{X}(t_k), T - t_k\big)\big].$$

The above amount corresponds to the share of the total cooperative outcome achievable in the time interval under consideration that Player 1 is willing to give to his neighbor if he plays cooperatively. Afterward, Player 2 moves and has the alternative of accepting the offer of Player 1 or rejecting it. If Player 2 agrees, both players implement, during this time interval $[t_{k-1}, t_k[$, the strategies prescribed by the cooperative differential game as if it were played from $t_{k-1}$ to $t_n = T$. Therefore, the system evolves along $\hat{X}(t) = \big(\hat{S}(t), \hat{K}(t)\big), t \in [t_{k-1}, t_k[$, and the supergame moves to the next stage $k+1$, if $k \neq n$, in which the players behave in the same manner, Player 1 moving first by selecting $\gamma_{k+1}$. If Player 2 rejects the offer of Player 1, then they play noncooperatively until the end of the horizon $T$.

Suppose that the following sequence $(\hat{\Gamma}^k\tilde{\Gamma})$ is realized in the supergame $G\big(X(t_0), T - t_0\big)$, which means that the cooperative game $\hat{\Gamma}$ is played on the first $k$ stages, $0 < k < n$, and the noncooperative one afterward. Player 1 will get the payoff $w_1[\hat{\Gamma}^k\tilde{\Gamma}]$ which corresponds to the sum of payoffs earned on the stages of the supergame and is given by

$$w_1[\hat{\Gamma}^k\tilde{\Gamma}] = \sum_{\ell=1}^{k} \left\{ \frac{\hat{v}\big(\hat{X}(t_{\ell-1}), T - t_{\ell-1}\big) - \gamma_\ell}{\hat{v}\big(\hat{X}(t_{\ell-1}), T - t_{\ell-1}\big)} \times \Big[\hat{v}\big(\hat{X}(t_{\ell-1}), T - t_{\ell-1}\big) - \hat{v}\big(\hat{X}(t_\ell), T - t_\ell\big)\Big] \right\} + \tilde{w}_1[\hat{X}(t_k), T - t_k]$$

and Player 2 will obtain

$$w_2[\hat{\Gamma}^k\tilde{\Gamma}] = \sum_{\ell=1}^{k} \left\{ \frac{\gamma_\ell}{\hat{v}\big(\hat{X}(t_{\ell-1}), T - t_{\ell-1}\big)} \times \Big[\hat{v}\big(\hat{X}(t_{\ell-1}, T - t_{\ell-1}\big) - \hat{v}\big(\hat{X}(t_\ell), T - t_\ell\big)\Big] \right\} + \tilde{w}_2[\hat{X}(t_k), T - t_k],$$

where as already defined $\tilde{w}_i\big[\hat{X}(t_k), T - t_k\big]$ is the payoff of Player $i$ if the mode of play is noncooperative from the moment $t_k$ until $T$.

### 4.1 The Last Stage Supergame

Consider now the last one-stage supergame starting at $t_{n-1}$, with duration $T - t_{n-1}$, which we denote by $G_1\big(\hat{X}(t_{n-1}), T - t_{n-1}\big)$. This game can be viewed as a game in extensive form, with the following players' strategy sets

$$P = \Big\{\gamma : 0 \le \gamma \le \hat{v}\big(\hat{X}(t_{n-1}), T - t_{n-1}\big)\Big\}$$

for Player 1, and, for Player 2, the set $R$ of all possible functions $\phi(\gamma), \gamma \in P$, taking its values in a two-point set $\{c, nc\}$, where $c$ stands for cooperation and $nc$

for noncooperation. This one-stage supergame is played as follows. Players 1 and 2 choose their strategies in the sets $\gamma \in P$ and $\phi \in R$. If the choice is $(\gamma, \phi)$, then the payoffs are given by $\tilde{w}_i\big(\hat{X}(t_{n-1}), T - t_{n-1}\big), i = 1, 2$, if $\phi(\gamma) = nc$, or by $\hat{v}\big(\hat{X}(t_{n-1})\big) - \gamma$ for Player 1, and $\gamma$ for Player 2, if $\phi(\gamma) = c$.

Define the set $\overline{P} \subset P$ as follows:

$$\overline{P} = \Big\{\gamma : \tilde{w}_2[\hat{X}(t_{n-1}), T - t_{n-1}] \leq \gamma \leq \hat{v}\big(\hat{X}(t_{n-1}), T - t_{n-1}\big) - \tilde{w}_1[\hat{X}(t_{n-1}), T - t_{n-1}]\Big\}.$$

Introduce the strategy of the second player $\overline{\phi}_M(\gamma)$, $M \subset P$, where $M$ is a closed set, of the form

$$\overline{\phi}_M(\gamma) = \begin{cases} c & \text{if } \gamma \in M, \\ nc & \text{if } \gamma \notin M. \end{cases}$$

**Proposition 4.1.** *Every pair of strategies*

$$(\overline{\gamma}, \overline{\phi}_M(\gamma)) \tag{5}$$

*is a Nash equilibrium in the game* $G_1\big(\hat{X}(t_{n-1}), T - t_{n-1}\big)$ *if* $\overline{P} \cap M$ *is a nonempty set,* $\overline{\gamma} \in \overline{P} \cap M$, *and* $\overline{\gamma}$ *is the minimal element of the set* $M$.

**Proof.** Denote by $w_i\{\gamma, \phi_M(\gamma)\}$ the payoff function of Player $i = 1, 2$. It is easy to see that the payoffs of players 1 and 2, when playing the pair of strategies $(\overline{\gamma}, \overline{\phi}_M(\gamma))$ are, respectively, given by

$$\begin{aligned} w_1\{\overline{\gamma}, \overline{\phi}_M(\gamma)\} &= \hat{v}\big(\hat{X}(t_{n-1}), T - t_{n-1}\big) - \overline{\gamma} \geq \tilde{w}_1\big(\hat{X}(t_{n-1}), T - t_{n-1}\big) \\ &= w_1\{\gamma, \overline{\phi}_M(\gamma)\} \qquad \text{if} \quad \gamma \notin M, \\ w_2\{\overline{\gamma}, \overline{\phi}_M(\gamma)\} &= \overline{\gamma} \geq \tilde{w}_2\big(\hat{X}(t_{n-1}), T - t_{n-1}\big) \\ &= w_2\{\overline{\gamma}, \phi_M(\gamma)\} \qquad \text{if} \quad \phi(\overline{\gamma}) \neq c. \end{aligned}$$

In all other cases the inequalities above turn out to be equalities. The statement of the proposition follows from the above inequalities. □

Consider the set $\overline{P} \backslash \{\tilde{w}_2[\hat{X}(t_{n-1}), T - t_{n-1}]\}$, i.e., the set $\overline{P}$ without the left end point. The following corollary gives the condition for Nash equilibrium to be subgame perfect in $G_1\big(\hat{X}(t_{n-1}), T - t_{n-1}\big)$.

**Corollary 4.1.** *The equilibrium* $\big(\overline{\gamma}, \overline{\phi}_M(\gamma)\big)$ *is subgame perfect if and only if* $M = \overline{P}$.

**Proof.** If $M = \overline{P}$, then subgame perfectness is evident because in every subgame Player 2 chooses $c$ which is prescribed by the strategy $\overline{\phi}_M(\gamma)$ and gets the payoff $\gamma \geq \tilde{w}_2\big(\hat{X}(t_{n-1}), T - t_{n-1}\big)$. Suppose that $M \neq \overline{P}$, then there exists $\gamma' \in \overline{P}$, $\gamma' > \tilde{w}_2\big(\hat{X}(t_{n-1}), T - t_{n-1}\big)$, $\gamma' \notin M$. This means that in the one move subgame following the choice $\gamma'$ by Player 1, Player 2, using the strategy $\overline{\phi}_M(\gamma)$, chooses $nc$ which is not optimal in this subgame since $\gamma' > \tilde{w}_2\big[\hat{X}(t_{n-1}), T - t_{n-1}\big]$. □

The best Nash equilibrium for Player 1 is

$$\left(\tilde{w}_2\left[\hat{X}(t_{n-1}), T - t_{n-1}\right], \overline{\phi}_M(\gamma)\right), \qquad \tilde{w}_2\left[\hat{X}(t_{n-1}), t_{n-1}\right] \in M,$$

and the best one for Player 2 is

$$\left(\hat{\mathrm{v}}\left(\hat{X}(t_{n-1}), T - t_{n-1}\right) - \tilde{w}_1\left[\hat{X}(t_{n-1}), T - t_{n-1}\right], \overline{\phi}_M(\gamma)\right),$$
$$\left\{\hat{\mathrm{v}}\left(\hat{X}(t_{n-1}), T - t_{n-1}\right) - \tilde{w}_1\left[\hat{X}(t_{n-1}), T - t_{n-1}\right]\right\} \in M.$$

The following corollary shows that the Nash bargaining outcomes are self-enforcing because they can be obtained as payoffs of a Nash equilibrium:

**Corollary 4.2.** *The Nash bargaining outcomes can be obtained as the payoffs of a Nash equilibrium.*

**Proof.** The Nash bargaining shares are obviously given by

$$\overline{\beta} = \frac{1}{2}\left\{\hat{\mathrm{v}}\left(\hat{X}(t_{n-1}), T - t_{n-1}\right) - \tilde{w}_1\left[\hat{X}(t_{n-1}), T - t_{n-1}\right]\right.$$
$$\left. - \tilde{w}_2\left[\hat{X}(t_{n-1}), T - t_{n-1}\right]\right\}.$$

To obtain the Nash bargaining outcomes, it suffices to construct a Nash equilibrium with the payoffs:

$$\hat{\mathrm{v}}\left(\hat{X}(t_{n-1}), T - t_{n-1}\right) - \overline{\gamma} = \tilde{w}_1\left[\hat{X}(t_{n-1}), T - t_{n-1}\right] + \overline{\beta},$$
$$\overline{\gamma} = \tilde{w}_2\left[\hat{X}(t_{n-1}), T - t_{n-1}\right] + \overline{\beta}.$$

According to the above proposition this equilibrium will be the pair $\left(\tilde{w}_2\left[\hat{X}(t_{n-1}), T - t_{n-1}\right] + \overline{\beta}, \overline{\phi}_M(\gamma)\right)$, where $\tilde{w}_2\left[\hat{X}(t_{n-1}), T - t_{n-1}\right] + \overline{\beta} \in M$, and is a minimal element of $M$. □

**Corollary 4.3.** *Any pair of payoffs of the form* $\left(\hat{\mathrm{v}}\left(\hat{X}(t_{n-1}), T - t_{n-1}\right) - \gamma, \gamma\right)$, $\gamma \in \overline{P}$, *could be achieved by a Nash equilibrium.*

**Proof.** It is sufficient to consider the strategy pair $(\gamma, \overline{\phi}_M(\gamma))$ where $\gamma = \min \gamma' \{\gamma' : \gamma' \in M\}$. □

To get the economic intuition behind the proposition and its corollaries, it may be useful to think about the problem as one of sharing of a rent (or the fruit of a cooperation between two players). Indeed, the strategy set of the vulnerable player is formed of elements that correspond to amounts of money that he can offer to the nonvulnerable player to, hopefully, reach an agreement to reduce emissions. These numbers range from a lower bound which corresponds to the minimum amount that can induce a cooperative attitude from the nonvulnerable player (i.e., the latter will be better off if he accepts this offer than otherwise) to an upper bound which equals the total cooperative rent. The preferred solution by the vulnerable player is the one where his neighbor accepts to cooperate while keeping the whole rent minus the lower bound. Conversely, the nonvulnerable player's preferred solution is certainly the one where he catches the whole rent. The proposition tells us that this lower bound and the acceptance by the nonvulnerable player is a Nash equilibrium. Recall that the non vulnerable player strategy set is restricted to two options, namely accept the offer or reject it (put differently cooperate or not). Corollary 1 states the condition under which this equilibrium is perfect.

One of the amounts that the vulnerable player can offer is the one which precisely corresponds to what would have been prescribed by the application of the symmetry (fairness) axiom of the Nash bargaining procedure. This outcome is proven to be also a Nash equilibrium which means that it is self-enforceable. This offer requires a high degree of altruism from the vulnerable player. However, one can eventually think about reversing the roles of the two players. Indeed, one can assume that the moves would be that the nonvulnerable player asks for a compensation to cooperate and the vulnerable player could accept or reject. In this case, to achieve the Nash bargaining outcomes, it will require the nonvulnerable player to be altruistic. This is to say that the share of the rent that each player can capture depends heavily on the definition of the strategy sets and sequence of moves.

### 4.2 Time Consistency in the One-Stage Supergame

If the players decide to play the Nash equilibrium strategies, with the payoffs $(\hat{v}(\hat{X}(t_{n-1}), T - t_{n-1}) - \beta, \beta)$, $\beta \in \overline{P}$, then they are playing a cooperative game $\hat{\Gamma}(\hat{X}(t_{n-1}), T - t_{n-1})$ on the time interval $[t_{n-1}, T]$. A natural question is how to distribute the payoffs on this time interval in order to get the desirable time consistency property which ensures that the players will not switch to noncooperative strategies during this interval. Consider the functions $\zeta_1(\tau), \zeta_2(\tau)$, $\tau \in [t_{n-1}, T]$, such that

$$\hat{v}(\hat{X}(t_{n-1}), T - t_{n-1}) - \beta = \int_{t_{n-1}}^{T} \zeta_1(\tau)\,d\tau,$$

$$\beta = \int_{t_{n-1}}^{T} \zeta_2(\tau)\,d\tau,$$

$$\int_{t}^{T} (\zeta_1(\tau) + \zeta_2(\tau))\,d\tau = \hat{v}(\hat{X}(t), T - t), \qquad t \in [t_{n-1}, T],$$

where $\hat{v}(\hat{X}(t), T - t)$ is the cooperative outcome in the subgame $\hat{\Gamma}(\hat{X}(t), T - t)$, with the initial conditions on the optimal cooperative trajectory in $\hat{\Gamma}(\hat{X}(t_{n-1}), T - t_{n-1})$. The functions $\zeta_1(\tau)$, $i = 1, 2$, which correspond to a payment rule on the time interval $[t_{n-1}, T]$, are called "payoff distribution procedure" (PDP). Intuitively, the PDP must be time consistent which means that no one player would find it preferable to switch to a noncooperative regime during the time interval under consideration. This switch will not occur if, at any intermediate time instant $t \in [t_{n-1}, T]$, the payoff obtained by each player, if both players continue to follow the cooperative strategies on the time interval $[t, T]$, is higher than what he would obtain by playing the noncooperative one. (For a complete treatment of time consistency in dynamic games one can refer to Petrosjan [12].) To state the time-consistency condition, let us introduce the following notations:

$$m_i(t) = \int_{t_{n-1}}^{t} \zeta_i(\tau)\, d\tau, \qquad i = 1, 2,$$
$$m_1^t = \hat{v}\big(\hat{X}(t_{n-1}), T - t_{n-1}\big) - \gamma - m_1(t),$$
$$m_2^t = \gamma - m_2(t).$$

The $m_i(t)$, $i = 1, 2$, correspond to the amount obtained by the players on the time interval $[t_{n-1}, t]$ and $m_i^t$, $i = 1, 2$, to future earnings under cooperation.

**Theorem 4.1.** *The PDP $\zeta_i(t)$, $t \in [t_{n-1}, T]$, is time consistent if*

$$\zeta_i(t) \leq -\frac{d}{dt}\tilde{w}_i\big[\hat{X}(t), T - t\big], \qquad t \in [t_{n-1}, T], \quad i = 1, 2.$$

**Proof.** Obviously, Player $i$, $i = 1, 2$, will not switch to noncooperative strategy at any intermediate moment $t \in [t_{n-1}, T]$ if its future earnings in the noncooperative game $\overline{\Gamma}\big(\hat{X}(t), T - t\big)$ are lower than those in the cooperative one, i.e.,

$$m_i^t \geq \tilde{w}_i\big[\hat{X}(t), T - t\big], \qquad t \in [t_{n-1}, T], \quad i = 1, 2. \tag{6}$$

The above condition holds at $t = t_{n-1}$, since the players have agreed to cooperate in $G_1\big(X(t_{n-1}), T - t_{n-1}\big)$. By definition, we have

$$m_1^{t_{n-1}} = \hat{v}\big(\hat{X}(t_{n-1}), T - t_{n-1}\big) - \gamma,$$
$$m_2^{t_{n-1}} = \gamma,$$

and thus (6) holds since $\gamma \in \overline{P}$. Therefore, (6) holds at any intermediate time instant if

$$\frac{d}{dt}(m_i^t) \geq \frac{d}{dt}\tilde{w}_i\big[\hat{X}(t), T - t\big], \qquad t \in [t_{n-1}, T], \quad i = 1, 2,$$

which can be written equivalently

$$\zeta_i(t) \leq -\frac{d}{dt}\tilde{w}_i\big[\hat{X}(t), T - t\big], \qquad t \in [t_{n-1}, T], \quad i = 1, 2.$$ □

Theorem 4.1 provides a condition under which the distribution of payoffs over time would be time consistent. This condition refers to speeds (or time derivatives) of the noncooperative outcomes. An economic interpretation of this condition is simply that the amount distributed at any instant of time should be less than minus the time variation of the noncooperative outcome. What remains to be seen is under which conditions on data and parameters, this condition would be fulfilled.

## 5 The Solution of the Supergame

If we consider the supergame $G(X(t_0), T - t_0)$ as a game in extensive form, then we can easily construct the players' strategy sets. Indeed, the strategy set of Player 1, denoted $P$, consists of $n$ tuples

$$\gamma = (\gamma_1, \dots, \gamma_n), \qquad \gamma_k \in \left[0, \hat{v}\big(\hat{X}(t_{k-1}), T - t_{k-1}\big)\right] = P_k, \qquad k = 1, \dots, n,$$

where $\hat{X}(t), t \in [t_0, T]$, is a cooperative trajectory in $G(X(t_0), T - t_0)$. The strategy set $R$ of Player 2 consists of all possible two valued vector functions $\phi(\gamma) = (\phi^1(\gamma), \dots, \phi^k(\gamma), \dots, \phi^n(\gamma))$ where each component $\phi^k(\gamma)$ is defined for all $\gamma \in P_k = \left[0, \hat{v}\left(\hat{X}(t_{k-1}), T - t_{k-1}\right)\right]$ and may take one of its two possible values in $\{c, nc\}$.

The supergame $G(X(t_0), T - t_0)$ is played as follows. Players 1 and 2 simultaneously choose their strategies $\gamma \in P, \phi \in R$. Let $\ell$, $1 \le \ell \le n$, be the first index for which $\phi^\ell(\gamma_\ell) = nc$, which means that the players play cooperatively on the time interval $[t_0, t_{\ell-1}]$ and noncooperatively afterward till $t_n$.

Let the stage payoffs be defined as follows:

$$\gamma_k^1(t_k) = \frac{\hat{v}(\hat{X}(t_k), T - t_k) - \gamma_k}{\hat{v}(\hat{X}(t_k), T - t_k)} \left[\hat{v}\left(\hat{X}(t_{k-1}), t - t_{k-1}\right) - \hat{v}\left(\hat{X}(t_k), T - t_k\right)\right], \quad (7)$$

$$\gamma_k^2(t_k) = \frac{\gamma_k}{\hat{v}(\hat{X}(t_k), T - t_k)} \left[\hat{v}\left(\hat{X}(t_{k-1}), t - t_{k-1}\right) - \hat{v}\left(\hat{X}(t_k), T - t_k\right)\right]. \quad (8)$$

The payoffs in the case where the players switch to noncooperation at instant $t_\ell$ are given by

$$w_i\{\gamma, \phi\} = \sum_{k=1}^{\ell-1} \gamma_k^i(t_k) + \tilde{w}_i[\hat{X}(t_\ell), T - t_\ell], \qquad i = 1, 2. \quad (9)$$

If $\phi^\ell(\gamma_\ell) = c$, for $\ell = 1, \dots, n$, then, obviously, the payoffs are

$$w_i\{\gamma, \phi\} = \sum_{k=1}^{n-1} \gamma_k^i(t_k), \qquad i = 1, 2.$$

For any game starting at position $\hat{X}(t_\ell)$, the total cooperative outcome is higher than the sum of noncooperative ones and therefore we have the following inequality:

$$\sum_{i=1}^{2} \sum_{k=\ell+1}^{n} \gamma_k^i(t_k) = \hat{v}(\hat{X}(t_\ell), T - t_\ell) \ge \sum_{i=1}^{2} \tilde{w}_i[\hat{X}(t_\ell), T - t_\ell], \qquad \ell = 0, \dots, n-1.$$

Thus there always exists such a sequence $\gamma_1, \ldots, \gamma_n$ that for corresponding $\gamma_1^i(t), \ldots, \gamma_n^i(t), i = 1, 2$, the following conditions hold:

$$\sum_{k=\ell+1}^{n} \gamma_k^i(t_k) \geq \tilde{w}_i[\hat{X}(t_\ell), T - t_\ell], \qquad \ell = 0, \ldots, n-1, \quad i = 1, 2.$$

Consider the set $\overline{P}$ of vectors $\nu = (\gamma_1, \ldots, \gamma_{n-1})$ such that for each $\gamma_k, k = 1, \ldots, n$, the following inequality takes place:

$$\tilde{w}_2[\hat{X}(t_{k-1}), T - t_{k-1}] \leq \gamma_k \leq \hat{v}\left(\hat{X}(t_{k-1}), T - t_{k-1}\right) - \tilde{w}_1[\hat{X}(t_{k-1}), T - t_{k-1}].$$

Let $\overline{\gamma} \in \overline{P}, M_k \subset \left[0, \hat{v}\left(\hat{X}(t_k), T - t_k\right)\right] = P_k, k = 0, \ldots, n-1$, and $M = \prod_{k=0}^{n-1} M_k$, the Cartesian product of the closed sets $M_k$. Define the strategy $\overline{\phi}_M(\gamma) \in R, \overline{\phi}_M(\gamma) = \{\overline{\phi}_M^1(\gamma), \ldots, \overline{\phi}_M^n(\gamma)\}$, by

$$\overline{\phi}_M^k(\gamma) = \begin{cases} c & \text{if } \gamma_k \in M_{k-1}, \\ nc & \text{if } \gamma_k \notin M_{k-1}. \end{cases}$$

**Theorem 5.1.** *If $\overline{P} \cap M$ is a nonempty set, $\overline{\gamma} \in \overline{P} \cap M$, and $\overline{\gamma} = \min \gamma\{\gamma : \gamma \in M\}$, then the pair of strategies $(\overline{\gamma}, \overline{\phi}_M(\gamma))$ is a Nash equilibrium in $G(\hat{X}(t_0), T - t_0)$ if the following conditions are satisfied:*

$$\sum_{k=\ell+1}^{n} \overline{\gamma}_k^i(t_k) \geq \tilde{w}_i[\hat{X}(t_\ell), T - t_\ell], \qquad \ell = 0, \ldots, n-1, \quad i = 1, 2, \tag{10}$$

*where $\overline{\gamma}_k^i(t_k)$ are computed from (7)–(8) for $\gamma = \overline{\gamma}$, and $\gamma_k$ is a minimal element of $M_{k-1}$.*

**Proof.** As we have already seen, there always exists such a sequence $\gamma = (\gamma_1, \ldots, \gamma_n) \in P$ and a corresponding sequence $\gamma_1^i(t_1), \ldots, \gamma_n^i(t_n), i = 1, 2$, for which (10) is satisfied. At the same time, the conditions for which (10) is satisfied for some $\gamma \in \overline{P}$ depend on the behavior of $\tilde{w}_i[\hat{X}(t_\ell), T - t_\ell], \ell = 0, \ldots, n-1, i = 1, 2$. Denote by $w_i\{\gamma, \phi_M(\gamma)\}$ the payoff function of Player $i, i = 1, 2$. For the pair $(\overline{\gamma}, \overline{\phi}_M(\gamma))$ we have

$$w_i\{\overline{\gamma}, \overline{\phi}_M(\gamma)\} = \sum_{k=1}^{n} \overline{\gamma}_k^i(t_k), \qquad i = 1, 2,$$

if the strategy $\gamma$ differs from $\overline{\gamma}$ for the first time on some stage $\ell$ by choosing $\gamma \neq \overline{\gamma}_\ell$ and $\gamma \notin M_{\ell-1}$, then the strategy $\overline{\phi}_M(\gamma)$ is constructed in such a manner that the players will not cooperate from this stage until the end of the game. In this case, from (9), the total payoff of Player 1 will be

$$\begin{aligned} w_1\{\gamma, \overline{\phi}_M(\gamma)\} &= \sum_{k=1} \overline{\gamma}_k^1(t_k) + \tilde{w}_1[\hat{X}(t_\ell), T - t_\ell] \\ &\leq \sum_{k=1}^{\ell} \overline{\gamma}_k^1(t_k) + \sum_{k=\ell+1}^{n} \overline{\gamma}_k^1(t_k) = w_1\{\overline{\gamma}, \overline{\phi}_M(\gamma)\}. \end{aligned} \tag{11}$$

If Player 2 does not agree for the first time at $\ell$ (i.e., $\phi(\overline{\gamma}_\ell) = nc$), then they will not cooperate any more and we have the following

$$\begin{aligned} w_2\{\overline{\gamma}, \overline{\phi}_M(\gamma)\} &= \sum_{k=1}^{\ell} \overline{\gamma}_k^2(t_k) + \sum_{k=\ell+1}^{n} \overline{\gamma}_k^2(t_k) \\ &\geq \sum_{k=1}^{\ell} \overline{\gamma}_k^2(t_k) + \tilde{w}_2[\hat{X}(t_\ell), T - t_\ell] = w_2\{\overline{\gamma}, \phi(\gamma)\}. \end{aligned} \tag{12}$$

For all other type of strategies $\gamma$ and $\phi(\gamma)$ we shall have in (11) and (12) equalities. The inequalities in (11) and (12) prove that $(\overline{\gamma}, \overline{\phi}_M(\gamma))$ is a Nash equilibrium. □

The economic interpretation of Theorem 5.1 follows the same reasoning as for Proposition 1.

Now, if we take

$$\overline{\beta}_k = \frac{1}{2}\{\hat{v}(\hat{X}(t_{k-1}, T - t_{k-1}) - \sum_{i=1}^{2} \tilde{w}_i[\hat{X}(t_{k-1}), T - t_{k-1}]\},$$

then we can get the Nash equilibrium with Nash bargaining outcomes, which would be equal to

$$\sum_{k=1}^{n-1} \tilde{w}_i[\hat{X}(t_{k-1}), T - t_{k-1}] + \overline{\beta}_k, \qquad i = 1, 2.$$

But in this case it has to be investigated to see whether or not the conditions (10) of Theorem 5.1 are satisfied.

The time consistency of the supergame $G\left(\hat{X}(t_0), T - t_0\right)$ has to be investigated for each stage of the game in a way similar to the one used in the one stage supergame $G_1\left(\hat{X}(t_{n-1}), T - t_{n-1}\right)$. Consider the functions $\zeta_i(t)$, $i = 1, 2$, defined on $[t_0, T]$, such that

$$\int_t^T (\zeta_1(\tau) + \zeta_2(\tau))\, d\tau = \hat{v}\left(\hat{X}(t), T - t\right)$$

$$\int_{t_{k-1}}^{t_n} \zeta_i(\tau)\, d\tau = \overline{\gamma}_k^i(t_k), \qquad k = 1, \ldots, n-1, \quad i = 1, 2,$$

where $\overline{\gamma}_k^i(t_k)$ are computed using (7)–(8) for $\gamma = \overline{\gamma}$.

The functions $\zeta_i(t), t \in [t_0, T], i = 1, 2$, are called the payoff distribution procedures (PDP) in the supergame $G(X(t_0), T - t_0)$.

The following theorem is given without proof which would be very similar to the proof of Theorem 4.1:

**Theorem 5.2.** *The PDP $\zeta_i(t)$, $i = 1, 2$, $t \in [t_0, T]$, is time consistent if*

$$\zeta_i(t) \leq -\frac{d}{dt}\, \tilde{w}_i[\hat{X}(t), T - t], \qquad t \in [t_0, T], \quad i = 1, 2.$$

Time consistency means that no one player will find it better to switch to noncooperation at any intermediate time in the game $G(X(t_0), T - t_0)$. Again, it remains to be investigated under which conditions on data and parameters, the condition involved in this theorem is likely to hold.

## 6 Concluding Remarks

We have shown in this paper that it is possible to obtain a cooperative solution in the supergame which is self-enforced, i.e., achieved as a Nash equilibrium. To achieve this, we provided a time consistency condition that ensures that the players will not switch to noncooperative strategy at any intermediate date in any stage.

The impact of the two assumptions made in this paper need to be discussed. First, we have assumed that the noncooperative game is played under an open-loop information structure. It is obvious to say that the adoption of a different information structure, e.g., a feedback information structure, would lead to different outcomes. What we would like to stress here is that, although the numbers would be different and the computation much more complex, our approach would have remained the same. Indeed, the main idea of this paper is to design a mechanism to distribute the fruit of cooperation over time which has the desirable property of being time consistent. The sufficient condition ensuring this property (see Theorem 4.1) involves the speed at which noncooperative outcomes will evolve. The condition would not change if the noncooperative outcomes are those resulting from the implementation of feedback strategies.

Second, we required in the model that investment in abatement capacity be nonnegative, which means that investment is irreversible. One may wonder if the vulnerable player would not revert to his noncooperative strategy once the nonvulnerable player has built up a certain abatement capacity. Actually, under the time consistency condition this should not happen because it will come at the expense of a welfare loss. The same argument can be invoked the other way around. Indeed, if investment were reversible, one can wonder if the nonvulnerable player would not switch to his noncooperative strategy once he has collected enough dividend from cooperation. Again, the answer is that he will not because future earnings under cooperation are higher than those that would be secured under noncooperation.

It would be valuable to extend our model and approach in different directions. The case where both players are vulnerable and the one where they can return to a cooperative mode of play after a noncooperative stage are of interest. Discounting of future earnings also needs to be considered. It is safe to claim that if both players discount their welfare streams at the same rate, then the results can be extended easily. It is considerably harder to do it if discount rates are different. Further, the extension to $n$-person setting to treat global environmental problems, e.g., global warming, is obviously needed. Finally, it is clearly of empirical interest to investigate under which circumstances (e.g., functional forms, data, ...) the time consistency conditions of Theorems 4.1 and 5.2 are likely to hold.

## Appendix A

To obtain the optimal production and investment schedules, one has to solve the standard optimal control problem corresponding to the maximization of the joint total welfare, subject to (1) and (2). Introduce the Hamiltonian:

$$H = \sum_{i=1}^{2} \{U_i(q_i) - F_i(I_i) + \pi_{K_i}[I_i - \mu_i K_i] + \gamma_i I_i + \nu_i Y_i\}$$

$$- D_1(S) + \psi\left[\sum_{i=1}^{2} \alpha_i(K_i) q_i - \delta S\right], \tag{13}$$

where $\pi_{k_i}$ and $\psi$ are the costate variables associated, respectively, with the state variables $K_i, i = 1, 2$, and $S$ and $\gamma_i$ and $\nu_i$ represent the multipliers associated with the nonnegativity constraints.

The sufficient optimality conditions for this concave control problem lead to the following relations:

$$\dot{\psi} = \delta\psi + \frac{\partial D_1}{\partial S}, \tag{14}$$

$$\dot{\pi}_{K_i} = \mu_i \pi_{K_i} - \psi \frac{\partial \alpha_i(K_i)}{\partial K_i} q_i, \qquad i = 1, 2, \tag{15}$$

$$\frac{\partial U_i}{\partial q_i} = -\nu_i - \psi \alpha_i(K_i), \qquad i = 1, 2, \tag{16}$$

$$\frac{\partial F_i}{\partial I_i} = \pi_{K_i} + \gamma_i, \qquad i = 1, 2, \tag{17}$$

$$\gamma_i I_i = \nu_i Y_i = 0, \qquad i = 1, 2, \tag{18}$$

$$\pi_{K_i}(T) = \psi(T) = 0, \qquad i = 1, 2, \tag{19}$$

$$K_1(0) = K_1^0, \qquad K_2(0) = 0, \qquad S(0) = S^0, \tag{20}$$

From these relations one sees that:

(a) $\psi$ can be interpreted as the marginal loss in welfare arising from a unit increase in the concentration level of pollutants. Therefore, (14) states that the appreciation of $\psi$ plus its marginal productivity (the rent plus the improvement in the quality of the environment) equal the marginal damage. Equation (15) is interpreted in a similar manner.
(b) The marginal utility of production equals the marginal cost of emissions (16).
(c) The investment decision in clean technology is determined by the condition that the marginal investment cost equals the marginal value of $K_i$ (17).
(d) Equations (19)–(20) are the usual transversality and initial conditions.

## Acknowledgments

Research supported by NSERC-Canada and CETAI, HEC, Montréal. We are grateful to an anonymous referee for very helpful comments.

## REFERENCES

[1] Barrett, S. Self Enforcing International Environmental Agreements. *Oxford Economic Papers*, **46**, 878–894, 1994.

[2] Carraro, C. and D. Siniscalo. Environmental Innovation Policy and International Competition. *Environmental and Resource Economics*, **2**, 183–200, 1992.

[3] Carraro, C. and D. Siniscalco. Strategies for the International Protection of the Environment. *Journal of Public Economics*, **52**, 309–328, 1993.

[4] Chander, P. and H. Tulkens. Theoretical Foundations of Negociations and Cost Sharing in Transfrontier Pollution Problems. *European Economic Review*, **36**, 288–299, 1992.

[5] Dockner, E. J. and F. Wirl. $CO_2$ Emissions Taxes: Energy Producers (OPEC) Versus Consumers, a Differential Game. *Proceedings of the Fifth International Symposium on Dynamic Games and Applications, Grimentz, Switzerland*, July 15-17, 1992.

[6] Filar, J. A. and P. S. Gaertner. Global Allocation of Carbon Dioxide Emission Reductions; A Game Theoretic Perspective. *Proceedings of the Seventh International Symposium of Dynamic Games and Applications*, vol. 1, (edited by J.A. Filar, V. Gaitsgory and F. Imado), Japan, December 16-18, 1996.

[7] Haurie, A. and G. Zaccour. Differential Game Models of Global Environmental Management. *Annals of the International Society of Dynamic Games*, **2**, 3–24, 1995.

[8] Kaitala, V. and M. Pohjola. Sustainable International Agreements on Green House Warming: A Game Theory Study. *Annals of the International society of Dynamic Games*, **2**, 67–88, 1995.

[9] Kaitala, V., M. Pohjola, and D. Tahvonen. Transboundary Air Pollution and Soil Acidification: A Dynamic Analysis of an Acid Rain Game between Finland and the USSR. *Environmental and Resource Economics*, **2**, 161–181, 1992.

[10] Krawczyk, J. B. Management of Effluent Discharges: A Dynamic Game Model. *Annals of the International Society of Dynamic Games*, **2**, 337–356, 1995.

[11] Krawczyk, J. B. and G. Zaccour. Management of Pollution from Decentralised Agents by the Local Government. Les Cahiers du GERAD **G–97–47**, July 1997.

[12] Petrosjan, L. A. *Differential Games of Pursuit*. World Scientific, 1993.

[13] Petrosjan, L. A. and V. V. Zakharov. *Introduction to Mathematical Ecology* (in Russian). Leningrad University Press, Leningrad, 1986.

[14] Tolwinski, B. and W. E. Martin. International Negotiations on Carbon Dioxide Reductions: A Dynamic Game Model, Mimeo, 1992.

[15] Tulkens, M., K. G. Mäler, and V. Kaitala. The Acid Rain Game as a Resource Allocation Process. *Proceedings to the Fifth International Symposium on Dynamic Games and Applications*, Grimentz, Switzerland, July 15–17, 1992.

[16] van der Ploeg, F. and A. J. de Zeeuw. International Aspects of Pollution Control. *Environmental and Resource Economics*, **2**, 117–139, 1992.

[17] van der Ploeg, F. and A. J. de Zeeuw. *Investment in Clean Technology and Transboundary Pollution Control*. Fondazione Eni Enrico Mattei, Nota di Lavoro 27.93, 1993.

# Solution and Stability for a Simple Dynamic Bottleneck Model

André de Palma
Thema
University of Cergy-Pontoise
Cergy-Pontoise, France

**Abstract**

Previous authors have considered the basic bottleneck model with one origin, one destination, and one link and shown that an equilibrium exists for the time of usage. In this paper we briefly review these results and introduce several intuitive day-to-day adjustment processes. Using numerical examples, we show that these adjustment processes never converge toward the solution of the basic bottleneck model. However, convergence occurs when the amount of heterogeneity in the driver's behavior is large enough. We hope that these results will provide some useful insights to researchers developing large-scale dynamic network equilibrium models.

## 1 Introduction

Perhaps J. M. Keynes was the first economist who discussed externality and congestion. He considered a population of individuals attending a theater show. One of these individuals may decide to stand up to see the show better; by doing so she increases her comfort, but at the same time decreases the comfort of all persons located behind her. Because of externalities, the individual interest is therefore in conflict with the social interest. Without regulation, at equilibrium everybody could be standing up, with the same visibility as if they were seated. This equilibrium is clearly dominated by the social optimum solution that entails everybody being seated. This is a simple example of the prisoner dilemma.

We focus in this article on congestible facilities for which the quality of service deteriorates as the number of individuals using it increases. Generally, given a choice of available activities, each potential user must decide the *intensity* of usage, the *time* of use, as well as *how* to use the facility, and *which activities* to perform. The individual's transportation demand is derived from the complex set of individual choices that includes the scheduling issue as an essential dimension.

The scope of this paper is

(1) to present the key ingredients of the two major generations of transportation models used to describe commuting behavior; and
(2) to discuss the stability of the solutions using simple adjustment processes.

In the first generation, the static models, individuals decide only usage and route choice. These are the most popular tools among practitioners in transportation. However, the array of decisions that an individual is facing is far more complex. The second generation, the dynamic models, consider another margin of adjustment: the time of usage decision. Although all our exposition will be performed in the transportation context, it should be clear that several other applications are possible, such as telecommunication, airline industry, recreational areas, etc.

This paper is organized as follows. In Section 2, we review the static network equilibrium models. We start with a simple model with two routes in parallel and then consider the extension to a general network. We discuss the two types of approaches that have been used, one based on a standard mathematical formulation and the other based on the theory of variational inequalities. In Section 3, we discuss the basic one-origin, one-destination, and one-bottleneck dynamic model as originally introduced by Vickrey. We outline previous results that give the user equilibrium and the social optimum solutions, and discuss various toll policies. In Section 4, we describe the computer implementation of the within-the-day dynamic process. In Section 5, we introduce four day-to-day adjustment processes to compute an equilibrium which approximates the user equilibrium. We also identify key parameters which explain the stability of the convergence processes. Finally, in the concluding section, we propose a brief research agenda for future research.

## 2 Static Models

The standard economic models of urban traffic congestion are static and treat just two margins of the user's choices: how many trips to perform and which road to select. Static equilibrium models were introduced in economics at the beginning of this century and have been applied in urban planning for more than 20 years. The basic model with one origin, one destination, and two parallel routes was introduced by Knight [20]. At *equilibrium*, each driver minimizes its travel time (or travel cost); at the (first-best) *social optimum*, the total travel cost is minimized. With inelastic (constant) demand and with linear congestion functions, it is easy to show that equilibrium road usage it not socially optimum. In general, the average equilibrium travel cost is strictly larger than the average social optimum cost: this is because drivers (who ignore the congestion externalities they induce on the other drivers using the same road) tend to overuse the shortest free-flow travel time route which therefore becomes too congested. This implies that usually, the total travel cost could be reduced by imposing a toll on the shortest route. For one route and elastic demand, it can be shown that too many individuals use their car: this is so because individuals base their decision on their user cost and not on the total social cost (as in Keynes's example). By doing so, individuals ignore the externality they induce, that is, the impact of their usage decision on the cost of other drivers. These simple examples show that

(1) commuters tend to overuse the shortest free-flow travel time route; and that
(2) too many commuters use their cars.[1]

This basic static transportation model could be extended to a general network. Consider an origin–destination matrix and a general transportation network defined by links, nodes, centroids (origin or destination), and link-specific congestion functions. The congestion functions define the relation between the number of users on a link and the average travel time to cross this link. The equilibrium concept is due to Wardrop [29]: at equilibrium no user can change her route choice in order to strictly reduce her travel time. The mathematical formulation of the static network equilibrium problem is due to Beckmann et al. [9]. They show that this problem could be formulated as a standard minimization problem of a convex function, under flow conservation constraints and nonnegativity constraints. In particular, they provide conditions which guarantee existence and uniqueness of a Wardrop equilibrium.

This initial analysis relies on several restrictive assumptions:

1. demand is inelastic (the number of users from one origin to a destination is fixed);
2. the travel cost function on each link is a continuous and nondecreasing function of the flow on this link; this implies in particular that there is no interaction between the travel costs on the different links (in particular, there is no treatment of the intersections);
3. there is only one type of user; this implies in particular that there are no trucks, that the route choices are deterministic, etc.

Several extensions of this mathematical programming formulation have since been developed. In particular, elastic demand and multicommodity flows have been addressed. Further extensions using stochastic route choices (based on the logit and probit models) have also been developed (see, e.g., Sheffi [25]).

More recently, various researchers have proposed a variational inequality formulation of the static problem (see Nesterov and Nemirovsky [23] and the reference book of Nagurney [22]). This reformulation of the static network equilibrium approach is very powerful both from the analytical point of view (to show existence and uniqueness under mild conditions) and from the numerical point of view (to develop fast algorithms to compute an equilibrium). It is more general and allows the modeling of complex interactions between flows traveling on different links: the conditions which guarantee existence and uniqueness of an equilibrium (the monotonicity of the travel cost operators) are much milder than in the standard approach.

---

[1]The same bias occurs for a firm producing an output and some (undesirable) pollutants: if not taxed efficiently, it will tend not to use the socially optimum production technology and to produce too much.

Recently, de Palma and Nesterov [15] have extended the variational inequalities approach mentioned above. The travel cost operators considered are monotone but not necessarily bounded and continuous (as in the standard variational inequality approach in transportation). They introduce the concept of *normal equilibrium* which generalizes both concepts of Wardrop and user equilibrium previously used in the static network equilibrium literature. This allows the treatment of new applications: road pricing (it can be shown easily that optimal congestion tolls are not differentiable), signalized intersections, and unbounded travel–density relationships (which are currently used in standard traffic planning).

A major topic of this paper is: how could a practical transport system reach an equilibrium? De Palma and Nesterov [15] consider a simple static network with two routes in parallel and discontinuous travel cost functions. They introduce a simple day-to-day adjustment process of the logit type (see details in Section 5.4). Unfortunately, this process does not converge. However, the oscillations become periodic and the center of those oscillations is the normal equilibrium. To economize space, and because we believe that dynamic models are more important in transportation, we will present the discussion of the day-to-day adjustment processes for the case of dynamic models.

Moreover, de Palma and Nesterov provide an interpretation of the original Beckmann's formulation using the mathematical concept of *potential*.

Static models are very popular tools for practical applications. In particular, the Emme/2 program [17] is used in several hundred cities for medium- and long-range planning. Let consider the standard stages of trip planning (see Bonsall et al. [10]). Trip assignment is the essence of static transportation systems. Trip generation and trip distribution are treated by static models with the elastic demand formulation. Modal choice (e.g., automobile versus transit) is also treated by static models since it is a particular case of elastic demand. However, the trip timing, which is an essential component of travel choice is ignored in static models. As we will discuss in Section 3, about half of the travel cost is due to penalties corresponding to early or late arrivals. By construction, this component of trip cost is missing in the static model. The time of use decision could only be treated within a dynamic model, i.e., a model for which the congestion level, and therefore the travel time, depend on the time of day. In static models, usage is implicitly assumed to be uniformly distributed over a fixed time period. Dynamic models, which explicitly treat the time of use dimension, will be discussed in the rest of this paper.

## 3 The Simple Bottleneck Model

The simple bottleneck model is the first dynamic model introduced in transportation. It can describe many congestible facilities (such as roads, of course, but also airline and telephone, inter alia). It has been originally introduced by Vickrey [28] and extensively studied (see, e.g., Arnott et al. [6] and [7]). Most results mentioned in this section could be found in Arnott et al. [5] and we refer the interested reader to those papers for the detailed proofs of the statement provided in this section. In the bottleneck model, the arrival rates are *endogenous* contrarily to the Operation Research literature where they are usually considered to be fixed (see, however, the analysis of Glazer and Hassin [19]).

Consider a continuous dynamic congestion model for a simple network, which consists of a single origin–destination pair connected by a single link. Let $N$ be the number of commuters. We assume that a bottleneck is located at the beginning of the link and denote by $s > 0$ its capacity. We also assume that the queue has no physical length and that the free traffic travel time on the link is constant (in the following we will set this value equal to zero w.l.o.g.).

It can be proved easily, that all departures occur in a finite interval. Let us denote by $[0, T]$ a closed interval; the departure rate, $r(t)$, satisfies

$$r(t)_2 \geq 0, \qquad t \in [O, T_3], \qquad \int_0^T r_1(u)\, du = N. \tag{1}$$

The departure rate defines a dynamic queue length $Q(t)$, $t \in [0, T]$, in accordance with the following differential equation:

$$Q(0) = 0, \tag{2}$$

$$\frac{dQ(t)}{dt} = \begin{cases} r(t) - s, & \text{if } Q(t) > 0, \\ (r(t) - s)^+, & \text{otherwise,} \end{cases} \tag{3}$$

where $(a)^+ \equiv \max\{0, a\}$. In other words, if there is no queue and if the departure rate is less than the capacity, then the queue length remains zero; otherwise, the queue length changes at a rate equal to the departure rate minus the capacity.

Further, we define the dynamic travel time $tt(t)$ for a departure time $t \in [0, T]$ as follows:

$$tt(t) = \frac{Q(t)}{s}. \tag{4}$$

An individual's travel cost function depends on her travel time and also on her schedule delay—time early or time late in arriving at work. To simplify, we assume a linear travel cost function (see Small [26] for an empirical estimation of travel cost functions). All individuals are assumed to have the same official work starting time, $t^*$. We assume a linear-additive specification for the travel cost function. The travel cost, $c(t)$, for a driver leaving home at time $t \in [0, T]$, is

$$c(t) = \alpha\, tt(t) + \beta \left[t^* - tt(t) - t\right]^+ + \gamma \left[t + tt(t) - t^*\right]^+, \tag{5}$$

where $\alpha$ is the unit cost of time spent traveling, $\beta$ the unit cost of time early, and $\gamma$ the unit cost of time late. For example, if an individual who starts her journey at time $t$ arrives early, her schedule delay is $\beta\,(t^* - tt(t) - t)$ and if she arrives late, her schedule delay is $\gamma\,(t + tt(t) - t^*)$. We assume that $\gamma > \alpha > \beta$: see Small [26], for an empirical support of this hypothesis. If $\alpha < \beta$, the solution is not unique and involves a mass of drivers leaving at the beginning of the departure period (this solution is clearly not realistic and will not be discussed here; see [3] for details).

### 3.1 No Toll Equilibrium

A no toll (or user) equilibrium is achieved when no user can modify her departure time in order to strictly decrease her travel cost $c(t)$; this definition is the dynamic counterpart of the Wardrop equilibrium concept introduced in the previous section.[2] Simple algebra shows that the equilibrium is characterized by

$$r(t) = \begin{cases} \frac{\alpha}{\alpha-\beta}s, & \text{for } t \in (t_0, t_n), \\ \frac{\alpha}{\alpha+\gamma}s, & \text{for } t \in (t_n, t_e), \\ 0, & \text{otherwise,} \end{cases} \tag{6}$$

where $t_n$ is the departure time of an individual arriving at destination at $t^*$: $t_n + tt(t_n) = t^*$.[3] One can prove that $t_n = t^* - (\delta/\alpha)(N/s)$ where $\delta = \beta\gamma/(\beta+\gamma)$. The beginning of the period starts at $t_o$ and ends at $t_e$ given by $t_0 = t^* - (\delta/\beta)(N/s)$ and $t_e = t^* + (\delta/\gamma)(N/s)$. At equilibrium, the total travel costs $TC^e = Nc^e$ (where $c^e$ is the individual travel cost) are equal to

$$TC^e = \delta\frac{N^2}{s}. \tag{7}$$

Interestingly, it can be shown that half of the equilibrium travel costs corresponds to congestion costs and the other half corresponds to schedule delay costs. Note that this result is no more valid if the cost function is not linear, as in (5). The marginal costs are computed at equilibrium, that is, assuming that all users readjust their departure time decisions, once a new driver decides to join the transport system. From (7), the marginal cost of a traveler is $MSC^e = 2\delta(N/s) = 2c^e$; this shows that the congestion externality is equal to the individual equilibrium travel costs $c^e$. For large systems, with a few thousand links, numerical simulations suggest that this property remains qualitatively valid (see [13] and [8]).

The equilibrium solution can be explained intuitively as follows. The user equilibrium equalizes the travel costs for all commuters (since there is a continuum of commuters). The queue length is such that at equilibrium the sum of travel time costs and schedule delay costs is constant over time. This implies that the queue builds up at a constant rate from $t_o$ to $t_n$ and dissipates, again at a constant rate, from $t_n$ to $t_e$ (this corresponds to the morning or evening peak).

---

[2]We provide in this paper the standard approach used in transportation. Another interpretation is as follows. The drivers select their departure time choice independently according to a mixed strategy whose density is $r(t)/s$, where $r(t)$ is given by (6). Note that these two interpretations are equivalent here. However, they are not the same when a day-to-day adjustment is considered (see Section 5). We would like to thank a referee for providing us with this alternative interpretation.

[3]The commuter who leaves at $t_n$ has no schedule delay costs (since she arrives at her destination at $t^*$) but incurs maximum congestion costs.

The timing of the rush hour and the costs are determined as follows. First note that the first and last commuter face no congestion costs and the same (maximum) schedule delay costs. Then use the conservation law (which implies that the length of the period in which departure rates are positive is $N/s$, since the bottleneck is used at full capacity during the peak period) to get expressions for $t_o$ and $t_e$. The equilibrium costs are given by $\beta(t^* - t_o) = \gamma(t_e - t^*) = TC^e$ (see (7)). This line of reasoning does not depend on the linearity of the cost function. Interestingly the value of time, $\alpha$, does not play any role in the above reasoning (and therefore, $t_o$, $t_e$, and $TC^e$ are all independent of the value of time, $\alpha$).

Note also that in the dynamic models the range of departures $(t_o, t_e)$ is computed endogenously (while in static models, the length of the peak period is given).

### 3.2 Social Optimum and Pricing

Clearly congestion is not optimal. Therefore the first-best (or social) optimum entails a departure rate equal to $s$, since this policy totally eliminates congestion (which represents half of the user costs). It can also be shown that the first and last departures occur at the same time for the user equilibrium and for the social optimum. Hence, the first-best social optimum aggregate costs, $TC^o$, are equal to the total schedule delay costs at equilibrium. One has

$$TC^o = \frac{\delta}{2}\frac{N^2}{s} = \frac{TC^e}{2}. \tag{8}$$

An optimal toll which is increasing at rate $\beta$ for early arrivals and decreasing at rate $\gamma$ for late arrivals, can decentralize the social optimum. In this case the sum of schedule delay costs and of the toll is the same for each driver (recall congestion is null at the first-best optimum). Therefore, $TC^e/2$ corresponds to the social saving from an optimal toll.[4] A more realistic toll is a coarse (or step) toll, that is, a toll which is piecewise constant. The total travel costs for a coarse toll with $n$ steps are $TC^n = f(n)\delta(N^2/s)$, $n \geq 1$, where $f(n) < 1$ is a decreasing function of $n$ with $\lim_{n\to\infty} f(n) = \frac{1}{2}$ the total travel costs reduce to the first-best social optimum aggregate costs as the number of steps tends to infinity (for details, see Laih [21]).

We could synthesize and extend the previous results using the following structural form (see [5] for details):

$$TC^i = \Gamma^i \delta \frac{N^2}{s}, \tag{9}$$

[4]Recall that the travel costs, $TC^e$, do not depend on the value of time. Therefore a usage dependent toll such that drivers face a toll proportional to the amount of congestion they incur would be able to decentralize the social optimum, if it is large enough. In this case, the drivers would pay a toll but their individual travel costs (sum of travel time, schedule delay, and toll) would remain the same.

where $i$ denotes the toll regime ($i = e, o$, or $n$). For no toll, we have $\Gamma^e = 1$; for the optimal toll, we have $\Gamma^o = \frac{1}{2}$, and for a optimal coarse toll we have $\Gamma^n \in (\frac{1}{2}, 1)$, $n = 1, 2, \ldots$. This shows that the reduced form (or structural form) of the equilibrium regime has a closed-form solution which only depends on the parameter $(N/s)$ usage over capacity ratio and on the behavioral parameters $\beta$ and $\gamma$. Interestingly, the equilibrium cost is *independent* of the value of time $\alpha$. This is because, as the value of time increases, on the one hand, the minute spent driving is more costly but, on the other hand, the peak is more uniform (so that the amount of congestion decreases).[5] For a deterministic bottleneck model but with linear or nonlinear cost functions, these two opposing forces exactly cancel out.

Using METROPOLIS, we performed simulations for large-scale systems, which suggest that similar expressions approximately hold. For example, in [14], it is shown that travel costs (net of free-flow travel time costs) are approximately independent of $\alpha$. Moreover, in [8], simulation results suggest that the individual travel costs function is still a *linear* function of usage in large-scale networks. This last finding has important consequences for practical applications.

### 3.3 Optimum Capacity

The social optimum capacity can be easily computed for each toll regime. The planner wishes to minimize the sum of total travel costs plus construction costs (toll revenues are ignored since they are pure transfers). Assume a unit construction cost constant, and equal to $\chi$. With no toll, the welfare function is $W = TC^i + \chi s$ ($i = e, o$, or $n$). The optimum capacity $\widehat{s^i}$, for toll regime $i$, is

$$\widehat{s^i} = N\sqrt{\frac{\delta}{\Gamma^i\,\chi}}, \qquad i = e,\, o, \text{ or } n. \tag{10}$$

Therefore, the coarser the toll, the smaller the optimum capacity. That is, the optimal capacity for the social optimum toll (fine toll) is smaller than for the step toll, which is itself smaller than for the no toll equilibrium. Note also that the *self-financing* property holds: assume a constant unit construction cost; then, if an optimal toll is implemented, the toll revenue is equal to the construction costs.

### 3.4 A Simple Numerical Example

To fix ideas about the order of magnitude, consider a single bottleneck model with 420,000 commuters. Following Small [26], we assume that $\beta = \$5/\text{hr}$ and $\gamma = \$20/\text{hr}$. Typically, in Paris, the rush hour is 6 hours long so that $N/s = 6$. The total equilibrium travel costs are then $TC^e = \$10.08 \times 10^6$. With 200 commuting days per year and an annual discount rate of 3%, the present value of the social saving from an optimal congestion toll is $\$67.2 \times 10^9$, which amounts to $\$1.6 \times 10^5$

[5]That is, as $\alpha$ increases, early and late departure rates converge toward $s$ (see (6)).

per commuter. Assume that the average length of the highways is 1 kilometer with a capacity of 2,000 veh/hr. The construction costs are $\$1.8 \times 10^9$ per lane and per kilometer (according to the French Ministry of Transportation). This corresponds to $\$1.5 \times 10^5$ per commuter. The fact that these two figures ($\$1.6 \times 10^5$ per commuter and $\$1.5 \times 10^5$ per commuter) are similar, is not coincidental. According to the self-financing property, when the capacities are computed optimally, the toll revenue exactly covers the construction costs when unit construction costs are constant.

## 4 Within-the-Day Dynamic Process

We now discuss the computer implementation of the basic dynamic model. We restrict our analysis to the departure time choices (ignoring the route choices). In this section, we describe the *dynamic process* during 1 day (we restrict this analysis to the morning commute). In the next section, we describe the day-to-day *adjustment process*.

The computer implementation of the dynamic process requires a partial discretization scheme. The term *partial* means that we restrict ourselves to the piecewise linear representation of the departure rate $r(t)$, but keep continuous time and queue length. We performed several numerical experiments to check that the results we present do not depend on the accuracy of the discretization scheme used in the simulations.

### 4.1 Discretization Scheme

We consider the piecewise constant departure rate. Let us choose a large enough integer number $N$ and let $t_i = ih$, $i = 0, \ldots, N$, with $h = T/N$. Let us choose a sequence of nonnegative real numbers $\{r_i\}_{i=1}^N$ such that $\sum_{i=1}^N r_i = 1$. Then the piecewise constant departure rate $r(t)$ is defined as $r(t) = r_i$ if $t \in [t_i, t_{i+1}]$.

Using this discretization strategy we can form the dynamic queue, travel time, and costs in accordance with (3), (4), and (5), respectively. This completely characterizes the simulation of the (within-the-day) dynamic process.

### 4.2 Statistics

At the end of each simulation day, we can define different statistics which will be used for two different tasks. First, those statistics are used to compute a convergence criteria; second, they are used to determine the initial values that will be used for the next simulation day. In the adjustment processes we will use the average costs for each time period. Namely, the average cost $c_i$ for time period $i$ is defined as

$$c_i \equiv \frac{1}{h} \int_{t_{i-1}}^{t_i} c(u)\, du, \qquad i = 1, \ldots, N. \tag{11}$$

For a given departure rate sequence, $\{r_i\}$, let us define the minimum cost $c^*$ as

$$c^* \equiv \min_{1 \leq i \leq N} c_{i,} \tag{12}$$

and the average cost $\overline{c}$ as

$$\overline{c} \equiv \int_0^T c(u) r(u)\, du. \tag{13}$$

It is easy to see that for a departure rate which is close to a user equilibrium, the value $c^*$ is close to $\overline{c}$ (in pure continuous models the equilibrium departure rates provide us with $c^* = \overline{c}$). Therefore, we can use the ratio

$$\rho \equiv \frac{\overline{c}}{c^*} \geq 1, \tag{14}$$

as a measure of the distance between the current departure rate and the user equilibrium departure rate (the equilibrium departure rate corresponds to $\rho = 1$).

## 5 Day-to-Day Adjustment Process

### 5.1 General Comments

The user equilibrium is well defined from a mathematical point of view, but for practical applications, this definition is incomplete for two (related) reasons. First, nothing is said about the stability of the process. The question is in this case: what is the consequence of the fact that an individual may make a slight mistake and select an erroneous departure time (see the ongoing discussion in evolutionary games in the excellent book by Samuelson [24])? The second question is related to the way drivers could compute a dynamic user equilibrium.[6] One way would be to develop a rigorous algorithm. However, it is unlikely that such an algorithmic approach would be useful for general and realistic networks and that it would be behaviorally sound.[7] We adopt here a second approach, and analyze several behavioral day-to-day adjustment processes.

At the beginning of each day, each driver has some information about the past performances of the transportation system. Based on this experience, she can use an intuitively sound forecasting process to determine the travel times she is likely to experience that day. Accordingly, each driver selects a departure time using some simple laws (departure rates) defined below. The departure rates are such that a driver modifies her departure times in order to decrease her expected travel

[6]See the recent work in Economics by Anderson et al. [2], where agents' dynamic decisions processes are subject to errors, and the concept of logit equilibrium.

[7]Note, however, that various researchers (see, e.g., [18] and [30]) have developed algorithms, based on the theory of variational inequalities, for the dynamic assignment problem (these authors solve for the dynamic route choice problem, in a general network, assuming that the departure rates are exogenously given).

costs. Of course, since the forecasting methods used are not perfect, there is no reason for all drivers to select the departure time corresponding to the minimum expected travel cost.

We restrict ourselves to a very particular set of hypotheses. Our main motivation is to keep a simple model, that is, a model that involves small cognitive processes as well as small memory requirements. This is necessary since casual observation strongly suggests that most tasks performed by drivers are simple, in practice. Our analysis is based on the following hypotheses:

1. Drivers use open-loop strategies, that is, they make at the beginning of the day their departure time decision: this implies that any new information acquired during the day does not modify the drivers' forecasts.
2. Information is common knowledge: this means that all drivers are assumed to have access to the same (off-line) information set.
3. The learning process is Markovian of order 1: that is, only the information collected on the previous day is used to perform travel time forecasts for the subsequent day.

In this section we introduce four adjustment strategies $A$, $B$, $C$, and $D$ (day-to-day adjustment processes) which will, under some conditions, converge toward the equilibrium solution. (We refer the reader to Ben-Akiva et al. [11] for preliminary work on day-to-day adjustment processes for dynamic network models.)

## 5.2 Modeling Approach

Adjustment strategies are defined by the rules of transformation of the departure rates $\{r_i^\omega\}$ of day $\omega$ into the departure rate$\{r_i^{\omega+1}\}$ of the next day $\omega + 1$. More generally, the set of departure rates on the previous day, $\{r_1^\omega, \ldots, r_i^\omega, r_N^\omega\}$, is used to compute the departure rates of the next day, $\omega + 1$: $\{r_i^{\omega+1}\}_{i=0}^N$ . An adjustment strategy is called *convergent* if $\rho_\omega \to 1$ when $\omega \to \infty$, where $\rho_\omega$ is the value of the convergence criteria (14) when the minimum cost and average cost are computed on day $\omega$.

At first sight it seems that the problem is very simple and the solution is obvious: if the cost $c_i^\omega$ of the $i$th interval on day $\omega$ is large enough (e.g., as compared with $\overline{c}_\omega$), the departure rate $r_i^{\omega+1}$ should be taken less than the departure rate of the previous day. Therefore, the following four adjustment strategies seem quite reasonable:

## 5.3 Strategy A (Additive Strategies)

Let $\phi(u)$, $u \geq 0$, be an increasing positive function. Let

$$\overline{c}_\phi^\omega = \frac{1}{N} \sum_{i=1}^{N} \phi(c_i^\omega - (c^*)^\omega), \tag{15}$$

where $(c^*)^\omega$ is the minimum cost on day $\omega$. Then, strategy A is defined by

$$\overline{r}_i^{\omega+1} = \left[r_i^\omega + R(\overline{c}_\phi^\omega - \phi(c_i^\omega - (c^*)^\omega))\right]^+, \tag{16}$$

where $R > 0$ is a step size. The departure rates are normalized as follows:

$$r_i^{\omega+1} = \frac{\overline{r}_i^{\omega+1}}{\sum_{j=1}^N \overline{r}_j^{\omega+1}}. \tag{17}$$

The idea of this strategy is as follows: if the cost of the $i$th departure interval is greater than an average cost $\overline{c}_\phi^\omega$ (computed by a weight function $\phi$), then the corresponding users are trying to change their departure time. If the cost is less than the average cost, then this departure time is attractive. The parameter $R$ here can be interpreted as a reviewing rate.

### 5.4 Strategy B (Multiplicative Strategy)

Let $\phi\ (u)$ be an increasing positive function. Let

$$\overline{c}_\phi^\omega = \sum_{i=1}^N \phi(-c_i^\omega). \tag{18}$$

Then, strategy B is defined by

$$r_i^{\omega+1} = (1-R)r_i^\omega + R\frac{\phi(-c_i^\omega)}{\overline{c}_\phi^\omega}, \tag{19}$$

where $R \in (0, 1)$ is a step size. The specification $\phi(u) = \exp\{u/\mu\}$ corresponds to the logit discrete-choice model for the fractions ($\exp\left\{-c_i^\omega/\mu\right\}/\overline{c}_\phi^\omega$), where $\mu > 0$ is an heterogeneity parameter. The larger the value of the parameter $\mu$, the more different will be the behavior of two drivers facing identical traffic conditions (that is, identical forecasts).

This strategy has a probabilistic interpretation: for each user the probability to choose a departure-time interval with small cost is greater than the probability to choose an interval with a higher cost. The parameter $R$ can be interpreted again as a reviewing rate. In this case, the departure rates can be interpreted as definitions of mixed strategies. As the number of drivers becomes large enough, we conjecture that both approaches provide the same results.

Note finally that the stationary state of the process (19) is given implicitly by

$$r_i^{ss} = \frac{\phi(-c_i^{ss})}{\overline{c}_\phi^{ss}}. \tag{20}$$

If this equation has a solution we referred to it as a *stochastic user equilibrium* (where the parameter $\mu$ measures the intensity of stochasticity). In the limit $\mu \to \infty$, departure-time choices are totally random. In the limit $\mu \to 0$, the stochastic-user equilibrium, given by (20), converges toward the deterministic-user equilibrium, described in Section 3 (see also Anderson et al. [1] and Ben-Akiva et al. [11]). That is, the implicit solution of (20) converges, for $\mu$ small enough, toward the deterministic-user equilibrium. Note, however, that the stochastic- (and the deterministic-) user equilibrium could never be reached if the adjustment processes are unstable.

### 5.5 Strategy C (Adaptive Strategy)

Strategy C is defined as

$$\bar{r}_i^{\omega+1} = \begin{cases} (1-\theta)r_i^{\omega} & \text{if } c_{i+1}^{\omega} > c_{i-1}^{\omega}, \\ (1+\theta)r_i^{\omega} & \text{if } c_{i+1}^{\omega} < c_{i-1}^{\omega}, \end{cases} \tag{21}$$

with

$$r_i^{\omega+1} = \frac{\bar{r}_i^{\omega+1}}{\sum_{j=1}^{N} \bar{r}_j^{\omega+1}}. \tag{22}$$

In this strategy $\theta \in (0, 1)$ is a parameter. A departure interval with positive $c'(t)$ corresponds to a higher departure rate and an interval with negative derivative corresponds to a lower departure rate.

### 5.6 Strategy D (Combined Adaptive Strategy)

Finally, strategy D is defined by

$$\bar{r}_i^{\omega+1} = \begin{cases} (1-\theta)r_i^{\omega} & \text{if } c_{i+1}^{\omega} > c_{i-1}^{\omega}, \\ (1+\theta)r_i^{\omega} & \text{if } c_{i+1}^{\omega} < c_{i-1}^{\omega}, \end{cases} \tag{23}$$

and

$$\text{if} \quad c_i^{\omega} < \bar{c}^{\omega} \quad \text{then} \quad \bar{r}_i^{\omega+1} = \bar{r}_i^{\omega+1} + R.$$

The departure rates are normalized as follows:

$$r_i^{\omega+1} = \frac{\bar{r}_i^{\omega+1}}{\sum_{j=1}^{N} \bar{r}_j^{\omega+1}}. \tag{24}$$

In this strategy $\theta \in (0, 1)$ and $R > 0$ are two adjustment parameters. This strategy is a modification of the strategy C. This modification allows the support of the departure rate distribution to change more quickly.

### 5.7 Numerical Results

In our experiments we choose a uniform distribution for $r(t)$ as a starting one (i.e., on the first simulated day). Surprisingly, the behavior of almost all adjustment strategies, described in the above, is rather bad. The strategies A and B usually demonstrate a stable two-level oscillation (we use the same exponential specification for $\phi$, for strategies A and B). The lower level of the oscillation corresponds to the smallest value of $\rho$ (1.07–1.10) and to the highest values of $c^*$ and $\overline{c}$. The highest level of $\rho$ usually corresponds to 1.50–1.60. However, if the heterogeneity parameter $\mu$ (of the logit model) in strategy B is large enough, we observe a convergence toward a stochastic-user equilibrium.

As far as process B is considered as a process for finding a stochastic-user equilibrium, we come to the following conclusion. The results of the simulation suggest that:[8]

1. The process exhibits oscillatory behavior if the heterogeneity parameter $\mu$ is small enough or if the reviewing rate $R$ is large enough. The period of the oscillations increases when $R$ decreases.
2. When $\mu$ is not too small ($\mu > 0.015$), it is possible to stabilize the process by decreasing the value of $R$ (which corresponds to the time step, i.e., the day, of the discretization procedure). However, when the value of $\mu$ is too small (i.e., $\mu < 0.01$), the oscillatory behavior persists for small values of the reviewing rate ($R = 0.01$), although the amplitude of the oscillation decreases.
3. When $\mu$ is large enough ($\mu > 0.025$), the process converges toward a stationary solution—the stochastic-user equilibrium—which is not the deterministic-user equilibrium.
4. The dynamic pattern of the evolution of costs from day to day appears to be independent of the initial condition. We use either the uniform distribution or the deterministic-user equilibrium as our starting point.

In Figures 1, 2, and 3, we provide examples of simulation results for strategy B: the average costs are displayed as a function of the iteration day. The initial

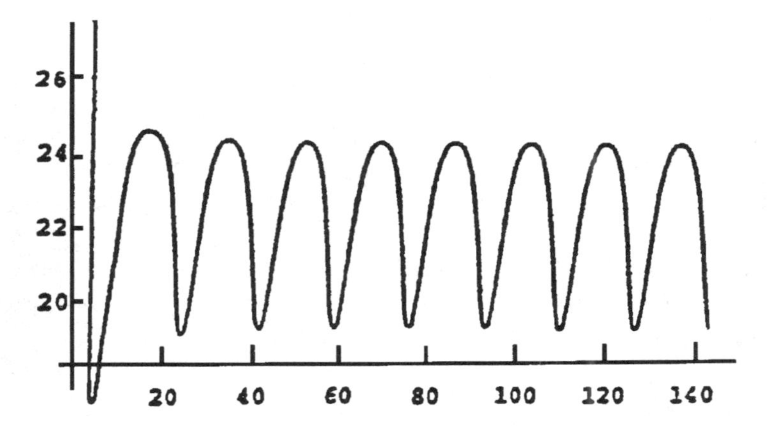


**Figure 1:** $\mu = 0.015$, $R = 0.2$. Equilibrium initial departure rate.

[8]The main criteria which was observed is the average cost. Other stopping criteria are being investigated elswhere for various types of dynamic models with day-to-day adjustment processes (see [8]).

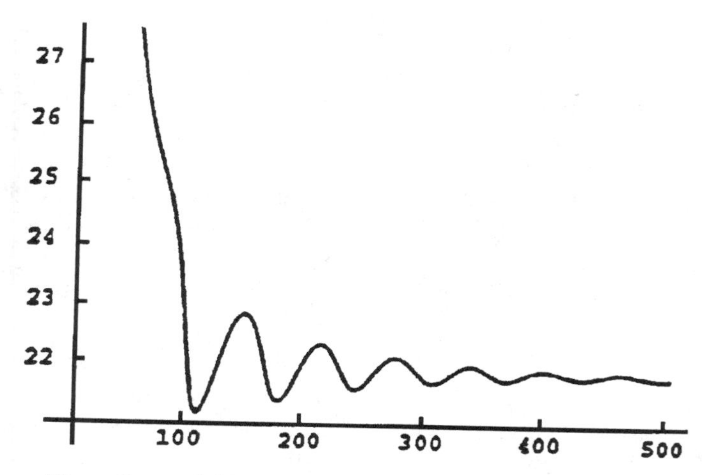


**Figure 2:** $\mu = 0.015$, $R = 0.04$. Uniform initial departure rate.

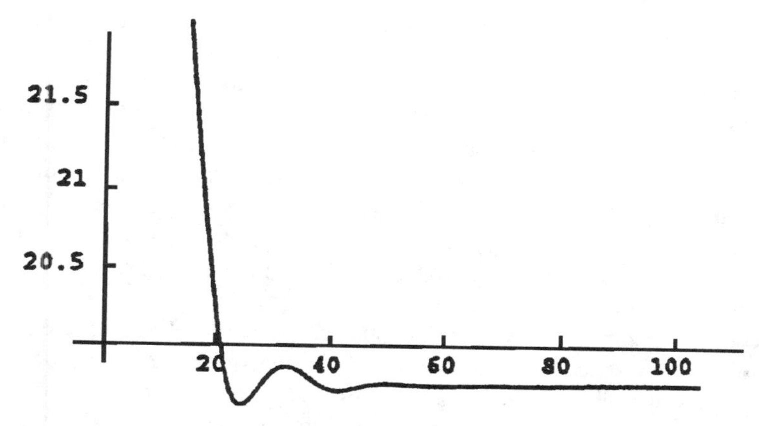


**Figure 3:** $\mu = 0.025$, $R = 0.2$. Uniform initial departure rate.

departure rate distribution is uniform. For $\mu = 0.015$ and $R = 0.2$, the system displays stable oscillations (see Figure 1). These oscillations could be damped, either by decreasing the reviewing rate $R$ (see Figure 2, for $\mu = 0.015$ and $R = 0.04$), or by increasing the heterogeneity parameter $\mu$ (see Figure 3, for $\mu = 0.025$ and $R = 0.2$). Note that when the value of $\mu$ is too small (equal to zero in particular, as for the deterministic process described in Section 3) the adjustment process does not converge. That is, a user equilibrium exists but is not stable.

Strategy C can work properly if it is used with the right support for the equilibrium departure rate. In this situation it can reach $\rho = 1.005$, but after this achievement the value of $\rho$ increases and starts to oscillate around the level 1.02–1.03. The record value of $\rho$ is usually reached after 40–60 days.

Strategy D seems more reliable. It can work with the wrong starting support for the departure rate. It can reach the same level of $\rho$, as strategy C: 1.006–1.008. However, the rate of convergence of this strategy is much slower: 200–300 iterations to reach the record value. After that, one can see the same picture: oscillation around the value 1.02–1.03.

Thus, our results suggest that any adjustment process is stable and converges toward the deterministic-user equilibrium only if the value of $\mu$ is large enough (see the discussion below).

## 5.8 Discussion

The reader may be perplexed at this point and wonder why, for continuous travel cost functions, convergence is guaranteed in a static model if $R$ is small enough (see [27]), while the results derived for the bottleneck model show that even when the reviewing rates $R$ or $\theta$ are arbitrary small, convergence may not occur. To understand this a priori paradoxical result we should question ourselves on the essence of congestion. Congestion occurs in a system when the individual costs are smaller than the social costs. (As a consequence, the user equilibrium and the social optimum solution differ.) *Congestion externality explains why several intuitive adjustment processes do not converge.* This is because, when an individual moves from one state to another (i.e., selects a new departure time), she attempts to reduce her own disequilibrium value, that is, she tends to decrease her own cost (simple adjustment process). By doing so, she induces a supplementary social cost to the other drivers which, in aggregate, could be larger in amplitude than her own cost reduction. That is why there is absolutely no guarantee that she will reduce the aggregate disequilibrium value (measured here by $\rho - 1$), given she does not take into account the impact of her move on other driver costs. Moreover, the system may well reach a stationary solution that is different from the user equilibrium as, in a similar way, the user equilibrium is different from the social optimum. We refer to this stationary solution as a *dynamic user equilibrium*. This is likely to be the relevant equilibrium concept (rather than the standard Wardrop assumption) to be taken into consideration.

The nature of the adjustment processes also implies that they are not convergent in general. However, strategy B converges when $\mu$ is large enough. This impact of $\mu$ is similar to that observed in some game-theoretic models. If $\mu$ is small, individual behavior is almost deterministic (the choice probabilities $\phi(-c_i^{\omega})/\overline{c}_{\phi}^{\omega}$ are equal to 1 or 0). Larger values of $\mu$ tend to smooth choice probabilities and as a result the equilibrium process converges.[9]

[9]This is reminiscent of de Palma et al. [12], where it is shown that a Nash equilibrium price is restored when $\mu$ is large enough. In the case $\mu > 0$, the reaction functions become continuous and incentive to undercut the competitor's price—which is the source of lack of equilibrium—is reduced.

Similarly, here, when $\mu$ is increasing, it follows that user decisions are more and more driven by idiosyncratic criteria (other than by travel times and schedule delays). This is likely to decrease the impact of congestion. The situation is very clear as $\mu \to \infty$. In this case, individuals ignore travel times and schedule delays and therefore the congestion is: (a) very small in the system and (b) in any way it is ignored by the individual (this is, in some sense, a paradoxical result since when drivers are less affected by congestion, congestion tends to disappear).

Note finally that the origin–destination case is the most difficult case. We have observed that the minimum value of $\mu$ required for convergence is decreasing as the size of the system is increasing (see de Palma et al. [13]). For small values of $\mu$, as discussed in Section 5.4, a stationary equilibrium exists; however, the analysis of the day-to-day adjustment processes suggests that it is meaningless from a practical point of view since it is *not stable* (for several other adjustment processes, not reported here, we have reached the same conclusion).

A stability analysis is not possible at this stage. As we have seen in the static analysis, oscillatory behavior does occur and could be interpreted only by introducing a new equilibrium concept (the normal equilibrium, in the case of the static model). The analogue concept, the dynamic-user equilibrium, has still to be defined rigorously in dynamic models. Here we propose another way to circumvent nonconvergence, which appears to be generic in transportation models with congestion effects. We suggest that, adding enough heterogeneity (either objective with $\alpha$, $\beta$, $\gamma$, or subjective with $\mu$) at the same times gives rise to a more realistic model and restores convergence.

## 6 Research Agenda for Dynamic Network Models

We have presented the basic congestion model used to describe commuters' behavior. The static models are still very much used for practical applications but they are limited since they fail to address one essential dimension of the commuters' behavior: the time of usage decision. On the other hand, the dynamic models are more difficult, but provide a much more accurate picture of traffic patterns.

We briefly describe below some major extensions of the dynamic transport models (see also Arnott et al. [6] and [7]).

*Elastic demand.* We have assumed so far that the total number of users is constant. The model could be easily extended to take into account situations where the number of commuters depends on the user costs (this could be used, for example, to model mode choice). For the isoelastic demand function, where $\epsilon$ is the elasticity of the demand, it can be shown that when $\varepsilon < 1$ ($\varepsilon > 1$), the optimal capacity is larger (smaller) with no toll other than with any optimal toll. More specifically, for $\varepsilon < 1$, the coarser the toll, the lower the optimal capacity; the reverse is true when $\varepsilon > 1$ (see Arnott et al. [5]).

*Simple networks.* It is possible (but difficult) to compute the equilibrium for simple networks with several routes in parallel, or two origins and one destination, for example. Unfortunately, the closed-form solutions are rather intricate so that we believe that this type of computation could not be easily (or at all) extended for slightly more complicated networks. Interestingly, it can be shown that

(a) counterintuitive results, such as the Braess paradox may occur (i.e., it is possible that the total user costs increase when the capacity of a road is increased); and
(b) that the social optimal solution entails congestion (which is not true for the single bottleneck models).

*Traffic lights*. It is possible to treat the case of two or several intersecting roads which are managed by traffic lights and to compute the optimal green time of these traffic lights (as well as the optimal capacity given this optimal green time). This computation allows several regimes to be compared: no traffic lights, one traffic light, and one bridge (or tunnel), which avoids the conflict between the flows of cars going in different directions.

*Heterogeneous drivers*. Distributive and antidistributive issues, related to congestion pricing, can be examined when the commuters differ in their unit travel costs (i.e., when they differ in their values of $\alpha$, $\beta$, and $\gamma$). It is also worth noting that when drivers are more heterogeneous, the minimum value of the parameter $\mu$, which guarantees convergence, is smaller.

*Stochastic demand and capacity*. When demand and/or capacity are fluctuating during the time of day or from day to day, it is not correct to use average values. Stochastic models have been formulated in order to study the impact of variable weather conditions, special events, strikes, etc. For this type of study, it is possible to examine the impact of the provision of information to drivers. It can be shown that under some conditions, the travel costs may increase when drivers are provided with unbiased information. This type of result has important insights for the design of driver information systems such as ATIS (Advanced Drivers Information Systems, see, e.g., Emmerink [16] ).

*Alternative functional forms*. Alternative models using different congestion functions (with flow rather than queue congestion) can be derived. Similarly, other types of demand functions could be used. However, there is still little empirical evidence to justify these types of extensions.

*General network*. The extension of the basic model to general networks has been studied by de Palma et al. [13]. They consider a closed-loop equilibrium game to describe the stochastic-user's route choice behavior (this is indeed the most expensive part of the commuter's choice). The departure time choice is modeled using a logit model (see strategy B in Section 5.4). Note that the adjustment processes converge under much milder conditions (on the values of the parameters) as the size of the network increases (i.e., the one origin–destination model described in Section 5 is the most difficult case for the convergence processes). Simulation experiments were performed for a large network, the Geneva network.[10] This model still needs to be estimated. However, using rough calibration methods of the parameters, the results are comparable with the empirical data available for the city of Geneva. A mathematical formulation of this model is still under progress as well as existence and uniqueness proofs of an equilibrium (using the variational inequalities approach). As mentioned above, using the same type of day-to-day adjustment processes we find that convergence is easier for larger networks (the value of $\mu$ required for convergence is smaller).

Much research has still to be performed both from the analytical point of view and from the algorithmic point of view for dynamic models. We hope that we have convinced the reader that the several issues raised are worthwhile to pursue both from the mathematical and from the practical point of view.

## Acknowledgments

Part of this work was developed with Moshe Ben-Akiva, Richard Arnott, and Robin Lindsey. The discussion on adjustment processes is based on joined work with Yurii Nesterov. I also benefited from several discussions with Jean-Louis Deneubourg, Alain Haurie, and Larry Samuelson. The very detailed comments of an Associate Editor and two referees are gratefully acknowledged. Finally we would like to thank PREDII, Ministere des Transports, France, for financial support.

## REFERENCES

[1] Anderson, S., A. de Palma, and J.-F. Thisse. *Discrete Choice Theory of Product Differentiation.* MIT Press, Cambridge, MA, 1992.

[2] Anderson, S., J. Goeree, and C. Holt. Stochastic Game Theory: Adjustment to Equilibrium Under Bounded Rationality. Mimeo, Department of Economics, University of Virginia, 1997.

[3] Arnott, R., A. de Palma, and R. Lindsey. Economics of a Bottleneck. Queen's University, Institute of Economic Research, Discussion paper 636, 1985.

[4] Arnott, R., A. de Palma, and R. Lindsey. Economics of a Bottleneck. *Journal of Urban Economics*, **27**, 111–130, 1993.

[5] Arnott, R., A. de Palma, and R. Lindsey. A Structural Model of Peak-Period Congestion: A Traffic Bottleneck with Elastic Demand. *American Economic Review*, **83**, 161–179, 1993.

[6] Arnott, R., A. de Palma, and R. Lindsey. Recent Developments in the Bottleneck Model. In: *Road Pricing, Traffic Congestion and the Environment* (K. Button and E. Verhoef, eds.). Elgar's Economics. 79–110, 1998.

[10] 1072 nodes, 3179 links, 81,000 users, 160 cpu, 20 Mbytes, for a pentium 100.

[7] Arnott, R., A. de Palma, and R. Lindsey. *Congestion: A Dynamic Approach*. MIT Press, Cambridge, MA, in press.

[8] Arnott, R., A. de Palma, and F. Marchal. From W. Vickrey to Large-Scale Dynamic Models. Mimeo, ITEP, Ecole Polytechnique Fédérale de Lausanne, Switzerland, 1998.

[9] Beckmann, M., C. McGuire, and C. Winsten. *Studies in Economics of Transportation*. Yale University Press, New Haven, 1956.

[10] Bonsall, P., M. Taylor, and W. Young. *Understanding Traffic systems: Data, Analysis and Presentation*. Cambridge University Press, Cambridge, 1996.

[11] Ben-Akiva, M., A. de Palma, and P. Kanaroglou. Dynamic Model of Peak Period Traffic Congestion with Elastic Arrival Rates. *Transportation Science*, **20** (2), 164–181, 1986.

[12] de Palma, A., V. Ginsburgh, Y.Y. Papageorgiou, and J.F. Thisse. The Principle of Minimum Differentiation Holds under Sufficient Heterogeneity. *Econometrica*, **53**, 767–781, 1985.

[13] de Palma, A., F. Marchal, and Yu. Nesterov. METROPOLIS: a Modular System for Dynamic Traffic Simulation. *Transportation Research Record*, **1607**, 178–184, 1997.

[14] de Palma, A. and F. Marchal. Evaluation of Activity Schedule Policies with the Use of Innovative Dynamic Traffic Models. In: *Traffic and Transportation Studies*, Proceedings of ICTTS'98, American Society of Civil Engineers, pp. 791–801, 1998.

[15] de Palma A. and Yu. Nesterov. Optimization Formulations and Static Equilibrium in Congested Transportation Networks. Mimeo, CORE, Université Catholique de Louvain, Belgium, 1998.

[16] Emmerink, R. *Information and Pricing in Road Transportation*. Advance in Spatial Science. Springer-Verlag, Berlin, 1998.

[17] *EMME/2: User's Manual. Software Release 7*. INRO, Montréal, 1994.

[18] Friesz, T., D. Bernstein, D. Smith, T. Tobin, and B. Wei. A Variational Inequality Formulation of the Dynamic Network User Equilibrium Problem. *Operations Research*, **41**, 179–191, 1993.

[19] Glazer, A. and R. Hassin. $?/M/1$: On the Equilibrium Distribution of Customers Arrivals. *European Journal of Operational Research*, **13**, 146–150, 1983.

[20] Knight, F. Some Fallacies in the Interpretation of Social Costs. *Quarterly Journal of Economics*, **38**, 582–606, 1924.

[21] Laih, C. H. Queuing at a Bottleneck with Single and Multiple Step Tolls. *Transportation Research*, **28 A**, 197-208, 1994.

[22] Nagurney, A. *Network Economics: A Variational Inequality Approach*. Kluwer Academic, Dordrecht, 1993.

[23] Nesterov, Yu. and A. Nemirovsky. *Interior Point Algorithms in Nolinear Optimization*. SIAM, Philadelphia, 1994.

[24] Samuelson, L. *Evolutionary Games and Equilibrium Selection.* MIT Press, Cambridge, MA, 1997.

[25] Sheffi, Y. *Urban Transportation Networks: Equilibrium Analysis with Mathematical Programming Methods.* Prentice Hall, Engelwood Cliffs, NJ, 1985.

[26] Small, K. A. The Scheduling of Consumer Activities: Work Trips. *American Economic Review*, **72**, 467–479, 1982.

[27] Smith, M. J. Existence and Calculation of Traffic Equilibria. *Transportation Research*, **17B**, 291–303, 1983.

[28] Vickrey, W. S. Congestion Theory and Transport Investment. *American Economic Review (Papers and Proceedings)*, **59**, 414–431, 1969.

[29] Wardrop, J. G. Some Theoretical Aspects of Road Traffic Research. *Proceedings of the Institution of Civil Engineers*, Part II (1), pp. 325–378, 1952.

[30] Wu, J. H., M. Florian, Y. W. Xu, and J. M. Rubio-Ardanaz. A Projection Algorithm for the Dynamic Network Equilibrium Problem. In: *Traffic and Transportation Studies.* Proceedings of ICTTS'98, American Society of Civil Engineers, pp. 379–390, 1998.

# Cumulants and Risk-Sensitive Control: A Cost Mean and Variance Theory with Application to Seismic Protection of Structures

Michael K. Sain
University of Notre Dame
Department of Electrical Engineering
Notre Dame, Indiana, USA

Chang-Hee Won
ETRI
TT & C Section
Taejon, Korea

B.F. Spencer, Jr.
University of Notre Dame
Department of Civil Engineering and Geological Sciences
Notre Dame, Indiana, USA

Stanley R. Liberty
University of Nebraska-Lincoln
Department of Electrical Engineering
Lincoln, Nebraska, USA

## Abstract

The risk-sensitive optimal stochastic control problem is interpreted in terms of managing the value of linear combinations of the cumulants of a traditional performance index. The coefficients in these linear combinations are fixed, explicit functions of the risk parameter. This paper demonstrates the possibility of controlling linear combinations of index cumulants with broader opportunities to choose the coefficients. In view of the considerable interest given to cumulants in the theories of signal processing, detection, and estimation over the last decade, such an interpretation offers the possibility of new insights into the broad modern convergence of the concepts of robust control in general. Considered in detail are the foundations for a full-state-feedback solution to the problem of controlling the second cumulant of a cost function, given modest constraints on the first cumulant. The formulation is carried out for a class of nonlinear stochastic differential equations, associated with an appropriate class of nonquadratic performance indices. A Hamilton–Jacobi framework is adopted; and the defining equations for solving the linear, quadratic case are determined. The method is then applied to a situation in which a building is to be protected from earthquakes. Densities of the cost function are computed,

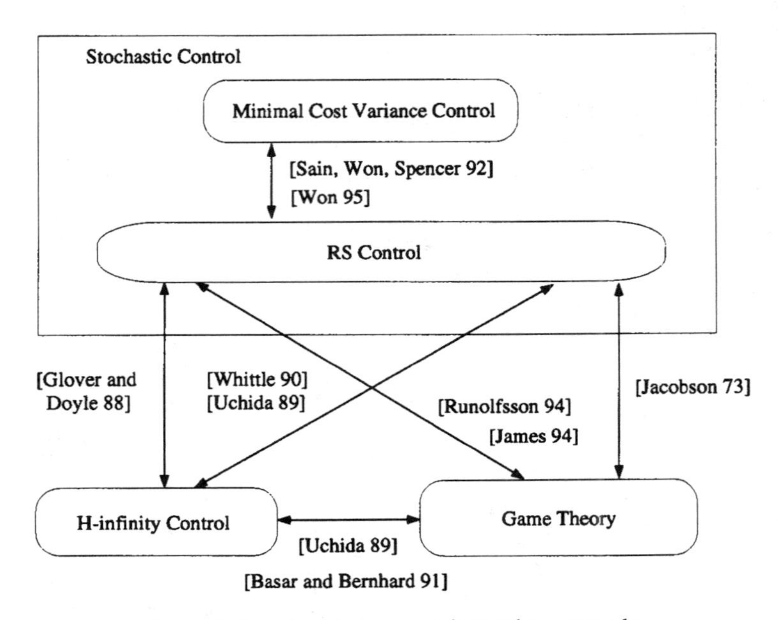


**Figure 1**: Relations between various robust controls.

so as to give insight into the question of how the first and second cumulants affect a cost considered as a random variable.

## 1 Introduction

Relationships among different areas of robust control, such as $H_\infty$ optimal control, game theory, and risk-sensitive (RS) stochastic control have been the subject of recent research; see, for example, [10] and [19]. Because they represent differing paradigms for thinking about the topic, and because these varying paradigms may fit one application area better than another, the overall investigative effort will certainly profit from such studies. By way of illustration, Figure 1 gives a partial overview of some of the connections between the different areas of robust control. The purpose of the present paper is not to develop these interrelationships, per se, but rather to point out the rather natural possibility of expanding the notions involved with the RS body of thought.

Consider an RS cost function [30],

$$J_{\text{RS}} = -\frac{2}{\theta} \log \left( E \left\{ \exp \left( -\frac{\theta}{2} J \right) \right\} \right),$$

where $\theta$ is a real parameter and $J$ is given by

$$J(t, x(t), k) = \int_t^{t_F} [L(s, x(s), k(s, x(s)))] ds + \psi(x(t_F)), \tag{1}$$

where $L : [t_0, t_F] \times \mathbb{R}^n \times \mathbb{R}^m \to \mathbb{R}^+$ and $\psi : \mathbb{R}^n \to \mathbb{R}^+$. Here we proceed intuitively, regarding $x(t)$ as a suitable random process, and $k(t, x(t))$ as a suitable feedback control mapping defined upon it. The moment generating function, or the first characteristic function of $J$, is $\phi_1(s) = E\{\exp(-sJ)\}$, and the cumulant generating function, or second characteristic function, follows by $\phi_2(s) = \log \phi_1(s)$. Suppose, for ease of conversation, that we may assume the existence of the cumulants associated with $J$. Then we have that $\phi_2(s) = \sum_{i=1}^{\infty}[(-1)^i/i!]\beta_i(J)s^i$, in which $\beta_i(J)$ is the $i$th *cumulant*, sometimes also called *semi-invariant*, of $J$. We remark that, if not all the cumulants exist, a truncated version, with remainder, can be written. It follows that

$$J_{\mathrm{RS}} = \left(-\frac{2}{\theta}\right)\left\{\sum_{i=1}^{\infty}\frac{(-1)^i}{i!}\beta_i(J)\left(\frac{\theta}{2}\right)^i\right\}. \tag{2}$$

Approximating to the second order,

$$\begin{aligned} J_{\mathrm{RS}} &= \beta_1(J) - \frac{\theta}{4}\beta_2(J) + O(\theta^2) \\ &= E\{J\} - \frac{\theta}{4}\,\mathrm{VAR}\{J\} + O(\theta^2). \end{aligned} \tag{3}$$

For small $\theta$, then, we see that the RS optimization problem leads to the well-understood task of optimizing the mean value $E\{J\}$ of $J$. If $\theta$ is not quite that small, then the next term to come into play involves $\mathrm{VAR}\{J\}$. In this situation we have in essence a weighted linear combination of $E\{J\}$ and $\mathrm{VAR}\{J\}$, which are in fact the first two cumulants of $J$. One can then address the optimization of the second cumulant $\mathrm{VAR}\{J\}$, under the restriction that the first cumulant $E\{J\}$ exists. We call this the minimal cost variance (MCV) control problem.

Therefore, in accordance with the number of cumulants that exist, RS problems associate naturally with a linear combination of cost cumulants. The particular coefficients which appear in the linear combination are of course fixed. Our purpose in this paper is to show that linear combinations of cumulants can be addressed directly, in their own right, without such restrictions on coefficients. We believe that this is quite in tune with the original motivations of the RS idea; and so we regard the present investigation as part of the broad notion of the RS technique, though it suggests perhaps a broadening of the class of problems that have been studied thus far. This way of proceeding might be given the name *cost cumulant control*. Linear Quadratic Gaussian (LQG) control, MCV control, and RS control certainly can be grouped into this paradigm. Investigating such a notion is clearly the work of multiple persons over extended time. For the present instance, we focus on the MCV problem.

Minimal cost variance (MCV) control was first examined for the open-loop situation by Sain in a dissertation [20] at the University of Illinois in 1965. Certain of those ideas appeared in journal form in 1966 [21]. In 1971, Sain and Liberty published an open-loop result on minimizing the performance variance while keeping the performance mean to a prespecified value [23]. In that paper, new mathematical representations were obtained, and for the first time analytical procedures were used to produce and display the cost densities associated with such control laws, as well as their effects upon the state and control variables of the system. Liberty continued to study characteristic functions of integral quadratic forms, further developing the MCV control idea. Some years later, with Hartwig, he published the results of generating cumulants in the time domain [16]. In 1992, Sain, Won, and Spencer showed that MCV control is related to RS control [24]. Cumulant control can also be viewed as a cost distribution shaping method. See [25, p. 358] for the RS case. In classical LQG control, only one cumulant is controlled, but in MCV control there are two cumulants, namely the mean and the variance, of the cost function that we can control. Thus, one has extra design freedom to shape the cost distribution. This point is demonstrated in Section 7 of this paper.

It is interesting to compare the time line of MCV, cost cumulants, and related topics, with that of RS optimization. RS optimal control seems to have started with Jacobson in 1973. In the 1970s Jacobson extended LQG results by replacing the quadratic criterion with the exponential of a quadratic cost functional, and related linear–exponential–quadratic–Gaussian (LEQG) control to differential games [12]. Many years later, Whittle noted Jacobson's results as an instance of RS control [31]. Speyer et al. [25] extended Jacobson's results to the noisy linear measurements case in discrete time. In [25], optimal control becomes a linear function of the smoothed history of the state, and the solutions are acquired by defining an enlarged state space composed of the entire state history. This enlarged state vector grows at every new stage but retains the feature of being a discrete linear system with additive white Gaussian noise. They also briefly discuss the continuous-time terminal LEQG problem, and the solutions are achieved by taking a formal limit of the discrete case solutions. In 1976, Speyer considered the noisy measurement case again in continuous time, but with zero state weighting in the cost function [26]. Unlike the previous work [25], the Hamilton–Jacobi–Bellman equation was used to produce the solutions. Kumar and van Schuppen derived the general solution of the partially observed exponential-of-integral (EOI) problem in continuous time with zero plant noise in 1981 [15]. Whittle then published his results for the general solution of the partially observed logarithmic-exponential-of-integral (LEOI) problem in discrete time [29]. Four years later, Bensoussan and van Schuppen reported the solution to the general case of a continuous-time partially observed stochastic EOI problem using a different method from Whittle [2]. In 1988, Glover and Doyle related $H_\infty$ and minimum entropy criteria to the infinite horizon version of LEOI theory in discrete time, thus establishing a relationship between RS control and $H_\infty$ optimal control [10]. This result was extended to continuous time by Glover [11]. In 1990, Whittle published the RS maximum principle in book form [30], and published a journal article about the RS maximum principle for the case of partially observed states using large deviation theory [31]. A year later, Bensoussan published a book with all solutions (including the partially observed case) of the exponential-of-integral problem [3]. Başar and Bernhard noted the relationship between deterministic dynamic games and $H_\infty$ optimal control in their book [1]. In 1992, James states that the RS optimal control problem with full-state-feedback information is equivalent to a stochastic differential game problem [13]. Fleming and McEneaney independently pointed out similar results in [8]. More recently, in 1994, James et al. published RS control and dynamic games solutions for partially observed discrete-time nonlinear systems [14]. Finally, Runolfsson presented the relationship between RS control and stochastic differential games in the infinite-horizon case using large deviation ideas [19]. The reader should have no difficulty with a further following of this line of work, as it remains active today in the major journals and conferences.

Viewed, therefore, in the cumulant control sense, we see that the literature on RS control actually extends backward further—perhaps a decade— from the work of Jacobson in 1973. On the other hand, it seems that these observations may serve to broaden the scope of RS investigations in the future. We intend this paper as a contribution toward that effort.

In Section 2, preliminaries needed to formulate the MCV control problem are given. Then in Section 3, MCV control is defined. A Hamilton–Jacobi–Bellman (HJB) equation and an associated verification theorem are derived in Section 4, and the solution structure for MCV control is found using that HJB equation in Section 5. Section 6 contains computer simulations of a representative civil structure being controlled by MCV methods, under seismic disturbances. Then in Section 7 various density and distribution graphs of the cost function for a simple MCV control example are given.

## 2 Preliminaries

This paper considers the Ito-sense stochastic differential equation (SDE)

$$dx(t) = f(t, x(t))dt + \sigma(t, x(t))\,dw(t), \qquad t \in T, \quad x(t_0) = x_0, \tag{4}$$

where $T = [t_0, t_F]$, $x(t) \in \mathbb{R}^n$ is the state, $x_0$ is a random variable that is independent of $w$, and $w(t)$ is a Brownian, not necessarily standard, motion of dimension $d$ defined on a probability space $(\Omega, \mathcal{F}, P)$. The above SDE is interpreted in terms of the stochastic integral equation

$$x(t) = x(t_0) + \int_{t_0}^{t} f(s, x(s))\,ds + \int_{t_0}^{t} \sigma(s, x(s))\,dw(s), \tag{5}$$

where the second integral is of the Ito type and where equality is w.p.1. In the sequel, $\mathbb{R}^n$ is equipped with an inner product $\langle\cdot,\cdot\rangle : \mathbb{R}^n \times \mathbb{R}^n \to \mathbb{R}$ with the usual action $\langle a, b\rangle = \sum_{i=1}^{n} a_i bi$. This then yields a norm in the customary manner $|x|^2 = \langle x, x\rangle$. The following conditions are sufficient for the existence and uniqueness of $x(t)$. See [7, p. 118]. Let $Q_0 = (t_0, t_F)\times\mathbb{R}^n$, and let $\bar{Q}_0 = T\times\mathbb{R}^n$ denote the closure of $Q_0$. Assume that $f : \bar{Q}_0 \to \mathbb{R}^n$ and $\sigma : \bar{Q}_0 \to \mathbb{R}^n \times \mathbb{R}^d$ are Borel measurable mappings [28, p. 50]. Suppose further that there exists a positive constant $C$ such that, for all $(t, x) \in \bar{Q}_0$ $|f(t, x)| \le C(1 + |x|)$, and $|\sigma(t, x)| \le C(1 + |x|)$. For any bounded $B \subset \mathbb{R}^n$ and $t_0 < t_1 < t_F$ assume that there exists a constant $K$, which may depend upon $B$ and $t_1$, such that, for all $x, y \in B$ and $t_0 \le t \le t_1$, $|f(t,x) - f(t,y)| \le K|x-y|$, and $|\sigma(t,x) - \sigma(t,y)| \le K|x-y|$. Then [7, p. 118] if $E\{|x(t_0)|^2\} < \infty$, the solution of (5) exists and is unique. Furthermore, if $E\{|x(t_0)|^m\} < \infty$, $m = 1, 2, \ldots$, then $E|x(t)|^m$ is bounded for $m = 1, 2, \ldots$, and $t_0 \le t \le t_F$. Moreover, the process is a Markov diffusion process; and transition functions become available [7, p. 123].

Now consider the SDE with control

$$dx(t) = f(t, x(t), u(t))\, dt + \sigma(t, x(t))\, dw(t), \qquad t \in T, \quad x(t_0) = x_0, \tag{6}$$

when $u(t) \in U$ is the control action. Let $p$ and $q$ be natural numbers. Assume that $f : \bar{Q}_0 \times U \to \mathbb{R}^n$ is $C^1(\bar{Q}_0 \times U)$, and that $\sigma : \bar{Q}_0 \to \mathbb{R}^{n \times d}$ is $C^1(\bar{Q}_0)$. Furthermore, assume $|f(t,0,0)| \le c$, $|\sigma(t,0)| \le c$,

$$\left|\frac{\partial f(t,x,u)}{\partial x}\right| + \left|\frac{\partial f(t,x,u)}{\partial u}\right| \le \bar{c},$$

and

$$\left|\frac{\partial \sigma(t,x)}{\partial x}\right| \le \bar{c} \qquad \text{for} \qquad (t,x,u) \in \bar{Q}_0 \times U, \quad (t,x) \in \bar{Q}_0,$$

and constants $c$ and $\bar{c}$.

In order to control the performance of (6), a memoryless feedback *control law* is introduced in the manner $u(t) = k(t, x(t))$, $t \in T$, where $k$ is a nonrandom function with random arguments. Then (6) can be written as

$$dx(t) = f^k(t, x(t))\, dt + \sigma(t, x(t))\, dw(t), \qquad t \in T, \quad x(t_0) = x_0, \tag{7}$$

where $f^k(t,x)$ denotes $f(t,x,k(t,x))$. This is in a form similar to (4).

Now we admit only bounded, Borel measurable feedback control laws, $k(t,x) : \bar{Q}_0 \to U$ which satisfy a local Lipschitz condition. Thus, for any bounded $B \subset \mathbb{R}^n$ and $t_0 < t_1 < t_F$ there exists a constant $c_1$, which may depend upon $B$ and $t_1$, such that, for all $x, y \in B$ and $t_0 \le t \le t_1$ $|k(t,x) - k(t,y)| \le c_1|x-y|$. Moreover, we require that $k(t,x)$ satisfy the linear growth condition $|k(t,x)| \le c_2(1+|x|), \forall (t,x) \in \bar{Q}_0$, for a constant $c_2$. A feedback control law $k$ which satisfies both of these conditions is called *admissible*. Then the previous existence result of (4) is applicable, and a unique solution process $x(t)$ of (7) exist [7, p. 156].

Now let $\hat{P}(t, x; s, B; k)$ be the transition function for (6), which is defined as

$$\hat{P}(t, x; s, B; k) = P[x(s) \in B \mid x(t) = x; u(\alpha) = k(\alpha, x(\alpha)), t \le \alpha \le s], \qquad \forall B \in \mathcal{B}(\mathbb{R}^n), \tag{8}$$

where $t_0 \le t < s$, and $\mathcal{B}(\mathbb{R}^n)$ denotes the Borel $\sigma$-algebra, namely the least $\sigma$-algebra containing all open subsets of $\mathbb{R}^n$ [9, p. 126]. Let $p(t, x, s, y; k)$ be the probability density corresponding to (8) so that $\hat{P}(t, x; s, B; k) = \int_B p(t, x, s, y; k)\, dy$, $\forall B \in \mathcal{B}(\mathbb{R}^n)$. This density satisfies the backward Fokker–Planck (or Kolmogorov) equation [7], [18], [33]

$$0 = \mathcal{O}(k)[p(t, x; s, y; k)], \qquad s > t,$$

where $\mathcal{O}(k)$ is the backward evolution operator given by

$$\mathcal{O}(k) = \frac{\partial}{\partial t} + \left\langle f(t, x, k(t, x)), \frac{\partial}{\partial x} \right\rangle + \frac{1}{2} \operatorname{tr} \left( \sigma(t, x) W(t) \sigma'(t, x) \frac{\partial^2}{\partial x^2} \right) \tag{9}$$

in which

$$\left\langle f(t, x, k), \frac{\partial}{\partial x} \right\rangle = \sum_{i=1}^{n} f_i(t, x, k) \frac{\partial}{\partial x_i} \triangleq \mathcal{O}^{(1)}(k),$$

$$\frac{1}{2} \operatorname{tr} \left( \sigma(t, x) W(t) \sigma'(t, x) \frac{\partial^2}{\partial x^2} \right) = \frac{1}{2} \sum_{i,j=1}^{n} (\sigma(t, x) W(t) \sigma'(t, x))_{ij} \frac{\partial^2}{\partial x_i \partial x_j}$$

$$\triangleq \mathcal{O}^{(2)}. \tag{10}$$

In (9), tr denotes the trace operator. The derivative operators are defined so that the $i$th element in the $n$-tuple $(\partial/\partial x)$ is $(\partial/\partial x_i)$, which we shall choose to regard as column vectors, and the $ij$th element in the $n \times n$ matrix operator $(\partial^2/\partial x^2)$ is $(\partial^2/\partial x_i \partial x_j)$. The matrix $W(t)$ characterizes $w(t)$ in the manner $E\{dw(t)dw'(t)\} = W(t)dt$ where the independent increments $dw(t)$ are assumed to be zero-mean Gaussian random variables, and the superscript ($'$) denotes transposition.

For all $(t, x) \in \bar{Q}_0$, a real valued function $\Phi(t, x)$ on $T \times \mathbb{R}^n$ satisfies a polynomial growth condition, if there exist constants $k_1$ and $k_2$ such that $|\Phi(t, x)| \leq k_1(1 + |x|^{k_2})$. Let $C^{1,2}(\bar{Q}_0)$ denote the space of $\Phi(t, x)$ such that $\Phi$ and the partial derivatives $\Phi_t$, $\Phi_{x_i}$, $\Phi_{x_i x_j}$ for $i,j = 1, \ldots, n$ are continuous on $\bar{Q}_0$. Also let $C_p^{1,2}(\bar{Q}_0)$ denote the space of $\Phi(t, x) \in C^{1,2}(\bar{Q}_0)$ such that $\Phi$, $\Phi_t$, $\Phi_{x_i}$, $\Phi_{x_i x_j}$ for $i, j = 1, \ldots, n$, satisfy a polynomial growth condition. Assumptions $\Phi(t, x) \in C_p^{1,2}(\bar{Q}_0)$, $k$ admissible, and $E\{|x(s)|^m | x(t) = x\}$ bounded for $m = 1, 2, \ldots$ and $t \leq s \leq t_F$ ensure existence of the terms in the right member of the Dynkin formula (see [9, p. 128, 135, 161]),

$$\Phi(t, x) = E\left\{ \int_t^{t_F} -\mathcal{O}(k)\Phi(s, x(s))\, ds + \Phi(t_F, x(t_F)) \mid x(t) = x \right\}. \tag{11}$$

In the sequel, we shorten this expectation notation to $E_{tx}$. In order to assess the performance of (6), consider the cost function (1)

$$J(t, x(t), k) = \int_t^{t_F} [L(s, x(s), k(s, x(s)))]\, ds + \psi(x(t_F)). \tag{12}$$

Assume that $L$ and $\psi$ satisfy the polynomial growth conditions $|L(t, x, u)| \leq c_3(1 + |x| + |u|)^{c_4}$, $\forall (t, x, u) \in \bar{Q}_0 \times U$, $|\psi(x)| \leq c_3(1 + |x|)^{c_4}$, $\forall x \in \mathbb{R}^n$, for constants $c_3$ and $c_4$. Fleming and Rishel show that a process $x(t)$ from (6), having an admissible controller $k$, with the assumptions above, is such that $E_{tx}\{J(t, x(t), k)\}$ is finite [7, p. 157]. Lastly, for real symmetric matrices $A$ and $B$, we denote $A \geq B$ if $A - B$ is positive semidefinite, and $A > B$ if $A - B$ is positive definite.

## 3 Minimal Cost Variance Control Problem

This section deals with the definition of the minimal cost variance (MCV) control problem in the completely observed, or full-state-feedback, case. MCV control is a type of cumulant control where we minimize the variance of the cost function while keeping the mean of the cost function at a specified level.

The class of admissible control laws, and comparison of control laws within the class, is defined in terms of the first and second moments of the cost. Introduce the notation $V_1(t, x; k) = E_{tx}\{J(t, x(t), k)\}$ and $V_2(t, x; k) = E_{tx}\{J^2(t, x(t), k)\}$.

**Definition 3.1.** A function $M : \bar{Q}_0 \to \mathbb{R}^+$, which is $C^{1,2}(\bar{Q}_0)$, is an *admissible mean cost function* if there exists an admissible control law $k$ such that $V_1(t, x; k) = M(t, x)$ for $t \in T$ and $x \in \mathbb{R}^n$.

**Remark.** One can of course think of generating a plethora of admissible mean cost functions. All that is necessary, in principle, is to choose a stabilizing admissible control law and then to evaluate the mean cost. In practice, there is the task of representing this mean cost on the one hand, the task of computing it on the other, and the background question of existence. In the present paper, the approach taken is to solve the mean-cost constraint equation for all possible solutions, and then to use the remaining design freedom to achieve a minimization of the cost variance. For the explicit situations discussed in the sequel, this leads to a simultaneous solution for both admissible mean costs and cost variances. This, of course, is in the general spirit of multiplier methods and a type of multiplier function does appear in our later applications.

**Definition 3.2.** Every admissible $M$ defines a class $K_M$ of *control laws $k$ corresponding to $M$* in the manner that $k \in K_M$ if and only if $k$ is an admissible control law which satisfies Definition 3.1.

It is now possible to define an MCV control law $k^*_{V|M}$.

**Definition 3.3.** Let $M$ be an admissible mean cost function, and let $K_M$ be its induced class of admissible control laws. An MCV control law $k^*_{V|M}$ satisfies $V_2(t, x; k^*_{V|M}) = V_2^*(t, x) \leq V_2(t, x; k)$ for $t \in T, x \in \mathbb{R}^n$, whenever $k \in K_M$. The corresponding minimal cost variance is given by $V^*(t, x) = V_2^*(t, x) - M^2(t, x)$ for $t \in T, x \in \mathbb{R}^n$.

An MCV control problem, therefore, is quite general in its scope. It presupposes that a cost mean $M$, not necessarily minimal, has been specified; and it seeks a control law which minimizes the variance of the cost, about $M$.

## 4 Hamilton–Jacobi–Bellman Equation for $V^*$

To solve the MCV control problem, we derive a Hamilton–Jacobi–Bellman (HJB) equation, under the assumption that a sufficiently smooth solution exists. A full-state-feedback MCV control law is constructed in the following section, using this HJB equation, for a large class of problems. We first present a number of preliminary results.

**Lemma 4.1.** *Assume that $M \in C_p^{1,2}(\bar{Q}_0)$ is an admissible mean cost function, and let $k \in K_M$ be a control law corresponding to $M$. Then $\mathcal{O}(k)[M(t,x)] + L(t,x,k(t,x)) = 0$ for $t \in T$, $x \in \mathbb{R}^n$, where $M(t_F, x) = \psi(x)$.*

**Proof.** See [21] or [33, p. 117]. □

**Lemma 4.2** (Verification Lemma). *Let $M \in C_p^{1,2}(Q) \cap C(\bar{Q})$ be a solution to the partial differential equation*

$$\mathcal{O}(k)[M(t,x)] + L(t,x,k(t,x)) = 0 \qquad \forall (t,x) \in Q, \tag{13}$$

*with the boundary condition $M(t_F,x) = \psi(x)$. Then $M(t,x) = V_1(t,x;k)$ for every $k \in K_M$ and $(t,x) \in Q$.*

**Proof.** Because $x$ is a Markov diffusion process, and because $M \in C_p^{1,2}(Q)$, we may use the Dynkin formula (11). From the boundary condition and the Dynkin formula we obtain

$$M(t,x) = E_{tx}\left\{\int_t^{t_F} -\mathcal{O}(k)[M(s,x(s))]\,ds + \psi(x(t_F))\right\}. \tag{14}$$

Substitute from (13) for the expression $-\mathcal{O}(k)[M(s,x(s))]$ in (14). Then we obtain

$$M(t,x) = E_{tx}\left\{\int_t^{t_F} [L(s,x(s),k(s,x(s)))]\,ds + \psi(x(t_F))\right\} = V_1(t,x;k), \tag{15}$$

for each $k \in K_M$ and $(t,x) \in Q$. □

**Lemma 4.3.** *Assume $L : \bar{Q}_0 \times U \to \mathbb{R}^+$ and $L \in C(\bar{Q}_0 \times U)$, together with the linear growth condition on $k$, and the polynomial growth condition on $L$ and $\psi$. Let $x^*(t)$ be the solution of (6) when $k = k^*_{V|M}$. Then*

$$\int_t^{t+\Delta t} L(s,x(s),k(s,x(s)))\,ds \left[\int_{t+\Delta t}^{t_F} L(s,x^*(s),k^*_{V\,|\,M}(s,x^*(s)))\,ds + \psi(x^*(t_F))\right]$$

*is uniformly integrable.*

**Proof.** By Doob [6, p. 629], if (we use $L_s$ and $L^*$ to abbreviate the longer expressions above)

$$E_{tx}\left|\int_t^{t+\Delta t} L_s ds \left[\int_{t+\Delta t}^{t_F} L^* ds + \psi(x^*(t_F))\right]\right|^\alpha$$

is bounded in $t$ for some $\alpha > 1$, then the lemma is proved. We have

$$\begin{aligned}\int_t^{t+\Delta t} L_s ds &\left[\int_{t+\Delta t}^{t_F} L^* ds + \psi(x^*(t_F))\right]\\ &= \Delta t L(t^+,x^+,k^+)[(t_F - t - \Delta t)L(t_1,x_1,k_1) + \psi(x_{t_F})],\end{aligned}$$

where $t + \Delta t \leq t_1 \leq t_F$, $x_1 = x^*(t_1, \omega)$, $x_{t_F} = x^*(t_F, \omega)$, $k_1 = k^*(t_1, x_1)$, and $t \leq t^+ \leq t + \Delta t$.
Consequently, we obtain

$$\begin{aligned}&|\Delta t L(t^+, x^+, k^+)[(t_F - t - \Delta t)L(t_1, x_1, k_1) + \psi(x_{t_F})|]\\&\quad \leq |\Delta t|\,|L(t^+, x^+, k^+)|\left[|t_F - t - \Delta t|\,|L(t_1, x_1, k_1)| + |\psi(x_{t_F})\right]|.\end{aligned}$$

Now we use the polynomial growth conditions on $L$ and $\psi$ to obtain

$$\begin{aligned}&|\Delta t L(t^+, x^+, k^+)[(t_F - t - \Delta t)L(t_1, x_1, k_1) + \psi(x_{t_F})]|\\&\quad \leq |\Delta t| c_3(1 + |x^+| + |k^+|)^{c_4}\\&\qquad \times \left[c_3|t_F - t - \Delta t|(1 + |x_1| + |k_1|)^{c_4} + c_3(1 + |x_{t_F}|)^{c_4}\right].\end{aligned}$$

From the linear growth condition on $k$,

$$\begin{aligned}&|\Delta t L(t^+, x^+, k^+)[(t_F - t - \Delta t)L(t_1, x_1, k_1) + \psi(x_{t_F})]|\\&\quad \leq |\Delta t(t_F - t - \Delta t)| c_5(1 + \|x\|)^{2c_4},\end{aligned}$$

where $\|x\|$ is the sup norm on the process segment which is a concatenation of the two subsegments which interface at $t + \Delta t$. Now we have

$$|\Delta t L(t^+, x^+, k^+)[(t_F - t - \Delta t)L(t_1, x_1, k_1) + \psi(x_{t_F})]| \leq c_6(1 + \|x\|)^{2c_4},$$

and thus

$$E_{tx}|\Delta t L(t^+, x^+, k^+)[(t_F - t - \Delta t)L(t_1, x_1, k_1) + \psi(x_{t_F})]|^{\alpha} \leq E_{tx}[c_6(1 + \|x\|)]^{2c_4\alpha}.$$

It can be shown [9, App. D] that $E_{tx}\|x\|^m < \infty$ for $m = 1, 2, \ldots$. Take $\alpha = m/2c_2 > 1$. Then $E_{tx}|\Delta t L(t^+, x^+, k^+)[(t_F - t - \Delta t)L(t_1, x_1, k_1) + \psi(x_{t_F})]|^{\alpha}$ is bounded, and by Doob [6, p. 629] we have uniform integrability, as desired. □

Next we derive the HJB equation for the second moment of the cost function, in the following theorem, which, as we observed early in this section, assumes the existence of an optimal controller. Then we present the verification theorem which is a sufficient condition for constructing a minimum.

**Theorem 4.1.** *Let $M$ be an admissible mean cost function and let $K_M$ be the corresponding class of control laws. Assume the existence of an optimal controller $k^*_{V/M}$ and an optimum value function $V_2^* \in C_p^{1,2}(\bar{Q}_0)$. Then $k^*_{V|M}$ and $V_2^*$ satisfy the partial differential equation*

$$\mathcal{O}(k^*_{V|M})[V_2^*(t, x)] + 2M(t, x)L(t, x, k^*_{V|M}(t, x)) = 0 \tag{16}$$

*for $t \in T$, $x \in \mathbb{R}^n$, where*

$$\begin{aligned}&\mathcal{O}(k^*_{V|M})[V_2^*(t, x)] + 2M(t, x)L(t, x, k^*_{V|M}(t, x))\\&\quad = \min_{k \in K_M}\{\mathcal{O}(k)[V_2^*(t, x)] + 2M(t, x)L(t, x, k(t, x))\},\end{aligned} \tag{17}$$

*along with the boundary condition $V_2^*(t_F, x) = M^2(t_F, x) = \psi^2(x)$, $x \in \mathbb{R}^n$.*

**Proof.** Define a controller $k_1 \in K_M$ by the action

$$k_1(r,x) = \left\{ \begin{array}{ll} k(r,x), & t \le r \le t+\Delta t, \\ k^*_{V|M}(r,x), & t+\Delta t < r \le t_F, \end{array} \right\}; \tag{18}$$

then the second moment is given by

$$\begin{aligned} V_2(t,x;k_1) &= E_{tx}\{J^2(t,x(t),k_1)\} \\ &= E_{tx}\left\{\left[\int_t^{t+\Delta t} L_s ds + \int_{t+\Delta t}^{t_F} L^* ds + \psi(x^*(t_F))\right]^2\right\} \\ &= E_{tx}\left\{\left[\int_t^{t+\Delta t} L_s ds\right]^2\right\} + 2E_{tx}\left\{\left[\int_t^{t+\Delta t} L_s ds\right]\right. \\ &\quad \left. \times \left[\int_{t+\Delta t}^{t_F} L^* ds + \psi(x^*(t_F))\right]\right\} \\ &\quad + E_{tx}\left\{\left[\int_{t+\Delta t}^{t_F} L^* ds + \psi(x^*(t_F))\right]^2\right\}, \end{aligned}$$

where $L^* = L(s, x^*(s), k^*_{V|M}(s, x^*(s)))$, $x^*(t)$ is the solution of (6) when $k = k^*_{V|M}$ and $L_s = L(s, x(s), k(s, x(s)))$. By definition, $V_2^*(t,x) \le V_2(t,x;k_1)$. Now we can substitute for $V_2(t,x;k_1)$ and obtain

$$\begin{aligned} V_2^*(t,x) \le E_{tx}&\left\{\left[\int_t^{t+\Delta t} L_s ds\right]^2\right\} \\ &+ 2E_{tx}\left\{\left[\int_t^{t+\Delta t} L_s ds\right]\left[\int_{t+\Delta t}^{t_F} L^* ds + \psi(x^*(t_F))\right]\right\} \\ &+ E_{tx}\left\{\left[\int_{t+\Delta t}^{t_F} L^* ds + \psi(x^*(t_F))\right]^2\right\}. \end{aligned}$$

We can apply the mean value theorem for each sample function. Then

$$\begin{aligned} V_2^*(t,x) \le \Delta t^2 E_{tx}\{L^2(t^+,x^+,k^+)\} &+ 2\Delta t E_{tx}\left\{L(t^+,x^+,k^+)\right. \\ &\left. \times \left[\int_{t+\Delta t}^{t_F} L^* ds + \psi(x^*(t_F))\right]\right\} + E_{tx}\{V_2^*(t+\Delta t, x(t+\Delta t))\}, \end{aligned}$$

where the last term in the right side uses the Chapman–Kolmogorov equation. Because of the assumptions we have made, we can use the Dynkin formula (11),

$$E_{tx}\{V_2^*(t+\Delta t, x(t+\Delta t))\} - V_2^*(t,x) = E_{tx}\left\{\int_t^{t+\Delta t} \mathcal{O}(k)[V_2^*(r,x(r))]dr\right\}.$$

Thus, we obtain

$$V_2^*(t,x) \le (\Delta t)^2 E_{tx}\{L^2(t^+,x^+,k^+)\} + 2\Delta t E_{tx}\left\{L(t^+,x^+,k^+)\right.$$

$$\times \left[ \int_{t+\Delta t}^{t_F} L^* ds + \psi(x^*(t_F)) \right] \Bigg\} + E_{tx} \left\{ \int_t^{t+\Delta t} \mathcal{O}(k)[V_2^*(r, x(r))] dr \right\}$$
$$+ V_2^*(t, x). \tag{19}$$

We also have [9, p. 164]

$$\lim_{\Delta t \to 0} E_{tx} \left\{ L(t^+, x^+, k^+) \left[ \int_{t+\Delta t}^{t_F} L^* ds + \psi(x^*(t_F)) \right] \right\} = M(t, x) L(t, x, k(t, x)),$$

because of the uniform integrability condition given in Lemma 4.3, $k^*_{V|M} \in K_M$, and the fact that, as $\Delta t$ goes to zero, $L(t^+, x^+, k^+)$ approaches $L(t, x, k(t, x))$. The third term in the right member of (19) can be treated in a similar manner, but we omit details for reasons of brevity. Now divide (19) by $\Delta t$ and let $\Delta t \to 0$ to get $0 \le \mathcal{O}(k)[V_2^*(t, x)] + 2M(t, x)L(t, x, k(t, x))$. Equality holds if $k(t, x) = k^*_{V|M}(t, x)$ where $k^*_{V|M}$ is an optimal feedback control law. This concludes the proof. See also [22]. □

The verification theorem for the second moment case is now presented. This verification theorem states that if there exists a sufficiently smooth solution of the HJB equation, then it is the optimal cost of control, and by using this solution an optimal feedback control law can be determined.

**Theorem 4.2** (Verification Theorem). *Let $V_2^* \in C_p^{1,2}(Q) \cap C(\bar{Q})$ be a nonnegative solution to the partial differential equation*

$$0 = \min_{k \in K_M} \{2M(t, x)L(t, x, k(t, x)) + \mathcal{O}(k)[V_2^*(t, x)]\}, \qquad \forall (t, x) \in Q, \tag{20}$$

*with the boundary condition $V_2^*(t_F, x) = \psi^2(x)$. Then $V_2^*(t, x) \le V_2(t, x; k)$ for every $k \in K_M$ and any $(t, x) \in Q$. If, in addition, such a $k$ also satisfies the equation*

$$2M(t, x)L(t, x, k(t, x)) + \mathcal{O}(k)[V_2^*(t, x)]$$
$$= \min_{\bar{k} \in K_M} \{2M(t, x)L(t, x, \bar{k}(t, x)) + \mathcal{O}(\bar{k})[V_2^*(t, x)]\}$$

*for all $(t, x) \in Q$, then $V_2^*(t, x) = V_2(t, x; k)$ and $k = k^*_{V|M}$ is an optimal control law.*

**Proof.** From (20) for each $k \in K_M$ and $(t, x) \in Q$,

$$2M(t, x)L(t, x, k(t, x)) + \mathcal{O}(k)[V_2^*(t, x)] \ge 0. \tag{21}$$

Because $x$ is a Markov diffusion process, and because $V_2^* \in C_p^{1,2}(Q)$, we may use the Dynkin formula (11). From the boundary condition and the Dynkin formula we obtain

$$V_2^*(t, x) = E_{tx} \left\{ \int_t^{t_F} -\mathcal{O}(k)[V_2^*(s, x(s))] ds + \psi^2(x(t_F)) \right\}. \tag{22}$$

Let $L_s = L(s, x(s), k(s, x(s)))$ and let $L_r = L(r, \psi(r), k(r, x(r)))$, and substitute from (21) expression $-\mathcal{O}(k)[V_2^*(s, x(s))]$ in (22). Then we obtain

$$\begin{aligned}
V_2^*(t,x) &\le E_{tx}\left\{\int_t^{t_F} 2L_s M(s,x(s))ds + \psi^2(x(t_F))\right\}\\
&= E_{tx}\left\{\int_t^{t_F} 2L_s E_{sx}\left\{\int_s^{t_F} L_r dr + \psi(x(t_F))\right\} ds + \psi^2(x(t_F))\right\}\\
&= E_{tx}\left\{\int_t^{t_F} E_{sx}\left\{2L_s\int_s^{t_F} L_r dr + 2L_s\psi(x(t_F))\right\} ds\right\}\\
&\quad + E_{tx}\left\{\psi^2(x(t_F))\right\}\\
&= \int_t^{t_F} E_{tx}\left\{E_{sx}\left\{2L_s\int_s^{t_F} L_r dr + 2L_s\psi(x(t_F))\right\}\right\} ds\\
&\quad + E_{tx}\left\{\psi^2(x(t_F))\right\}.
\end{aligned}$$

For a justification of the interchange of the integral and the expectation in the last equation, see [6, p. 62] and [5, p. 65]. Consequently, we have

$$\begin{aligned}
V_2^*(t,x) &\le \int_t^{t_F} E_{tx}\left\{\left[2L_s\int_s^{t_F} L_r dr + 2L_s\psi(x(t_F))\right]\right\} ds + E_{tx}\left\{\psi^2(x(t_F))\right\}\\
&= E_{tx}\left\{\int_t^{t_F}\left[2L_s\int_s^{t_F} L_r dr + 2L_s\psi(x(t_F))\right] ds + \psi^2(x(t_F))\right\}\\
&= E_{tx}\left\{\int_t^{t_F} L_s ds\int_t^{t_F} L_r dr + 2\int_t^{t_F} L_s\psi(x(t_F))ds + \psi^2(x(t_F))\right\}\\
&= E_{tx}\left\{\left[\int_t^{t_F} L_s ds + \psi(x(t_F))\right]^2\right\} = V_2(t,x;k).
\end{aligned}$$

This proves the first part. For the second part, the inequality becomes equality. □

Note that if $M$ is given a priori, then the optimal second moment results in the optimal variance.

With these results in hand, it is then possible to transfer the results to the cumulant functions by means of the following pair of theorems, which make use of the notation $|a|_A^2 = a'Aa$:

**Theorem 4.3.** *Let $M \in C_p^{1,2}(\bar{Q}_0)$ be an admissible mean cost function, and let $M$ induce a nonempty class $K_M$ of admissible control laws. Assume the existence of an optimal control law $k = k_{V|M}^*$ and an optimum value function $V^* \in C_p^{1,2}(\bar{Q}_0)$. Then the MCV function $V^*$ satisfies the HJB equation*

$$\min_{k\in K_M} \mathcal{O}(k)[V^*(t,x)] + \left|\frac{\partial M(t,x)}{\partial x}\right|^2_{\sigma(t,x)W(t)\sigma'(t,x)} = 0, \tag{23}$$

*for $t, x) \in \bar{Q}_0$, together with the terminal condition, $V^*(t_F, x) = 0$.*

**Proof.** From Definition 3.3 and (20), it follows that

$$\min_{k\in K_M} \left\{\mathcal{O}(k)[V^*(t,x)+M^2(t,x)]+2M(t,x)L(t,x,k(t,x))\right\}=0.$$

To prove (23), it is necessary and sufficient to establish that

$$\mathcal{O}(k)[M^2(t,x)]+2M(t,x)L(t,x,k(t,x)) = \left|\frac{\partial M(t,x)}{\partial x}\right|^2_{\sigma(t,x)W(t)\sigma'(t,x)} \tag{24}$$

whenever $k\in K_M$. In order to see this, write $\mathcal{O}(k)$ as the sum $\mathcal{O}^{(1)}(k)+\mathcal{O}^{(2)}$ of two operators, where $\mathcal{O}^{(i)}$ involves the partial derivative of exactly the $i$th order; see (10). Then $\mathcal{O}^{(1)}(k)[M^2(t,x)] = 2M(t,x)\mathcal{O}^{(1)}(k)[M(t,x)]$, which by Lemma 4.1 becomes

$$\mathcal{O}^{(1)}(k)[M^2(t,x)] = -2M(t,x)\left[\mathcal{O}^{(2)}[M(t,x)]+L(t,x,k(t,x))\right]. \tag{25}$$

Substitute in (24) to get

$$\mathcal{O}^{(2)}[M^2(t,x)]-2M(t,x)\mathcal{O}^{(2)}[M(t,x)] = \left|\frac{\partial M(t,x)}{\partial x}\right|^2_{\sigma(t,x)W(t)\sigma'(t,x)} \tag{26}$$

for $t\in T$, $x\in\mathbb{R}^n$. Note that $k$ does not appear explicitly in (26). Note also that because $\mathcal{O}^{(2)}=\frac{1}{2}\operatorname{tr}(\sigma(t,x)W(t)\sigma'(t,x)(\partial^2/\partial x^2))$, (26) is equivalent to

$$\frac{1}{2}\operatorname{tr}\left(\sigma(t,x)W(t)\sigma'(t,x)\left[\frac{\partial^2(M^2(t,x))}{\partial x^2}-2M(t,x)\frac{\partial^2 M(t,x)}{\partial x^2}\right]\right) = \left|\frac{\partial M(t,x)}{\partial x}\right|^2_{\sigma(t,x)W(t)\sigma'(t,x)}. \tag{27}$$

However,

$$\frac{\partial^2 M^2(t,x)}{\partial x^2}-2M(t,x)\frac{\partial^2 M(t,x)}{\partial x^2} = 2\frac{\partial M(t,x)}{\partial x}\left(\frac{\partial M(t,x)}{\partial x}\right)',$$

which combines with (27) to establish (24). □

**Theorem 4.4** (Verification Theorem). *Let $M$ be an admissible mean cost function satisfying $M^2(t,x)\in C_p^{1,2}(Q)\cap C(\bar{Q})$, and let $K_M$ be the associated nonempty class of admissible control laws. Suppose that a nonnegative function $V^*\in C_p^{1,2}(Q)\cap C(\bar{Q})$ is a solution to the partial differential equation*

$$\min_{k\in K_M}\mathcal{O}(k)[V^*(t,x)]+\left|\frac{\partial M(t,x)}{\partial x}\right|^2_{\sigma(t,x)W(t)\sigma'(t,x)} = 0,\qquad \forall(t,x)\in Q, \tag{28}$$

*together with the boundary condition $V^*(t_F,x)=0$. Then $V^*(t,x)\le V(t,x;k)$ for every $k\in K_M$ and any $(t,x)\in Q$. If, in addition, such a $k$ satisfies the equation*

$$\mathcal{O}(k)[V^*(t,x)] = \min_{\bar{k}\in K_M}\left\{\mathcal{O}(\bar{k})[V^*(t,x)]\right\}$$

*for all $(t,x)\in Q$, then $V^*(t,x)=V(t,x;k)$ and $k=k^*_{V|M}$ is an optimal control law.*

**Proof.** From (28) for each $k \in K_M$ and $(t, x) \in Q$

$$\mathcal{O}(k)[V^*(t,x)] + \left|\frac{\partial M(t,x)}{\partial x}\right|^2_{\sigma(t,x)W(t)\sigma'(t,x)} \geq 0. \tag{29}$$

Because $x$ is a Markov diffusion process, and because $V^*, M^2 \in C_p^{1,2}(Q)$, we may use the Dynkin formula (11) to obtain

$$V^*(t,x) = E_{tx}\left\{\int_t^{t_F} -\mathcal{O}(k)[V^*(s,x(s))]\,ds\right\} \tag{30}$$

and

$$M^2(t,x) = E_{tx}\left\{\int_t^{t_F} -\mathcal{O}(k)[M^2(s,x(s))]ds + \psi^2(x(t_F))\right\}. \tag{31}$$

By (29) and (30), we get, with the aid of the notation $L_s = L(s, x(s), k(s, x(s)))$,

$$\begin{aligned} V^*(t,x) &\leq E_{tx}\left\{\int_t^{t_F} \left|\frac{\partial M(s,x(s))}{\partial x}\right|^2_{\sigma(s,x(s))W(s)\sigma'(s,x(s))} ds\right\} \\ &= E_{tx}\left\{\int_t^{t_F} \left[\mathcal{O}(k)[M^2(s,x(s))] + 2M(s,x(s))L_s\right] ds\right\}, \end{aligned}$$

where we used (24) to obtain the last expression. Use Definition 3.3 and (31) to get

$$V_2^*(t,x) \leq E_{tx}\left\{\int_t^{t_F} 2M(s,x(s))L_s ds + \psi^2(x(t_F))\right\}.$$

By the analysis in the proof of Theorem 4.2, the above inequality becomes $V_2^*(t,x) \leq V_2(t,x;k)$, which in turn implies $V^*(t,x) \leq V(t,x;k)$. □

Equation (28) in Theorem 4.4 differs from the classical HJB result for minimal mean controllers in that the integrand function $L$ does not appear explicitly. In order to compare $k^*_{V|M}$ with $k^*_M$, recall that $V_1^*$ satisfies (see, e.g., [33])

$$\min_k \left[\mathcal{O}(k)[V_1^*(t,x)] + L(t,x,k(t,x))\right] = 0 \tag{32}$$

together with the terminal condition $V_1^*(t_F, x) = \psi(x)$. Equation (23) may therefore be compared with a minimal mean problem in which

$$L(t,x,k(t,x)) = \left|\frac{\partial M(t,x)}{\partial x}\right|^2_{\sigma(t,x)W(t)\sigma'(t,x)}, \tag{33}$$

provided that it is realized in (23) that the control law is constrained to be in the class $K_M$, whereas in (32) it is constrained only to be an admissible function of its arguments. Thus an analogy between the MCV problem and a minimal mean problem with control law constraints, but no cost of control action, can be drawn immediately. The precise nature of this analogy depends upon the nature of $f$.

## 5 Solutions of MCV Control

In this section we derive the full-state-feedback solution of the MCV control problem for a linear system and a quadratic cost function. Here we will look for an admissible linear controller that minimizes the variance of the cost function. We consider the class of admissible controls that satisfy the equation $L(t, x, k(t, x)) + \mathcal{O}(k)[M(t, x)] = 0$. We assume that $L$, $f$, and $M$ are given, and we wish to find $k$. Henceforward we will write $\sigma(t, x) = E(t)$, and make the assumptions

$$\sigma(t, x) = E(t), \tag{34}$$

$$L(t, x, k(t, x)) = h(t, x) + k'(t, x)R(t)k(t, x), \tag{35}$$

$$\psi(x(t_F)) = x'(t_F)Q_F x(t_F), \tag{36}$$

$$f(t, x, k(t, x)) = g(t, x) + B(t)k(t, x), \tag{37}$$

where $k$ is an admissible feedback control law; $h : \bar{Q}_0 \to \mathbb{R}^+$ is $C(\bar{Q}_0)$ and satisfies the polynomial growth conditions assumed for $L$; and $g : \bar{Q}_0 \to \mathbb{R}^n$ is $C^1(\bar{Q}_0)$ and satisfies the linear growth condition and the local Lipschitz condition assumed for $f$. Moreover, $E(t)$, $R(t) > 0$, and $B(t)$ are continuous real matrices of appropriate dimensions for all $t \in T$.

Here we state the results of Liu and Leake [17]. Let $x \in \mathbb{R}^n$ be a real $n$-vector, let $z(x)$ and $y(x)$ be real $r$-vector functions, and let $\alpha(x)$ be a real function defined on $\mathbb{R}^n$.

**Lemma 5.1.** *Let $y(x)$ and $\alpha(x)$ be given. Then there exists $z(x)$ which satisfies the condition*

$$\langle z(x), z(x)\rangle + 2\langle z(x), y(x)\rangle + \alpha(x) = 0 \tag{38}$$

*if and only if $|y(x)|y(x)|^2 \geq \alpha(x)$. In such a case, the set of all solutions to (38) is represented by*

$$z(x) = \beta(x)a(x) - y(x), \tag{39}$$

*where $\beta(x) = (|y(x)|^2 - \alpha(x))^{1/2}$ and $a(x)$ is an arbitrary unit vector.*

**Proof.** The sufficiency follows by direct evaluation. To show that the conditions are necessary, note that $|y|^2 < \alpha$ implies that $|z+y|^2 < 0$, which is a contradiction. Let $w = z + y$, then (38) implies that $\langle w, w\rangle = \beta^2$; taking $a = w/|w|$, we have $w = \beta a$, then (39) follows. □

**Lemma 5.2** (Liu and Leake Lemma). *Let $X$ be a positive-definite symmetric-real matrix. Then there exists $z(x)$ which satisfies the condition*

$$\langle z(x), Xz(x)\rangle + 2\langle z(x), y(x)\rangle + \alpha(x) = 0 \tag{40}$$

*if and only if $\langle y(x), X^{-1}y(x)\rangle \geq \alpha(x)$. In this case, the set of all solutions to (40) is represented by*

$$z(x) = \beta(x)H^{-1}\alpha(x) - X^{-1}y(x), \tag{41}$$

*where*

$$\beta(x) = \left(\langle y(x), X^{-1}y(x)\rangle - \alpha(x)\right)^{1/2}, \tag{42}$$

*$H$ is a nonsingular matrix such that $X = H'H$, and $a(x)$ is an arbitrary unit vector.*

**Proof.** The existence of such an $H$ is well known. One instance is sometimes denoted $X^{(1/2)}$. No difficulty accrues due to the nonuniqueness of $H$, because it is subsumed into the unit vector $a(x)$. The proof of this lemma follows from Lemma 5.1 by a change of variables $\hat{z} = Hz$ and $\hat{y} = (H^{-1})'y$. □

**Remark.** The presence of the term involving the unit vector $a(x)$ in (39) is worthy of reflection. The only way in which this term can vanish is for $\beta(x)$ to vanish. Later, in the application of this result, the term will appear in a product with another vector $Gx$, for $G$ a matrix, and will be chosen to be in the direction opposite to that of $Gx$. This means that, whenever $x$ approaches the null space of $G$, it is not possible to define a unique limiting value for the unit vector. Therefore, in order for the term to remain continuous as $x$ approaches the null space of $G$, we shall simply require that $\beta(x)$ then approach zero.

**Lemma 5.3.** *Assume that $M \in C_p^{1,2}(Q) \cap C(\bar{Q})$ and that the above assumptions (34)-(37) are satisfied. Then we have a solution $k(t,x)$, which may or may not be admissible, if and only if,*

$$\begin{aligned}\frac{1}{4}\left(\frac{\partial M(t,x)}{\partial x}\right)' B(t)R^{-1}(t)B'(t)\left(\frac{\partial M(t,x)}{\partial x}\right)\\ \geq \frac{\partial M(t,x)}{\partial t} + \frac{1}{2}\operatorname{tr}\left(E(t)W(t)E'(t)\frac{\partial^2 M(t,x)}{\partial x^2}\right)\\ + h(t,x) + g'(t,x)\left(\frac{\partial M(t,x)}{\partial x}\right).\end{aligned}$$

**Proof.** Rewriting (13),

$$\begin{aligned}\frac{\partial M(t,x)}{\partial t} + L(t,x,k(t,x)) + \frac{1}{2}\operatorname{tr}\left(\sigma(t,x)W(t)\sigma'(t,x)\frac{\partial^2 M(t,x)}{\partial x^2}\right)\\ + f'(t,x,k(t,x))\frac{\partial M(t,x)}{\partial x} = 0.\end{aligned} \tag{43}$$

By substituting expressions for (34), (35), and (37) into (43) and suppressing the arguments, we obtain,

$$\frac{\partial M}{\partial t} + h + k'Rk + \frac{1}{2}\operatorname{tr}\left(EWE'\frac{\partial^2 M}{\partial x^2}\right) + g'\left(\frac{\partial M}{\partial x}\right) + k'B'\left(\frac{\partial M}{\partial x}\right) = 0. \tag{44}$$

One can then solve the above equation for $k$, using the method of Liu and Leake; see Lemma 5.2. We may identify from (44) $z \Leftrightarrow k$, $X \Leftrightarrow R$, $y \Leftrightarrow \frac{1}{2}B'(\partial M/\partial x)$, and $\alpha \Leftrightarrow \partial M/\partial t + \frac{1}{2}\operatorname{tr}(EWE'(\partial^2 M/\partial x^2)) + h + g'(\partial M \partial M/\partial x \partial x)$. Accordingly, we must satisfy

$$\frac{1}{4}\left(\frac{\partial M}{\partial x}\right)' BR^{-1}B'\left(\frac{\partial M}{\partial x}\right) \geq \frac{\partial M}{\partial t}+\frac{1}{2}\operatorname{tr}\left(EWE'\frac{\partial^2 M}{\partial x^2}\right)+h+g'\left(\frac{\partial M}{\partial x}\right), \quad (45)$$

which corresponds to the $\langle y(x), X^{-1}y(x)\rangle \geq \alpha(x)$ condition in Lemma 5.2. □

Now we are ready to characterize all controllers, $k \in K_M$.

**Theorem 5.1.** *Assume that the condition of Lemma 5.3 is satisfied. Then a control law $k$ is in $K_M$ if and only if (1) it is admissible and (2) it is of the form*

$$k(t,x) = \beta(x)H^{-1}a(x) - \frac{1}{2}R^{-1}(t)B'(t)\left(\frac{\partial M(t,x)}{\partial x}\right), \quad (46)$$

*where $a(x)$ is an arbitrary unit vector, $H'H = R$, and*

$$\beta(x) = \sqrt{\frac{1}{4}\left(\frac{\partial M}{\partial x}\right)' BR^{-1}B'\left(\frac{\partial M}{\partial x}\right) - \frac{\partial M}{\partial t} - \frac{1}{2}\operatorname{tr}\left(EWE'\frac{\partial^2 M}{\partial x^2}\right) - h - g'\left(\frac{\partial M}{\partial x}\right)}. \quad (47)$$

*Moreover, $\beta(x) = 0$ corresponds to the optimal mean cost law.*

**Proof.** See Lemmas 5.2 and 5.3. Now, for any admissible control law, we have from (44) the equation

$$\left[\frac{\partial M}{\partial t} + \frac{1}{2}\operatorname{tr}\left(EWE'\frac{\partial^2 M}{\partial x^2}\right) + h + g'\frac{\partial M}{\partial x}\right] + k'Rk + k'B'\frac{\partial M}{\partial x} = 0. \quad (48)$$

Incorporating (45), we obtain

$$\left(k'Rk + k'B'\left(\frac{\partial M}{\partial x}\right)\right) \geq -\frac{1}{4}\left(\frac{\partial M}{\partial x}\right)' BR^{-1}B'\left(\frac{\partial M}{\partial x}\right). \quad (49)$$

The left-hand side of the above inequality is $-\alpha$ (see (48)), and the right-hand side is $-\langle y, X^{-1}y\rangle$. Thus if $k$ is the minimal mean cost law then $\beta = 0$ and $M(t,x) = V_1(t,x;k) = V_1^*(t,x)$. If $k$ is not the minimal mean cost law, and (45) is satisfied, then $\beta > 0$. □

For a wide variety of problems, therefore, we have shown that suboptimal mean control introduces the possibility of reducing variance, that is, we have some freedom in selecting $a(x)$.

To find the solution of the MCV control problem, we rewrite the HJB equation (28) of Theorem 4.4 as

$$\frac{-\partial V^*(t,x)}{\partial t} = \min_{k\in K_M}\left[f'(t,x,k(t,x))\frac{\partial V^*(t,x)}{\partial x}\right.$$
$$\left. + \frac{1}{2}\operatorname{tr}\left(\sigma(t,x)W(t)\sigma'(t,x)\frac{\partial^2 V^*(t,x)}{\partial x^2}\right)\right.$$

$$+ \left| \frac{\partial M(t,x)}{\partial x} \right|^2_{\sigma(t,x)W(t)\sigma'(t,x)} \Bigg] \tag{50}$$

with boundary condition $V^*(t_F, x) = 0$.

**Theorem 5.2.** *Assume that the conditions of Theorems 4.4 and 5.1 are satisfied. Then a nonlinear optimal MCV control law is of the form*

$$k^*_{V|M}(t,x) = \frac{-\beta(x)R^{-1}(t)B'(t)(\partial V^*(t,x)/\partial x)}{|B'(t)[\partial V^*(t,x)/\partial x]|_{R^{-1}(t)}} - \frac{1}{2}R^{-1}(t)B'(t)\left(\frac{\partial M(t,x)}{\partial x}\right), \tag{51}$$

*provided that $B'(t)[\partial V^*(t,x)/\partial x]$ is nonzero, and that the optimal cost function $V^*$ satisfies the partial differential equation*

$$-\frac{\partial V^*}{\partial t} = -\beta \left| (H^{-1})'B'\frac{\partial V^*}{\partial x} \right| + g'\frac{\partial V^*}{\partial x} - \frac{1}{2}\left(\frac{\partial M}{\partial x}\right)' BR^{-1}B'\left(\frac{\partial V^*}{\partial x}\right)$$
$$+ \frac{1}{2}\operatorname{tr}\left(EWE'\frac{\partial^2 V^*}{\partial x^2}\right) + \left|\frac{\partial M}{\partial x}\right|^2_{EWE'}, \tag{52}$$

*with boundary condition $V^*(t_F, x) = 0$. Moreover, whenever $B'(t)[\partial V^*(t,x)/\partial x]$ is zero, then $\beta(x)$ is also zero; and $k^*_{V|M}(t,x)$ employs only the second term $-\frac{1}{2}R^{-1}(t)B'(t)(\partial M(t,x)/\partial x)$ in (51).*

**Remark.** When $B'(t)[\partial V^*(t,x)/\partial x]$ is zero, then $M$ satisfies the partial differential equation for the minimum mean cost problem, as can be seen from the form for $\beta(x)$ in (47). For pairs $(t,x)$ at which this occurs, then we may see the controller accumulating average cost at the same rate as would a minimum average cost controller.

**Proof.** From the HJB equation (50), we obtain

$$-\frac{\partial V^*}{\partial t} = \min_{k \in K_M}\left[(g' + k'B')\frac{\partial V^*}{\partial x}\right] + \frac{1}{2}\operatorname{tr}\left(EWE'\frac{\partial^2 V^*}{\partial x^2}\right) + \left|\frac{\partial M}{\partial x}\right|^2_{EWE'} \tag{53}$$

with boundary condition $V^*(t_F, x) = 0$. Substituting (46) into (53), we obtain

$$-\frac{\partial V^*}{\partial t} = \min_{|a|=1}\left[\beta a'(H^{-1})'B'\left(\frac{\partial V^*}{\partial x}\right)\right] + g'\frac{\partial V^*}{\partial x} - \frac{1}{2}\left(\frac{\partial M}{\partial x}\right)' BR^{-1}B'\left(\frac{\partial V^*}{\partial x}\right)$$
$$+ \frac{1}{2}\operatorname{tr}\left(EWE'\frac{\partial^2 V^*}{\partial x^2}\right) + \left|\frac{\partial M}{\partial x}\right|^2_{EWE'}. \tag{54}$$

The minimization is achieved by choosing

$$a = -a^*\left((H^{-1})'B'\left(\frac{\partial V^*}{\partial x}\right)\right), \tag{55}$$

where $a^*\left((H^{-1})'B'(\partial V^*/\partial x)\right)$ is a unit vector in the direction of $(H^{-1})'B'(\partial V^*/\partial x)$, when it is nonzero. Notice that, when this quantity is zero, the bracketed term in (54) vanishes. Therefore, we can rewrite (54) in the manner (51), and the result follows. □

Our next step is to examine $\beta(x)$ in the case in which the dynamical system is linear and the cost function accumulates at a quadratic rate. We do not yet assume that the controller is linear, but we shall assume that the average cost function is quadratic:

$$M(t,x) = x'\mathcal{M}(t)x + m(t), \tag{56}$$

so that we can obtain the explicit evaluations

$$\frac{\partial^2 M(t,x)}{\partial x^2} = 2\mathcal{M}(t), \qquad \frac{\partial M(t,x)}{\partial x} = 2\mathcal{M}(t)x, \qquad \frac{\partial M(t,x)}{\partial t} = x'\dot{\mathcal{M}}(t)x + \dot{m}(t), \tag{57}$$

with which one obtains the following lemma.

**Lemma 5.4.** *Invoke the assumptions of Lemma 5.3, together with (56). Then, for the quadratic cost rate accumulation and linear dynamical system cases,*

$$\begin{aligned} h(t,x) &= x'(t)Q(t)x(t), \qquad Q(t) \geq 0, \\ g(t,x) &= A(t)x(t), \end{aligned} \tag{58}$$

*of (35) and (37), we find $\beta(x)$ in (46) to be $\beta(x) = |x|_{\mathcal{R}} - \dot{m} - \operatorname{tr}(EWE'\mathcal{M})$, where*

$$\mathcal{R} \triangleq \mathcal{M}BR^{-1}B'\mathcal{M} - \dot{\mathcal{M}} - Q - A'\mathcal{M} - \mathcal{M}A. \tag{59}$$

*Moreover, if $B'(t)[\partial V^*(t,x)/\partial x]$ is zero when $x$ is zero, then*

$$\dot{m} = -\operatorname{tr}(EWE'\mathcal{M}). \tag{60}$$

*A particular case of this situation occurs when*

$$V^*(t,x) = x'\mathcal{V}(t)x + v(t), \tag{61}$$

*in which case the optimal MCV control law (51) can be rewritten as*

$$k^*_{V|M}(t,x) = \frac{-|x|_{\mathcal{R}} R^{-1}(t)B'(t)\mathcal{V}(t)x}{|B'(t)\mathcal{V}(t)x|_{R^{-1}(t)}} - R^{-1}(t)B'(t)\mathcal{M}(t)x, \tag{62}$$

*provided that $B'(t)\mathcal{V}(t)x$ is nonzero. When $B'(t)\mathcal{V}(t)x$ is zero, $|x|_{\mathcal{R}}$ is required to vanish, so that only the second term in the right member remains.*

**Proof.** We must return to (47) to find $\beta(x)$, which is rewritten as

$$\beta^2 = \frac{1}{4}\left|B'\frac{\partial M}{\partial x}\right|^2_{R^{-1}} - \frac{\partial M}{\partial t} - \frac{1}{2}\operatorname{tr}\left(EWE'\frac{\partial^2 M}{\partial x^2}\right) - h - g'\frac{\partial M}{\partial x}. \tag{63}$$

Now substitute (56) to obtain

$$\beta^2 = \frac{1}{4}|B'2\mathcal{M}x|^2_{R^{-1}} - x'\dot{\mathcal{M}}x - \dot{m} - \operatorname{tr}(EWE'\mathcal{M}) - h - g'2\mathcal{M}x. \tag{64}$$

From (64) and (58) we obtain

$$\beta^2 = |x|^2_{\mathcal{M}BR^{-1}B'\mathcal{M}-\dot{\mathcal{M}}-Q-(A'\mathcal{M}+\mathcal{M}A)} - \dot{m} - \operatorname{tr}(EWE'\mathcal{M}), \tag{65}$$

which is written as $\beta^2 = |x|^2_{\mathcal{R}} - \dot{m} - \operatorname{tr}(EWE'\mathcal{M})$. Then, if $(\beta(x)$ vanishes when $B'(t)[\partial V^*(t,x)/\partial x]$ is zero, we have $\dot{m} = -\operatorname{tr}(EWE'\mathcal{M})$ as a constraint on our choice of $m(t)$. □

Let us now restrict the class of controllers, $K_M$, to be vector space morphisms. To denote this, we replace the notation $K_M$ by $K_{ML}$. It follows from the work of Liberty and Hartwig [16] that $M$ and $V$ are then quadratic, which is consistent with the assumptions and results in the foregoing lemma. It is straightforward to see that (62) defines a homogeneous mapping, by $k^*_{V|M}(t, \alpha x) = \alpha k^*_{V|M}(t, x)$. Indeed, the result follows by the definition

$$f(x) = |x|_{\mathcal{R}}/|x|_{\mathcal{V}BR^{-1}B'\mathcal{V}}, \tag{66}$$

on the domain in which $|x|_{\mathcal{V}BR^{-1}B'\mathcal{V}}$ does not vanish, together with the observation that $f(\alpha x) = f(x)$ on this domain. We shall extend this to the point $x = 0$ momentarily. But the question of whether or not $k^*_{V|M}(t,x)$ is a morphism under the addition of vectors needs further examination. The basic idea which we require is given in the following lemma:

**Lemma 5.5.** *Let $f(x)$ be given by (66), and consider the controller term $-f(x)R^{-1}B'\mathcal{V}$. If this term is a morphism of vector addition, then $f(x)$ is constant for all $x$ such that $B'\mathcal{V}x$ is nonzero.*

**Proof.** Denote by $x_1$ and $x_2$ two values of $x$ satisfying the assumptions of the lemma. Examine first the case in which $x_1$ and $x_2$ are chosen so that $B'\mathcal{V}x_1$ and $B'\mathcal{V}x_2$ are linearly independent. Additivity implies that

$$-R^{-1}B'\mathcal{V}[x_1 + x_2]f(x_1 + x_2) = -R^{-1}B'\mathcal{V}x_1 f(x_1) - R^{-1}B'\mathcal{V}x_2 f(x_2), \tag{67}$$

so that

$$[f(x_1 + x_2) - f(x_1)]B'\mathcal{V}x_1 + [f(x_1 + x_2) - f(x_2)]B'\mathcal{V}x_2 = 0, \tag{68}$$

and it is necessary to conclude that $f(x_1 + x_2) = f(x_1) = f(x_2)$. Turn now to the case in which the two vectors $B'\mathcal{V}x_1$ and $B'\mathcal{V}x_2$ are not linearly independent. Let $\alpha$ be a real number such that $B'\mathcal{V}x_1 = \alpha B'\mathcal{V}x_2$. In this situation, $B'\mathcal{V}(x_1 - \alpha x_2) = 0$. By additivity, it follows that

$$0 = -R^{-1}B'\mathcal{V}x_1 f(x_1) + \alpha R^{-1}B'\mathcal{V}x_2 f(x_2) = -[f(x_1) - f(x_2)]R^{-1}B'\mathcal{V}x_{(1)} \tag{69}$$

so that $f(x_1) = f(x_2)$. □

With the preceding results in hand, we are now able to conclude how the constant character of $f(x)$ induces a corresponding relationship between the weighting matrices in $f(x)$.

**Lemma 5.6.** *Let $\mathcal{R}$ and $\mathcal{V}BR^{-1}B'\mathcal{V}$ have identical null spaces, and consider the function $f(x)$ defined by (66) on the domain in which $|x|_{\mathcal{V}BR^{-1}B'\mathcal{V}}$ does not vanish. Then $f(x)$ is equal to a (positive) constant $\gamma$ on this domain, if and only if $\mathcal{R} = \gamma^2\mathcal{V}BR^{-1}B'\mathcal{V}$ on the domain.*

**Remark.** The assumption that $\mathcal{R}$ has the same null space as $\mathcal{V}BR^{-1}B'\mathcal{V}$ is quite natural. Of course, the null space of $\mathcal{R}$ must contain that of $R^{-1}B'\mathcal{V}$, by our foregoing discussions. Moreover, if this containment were strict, then the only possible constant behavior of our function would be zero. This is of course unduly restrictive. So we set the two null spaces equal to each other.

**Proof.** Notice that $\mathcal{R}$ and $\mathcal{V}BR^{-1}B'\mathcal{V}$ are nonnegative semidefinite. Thus $\gamma$ must be seen as positive. Sufficiency of the lemma is of course straightforward. Necessity follows by construction of the gradient vector, which is given by

$$[(f(x))^{-1}\mathcal{R}x - f(x)\mathcal{V}BR^{-1}B'\mathcal{V}x]/|x|^2_{\mathcal{V}BR^{-1}B'\mathcal{V}}, \tag{70}$$

from which it follows that $[\gamma^{-1}\mathcal{R}x - \gamma\mathcal{V}BR^{-1}B'\mathcal{V}x] = 0$ on the domain, and a simple multiplication by $\gamma$ achieves the result, as desired. Notice that $\mathcal{R}x$ does not vanish on the domain, so that the function $f$ is nonzero for the values of $x$ considered. □

**Remark.** Inasmuch as this equation holds over the entire nonzero images of the two members, there is clearly no loss of generality in extending the domain of the equation to the whole space.

The solution to the full-state-feedback MCV control problem with linear controller is then given by the following theorem:

**Theorem 5.3.** *Assume $V^* \in C_p^{1,2}(Q) \cap C(\bar{Q})$ and the same assumptions as in Theorem 5.2 and Lemma 5.4. Then for $k \in K_{ML}$, there exists a linear MCV controller, if and only if there exist solutions $\mathcal{M}$ and $\mathcal{V}$ to the pair of matrix differential equations*

$$\dot{\mathcal{M}} + A'\mathcal{M} + \mathcal{M}A + Q - \mathcal{M}BR^{-1}B'\mathcal{M} + \gamma^2\mathcal{V}BR^{-1}B'\mathcal{V} = 0, \tag{71}$$

$$\begin{aligned}\dot{\mathcal{V}} + 4\mathcal{M}EWE'\mathcal{M} + A'\mathcal{V} + \mathcal{V}A - \mathcal{M}BR^{-1}B'\mathcal{V}&\\ - \mathcal{V}BR^{-1}B'\mathcal{M} - 2\gamma\mathcal{V}BR^{-1}B'\mathcal{V} &= 0,\end{aligned} \tag{72}$$

*with boundary conditions $\mathcal{M}(t_F) = Q_F$ and $\mathcal{V}(t_F) = 0$, for a suitable positive time function $\gamma(t)$. In such a case, the controller is given by*

$$k^*_{V|M}(t,x) = -R^{-1}(t)B'(t)[\mathcal{M}(t) + \gamma(t)\mathcal{V}(t)]x. \tag{73}$$

**Remark.** In these equations, the choice $\gamma$ equal to zero results in the classical minimum mean cost situation.

**Proof.** Using the development above, and (61), (52) is rewritten as

$$\begin{aligned}-|x|^2_{\dot{\mathcal{V}}} - \dot{v} = &-\sqrt{|x|^2_{\mathcal{R}} - \dot{m} - \mathrm{tr}(EWE'\mathcal{M})}(|2B'\mathcal{V}x|_{R^{-1}}) + \mathrm{tr}(EWE'\mathcal{V})\\ &+ |2\mathcal{M}x|^2_{EWE'} + 2x'A'\mathcal{V}x - \frac{1}{2}(2\mathcal{M}x)'BR^{-1}B'(2\mathcal{V}x)\end{aligned}$$

with boundary condition $V^*(t_F, x) = 0$. We first note that $\dot{v} = -\operatorname{tr}(EWE'\mathcal{V})$, $v(t_F) = 0$, so that

$$v(t) = \int_t^{t_F} \operatorname{tr}(E(\tau)W(\tau)E'(\tau)\mathcal{V}(\tau))d\tau.$$

We also have $m(t) = \int_t^{t_F} \operatorname{tr}(E(\tau)W(\tau)E'(\tau)\mathcal{M}(\tau))\,d\tau$ with $m(t_F) = 0$. Collecting $|x|^2$ terms we end up with

$$0 = |x|^2_{\dot{\mathcal{V}}+4\mathcal{M}EWE'\mathcal{M}+A'\mathcal{V}+\mathcal{V}A-\mathcal{M}BR^{-1}B'\mathcal{V}-\mathcal{V}BR^{-1}B'\mathcal{M}} - 2|x|_{\mathcal{R}}|x|_{\mathcal{V}BR^{-1}B'\mathcal{V}}, \quad (74)$$

which is denoted by $0 = |x|^2_{\mathcal{E}(t)} - 2|x|_{\mathcal{R}(t)}|x|_{\mathcal{S}(t)}$. Now rewrite (62) in the form

$$k^*_{V|M}(t, x) = -R^{-1}(t)B'(t)\left[\mathcal{M}(t) + \frac{|x|_{\mathcal{R}(t)}}{|x|_{\mathcal{V}(t)B(t)R^{-1}(t)B'(t)\mathcal{V}(t)}}\mathcal{V}(t)\right].$$

Because we are considering only linear controllers, we know from Lemma 5.5 that $f(x)$, defined in that lemma, must be constant on the domain in which $B'(t)\mathcal{V}x$ is nonzero. But then the proof of Lemma 5.6, under our assumptions, implies that $f(x)$ must then be constant for all $x$, provided that we agree to define it equal to the same constant value on the nullspaces discussed above. Then Lemma 5.6 gives (71). Moreover, (72) is then a consequence of (74). We can then write the expression for the controller in the stated form. □

## 6 Earthquake Application

A three-degree-of-freedom (3DOF) structure under seismic excitation is studied in this section. We illustrate how cost mean and minimal cost variance are related to constant parameters $\gamma$. Performance characteristics, such as the standard deviation of displacement and control forces are also illustrated. Finally, we indicate the way in which control energy is related to the selected structure energy, through the cost functional, as the parameter $\gamma$ varies.

Consider the 3DOF, single-bay structure with an active tendon controller as shown in Figure 2. The structure is subject to a one-dimensional earthquake excitation. If we assume a simple shear frame model for the structure, then we can write the governing equations of motion in state space form as

$$dx(t) = \begin{bmatrix} 0 & I \\ -M_s^{-1}K_s & -M_s^{-1}C_s \end{bmatrix} x(t)dt + \begin{bmatrix} 0 \\ M_s^{-1}B_s \end{bmatrix} u(t)dt + \begin{bmatrix} 0 \\ -\Gamma_s \end{bmatrix} dw(t),$$

where

$$M_s = \begin{bmatrix} m_1 & 0 & 0 \\ 0 & m_2 & 0 \\ 0 & 0 & m_3, \end{bmatrix}, \qquad B_s = \begin{bmatrix} -4k_c\cos\alpha \\ 0 \\ 0 \end{bmatrix},$$

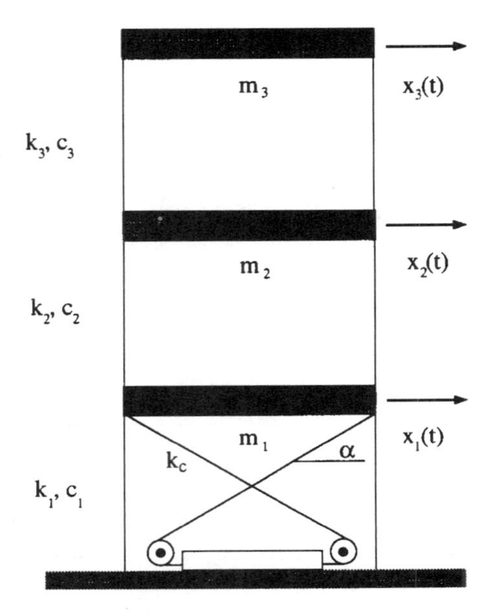


**Figure 2**: Schematic diagram for a three-degree-of-freedom structure.

$$C_s = \begin{bmatrix} c_1 + c_2 & -c_2 & 0 \\ -c_2 & c_2 + c_3 & -c_3 \\ 0 & -c_3 & c_3 \end{bmatrix}, \qquad \Gamma_s = \begin{bmatrix} 1 \\ 1 \\ 1 \end{bmatrix},$$

$$K_s = \begin{bmatrix} k_1 + k_2 & -k_2 & 0 \\ -k_2 & k_2 + k_3 & -k_3 \\ 0 & -k_3 & k_3 \end{bmatrix},$$

$m_i$, $c_i$, $k_i$ are the mass, damping, and stiffness, respectively, associated with the $i$th floor of the building, and $k_c$ is the stiffness of the tendon. The (nonstandard) Brownian motion term has $W = 1.00 \times 2\pi$ in$^2$/s$^3$. The parameters were chosen to match modal frequencies and dampings of the experimental structure in [4]. The cost function is given by

$$J = \int_0^{t_F} \left(z'(t) K_s z(t) + k_c u^2(t)\right) dt,$$

where $z$ is a vector of floor displacements and $x = (z, \dot{z})$.

Figure 3 shows that the average value of the cost function $E\{J\}$ increases as the MCV parameter $\gamma$ increases. On the other hand, the minimal associated variance of the cost function decreases; see Figure 4. Recall that $\gamma = 0$ point corresponds to the classical LQG case. Figure 5 shows the RMS displacement

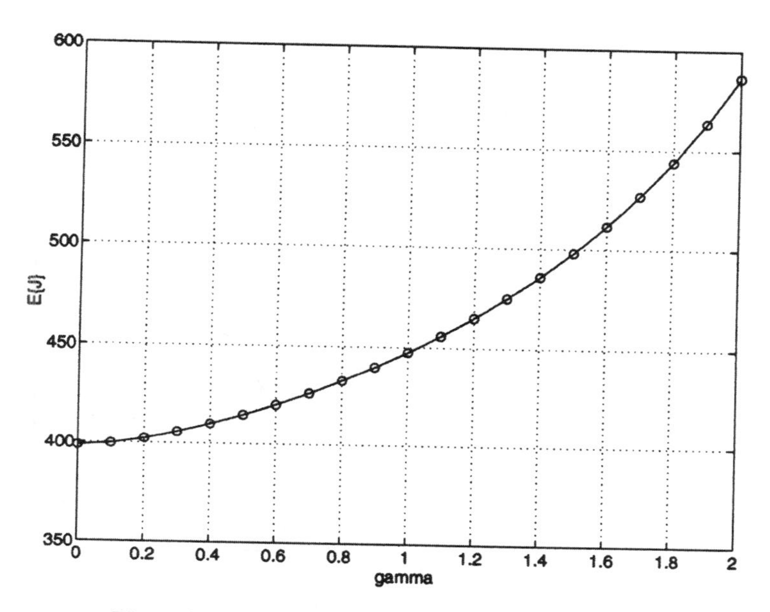


**Figure 3**: Cost mean; full-state-feedback, MCV, 3DOF.

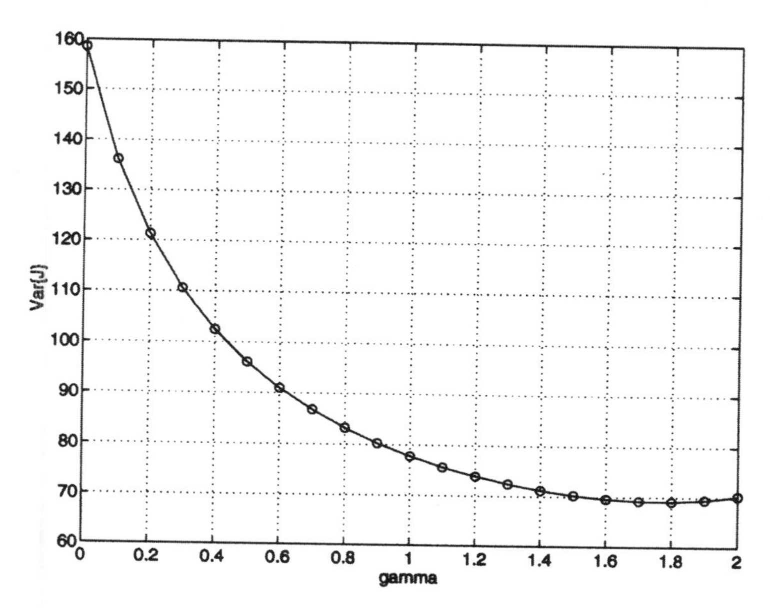


**Figure 4**: Optimal variance; full-state-feedback, MCV, 3DOF.

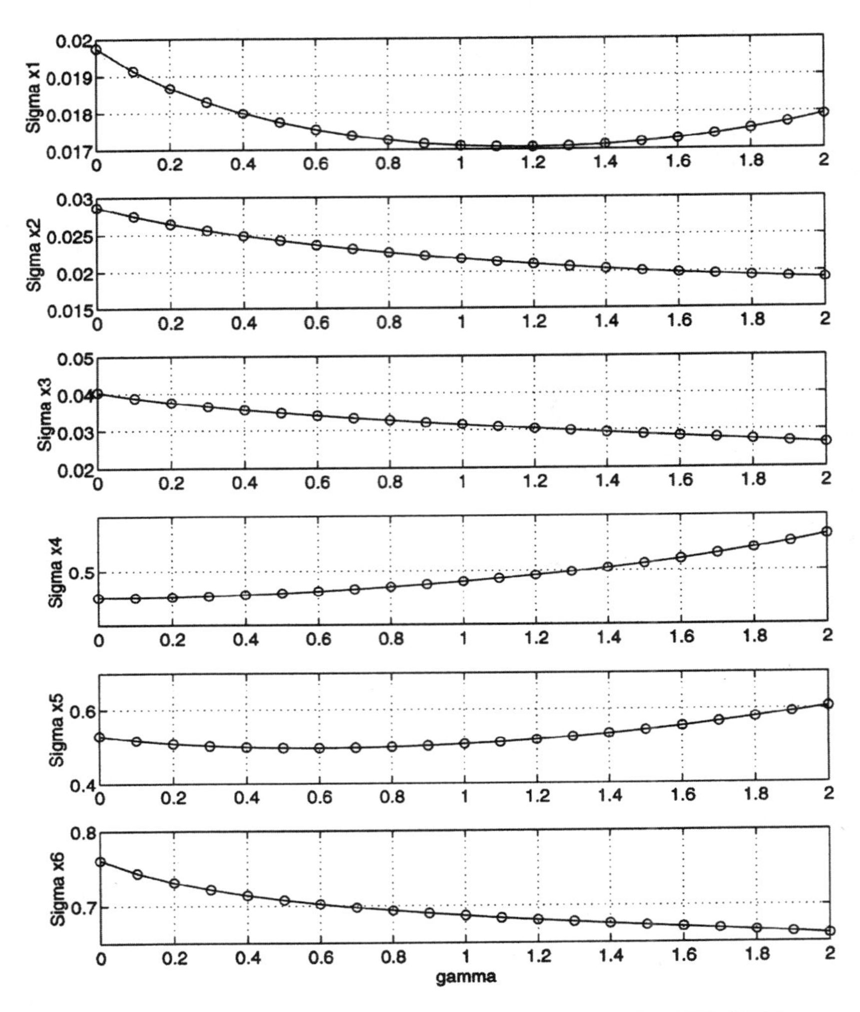


**Figure 5**: Displacements and velocities; full-state-feedback, MCV, 3DOF.

responses of first ($\sigma_{x_1}$), second ($\sigma_{x_5}$), and third ($\sigma_{x_3}$) floor; and the RMS velocity responses of first ($\sigma_{x_4}$), second ($\sigma_{x_3}$), and third ($\sigma_{x_6}$) floor, respectively, versus the MCV parameter, $\gamma$. It is important to note that both third floor RMS displacement and velocity responses can be decreased by choosing large $\gamma$. For larger $\gamma$, note that we require a larger control force, which means that more effort is needed to reduce the RMS displacement and velocity responses; see Figure 6.

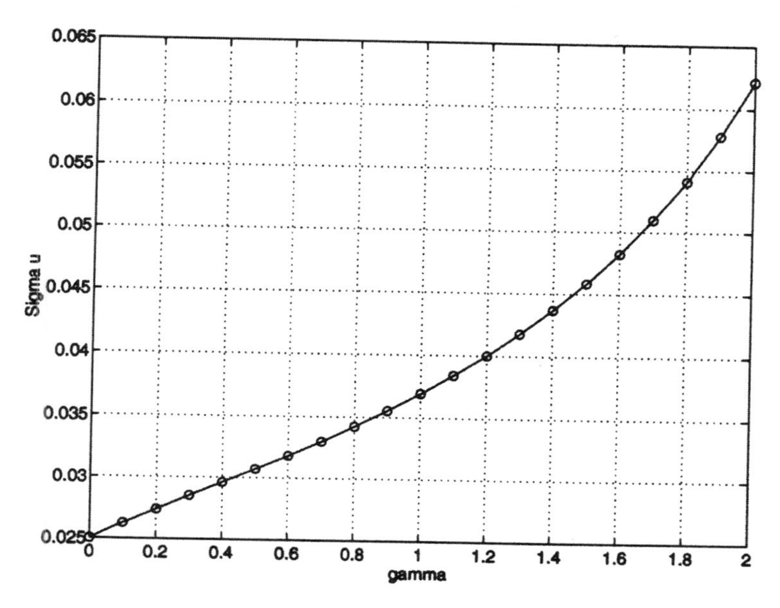


**Figure 6**: Control force; full-state-feedback, MCV, 3DOF.

## 7 A Simple Distribution Function Shaping Example

Consider a system described on the interval $T = [0, 1]$ by

$$dx(t) = x(t)dt + u(t)dt + e\,dw(t),$$

where the state $x(t) \in \mathbb{R}$, the control action $u(t) \in \mathbb{R}$, $w(t)$ is a standard Brownian motion with $e^2 = 0.25$, $x_0 = 1$, and incremental covariance of $x_0 = 0$. The cost function is given by $J = \int_0^1 [x^2(t) + u^2(t)]dt$. The MCV feedback controller for the full-state-feedback case is used for several values of $\gamma$. For each of these controllers we calculated the feedback control gain matrices and plotted the density and distribution graphs. We also graphed the same item for $J_x = \int_0^1 x^2(t)dt$, and $J_u = \int_0^1 u^2(t)dt$. Figure 7 shows the density graphs for $\gamma = 0, 2$ and 4. We note that the mean is smallest for $\gamma = 0$ and largest for $\gamma = 4$. From Figure 8, we notice that the probability of $J$ being smaller than a particular $J_0$ is largest for $\gamma = 0$ and smallest for $\gamma = 4$ with $\gamma = 2$ in between those two. If we are trying to find a controller that would give the best chance of giving smaller cost, we should choose $\gamma = 0$ in this example.

In Figures 9 and 10, we have density and distribution graphs for $J_x$, respectively. Then in Figures 11 and 12, we have density and distribution graphs for $J_u$. Notice that as $\gamma$ increases the density of $J_x$ shifts to the left while the $J_u$ graphs shift to the right. This corresponds to the tradeoff between the control effort and the state regulation.

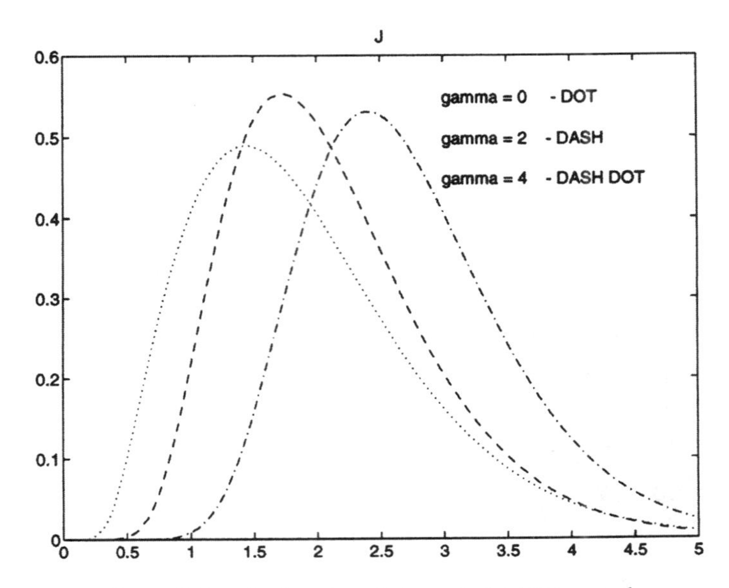


**Figure 7**: Density function of the cost, $J$; MCV control.

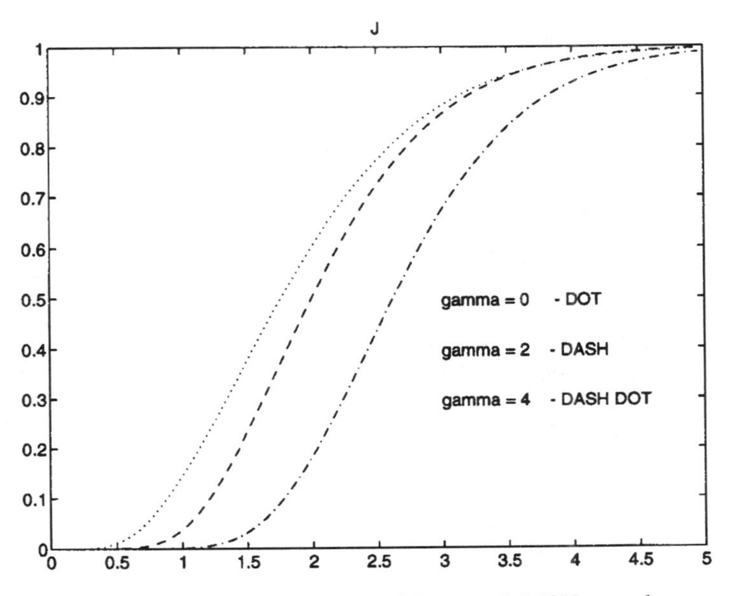


**Figure 8**: Distribution function of the cost, $J$; MCV control.

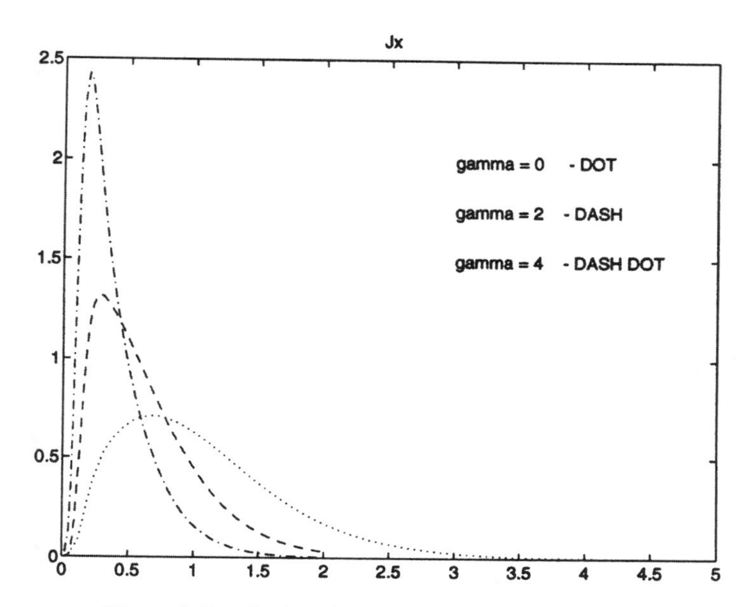


**Figure 9**: Density function of the cost, $J_x$; MCV control.

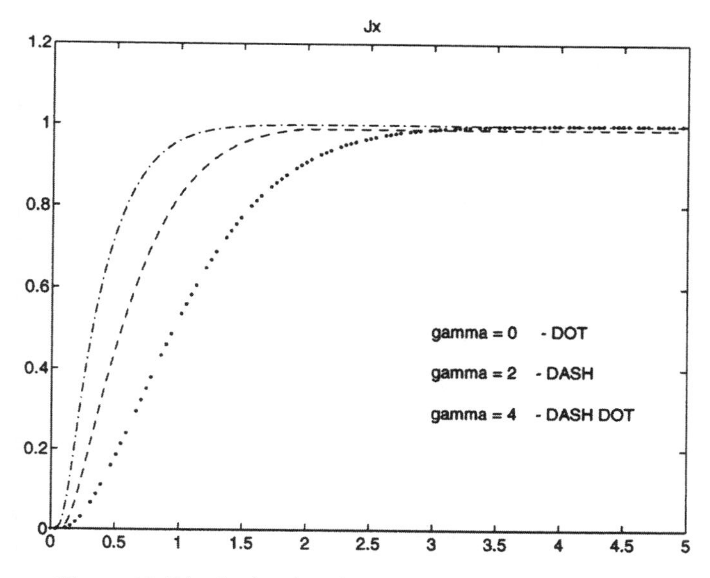


**Figure 10**: Distribution function of the cost, $J_x$; MCV control.

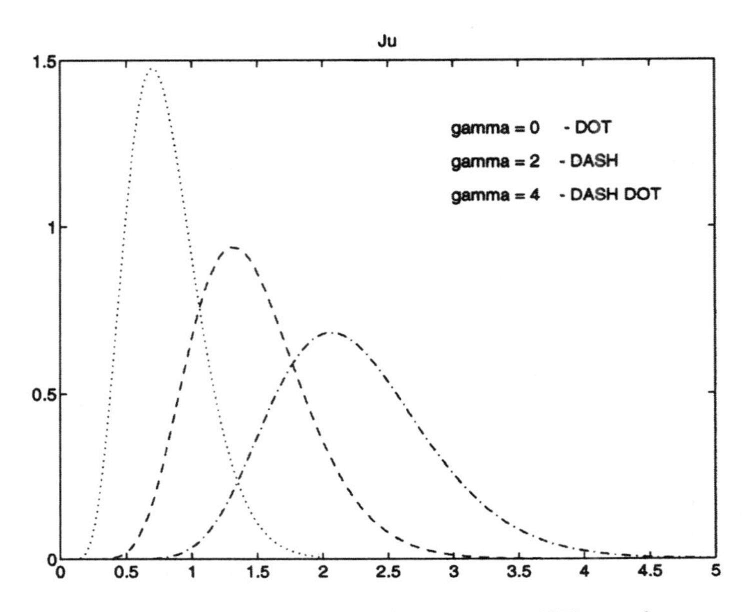


**Figure 11**: Density function of the cost, $J_u$; MCV control.

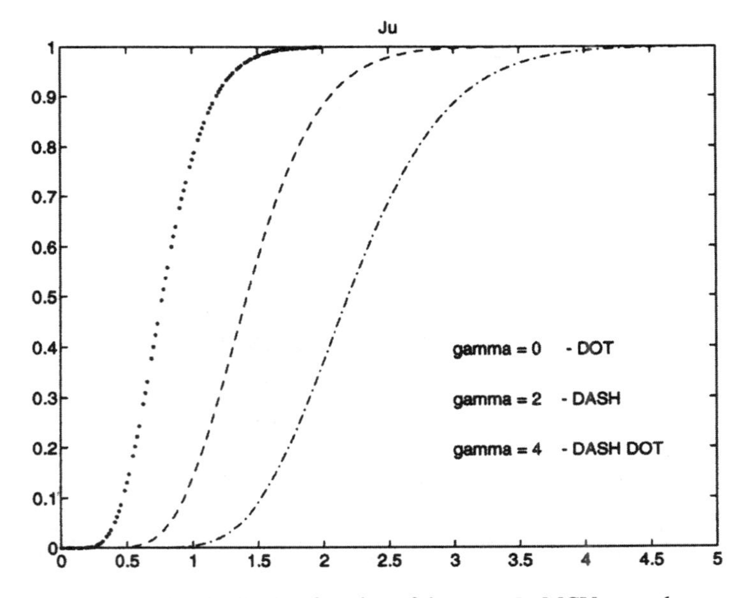


**Figure 12**: Distribution function of the cost, $J_u$; MCV control.

## Acknowledgments

This work was funded in part by the Frank M. Freimann Chair, in the Department of Electrical Engineering at the University of Notre Dame, and by the National Science Foundation under Grants CMS 93-01584, CMS 95-00301, and CMS 95-28083. Submitted to the *Annals of Dynamic Games*, March 24, 1997; revised January 8, 1998 and September 21, 1998.

## REFERENCES

[1] Başar, T. and P. Bernhard. *$H_\infty$-Optimal Control and Related Minimax Design Problems.* Birkhäuser, Boston, 1991.

[2] Bensoussan, A. and J. H. van Schuppen. Optimal Control of Partially Observable Stochastic Systems with an Exponential-of-Integral Performance Index. *SIAM Journal on Control and Optimization*, **23**, 599–613, 1985.

[3] Bensoussan, A. *Stochastic Control of Partially Observable Systems.* Cambridge University Press, London, 1992.

[4] Chung, L. L., T. T. Soong, and A. M. Reinhorn. Experiments on Active Control of MDOF Seismic Structures. *Journal of Engineering Mechanics*, ASCE, **115**, No. 8, 1609–1627, 1989.

[5] Davenport, W. B. Jr. and W. L. Root. *An Introduction to the Theory of Random Signals and Noise.* McGraw-Hill, New York, 1958.

[6] Doob, J. L. *Stochastic Processes.* Wiley, New York, 1953.

[7] Fleming, W. H. and R. W. Rishel. *Deterministic and Stochastic Optimal Control.* Springer-Verlag, New York, 1975.

[8] Fleming, W. H. and W. M. McEneaney. Risk Sensitive Optimal Control and Differential Games. *Stochastic Theory and Adaptive Control.* (T. E. Duncan and B. Pasik-Duncan, eds.). Lecture Notes in Control and Information Sciences 184. Springer-Verlag, New York, pp. 185–197, 1992.

[9] Fleming, W. H. and H. M. Soner. *Controlled Markov Processes and Viscosity Solutions.* Springer-Verlag, New York, 1992.

[10] Glover, K. and J. C. Doyle. State-Space Formulae for All Stabilizing Controllers That Satisfy an $H_\infty$-Norm Bound and Relations to Risk Sensitivity. *Systems and Control Letters*, **11**, 167–172, 1988.

[11] Glover, K. Minimum Entropy and Risk-Sensitive Control: The Continuous Time Case. *Proceedings of the 28th IEEE Conference on Decision and Control*, pp. 388–391, December 1989.

[12] Jacobson, D. H. Optimal Stochastic Linear Systems with Exponential Performance Criteria and Their Relationship to Deterministic Differential Games. *IEEE Transactions on Automatic Control*, **AC-18**, 124–131, 1973.

[13] James, M. R. Asymptotic Analysis of Nonlinear Stochastic Risk-Sensitive Control and Differential Games. *Mathematics of Control, Signals, and Systems*, **5**, 401–417, 1992.

[14] James, M. R., J. S. Baras, and R. J. Elliott. Risk-Sensitive Control and Dynamic Games for Partially Observed Discrete-Time Nonlinear Systems. *IEEE Transactions on Automatic Control*, **AC-39**, No. 4, 780–792, 1994.

[15] Kumar, P. R. and J. H. van Schuppen. On the Optimal Control of Stochastic Systems with an Exponential-of-Integral Performance Index. *Journal of Mathematical Analysis and Applications*, **80**, 312–332, 1981.

[16] Liberty, S. R. and R. C. Hartwig. On the Essential Quadratic Nature of LQG Control-Performance Measure Cumulants. *Information and Control*, **32**, No. 3, 276–305, 1976.

[17] Liu, R. W. and R.J.L. Leake. Inverse Lyapunov Problems. Technical Report No. EE-6510, Department of Electrical Engineering, University of Notre Dame, August 1965.

[18] Pontryagin, L. S., V. G. Boltyanski, R. V. Gamkriledze, and E. F. Mischenko. *The Mathematical Theory of Optimal Processes*, Interscience New York, 1962.

[19] Runolfsson, T. The Equivalence Between Infinite-Horizon Optimal Control of Stochastic Systems with Exponential-of-Integral Performance Index and Stochastic Differential Games. *IEEE Transactions on Automatic Control*, **39**, No. 8, 1551–1563, 1994.

[20] Sain, M. K. On Minimal-Variance Control of Linear Systems with Quadratic Loss. Ph.D. Dissertation, Department of Electrical Engineering, University of Illinois, Urbana, January 1965.

[21] Sain, M. K. A Sufficiency Condition for Minimum Variance Control of Markov Processes. *Proceedings of the Fourth Allerton Conference on Circuit and System Theory*, Monticello, IL, pp. 593–599, October 1966.

[22] Sain, M. K. Performance Moment Recursions, with Application to Equalizer Control Laws. *Proceedings of the Fifth Allerton Conference on Circuit and System Theory*, Monticello, IL, pp. 327–336, 1967.

[23] Sain, M. K. and S. R. Liberty. Performance Measure Densities for a Class of LQG Control Systems. *IEEE Transactions on Automatic Control*, **AC-16**, No. 5, 431–439, 1971.

[24] Sain, M. K., C.-H. Won, and B. F. Spencer, Jr. Cumulant Minimization and Robust Control. *Stochastic Theory and Adaptive Control*. (T. E. Duncan and B. Pasik-Duncan, eds.). Lecture Notes in Control and Information Sciences 184. Springer-Verlag, New York, pp. 411–425, 1992.

[25] Speyer, J. L., J. Deyst, and D. H. Jacobson. Optimization of Stochastic Linear Systems with Additive Measurement and Process Noise Using Exponential Performance Criteria. *IEEE Transactions on Automatic Control*, **AC-19**, No. 4, 358–366, 1974.

[26] Speyer, J. L. An Adaptive Terminal Guidance Scheme Based on an Exponential Cost Criterion with Application to Homing Missile Guidance. *IEEE Transactions on Automatic Control*, **AC-21**, 371–375, 1976.

[27] Uchida, K. and M. Fujita. On the Central Controller: Characterizations via Differential Games and LEQG Control Problems. *Systems and Control Letters*, **13**, 9–13, 1989.

[28] Wheeden, R. L. and A. Zygmund. *Measure and Integral, An Introduction to Real Analysis*. Marcel Dekker, New York, 1977.

[29] Whittle, P. Risk-Sensitive Linear/Quadratic/Gaussian Control. *Advances in Applied Probability*, **13**, 764–777, 1981.

[30] Whittle, P. *Risk Sensitive Optimal Control*, Wiley. New York, 1990.

[31] Whittle, P. A Risk-Sensitive Maximum Principle: The Case of Imperfect State Observation. *IEEE Transactions on Automatic Control*, **36**, No. 7, 793–801, 1991.

[32] Won, Chang-Hee, M. K. Sain, and B. F. Spencer, Jr. Performance and Stability Characteristics of Risk-Sensitive Controlled Structures under Seismic Disturbances. *Proceedings of the American Control Conference*, pp. 1926–1930, June 1995.

[33] Wonham, W. M. Stochastic Problems in Optimal Control. *1963 IEEE Int. Conv. Rec.*, part 2, pp. 114–124, 1963.

## Annals of the International Society of Dynamic Games

*Series Editor*

Tamer Başar
Coordinated Science Laboratory
University of Illinois
1308 West Main Street
Urbana, IL 61801
U.S.A.

This series publishes volumes in the general area of dynamic games and their applications. It is an outgrowth of activities of "The International Society of Dynamic Games," *ISDG*, which was founded in 1990. The primary goals of *ISDG* are to promote interactions among researchers interested in the theory and applications of dynamic games; to facilitate dissemination of information on current activities and results in this area; and to enhance the visibility of dynamic games research and its vast potential applications.

The *Annals of Dynamic Games* Series will have volumes based on the papers presented at the society's biannual symposia, including only those that have gone through a stringent review process, as well as volumes of invited papers dedicated to specific, fast-developing topics, put together by a guest editor or guest co-editors. More information on this series and on volumes planned for the future can be obtained by contacting the Series Editor, Tamer Başar, whose address appears above.

We encourage the preparation of manuscripts in LaTeX using Birkhäuser's macro.sty for this volume.

Proposals should be sent directly to the editor or to: Birkhäuser Boston, 675 Massachusetts Avenue, Cambridge, MA 02139, U.S.A.

Volumes in this series are:

Advances in Dynamic Games and Applications
*Tamer Başar and Alan Haurie, Eds.*

Control and Game-Theoretic Models of the Environment
*Carlo Carraro and Jerzy A. Filar, Eds.*

New Trends in Dynamic Games and Applications
*Geert Jan Olsder, Ed.*

Stochastic and Differential Games
*Martino Bardi, T.E.S. Raghavan, and T. Parthasarathy, Eds.*

Advances in Dynamic Games and Applications
*Jerzy Filar, Vladimir Gaitsgory, and Koichi Mizukami*